SALVAGE

LANDOLT-BÖRNSTEIN

Numerical Data and Functional Relationships
in Science and Technology

New Series

Editor in Chief: K.-H. Hellwege

Group I: Nuclear and Particle Physics

Volume 6

Properties and Production Spectra
of Elementary Particles

A. N. Diddens · H. Pilkuhn
K. Schlüpmann

Editor: H. Schopper

Springer-Verlag Berlin · Heidelberg · New York 1972

LANDOLT-BÖRNSTEIN

Zahlenwerte und Funktionen aus Naturwissenschaften und Technik

Neue Serie

Gesamtherausgabe: K.-H. Hellwege

Gruppe I: Kern- und Teilchenphysik

Band 6
Eigenschaften und Erzeugungsspektren von Elementarteilchen

A. N. Diddens · H. Pilkuhn
K. Schlüpmann

Herausgeber: H. Schopper

Springer-Verlag Berlin · Heidelberg · New York 1972

ISBN 3-540-06047-2 Springer-Verlag Berlin-Heidelberg-New York
ISBN 0-387-06047-2 Springer-Verlag New York-Heidelberg-Berlin

Vorwort

Der vorliegende Band ist der erste in der Neuen Serie des Landolt-Börnstein, in dem Daten über Elementarteilchen zusammengestellt sind. Zwei weitere Bände werden in Kürze folgen. Diese Bände unterscheiden sich in ihrer Zielsetzung weitgehend von den Tabellen, die als „Laborberichte" von mehreren großen Laboratorien laufend veröffentlicht werden. Diese „Laborberichte" beziehen sich meistens auf eine spezielle Reaktion, und ihr wesentlicher Zweck besteht darin, eine auf den neuesten Stand gebrachte Übersicht für die Spezialisten zu bringen.

Da die Elementarteilchenphysik bereits einen erheblichen Umfang angenommen hat, ist eine Spezialisierung schon innerhalb dieses Teilgebietes der Physik kaum zu vermeiden. Die Daten aus dem Gebiet der Elementarteilchenphysik sind aber auch für andere Forschungsgebiete von Interesse, so außer für die Kernphysik insbesondere für die Astrophysik und Kosmologie. In den hier veröffentlichten Tabellen wurde daher versucht, eine kritische Auswahl von Daten in einer solchen Form darzustellen, daß auch der Nichtspezialist die von ihm benötigten Werte schnell finden kann. Es ist geplant, im Laufe der Zeit alle wichtigen Daten der Elementarteilchenphysik systematisch zusammenzustellen, wobei genügend Hinweise auf detaillierte Tabellen für spezielle Fälle gegeben werden. Auf diese Weise sollte es möglich werden, durch Nachschlagen im Landolt-Börnstein schnell herauszufinden, ob bestimmte Daten überhaupt gemessen wurden und wenn ja, welches ihre besten Werte sind.

Das erste Kapitel dieses Bandes behandelt die Eigenschaften der Elementarteilchen und ist weitgehend in sich abgeschlossen. Es geht über vorhandene Tabellen wesentlich hinaus, da neben den eigentlichen Eigenschaften der Teilchen auch Kopplungskonstanten und Formfaktoren angegeben werden. Der Zusammenhang zwischen den experimentell gemessenen Größen und den fundamentalen Eigenschaften der Wechselwirkungen wird erläutert.

Die weiteren Kapitel dieses und der folgenden Bände enthalten Daten über Streuprozesse. Um den Gebrauch aller dieser Tabellen zu erleichtern, wurde soweit wie möglich eine einheitliche Nomenklatur benutzt. Sie wird in Kapitel 2 eingeführt und erläutert.

Die Erzeugung von Teilchen in sogenannten inklusiven Reaktionen gewinnt laufend an Interesse. Aber auch für praktische Anwendungen ist die Kenntnis der Erzeugungsraten von Teilchen wichtig. In Kapitel 3 sind daher die Spektren von Teilchen zusammengestellt, die bei der Proton-Proton-Wechselwirkung erzeugt werden.

Den Autoren ist für ihre Mühe bei der Zusammenstellung der Tabellen zu danken, dem Chefherausgeber, der Redaktion und dem Verlag für die Lösung vieler sachlicher und technischer Probleme, die bei der Bearbeitung des Bandes auftraten. Besonderer Dank gebührt Herrn Dr. G. Craig, der nicht nur zu einer einwandfreien Übersetzung, sondern auch zur Präzision des Ausdrucks wesentlich beigetragen hat. Dieser Band wurde, wie alle Bände des Landolt-Börnstein, ohne finanzielle Unterstützung von anderer Seite veröffentlicht.

Genf, Oktober 1972   **Der Herausgeber**

Preface

This volume is the first in the New Series of Landolt-Börnstein which contains a compilation of data on elementary particles. Two further volumes will follow in the near future. The purpose of these volumes is rather different from that of the tables published continuously by several big laboratories. These "laboratory reports" deal mostly with a special reaction and their main purpose is to provide an up-to-date review for the specialists.

Since elementary particle physics has become rather extensive, a specialization already inside this field of physics can hardly be avoided. But the data on elementary particle physics are also of interest for other areas of research, as for nuclear physics and in particular for astrophysics and cosmology. In the tables published here it has, therefore, been tried to present a critical selection of data in such a form that also the non-specialists can quickly find what they need. In due time we plan to compile all important data on elementary particle physics systematically giving at the same time sufficient references to more detailed tables for special cases. In this way it should be possible to find out quickly by consulting Landolt-Börnstein whether certain data have been measured at all and if so, which are the best values.

The first chapter of this volume deals with the properties of elementary particles and is largely selfconsistent. It exceeds existing tables considerably since besides the properties of elementary particles also coupling constants and form factors have been compiled. The relations between experimentally measured quantities and the fundamental properties of interactions are explained.

The consecutive chapters of this and the two following volumes contain data of scattering processes. In order to facilitate the use of all these tables a uniform nomenclature has been used as far as possible. It is introduced and explained in chapter 2.

The production of particles in so-called inclusive reactions finds increasing interest. In chapter 3 spectra of particles produced in proton-proton-interactions have been compiled.

I should like to thank the authors for their work of compiling the data, and the editor in chief, the editorial office and the publishers for their effort solving the many pertinent and technical problems which arose during the preparation of the volume. Special thanks are due to Dr. G. Craig who contributed not only to an impeccable translation but also to the precision of the style. This volume like all other volumes of the Landolt-Börnstein has been published without financial support from any outside source.

Geneva, October 1972   **The Editor**

Inhaltsverzeichnis

1 Teilcheneigenschaften, Kopplungskonstanten und Formfaktoren

Von H. PILKUHN, Universität Karlsruhe/Deutschland

Table of contents

1 Particle properties, coupling constants and form factors

By H. PILKUHN, University of Karlsruhe/Germany

2 Bezeichnung, Konstanten und allgemeine Beziehungen

Von P. J. Carlson, A. N. Diddens, G. Giacomelli, K. Schlüpmann und H. Schopper

3 Teilchenerzeugung bei der Proton-Proton-Wechselwirkung

Von A. N. Diddens, CERN, Genf/Schweiz, und K. Schlüpmann, Freie Universität Berlin/Deutschland

2 Notation, constants and general relations

By P. J. Carlson, A. N. Diddens, G. Giacomelli, K. Schlüpmann and H. Schopper

3 Particle production in proton-proton interactions

By A. N. Diddens, CERN, Geneva/Switzerland, and K. Schlüpmann, University (FU) of Berlin/Germany

1 Particle properties, coupling constants and form factors

1.1 Introduction: The basic interactions

1.1.1 Units and conventions

The following units are normally used in particle physics

$$\hbar = c = 1 = 6.582183 \cdot 10^{-22} \text{ MeV sec} = 197.3289 \text{ MeV fm}, \tag{1}$$

where 1 MeV is 10^6 eV or 10^{-3} GeV (or BeV) and 1 fm (fermi) is 10^{-13} cm. The fine structure constant $\alpha \approx 137^{-1}$ is written as $e^2/4\pi$.

For cross sections one uses

$$\text{GeV}^{-2} = 0.38935 \text{ mb}, \quad 1 \text{ mb (millibarn)} = 10^{-27} \text{ cm}^2, \quad 1 \,\mu\text{b} = 10^{-3} \text{ mb}. \tag{2}$$

Other fundamental constants are collected in 1.4.1. The conventions about metric, state normalization, Dirac matrices etc., are not uniform. We shall use the 4-momentum vector P, energy E, particle mass m, and metric tensor $g_{\mu\nu}$ as follows:

$$P = (P^\mu) = (P_0, \vec{p}) = (E, p_x, p_y, p_z), \quad P^2 = E^2 - p^2 = m^2,$$

$$g_{\mu\nu} = \begin{pmatrix} 1 & & & 0 \\ & -1 & & \\ & & -1 & \\ 0 & & & -1 \end{pmatrix}, \quad P_a \cdot P_b = P_a^\mu P_b^\nu g_{\mu\nu} = P_a^\mu P_{b\mu} = P_{a\mu} P_b^\mu. \tag{3}$$

We define the connection between the S- and T-matrices and the normalization of 1-particle states in a Lorentz-invariant way:

$$S_{if} = \langle f|i \rangle + i(2\pi)^4 \, \delta^4(P_i - P_f) \, T_{if}, \quad \langle \vec{p}_f | \vec{p}_i \rangle = (2\pi)^3 \, 2E_f \, \delta(\vec{p}_f - \vec{p}_i). \tag{4}$$

The Dirac equation, γ-matrices and spinor normalizations are taken as

$$P \cdot \gamma u = mu, \quad \bar{u} = u^+ \gamma^0, \quad \bar{u}(M') \, u(M) = \delta_{MM'} \cdot 2m, \tag{5}$$

$$\gamma^0 = \gamma_0 = \begin{pmatrix} 1 & 0 \\ 0 & -1 \end{pmatrix}, \quad -\gamma^i = \gamma_i = \begin{pmatrix} 0 & \sigma_i \\ -\sigma_i & 0 \end{pmatrix}, \quad \gamma_5 = i\gamma_0\gamma_1\gamma_2\gamma_3 = \begin{pmatrix} 0 & 1 \\ 1 & 0 \end{pmatrix}, \tag{6}$$

where M denotes the magnetic quantum number. The relation between the Dirac spinor u and the Pauli spinor χ is

$$u = (E+m)^{\frac{1}{2}} \begin{pmatrix} \chi \\ \dfrac{\vec{\sigma}\vec{p}\chi}{E+m} \end{pmatrix} = (E+m)^{-\frac{1}{2}} (P\cdot\gamma + m)\begin{pmatrix} \chi \\ 0 \end{pmatrix}, \quad \chi(\tfrac{1}{2}) = \begin{pmatrix} 1 \\ 0 \end{pmatrix}, \quad \chi(-\tfrac{1}{2}) = \begin{pmatrix} 0 \\ 1 \end{pmatrix}. \tag{7}$$

For an outgoing antiparticle of 4-momentum P and helicity λ ($=$ spin component along $\vec{p}$), we simply use the negative-energy spinor $u(-P, \lambda)$. For photons and vector mesons, we need the spin-1 states $\varepsilon(\lambda)$:

$$P \cdot \varepsilon(\lambda) = 0, \quad \varepsilon(\lambda) \cdot \varepsilon^*(\lambda) = -1, \tag{8}$$

$$\varepsilon(\pm 1) = 2^{-\frac{1}{2}}(0, \mp 1, -i, 0), \quad \varepsilon(0) = \frac{1}{m}(p, 0, 0, E). \tag{9}$$

Textbooks using the same metric are: [Schweber 61; Muirhead 65; Björken 65, 67; Gasiorowicz 66; Pilkuhn 67]. Other metrics are used by [Martin 70; Källen 64, 65; Marshak 69]. Some books take $-i$ instead of $+i$ in (4), some omit the factor $2E_f$ in (4) and some the factor $2m$ in (5). The latter factor is used to make the momentum space (see e.g. Eq. (4) in 1.1.5 below) the same for fermions and bosons. Some books take all γ-matrices hermitean. The γ_5 in (6) is sometimes called $i\gamma_5$, or sometimes even $-\gamma_5$ (for example in [Marshak 69]).

For spins $\tfrac{3}{2}$ and 2, the Rarita-Schwinger spinor $u_\mu(M)$ and tensor $\varepsilon_{\mu\nu}(M)$ will be used:

$$u_\mu(\pm\tfrac{3}{2}) = u(\pm\tfrac{1}{2})\,\varepsilon_\mu(\pm 1), \quad u_\mu(\pm\tfrac{1}{2}) = 3^{-\frac{1}{2}}u(\mp\tfrac{1}{2})\,\varepsilon_\mu(\pm 1) + (\tfrac{2}{3})^{\frac{1}{2}}u(\pm\tfrac{1}{2})\,\varepsilon_\mu(0). \tag{10}$$

$$\varepsilon_{\mu\nu}(\pm 2) = \varepsilon_\mu(\pm 1)\,\varepsilon_\nu(\pm 1), \quad \varepsilon_{\mu\nu}(\pm 1) = 2^{-\frac{1}{2}}[\varepsilon_\mu(\pm 1)\,\varepsilon_\nu(0) + \varepsilon_\mu(0)\,\varepsilon_\nu(\pm 1)],$$

$$\varepsilon_{\mu\nu}(0) = 6^{-\frac{1}{2}}[\varepsilon_\mu(1)\,\varepsilon_\nu(-1) + \varepsilon_\mu(-1)\,\varepsilon_\nu(1) + 2\varepsilon_\mu(0)\,\varepsilon_\nu(0)]. \tag{11}$$

Pilkuhn

The resulting propagators are collected in 1.3.3. Finally, we shall frequently need the triangular function λ for the calculation of momenta in the a, b, or c rest frames. For $\vec{p}_a \pm \vec{p}_b \pm \vec{p}_c = 0$,

$$p^2(p_i = 0) = (4m_i^2)^{-1}\, \lambda(m_a^2, m_b^2, m_c^2), \quad i = a, b, c:$$
$$\lambda(a, b, c) = a^2 + b^2 + c^2 - 2(ab + bc + ca) = [a - (\sqrt{b} + \sqrt{c})^2]\,[a - (\sqrt{b} - \sqrt{c})^2]\,. \tag{12}$$

Constants which are closely related to experiment are normally adorned with a statistical error. This error is the 1-standard deviation uncertainty in the last digits of the constant and will be given in parentheses immediately after the digits, for example $0.72(5) = 0.72 \pm 0.05$, $0.72(11) = 0.72 \pm 0.11$, $7.2(1.1) \cdot 10^{-4} = (7.2 \pm 1.1) \cdot 10^{-4}$. Where experimental information is lacking (e.g. Σ^0 lifetime) or obviously less accurate than the theoretical prediction (e.g. branching ratio $\Lambda e\bar{e}/\Lambda\gamma$ in Σ^0 decay, or some branching ratios of resonance decays that follow from isospin-invariance + phase space), theoretical values are given, followed by the abbreviation "(th)" to distinguish them from experimental values. Whenever possible, I have adopted the recommended values of the "Compilation of coupling constants and low-energy parameters" [Ebel 71], the "Review of Particle Properties" [Particle 71, 72] and other review articles. In these review articles the reader will find further references to the original literature. Also, in this very compact tabulation it is not possible to enumerate all the additional effects which may occur for certain reactions or decays. For this reason, I have quoted relatively many reviews. I have used the reference system of [Ebel 71] and [Particle 71], which gives the name of the first author and the year of publication as reference key, and the journal abbreviation, volume and page numbers, and list of authors in the list of references. The journals are abbreviated as follows:

AP	Annals of Physics
FP	Fortschritte der Physik
JETP	English translation of Soviet Physics JETP
JETL	JETP Letters
NC	Nuovo Cimento
NCL	Nuovo Cimento Letters
NP	Nuclear Physics
PL	Physics Letters
PPSL	Proceedings of the Physical Society of London
PR	Physical Review
PRep	Physics Reports
PRL	Physical Review Letters
PRSL	Proceedings of the Royal Society of London
PTP	Progress of Theoretical Physics
RMP	Reviews of Modern Physics
SJNP	Soviet Journal of Nuclear Physics
STMP	Springer Tracts in Modern Physics
ZP	Zeitschrift für Physik.

Finally, I wish to acknowledge the help of my colleagues at Karlsruhe, particularly of D. Wegener in connection with the electromagnetic for factors.

1.1.2 The classification of particles and resonances. State mixing

Particles may be classified experimentally according to their lifetimes. Tab. a shows the classes of stable particles, weakly decaying particles (lifetimes $> 0.8 \cdot 10^{-10}$ sec) and electromagnetically decaying particles (lifetimes $10^{-16} \ldots 10^{-19}$ sec). Our knowledge of these three groups has reached a certain completeness in recent years. There is a forth group of extremely shortlived states, normally called "resonances". Here only the lower-lying states are known with some accuracy, and a complete tabulation is premature. Among these resonances, I have included the nonets of vector and tensor mesons and the baryon decuplet, which form complete SU(3)-multiplets, and four additional baryon multiplets which are nearly complete. From the remaining resonances I have included only a few low-lying states with not too broad widths. An exception is the ε-meson, which is presently not even confirmed as a resonance, but which is used in the description of πN and NN scattering. Complete tabulation of all resonances is given in the "review of particle properties", which is published at regular intervals [Particle 71]. On the other hand, I have included some "elementary particle" properties of the nuclei d, t, ^{3}He and α in the appropriate places.

Pilkuhn

Tab. *a*). Classification of particles according to their lifetimes

| Stable particles | | | | Weakly decaying particles[2]) | | | |
Name	Symbol	Spin	Antiparticle	Name	Symbol	Spin	Antiparticle
photon	γ	1	γ	muon	μ^- or μ	$\frac{1}{2}$	μ^+ or $\bar{\mu}$
electron	e^- or e	$\frac{1}{2}$	e^+ or $\bar{e}$	neutron	n	$\frac{1}{2}$	$\bar{n}$
electron-neutrino	ν_e	$\frac{1}{2}$	$\bar{\nu}_e$	lambda-hyperon	Λ	$\frac{1}{2}$	$\bar{\Lambda}$
muon-neutrino[1])	ν_μ	$\frac{1}{2}$	$\bar{\nu}_\mu$	sigma-hyperons	Σ^+, Σ^-	$\frac{1}{2}$	$\bar{\Sigma}^-, \bar{\Sigma}^+$
proton	p	$\frac{1}{2}$	$\bar{p}$	Xi or cascade particle	Ξ^0, Ξ^-	$\frac{1}{2}$	$\bar{\Xi}^0, \bar{\Xi}^+$
deuteron	d	1	$\bar{d}$	omega-particle	Ω^-	$\frac{3}{2}$	$\bar{\Omega}^+$
helion or helium-3	^{3}He	$\frac{1}{2}$	$^3\overline{\text{He}}$ [3])	charged pi-meson or pion	π^+	0	π^-
α-particle (helium)	^{4}He or α	0	$^4\overline{\text{He}}$ [4])	K-mesons or kaons	K^+, K^0	0	$K^-, \bar{K}^0$

Electromagnetically decaying particles:
Σ^0 and $\bar{\Sigma}^0$, π^0 and η (η-meson). Limiting cases: X (X-meson or η', see 1.4.2) and ω (omega-meson, see 1.4.3).

[1]) The distinction between ν_e and ν_μ is of importance only in neutrino reactions: ν_e produces electrons, ν_μ produces muons.
[2]) The lightest weakly decaying nuclei are the triton (t or ^{3}H, see table in 1.4.6*a*) and the hypertriton $^3_\Lambda$H. The latter has a binding energy of only 100 keV and a lifetime of $2 - 3 \cdot 10^{-10}$ sec (see [Ram 71] for example).
[3]) The first observation of this particle is published in SJNP 12, 171 (1971).
[4]) Not yet observed.

A more theoretical classification of particles and resonances is provided by their quantum numbers. The *lepton group* consists of e, μ, ν_e, ν_μ and their antiparticles; the *photon* is in a group by itself; the remaining states have strong interactions and are called *hadrons*. The hadrons are subdivided into baryons (which are fermions) and mesons (which are bosons). They occur in multiplets of $2I+1$ states ($I =$ isospin), the members of which are distinguished by their charge Q, which is related to the third component of isospin I_3 as follows:

$$Q = I_3 + \tfrac{1}{2}Y \quad (Y = S + B).\tag{1}$$

The hypercharge Y ($=$ twice the average charge of the isospin multiplet) is a new quantum number which is conserved both in strong and electromagnetic interactions. It is equivalent to the older quantum number S (strangeness), which is defined such that nucleons and pions have $S = 0$ ($B =$ baryon number: $+1$ for baryons, -1 for antibaryons and 0 for mesons). Hadrons with $Y \neq 0$ will be called "hypercharged", the others "hyperneutral" (the corresponding names with respect to S are "strange" and "nonstrange"). Hadrons also form approximate SU(3)-multiplets of multiplicity 1 (singlet: $I = 0$), 8 (octet: $I = 0$, $2 \times (I = \frac{1}{2})$, $I = 1$), and 10 (decuplet: $I = 0, \frac{1}{2}, 1, \frac{3}{2}$). This classification is shown in Fig. 1. For each singlet, there exists an octet of identical spin and parity and similar masses. The SU(3)-breaking part of the strong interaction induces a certain mixing between the two $I = 0$ states, i.e. the physical states η and $X = \eta'$ of the pseudoscalar mesons[1]), ϕ and ω of the vector mesons, f and f' of the tensor mesons become linear combinations of the octet and singlet states (denoted by 8 and 1):

$$\eta = \eta_8 \cos\theta_p - X_1 \sin\theta_p, \quad \phi = \phi_8 \cos\theta_v - \omega_1 \sin\theta_v, \quad f' = f_8 \cos\theta_t - f_1 \sin\theta_t,$$
$$X = \eta_8 \sin\theta_p + X_1 \cos\theta_p, \quad \omega = \phi_8 \sin\theta_v + \omega_1 \cos\theta_v, \quad f = f_8 \sin\theta_t + f_1 \cos\theta_t \tag{2}$$

with approximate mixing angles of $10°$, $40°$ and $32°$ (see tables in 1.4.2 ... 1.4.4). These angles are obtained from the Gell-Mann-Okubo mass formulas. For the pseudoscalar mesons,

$$4m_K^2 = m_\pi^2 + 3(m_\eta^2 \cos^2\theta_p + m_X^2 \sin^2\theta_p).\tag{3}$$

For baryons, linear mass formulas are used, e.g. $2m_N + 2m_\Xi = m_\Sigma + 3m_\Lambda$. We shall call the mixed singlet-octet a "nonet", although this term is frequently reserved for meson nonets in the quark model[2]) [Kokkedee 69]. The experimental determination of particle quantum numbers is explained for example in the books of [Källen 64, 65], [Cool and Marshak 68], or [Miller 69]. Special books on SU(3)-symmetry are [Gourdin 67] and [Carruthers 66]. A treatment in terms of U-spin (U-spin multiplets have fixed Q instead of fixed I) is given by [Pilkuhn 67].

[1]) At the time of writing, the spin of X could also be 2, in which case $\theta_p = 0$.
[2]) The signs in Eq. (2) are chosen such that $\sin\theta = +3^{-\frac{1}{2}}$ in the quark model. See also 1.3.4*d*.

Pilkuhn

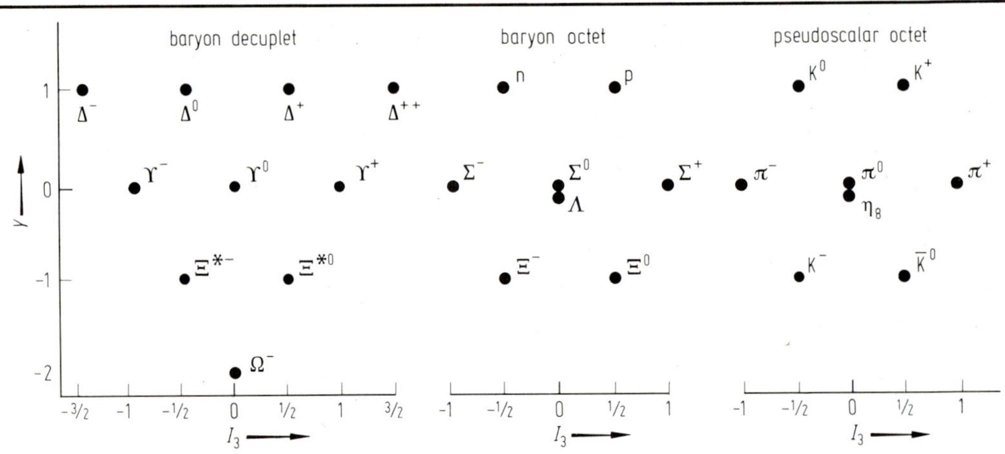

Fig. 1. (Y, I_3)-diagrams for the baryon decuplet, baryon octet and pseudoscalar meson octet. The octet state η_8 is defined in Eq. (2) in 1.1.2. The octets of the vector (K^*, ϱ, ϕ_8, $\bar{K}^*$) and tensor (K_N, A_2, f_8, $\bar{K}_N$) mesons are analogous to the pseudoscalar meson octet. The symbol Y represents the $\Sigma^*(1385)$ resonance, and Ξ^* stands for the $\Xi^*(1530)$ resonance. The $Y = 0$ row contains the hyperneutral particles.

The neutral kaon states K^0 and $\bar{K}^0$ have opposite hypercharge but otherwise identical quantum numbers. As hypercharge is not conserved in weak interactions, the states having definite lifetimes in a sufficiently thin medium (K_S = shortlived kaon, K_L = longlived kaon) are linear combinations of K^0 and $\bar{K}^0$ of the type

$$K_S = 2^{-\frac{1}{2}}[(1+\varepsilon)\,K + (1-\varepsilon)\,\bar{K}], \qquad K_L = 2^{-\frac{1}{2}}[(1-\varepsilon)\,K - (1+\varepsilon)\,\bar{K}]\,, \tag{4}$$

with $|\varepsilon| \ll 1$ (see Tab. 1 in 1.4.2, notes f and g). This mixing is physically different from (2) because it takes a time interval of $> 10^{-10}$ sec to convert a state with definite hypercharge into a state with definite lifetime. Otherwise, state mixing by weak interactions is negligible except for the theory of weak decays. State mixing by electromagnetic interactions occurs between Λ and Σ^0 and between η and π^0:

$$\begin{aligned}
\Lambda &= \Lambda_1 \cos\theta_{\Lambda\Sigma} + \Sigma_3 \sin\theta_{\Lambda\Sigma}, & \eta &= \eta_1 \cos\theta_{\pi\eta} + \pi_3 \sin\theta_{\pi\eta}\,, \\
\Sigma^0 &= -\Lambda_1 \sin\theta_{\Lambda\Sigma} + \Sigma_3 \cos\theta_{\Lambda\Sigma}, & \pi^0 &= -\eta_1 \sin\theta_{\pi\eta} + \pi_3 \cos\theta_{\pi\eta}\,,
\end{aligned} \tag{5}$$

where the subscripts 1 and 3 denote the pure SU(3) singlet and triplet states. As these mixing angles are small, one can replace $\sin\theta$ by θ and $\cos\theta$ by 1. Then the SU(3) formulas for electromagnetic mass splitting give (see [Dalitz 64]; [Carruthers 66] or [Gourdin 67])

$$\theta_{\Lambda\Sigma} = \frac{m_p - m_n + m_{\Sigma^0} - m_{\Sigma^+}}{\sqrt{3}(m_\Lambda - m_\Sigma)}, \qquad \theta_{\pi\eta} = \frac{m_{K^+}^2 - m_{K^0}^2 + m_{\pi^0}^2 - m_{\pi^+}^2}{\sqrt{3}(m_\eta^2 - m_\pi^2)}\,. \tag{6}$$

Particles that carry no charge whatsoever are eigenstates of the charge conjugation operator C. The neutral vector mesons ϱ^0, ω and ϕ have $C = -1$ like the photon, and π^0, η, X, A_2^0, f and f have $C = 1$. As a consequence of charge conjugation and isospin invariance, all hyperneutral mesons are eigenstates of the G-parity operator

$$G = C \cdot \exp(i\pi I_2) \tag{7}$$

with eigenvalue $+1$ for $\varrho^\pm$, ϱ^0, η, X, f and f and -1 for $\pi^\pm$, π^0, ω, ϕ, $A_2^\pm$ and A_2^0.

1.1.3 Electromagnetic interactions

The Hamiltonian of the electromagnetic interaction is

$$\mathcal{H}_{em} = e A_\mu j_{em}^\mu, \qquad j_{em}^\mu = \bar{\Psi}_{(e)} \gamma^\mu \Psi_{(e)} + \bar{\Psi}_{(\mu)} \gamma^\mu \Psi_{(\mu)} + j_{em}^\mu(\text{hadr}) \tag{1}$$

where A_μ is the 4-potential and j_{em}^μ the electromagnetic current operator. The electron and muon components of j_{em}^μ are used in quantum electrodynamics. The hadronic part $j_{em}^\mu(\text{hadr})$ is not known, but there is some evidence that it has the isospin- and SU(3)-properties of the electric charge operator Q, i.e. it is a mixture of an isovector operator (I_3) and an isoscalar operator (Y). One may thus say that the photon behaves as a mixture of an isovector and an isoscalar vector meson. The vector-meson-dominance model (VMD) explains this by a dominance of the low-lying vector meson states ($\varrho_\mu(x) = \varrho$-meson field etc.):

$$j_{em}^\mu(\text{hadr}) \approx -\frac{m_\varrho^2}{f_\varrho}\,\varrho^\mu(x) - \frac{m_\omega^2}{f_\omega}\,\omega^\mu(x) - \frac{m_\phi^2}{f_\phi}\,\phi^\mu(x)\,. \tag{2}$$

SU(3)-symmetry predicts the photon to behave like the U-spin singlet member of an octet (i.e. like $\sqrt{3}\varrho + \phi_8$). In practice, field theory is only needed for radiative corrections. We shall be concerned directly with the T-matrix element (Eq. (4) in 1.1.1). For the emission of a photon, T is of the form

$$T(\gamma) = e\varepsilon_\mu^*(\lambda)\, T^\mu \quad \text{with} \quad P_{\gamma\mu} T^\mu = 0, \tag{3}$$

with $P_\gamma =$ 4-momentum of the photon and ε_μ^* the complex conjugate of Eq. (9) in 1.1.1. The second equation in (3) may be called gauge invariance; it is related to the conservation of the electromagnetic current. For the decay $\Sigma^0 \to \Lambda\gamma$, gauge invariance and parity conservation restrict T^μ to the form

$$T_\mu(\Sigma^0 \to \Lambda) = \bar{u}_\Lambda \left[F_3(t)\left(\frac{\gamma_\mu t}{m_\Sigma^2 - m_\Lambda^2} - \frac{P_{\gamma\mu}}{m_\Sigma - m_\Lambda} \right) + \frac{\kappa^{\Lambda\Sigma}}{m_\Sigma + m_\Lambda} \frac{1}{2}(\gamma_\mu\gamma_\nu - \gamma_\nu\gamma_\mu) P_\gamma^\nu F_2(t) \right] u_\Sigma \tag{4}$$

with $t = P_\gamma^2 = 0$ for a real photon and $F_2(t) = 1$. Similarly, the matrix element of the electromagnetic current between identical baryons b of 4-momentum P in the initial state and 4-momentum P' in the final state is given by

$$T_\mu(\mathrm{b}) = \bar{u}_\mathrm{b}' \left[F_1(t)\gamma_\mu + \frac{\kappa}{2m} F_2(t)\frac{1}{2}(\gamma_\mu\gamma_\nu - \gamma_\nu\gamma_\mu)(P - P')^\nu \right] u_\mathrm{b} \tag{5}$$

$$= \bar{u}_\mathrm{b}'[(F_1 + \kappa F_2)\gamma_\mu - \kappa F_2(2m)^{-1}(P + P')_\mu] u_\mathrm{b},$$

$$t = (P - P')^2, \quad F_1(0) = Q, \quad F_2(0) = 1, \quad e(Q + \kappa)/2m = \mu.$$

where μ is the magnetic moment. One frequently writes $t = -\vec{q}^2$, where q is the momentum transfer in the cms, in the case $m_\mathrm{b} = m_{\mathrm{b}'}$.

In general, for processes having only one hadron in the initial and final state, T_μ is determined by a few "coupling constants" (eQ and $e\kappa$ in our example) for a real photon. When the photon is taken off the mass shell, these coupling constants are modified by "form factors" $F_i(t)$.

It should be noted that (5) is the relativistically correct form for electron scattering on any target of spin $\frac{1}{2}$. In particular, the formula also applies to $^3\mathrm{He}$, with $m = m(^3\mathrm{He})$ in (5), and in Eq. (7) in 1.1.1. On the other hand, the interpretation of a form factor as the Fourier transform of a spatial charge distribution is a nonrelativistic concept which becomes useless for large q^2.

For pions, we have

$$T_\mu(\pi) = Q_\pi e F_\pi(t)(P_\pi + P_\pi')_\mu, \quad F_\pi(0) = 1. \tag{6}$$

In particular, the π^0 meson (and also η and X) cannot emit or absorb photons, due to charge conjugation invariance, except by transforming themselves into ϱ^0, ω or ϕ mesons (see 1.4.3c).

The complete matrix element of electron-hadron-scattering is (see Eq. (1) in 1.3.3 for $m_e = 0$, and Eq. (3) in 1.3.3 with gauge invariance (3))

$$T(\mathrm{eb} \to \mathrm{e'b'}) = e^2 \bar{u}_e' \gamma^\mu u_e\, T_\mu/t. \tag{7}$$

In addition to interactions obviously involving photons, there are decays such as $\omega \to \pi^+\pi^-$ or $\eta \to \pi\pi\pi$ which do not conserve isospin but do conserve hypercharge and parity. Their matrix elements are roughly a factor $\alpha = e^2/4\pi = 1/137$ smaller than those of strong interactions; for this reason they are also called electromagnetic. They could be induced by emission and reabsorption of a virtual photon. In this case the isospin could change by as much as 2 units, but it appears that the $\Delta I = 2$ piece is suppressed. This is analogous to the nonleptonic weak interactions (see the following section). For a recent review of the electromagnetic interactions of hadrons, see [Cumming 71].

1.1.4 Weak and superweak interactions

The Hamiltonian of the weak interaction is probably

$$\mathscr{H}_{weak} = 2^{-\frac{1}{2}} g j_\mu j^{\mu+}, \quad j_\mu = \bar{\Psi}_{(e)}\gamma_\mu(1 - \gamma_5)\, \Psi_{v_e} + \bar{\Psi}_{(\mu)}\gamma_\mu(1 - \gamma_5)\, \Psi_{v_\mu} + j_\mu(\mathrm{hadr}), \tag{1}$$

where g is the weak interaction constant measured in muon decay. The word "probably" must be added for three reasons: Firstly, neutrino-electron scattering as predicted by (1) remains to be detected. Secondly, nonleptonic decays such as $\Lambda \to \mathrm{p}\pi^-$ cannot be shown to be of the form $j_\mu j^{\mu+}$ (see below). Finally (1) does not allow the calculation of higher order corrections. Therefore, all the physics is already contained in the form of the matrix element for emission of an e, $\bar{v}_e$ or μ, $\bar{v}_\mu$ pair:

$$T(l\bar{v}_l) = \bar{u}_l(P_l\lambda_l)\gamma_\mu(1 - \gamma_5)u_{v_l}(-P_{\bar{v}}, \lambda_{\bar{v}})(V^\mu + A^\mu) \tag{2}$$

where V^μ is a vector like T^μ, and A^μ is an axial vector.

Pilkuhn

For baryon decays $b \to d l \bar{v}$, V^μ and A^μ are given by

$$V_\mu = 2^{-\frac{1}{2}} \bar{u}_d \left[g_V(t) \gamma_\mu + g_T(t) \frac{\gamma_\mu \gamma_v - \gamma_v \gamma_\mu}{2(m_b + m_d)} (P - P')^v + g_S(t) \frac{(P - P')_\mu}{m_b + m_d} \right] u_b , \tag{3}$$

$$A_\mu = 2^{-\frac{1}{2}} \bar{u}_d [-g_A(t) \gamma_\mu + g_P(t) (P - P')_\mu + g_3(t) (P + P')_\mu] \gamma_5 u_b , \tag{4}$$

where (3) is simply the non-gauge-invariant generalization of T^μ. Actually, it appears that at least for hypercharge conserving decays ("hyperneutral decays"), V_μ is in fact also gauge invariant. This is called the conserved vector current hypothesis (CVC). Also, the isospin properties of the charged vector current are identical with those of the $\varrho^\pm$ and $K^{*\pm}$ states. It thus appears that the vector mesons dominate both the electromagnetic and the weak vector currents.

The relative strength of hyperneutral and hypercharged couplings is fixed by Cabibbo's universality hypothesis (θ_V = Cabibbo angle):

$$g_V^{\Delta Y = 0} = g \cos \theta_V, \qquad g_V^{|\Delta Y| = 1} = g \sin \theta_V \tag{5}$$

apart from SU(3)-coefficients (see 1.4.6 b). Similarly, the axial currents form the charged components of an octet. Consequently, the following selection rules hold both for V_μ and for A_μ:

$$\Delta I = 1 \quad \text{for} \quad \Delta Y = 0, \quad \Delta I = \tfrac{1}{2} \quad \text{and} \quad \Delta Y = \Delta Q \quad \text{for} \quad |\Delta Y| = 1 . \tag{6}$$

The relative strength of the axial $|\Delta Y| = 1$ and $\Delta Y = 0$ couplings is $\tan \theta_A \approx \tan \theta_V$, but $g_A \neq g_V$ (see 1.4.6 a).

The matrix elements A_μ are not gauge invariant, but $(P - P')_\mu A^\mu$ (the matrix elements of the divergence of the axial current) are dominated by the pseudoscalar states $\pi^\pm$ and $K^\pm$. This leads to the Goldberger-Treiman relation

$$-g_\pi f_{bb'\pi} = (8\pi)^{-\frac{1}{2}} m_\pi g_{A, bb'} \tag{7}$$

between the pion decay constant g_π (see 1.4.2 d), the axial decay constant g_A, and the strong (pseudovector) coupling constant $f_{bb'\pi^\pm}$ (see 1.3.3 and 1.3.4). The corresponding relation for kaons is much more uncertain. The full axial hyperneutral current has negative G-parity like the pion. Hyperneutral currents with positive G-parity in the axial part and negative G-parity in the vector part, socalled "second-class" currents [Weinberg 58] appear to be absent except for secondary effects in nuclei [Wilkinson 71; Delorme 71].

The matrix elements of an octet operator, taken between two octet states such as the baryons, contain two reduced matrix elements g_F and g_D in addition to SU(3)-coefficients. We shall use

$$g_F + g_D = g \cos \theta, \qquad g_D [g_F + g_D]^{-1} = \alpha_{weak} . \tag{8}$$

At $t = 0$, however, the vector current couples to the total charge and has therefore α(weak, vector, $t = 0$) $= 0$. Also, it is so far not possible to measure any t-dependence of α. Therefore, α will denote the average fraction of D-type coupling of the axial current. The precise factors are given in 1.4.6 b.

Nonleptonic weak decays such as $\Lambda \to \pi N$, $\Sigma \to \pi N$ (1.4.6 c) or $K \to \pi\pi$, $K \to \pi\pi\pi$ (1.4.2g and h) follow the $\Delta I = \tfrac{1}{2}$ rule (the decay $K^\pm \to \pi^\pm \pi^0$ being the most notable violation). If these decays also originate from the interaction $j_\mu j^{\mu+}$ with $\Delta I = 1$ for the hyperneutral and $\Delta I = \tfrac{1}{2}$ for the hypercharged current, then $\Delta I = \tfrac{3}{2}$ should also result. The suppression of the $\Delta I = \tfrac{3}{2}$ piece is not understood, similar to the $\Delta I = 2$ suppression in the corresponding electromagnetic decays.

Interactions which do not conserve CP (P = parity, C = charge conjugation) are so far restricted to the neutral kaon decays. Even there, the whole effect seems to be due to the fact that the states K_S and K_L in Eq. (4) in 1.1.2 are not CP-eigenstates (they are so for $\varepsilon = 0$). If this is true, one can divide the weak interactions into the normal weak interactions which conserve CP, and into a "superweak", CP-violating interaction which gives measurable effects only in $K^0 - \bar{K}^0$-transitions [Wolfenstein 64].

For the weak interactions in particle physics, the book of [Marshak, Riazuddin, and Ryan 69] can be recommended. (There is at present no comparable book on the electromagnetic interactions of hadrons.) Books on nuclear β-decay are [Konopinski 66], [Schopper 66] and [Wu 66]. Reviews on leptonic decays of hadrons are given by [Willis 68] and [Rubbia 69]. See also [Bailin 71].

In nuclei, two other forms of weak decays can be observed which will not be discussed here: stimulated Λ decay $\Lambda p \to pn$ in hyperfragments [Davies 67] and parity-violating, hypercharge conserving decays such as $^{16}O \to {}^{12}C + \alpha$ [Hättig 70] or the parity impurities in $np \to d\gamma$ reaction [Lobashov 70]. See also PRL 29, 518 (1972).

1.1.5 Strong interactions and resonances

The electromagnetic and weak interactions of hadrons are essentially characterized by a few coupling constants and form factors. For the strong interactions, the situation is more complicated. Suppose for example that in a reaction $ab \to cd$, the exchange of a stable meson e of mass m_e is possible. Taking all particles as spinless, there exists a unique decomposition of the scattering amplitude,

$$T(\mathrm{ab} \to \mathrm{cd}) = g_{\mathrm{ace}} g_{\mathrm{bde}} (m_\mathrm{e}^2 - t)^{-1} + T' \tag{1}$$

where T' is regular at $t = m_\mathrm{e}^2$. The g's are the coupling constants. The decomposition is accomplished by means of dispersion relations (1.3.5). We can also decompose

$$T(\mathrm{ab} \to \mathrm{cd}) = V_{\mathrm{ace}} V_{\mathrm{bde}} (m_\mathrm{e}^2 - t)^{-1} + T'', \qquad V = g \cdot F(t), \qquad F(t) = 1 + F' \cdot (t - m_\mathrm{e}^2) + \cdots \tag{2}$$

where V are the vertex functions and F the corresponding form factors. Since terms proportional to F' have no pole at $t = m_\mathrm{e}^2$, they are contained in T' and can be obtained only if a reliable model for T'' exists. The product $V_{\mathrm{ace}} \cdot V_{\mathrm{bde}}$ is called a residue function. In the Regge pole model, one tries to approximate T by a sufficiently large number of "Regge poles" (see 1.5.6 d). This is always possible, but in that case the residue functions will not necessarily factorize. If one insists on factorization, no strong reaction is known with $T'' = 0$. This should be a warning to anyone who tries to use the coupling constants with or without form factors in a single-particle exchange model. Strong form factors will be given only for π exchange (1.5.5) and for the Δ resonance (1.4.7 a). As the stable particle poles lie outside the physical region, other parameters are of equal or even greater importance in describing the physical amplitude. In particular, scattering lengths are included in the tabulation.

The situation is slightly better for the decay coupling constants of resonances. Consider first the weak decay of an unstable particle. The total decay rate Γ is obtained from the exponential decrease of intensity I in proper time $\tau = t_{lab} E_{lab}/m$

$$\mathrm{d}I/\mathrm{d}\tau = -\Gamma I, \qquad \Gamma = t_0^{-1} \ln 2 = 0.69315 \, t_0^{-1} \tag{3}$$

where t_0 is the half life (Γ^{-1} is the mean life or just lifetime). The partial rate $\Gamma(\mathrm{d} \to 1 \ldots \mathrm{n})$ for the decay of particle d into n particles of spins $S_1 \ldots S_n$ is related to the decay matrix element by the "golden rule"

$$\Gamma(\mathrm{d} \to 1 \ldots \mathrm{n}) = \frac{1}{2m} \int \mathrm{d}\,\mathrm{Lips}(m^2; P_1 \ldots P_\mathrm{n}) \sum_{\lambda_1 \cdots \lambda_\mathrm{n}} |T(\mathrm{d} \to 1 \ldots \mathrm{n})|^2,$$

$$\mathrm{d}\,\mathrm{Lips}(m^2; P_1 \ldots P_\mathrm{n}) = \prod_{i=1}^{n} \frac{\mathrm{d}^3 p_i}{2E_i} \, \delta\!\left(m - \sum_{i=1}^{n} E_i\right) \delta\!\left(\sum_{i=1}^{n} \vec{p}_i\right)(2\pi)^{4 - 3n}, \tag{4}$$

where m is the mass of particle d, and d Lips is the Lorentz invariant phase space element [Pilkuhn 67] in the cms ($\vec{p}_\mathrm{d} = 0$). The summation in front of $|T|^2$ extends over the helicities $\lambda_1 \ldots \lambda_n$ of the produced particles. Already for electromagnetic particle decays (π^0, η^0, Σ^0), (3) cannot be measured. For strongly decaying resonances, Γ can be measured from the invariant mass distribution of its decay products, which is given by

$$P(s_{1 \ldots \mathrm{n}}) = \mathrm{const} \cdot m\Gamma(\mathrm{d} \to 1 \ldots \mathrm{n}) \left[(m^2 - s_{1 \ldots \mathrm{n}})^2 + m^2 \Gamma^2\right]^{-1}, \qquad s_{1 \ldots \mathrm{n}} = \left(\sum_{i=1}^{n} P_i\right)^2, \tag{5}$$

provided the production matrix element for the resonance (in a reaction ab $\to$ cd say) does not vary between $s_{1 \ldots \mathrm{n}} = m^2$ and $m^2 \pm m\Gamma$. If Γ is not too large, Γ is the full width at half maximum of the distribution (5), and it does not matter whether one puts m or $\sqrt{s}$ in (5). If on the other hand Γ is large, (5) cannot be fitted with an s-independent Γ. In this case, the normally adopted prescription is to replace m by $\sqrt{s}$ everywhere in (4) except possibly in front of the integral. For a decay into two particles, the main change is in the "threshold factor", $\Gamma \approx p^{2l+1}$, where p is the actual decay momentum at invariant mass $\sqrt{s_{12}}$ instead of the decay momentum at the resonance position m_d (see 1.2.1). The resulting $P(s_{1 \ldots \mathrm{n}})$ is called a Breit-Wigner resonance. As to the factor $1/m$ in front of the integral, it appears to be most appropriate not to change it. If one writes (4) in the form $m\Gamma = \int \ldots$, one has exactly the combination $m\Gamma$ entering (5). (One may of course also write $\sqrt{s_{1 \ldots \mathrm{n}}}\,\Gamma$ in (4) *and* in (5), which is merely a redefinition of the symbol Γ.)

If Γ is large, (5) need no longer describe the invariant mass distribution. In this case, the simplest starting point is the decomposition of the T-matrix for the over-all process ab $\to 1 \ldots$ n,

$$T(\mathrm{ab} \to \mathrm{c}1 \ldots \mathrm{n}) = T(\mathrm{ab} \to \mathrm{cd})\,(m^2 - s_{1 \ldots \mathrm{n}} - im\Gamma)^{-1}\, T(\mathrm{d} \to 1 \ldots \mathrm{n}) + T'', \tag{6}$$

where the contribution T''_d of T'' to the resonating partial wave d is negligible around $s_{1 \ldots \mathrm{n}} = m^2 - im\Gamma$. One may then still determine the whole function $\Gamma(s_{1 \ldots \mathrm{n}})$. If on the other hand no decomposition (6) with negligible T''_d exists, one should read (6) analogous to (1), with $T(\mathrm{ab} \to \mathrm{cd}) \cdot T(\mathrm{d} \to 1 \ldots \mathrm{n})$ given by their values at $m^2 - im\Gamma$ (residues) and Γ given by $\Gamma(m)$ according to (4).

It is possible that in such cases the resonance approximation is not too useful altogether. It may be difficult to prove that the residue factorizes. For resonances other than s-wave resonances, $T(\mathrm{d} \to 1 \ldots \mathrm{n})$ is complex at the resonance position $m^2 - im\Gamma$ since the decay momentum is complex there.

In practice, $|T|^2$ is frequently approximated by a sum of distributions of the type (5) in the various invariant masses (for three particles this is done in the Dalitz plot) plus a constant, noninterfering "background". More elaborate prescriptions are used for example in the Veneziano model. Sometimes (5) is used with constant Γ although a Breit-Wigner width gives a better fit. In this case $m\Gamma(m)$ is somewhat smaller than the "experimental" $m\Gamma$. The coupling constants always refer to $m\Gamma(m)$. Sometimes the resonance form $\Gamma[(m-\sqrt{s})^2 + \Gamma^2/4]^{-1}$ is used. This agrees with (5) for $\sqrt{s} + m \approx 2m$ and is generally harmless for baryon resonances, except for the Δ resonance.

The vertices of strong interactions can also be translated to interaction Lagrangians (see 1.3.3).

1.1.6 Textbooks and review articles

Bailin, D.	71	in Reports on progress in physics, Vol. 34.
Bjorken, J. D., and S. D. Drell	65	Relativistic quantum fields (New York: McGraw-Hill).
Bjorken, J. D., and S. D. Drell	67	(Deutsche Übersetzung) Relativistische Quantenfeldtheorie, B. I.-Hochschultaschenbücher 101/101a (Mannheim: Bibliogr. Inst.).
Carruthers, P.	66	Introduction to unitary symmetry (New York: Wiley).
Cool, R. L., and R. E. Marshak	68	Advances in particle physics, Vol. 2 (New York: Wiley Interscience).
Cumming, J., and H. Osborn	71	Hadronic Interactions of electrons and photons (London: Academic Press).
Ebel	71	NP, B 33, 317: G. Ebel, A. Müllensiefen, H. Pilkuhn, F. Steiner, D. Wegener, M. Gourdin, C. Michael, J. L. Petersen, M. Roos, B. R. Martin, G. Oades, J. J. de Swart.
Gasiorowicz, S.	66	Elementary particle physics (New York: Wiley).
Gourdin, M.	67	Unitary symmetry (Amsterdam: North-Holland).
Källen, G.	64	Elementary particle physics (Reading: Addison-Wesley).
Källen, G.	65	(Deutsche Übersetzung) Elementarteilchenphysik, B. I.-Hochschultaschenbücher 100/100a/100b (Mannheim: Bibliogr. Inst.).
Konopinski, E. J.	66	The theory of beta radioactivity (Oxford: Clarendon Press).
Kokkedee, J. J. J.	69	The quark model (New York, Amsterdam: Benjamin).
Marshak, R. E., Riazuddin, C. P. Ryan	69	Theory of weak interactions in particle physics (New York: Wiley Interscience).
Martin, A. D., and T. D. Spearman	70	Elementary particle theory (Amsterdam: North-Holland).
Miller, D. H.	69	in High energy physics (Academic Press, ed. Burhop, E. H. S.), Vol. 2.
Muirhead, H.	65	The theory of elementary particles (Pergamon Press).
Particle	71, 72	RMP 43, S 1, and PL 39 B, 1: Particle data group.
Pilkuhn, H.	67	The interactions of hadrons (Amsterdam: North-Holland).
Rubbia, C.	69	in High energy physics (Academic Press, ed. Burhop, E. H. S.), Vol. 3.
Schopper, H. F.	66	Weak interactions and nuclear beta decay (Amsterdam: North-Holland).
Schweber, S. S.	61	An introduction to relativistic quantum field theory (Evanston: Row, Peterson & Co.).
Willis, W., and J. Thompson	68	in Advances in particle physics (New York: Wiley Interscience, ed. Marshak, R. E., R. L. Cool), Vol. 1.
Wu, C. S., and S. A. Moszkowski	66	Beta decay (New York: Wiley).

Further references for 1.1

For journal abbreviation see p. 2

Dalitz	64	PL 10, 153: R. H. Dalitz, F. von Hippel.
Davies	67	High En. Phys. (Academic Press, ed. Burhop). Vol. 2, 365: D. H. Davies, J. Sacton.
Delorme	71	NP B 34, 317: J. Delorme, M. Rho.
Hättig	70	PRL 25, 941: H. Hättig, K. Hünchen, H. Wäffler,
Lobashov	70	JETL 11, 76: V. M. Lobashov, A. E. Egorov, D. M. Kaminker, V. A. Nazarenko, L. F. Saenko, L. M. Smorotritskii, G. I. Kharkevich, V. A. Knyaz'kov.
Ram	71	NP B 28, 566: B. Ram, W. Williams.
Weinberg	58	PR 112, 1375: S. Weinberg.
Wilkinson	71	PRL 26, 1127: D. H. Wilkinson, D. E. Alburger.
Wolfenstein	64	PL 13, 562: L. Wolfenstein.

1.2 General formulas for decays and resonances

1.2.1 Decay angular distributions, phases, and penetration factors

For the two-body decay of a resonance d in its rest frame, Eq. (4) in 1.1.5 simplifies to

$$d\,\mathrm{Lips} = \frac{p\,d\Omega}{16\pi^2 s_{12}^{\frac{1}{2}}}, \qquad m\Gamma_{12} = \frac{p}{16\pi\sqrt{s_{12}}} \sum_{\lambda_1\lambda_2} \int_{-1}^{1} d\cos\vartheta\,|T_{M\lambda_1\lambda_2}(s_{12},\vartheta,\varphi)|^2, \tag{1}$$

where $p = 2s_{12}^{-\frac{1}{2}}\lambda^{\frac{1}{2}}(s_{12},m_1^2,m_2^2)$ is the decay momentum, $\sqrt{s_{12}}$ is the invariant mass of particles 1 and 2 [1]), λ_1 and λ_2 are their helicities, M is the magnetic quantum number of d (averaging over M is unnecessary), ϑ is the angle between the spin quantization axis of d and the momentum of particle 1, and φ is a possible rotation angle around the spin quantization axis (counted for example from the collision plane of the reaction $ab\to cd$). For identical particles in the final state ($\pi^0\to\gamma\gamma$), the upper limit of integration in (1) is 0. The angular dependence of T is ($S=$ spin of d):

$$T_{M\lambda_1\lambda_2}(\vartheta,\varphi) = \left(\frac{2S+1}{4\pi}\right)^{\frac{1}{2}} D^{S*}_{M\lambda}(\varphi,\vartheta,-\varphi)\,T_S(\lambda_1,\lambda_2)$$

$$= \left(\frac{2S+1}{4\pi}\right)^{\frac{1}{2}} e^{i\varphi(M-\lambda)}\,d^S_{M\lambda}(\vartheta)\,T_S(\lambda_1,\lambda_2), \tag{2}$$

$$d^S_{MM'}(\vartheta) = (\exp(-i\vartheta J_y^{(S)}))_{MM'} \qquad \lambda = \lambda_1 - \lambda_2.$$

Tab. 1. The functions $d^S_{MM'}(\theta)$ for $S=\frac{1}{2}\cdots\frac{5}{2}$. The sign convention of [Rose 57] is chosen, and $d^S_{MM'}=d^S_{-M'-M}=(-1)^{M-M'}d_{M'M}$

$M\backslash M'$	$+\frac{1}{2}$	$-\frac{1}{2}$
$+\frac{1}{2}$	$\cos\frac{1}{2}\theta$	$-\sin\frac{1}{2}\theta$
$-\frac{1}{2}$	$\sin\frac{1}{2}\theta$	$\cos\frac{1}{2}\theta$

$M\backslash M'$	$+1$	0	-1
$+1$	$\cos^2\frac{1}{2}\theta$	$-2^{-\frac{1}{2}}\sin\theta$	$\sin^2\frac{1}{2}\theta$
0	$2^{-\frac{1}{2}}\sin\theta$	$\cos\theta$	$-2^{-\frac{1}{2}}\sin\theta$
-1	$\sin^2\frac{1}{2}\theta$	$2^{-\frac{1}{2}}\sin\theta$	$\cos^2\frac{1}{2}\theta$

$M\backslash M'$	$+\frac{3}{2}$	$+\frac{1}{2}$	$-\frac{1}{2}$	$-\frac{3}{2}$
$+\frac{3}{2}$	$\cos^3\frac{1}{2}\theta$	$-3^{\frac{1}{2}}\cos^2\frac{1}{2}\theta\sin\frac{1}{2}\theta$	$3^{\frac{1}{2}}\cos\frac{1}{2}\theta\sin^2\frac{1}{2}\theta$	$-\sin^3\frac{1}{2}\theta$
$+\frac{1}{2}$	$3^{\frac{1}{2}}\cos^2\frac{1}{2}\theta\sin\frac{1}{2}\theta$	$\cos\frac{1}{2}\theta(1-3\sin^2\frac{1}{2}\theta)$	$\sin\frac{1}{2}\theta(1-3\cos^2\frac{1}{2}\theta)$	$3^{\frac{1}{2}}\cos\frac{1}{2}\theta\sin^2\frac{1}{2}\theta$
$-\frac{1}{2}$	$3^{\frac{1}{2}}\cos\frac{1}{2}\theta\sin^2\frac{1}{2}\theta$	$\sin\frac{1}{2}\theta(3\cos^2\frac{1}{2}\theta-1)$	$\cos\frac{1}{2}\theta(1-3\sin^2\frac{1}{2}\theta)$	$-3^{\frac{1}{2}}\cos^2\frac{1}{2}\theta\sin\frac{1}{2}\theta$
$-\frac{3}{2}$	$+\sin^3\frac{1}{2}\theta$	$3^{\frac{1}{2}}\cos\frac{1}{2}\theta\sin^2\frac{1}{2}\theta$	$3^{\frac{1}{2}}\cos^2\frac{1}{2}\theta\sin\frac{1}{2}\theta$	$\cos^3\frac{1}{2}\theta$

$M\backslash M'$	2	1	0	-1	-2
2	$\cos^4\frac{1}{2}\theta$	$-2\cos^3\frac{1}{2}\theta\sin\frac{1}{2}\theta$	$6^{\frac{1}{2}}\cos^2\frac{1}{2}\theta\sin^2\frac{1}{2}\theta$	$-2\cos\frac{1}{2}\theta\sin^3\frac{1}{2}\theta$	$\sin^4\frac{1}{2}\theta$
1	$2\cos^3\frac{1}{2}\theta\sin\frac{1}{2}\theta$	$\cos^2\frac{1}{2}\theta(\cos^2\frac{1}{2}\theta-3\sin^2\frac{1}{2}\theta)$	$-\frac{3^{\frac{1}{2}}}{2}\sin\theta\cos\theta$	$\sin^2\frac{1}{2}\theta(3\cos^2\frac{1}{2}\theta-\sin^2\frac{1}{2}\theta)$	
0	$6^{\frac{1}{2}}\cos^2\frac{1}{2}\theta\sin^2\frac{1}{2}\theta$	$\frac{3^{\frac{1}{2}}}{2}\sin\theta\cos\theta$	$\frac{1}{2}(3\cos^2\theta-1)$	$-\frac{3^{\frac{1}{2}}}{2}\sin\theta\cos\theta$	

$M\backslash M'$	$\frac{5}{2}$	$\frac{3}{2}$	$\frac{1}{2}$	$-\frac{1}{2}$	$-\frac{3}{2}$	$-\frac{5}{2}$
$\frac{5}{2}$	c^5	$-5^{\frac{1}{2}}c^4s$	$10^{\frac{1}{2}}c^3s^2$	$-10^{\frac{1}{2}}c^2s^3$	$5^{\frac{1}{2}}cs^4$	$-s^5$
$\frac{3}{2}$	$5^{\frac{1}{2}}c^4s$	$c^3(c^2-4s^2)$	$2^{\frac{1}{2}}sc^2(3s^3-2c^2)$	$2^{\frac{1}{2}}s^2c(3c^2-2s^2)$	$s^3(s^2-4c^2)$	$5^{\frac{1}{2}}cs^4$
$\frac{1}{2}$	$10^{\frac{1}{2}}c^3s^2$	$2^{\frac{1}{2}}sc^2(2c^2-3s^2)$	$c(c^4+3s^4-6c^2s^2)$	$s(6s^2c^2-s^4-3c^4)$	$c\equiv\cos\frac{1}{2}\theta,\ s\equiv\sin\frac{1}{2}\theta$	

The necessary d-functions are given in Tab. 1. $|T|^2$ in (1) is in fact independent of φ. The decay angular distribution for polarization density matrix q of d is (see [Pilkuhn 67], ref. 1.1.6):

$$W(\vartheta,\varphi) = \sum_{MM'} \varrho^S_{MM'} A^S_{MM'}(\vartheta,\varphi), \qquad \int d\cos\vartheta\,d\varphi\,W = 1, \tag{3}$$

---

[1]) λ is defined in Eq. (12) in 1.1.1, and s in Eq. (5) in 1.1.5.

Pilkuhn

$$A_{MM'}^S(\vartheta, \varphi) = \frac{2S+1}{4\pi} \sum_{\lambda_1 \lambda_2} |T_S(\lambda_1, \lambda_2)|^2 \, D_{M\lambda}^{S*} D_{M'\lambda}^S \left[\sum_{\lambda_1 \lambda_2} |T_S(\lambda_1, \lambda_2)|^2 \right]^{-1},$$

(4)

$$m\Gamma_{12} = p(32\pi^2 s_{12}^{\frac{1}{2}})^{-1} \sum_{\lambda_1 \lambda_2} |T_S(\lambda_1, \lambda_2)|^2, \quad \text{since} \quad \int d\cos\vartheta(2S+1)|D|^2 = 2.$$

(5)

When particle d has spin $\frac{1}{2}$, one normally puts $\varrho = \frac{1}{2}(1 + \vec{P} \cdot \vec{\sigma})$, where $\vec{P}$ is the "polarization vector". T_S is real except in weak and electromagnetic decays, where it has the elastic scattering phase of the final state. For the decay into two spinless particles, the simplest Lorentz-invariant matrix elements are given in Tab. 2. For arbitrary spin S, one has $(s = s_{12})$

$$T_M(s, \vartheta, \varphi) = \frac{2S+1}{4\pi} D_{M0}^{S*}(\vartheta, \varphi) \, T_S(s) = Y_M^S(\vartheta, \varphi) \, T_S(s).$$

(6)

Tab. 2. Matrix element T, width Γ and suppression factor N for the decay of a resonance of spin S into two spinless particles. $\varepsilon_{\mu\nu}$ is the tensor defined in Eq. (11) in 1.1.1. Only G_1 is dimensionless

S	$T_M(s, \vartheta, \varphi)$	T_S	$m\Gamma$ (Breit-Wigner)	N
0	G_0	$(4\pi)^{\frac{1}{2}} G_0$	$\dfrac{G_0^2}{4\pi} p s^{-\frac{1}{2}}$	1
1	$G_1 \varepsilon_\mu(M)(P_2 - P_1)^\mu$	$\left(\dfrac{4\pi}{3}\right)^{\frac{1}{2}} 2 G_1 p$	$\dfrac{G_1^2}{4\pi} \frac{2}{3} p^3 s^{-\frac{1}{2}}$	$1 + R^2 p^2$
2	$G_2 \varepsilon_{\mu\nu}(M)(P_2 - P_1)^\mu(P_2 - P_1)^\nu$	$\left(\dfrac{8\pi}{15}\right)^{\frac{1}{2}} 4 G_2 p^2$	$\dfrac{G_2^2}{4\pi} \frac{16}{15} p^5 s^{-\frac{1}{2}}$	$1 + R^2 p^2/3 + R^4 p^4/9$

It is noted that T gives rise to the "threshold behaviour" p^{2S+1} of Γ. This behaviour is reliable only for small p. For large p (i.e. for $p \gg p_{res}$), Γ certainly does not diverge like p^{2S+1}. Several modifications of Γ can be used here. A simple modification adopted from potential theory ($R \approx$ range of potential) is to put

$$m\Gamma = m\Gamma \,(\text{Breit-Wigner})/N, \qquad N = R^2 p^2 |h^{(1)}(Rp)|^2,$$

with N given in the last column of Tab. 2. It must be noted that this may change the coupling constant considerably. For example, for a p-wave resonance of given $\Gamma(m)$, $G_1^2/4\pi$ will be twice as large for $R^2 p^2(m) = 1$ as for $R^2 p^2(m) = 0$. This ambiguity is very pronounced for the Δ resonance (see 1.4.7a). When the decay products carry spin, one uses $N(L)$, where L is orbital angular momentum. See also [von Hippel 72].

For $S = 1$, the angular distribution for decays into two spinless particles is determined by 3 independent and real parameters of the density matrix ϱ. For the spin quantization axis of d in the production plane of reaction $ab \rightarrow cd$, one finds

$$W(\vartheta, \varphi) = \frac{1 - A}{4\pi}, \qquad A = (\varrho_{00} - \varrho_{11})(1 - 3\cos^2\vartheta) + 3\sqrt{2}\,\text{Re}\varrho_{10}\sin2\vartheta + 3\varrho_{1,-1}\sin^2\vartheta\cos2\varphi.$$

(7)

For $S = 2$, the corresponding distribution is [Dalitz 66]

$$W(\vartheta, \varphi) = \frac{15}{16\pi} \{3\varrho_{00}(\cos^2\vartheta - \tfrac{1}{3})^2 + \varrho_{11}\sin^2 2\vartheta + \varrho_{22}\sin^4\vartheta - 2A_1\sin2\vartheta\cos\varphi$$

$$-4A_2\sin^2\vartheta\cos2\varphi + 4\text{Re}\varrho_{2,-1}\sin^3\vartheta\cos\vartheta\cos3\varphi + \varrho_{2,-2}\sin^4\vartheta\cos4\varphi\},$$

(8)

$$A_1 = \text{Re}\varrho_{21}\sin^2\vartheta + \sqrt{6}\text{Re}\varrho_{10}(\cos^2\vartheta - \tfrac{1}{3}), \qquad A_2 = \varrho_{1,-1}\cos^2\vartheta - \sqrt{\tfrac{3}{2}}\text{Re}\varrho_{20}(\cos^2\vartheta - \tfrac{1}{3}).$$

(9)

For pseudoscalar and vector meson exchange alone, $\varrho_{22} = A_1 = A_2 = \varrho_{2,-2} = 0$.

Decays in flight. The lifetime Γ_{lab}^{-1} is $\Gamma^{-1} E_{lab}/m$ due to the time dilatation. The phase space element entering Eq. (1) can be transformed to $dE_{1\,lab}$:

$$d\text{Lips}(s_{12}; P_1, P_2) = \frac{p\,d\cos\vartheta}{8\pi s_{12}^{-\frac{1}{2}}} = \frac{dE_{1\,lab}}{8\pi p_{lab}}, \qquad p_{lab} = (E_{lab}^2 - s_{12})^{\frac{1}{2}}.$$

(10)

For spinless or unpolarized particles d, $\Sigma|T|^2$ is independent of both ϑ and φ. In these cases, the decay distribution is also flat in $E_{1\,lab}$, as $d\text{Lips}/dE_{1\,lab}$ is independent of $E_{1\,lab}$ according to (10). $E_{1\,lab}$ is related to the emission angle ϑ_{lab} of particle 1 as follows:

$$E_{1\,lab} = (E_{lab}^2 - p_{lab}^2\cos^2\vartheta_{lab})^{-1}[s_{12}^{\frac{1}{2}}E_{lab}E_1 \pm p_{lab}\cos\vartheta_{lab}(s_{12}p^2 - m_1^2 p_{lab}^2\sin^2\vartheta_{lab})^{\frac{1}{2}}],$$

(11)

Pilkuhn

where $E_1 = \frac{1}{2}s_{12}^{-\frac{1}{2}}(s_{12} + m_1^2 - m_2^2)$ is the cms energy of particle 1. The sign ambiguity in front of the square root arises only for $p_{lab}/E_{lab} > p/E_1$. In this case the maximum emission angle occurs for vanishing square root, $\sin\vartheta_{lab,max} = s_{12}^{\frac{1}{2}}pm_1^{-1}p_{lab}^{-1}$. The limits of E_{1lab} in (10) occur at $\cos^2\vartheta_{lab} = 1$ in (11):

$$E_{1lab}(^{max}_{min}) = s_{12}^{-\frac{1}{2}}(E_1 E_{lab} \pm pp_{lab}) . \tag{12}$$

1.2.2 Born terms for meson decays into particles with spin

A general decay matrix element can be decomposed into Born terms and form factors. The Born term contains the minimal number of 4-momenta which still allows the most general spin dependence and which is consistent with possible additional symmetries such as parity, gauge invariance and chiral symmetry. The T's of Tab. 2 in 1.2.1 are Born terms; the N's are form factors. Tab. 1 gives parity conserving Born terms between vector mesons or photons (spin and parity 1^-), pseudoscalar mesons (0^-) and tensor mesons (2^+). If parity is not conserved, one should add

$$T' = g'\varepsilon_1^*\varepsilon_2^* - h(P_1\varepsilon_1^*)(P_2\varepsilon_2^*)$$
$$T_0'(\lambda_1, \lambda_2) = \delta_{\lambda_1, \lambda_2}(4\pi)^{\frac{1}{2}}[g'\delta_{\lambda_1, \pm 1} + (hsp^2 - g'P_1P_2)\delta_{\lambda_1, 0}m_1^{-1}m_2^{-1}] \tag{1}$$

to the matrix element of $T(0^- \to 1^- 1^-)$, where g' and h are two independent coupling constants. Conversely, for $0^+ \to 1^- 1^-$ and $0^- \to 1^+ 1^-$ decays, T' is the parity-conserving amplitude. Contrary to the matrix elements which are given in Tab. 1, T' is not automatically gauge invariant. It becomes so for

$$g' = hP_1 P_2 . \tag{2}$$

Tab. 1. Born terms for decays involving particles of $(J^P) = 0^-$, 1^- and 2^+. $T(1^- \to 1^- 0^-)$ is obtained from $T(0^- \to 1^- 1^-)$ by crossing symmetry. $\varepsilon_{\alpha\beta\gamma\delta}$ is totally antisymmetric in its indices, and $\varepsilon_{0123} = 1$

Decay	T	$T_S(\lambda_1, \lambda_2)$	$m\Gamma$
$0^- \to 1^- 1^-$	$ig\varepsilon_{\alpha\beta\gamma\delta}P_1^\alpha\varepsilon_1^{*\beta}P_2^\gamma\varepsilon_2^{*\delta}$	$(4\pi)^{\frac{1}{2}}\delta_{\lambda_1\lambda_2}\cdot\lambda_1 gp\sqrt{s}$	$\dfrac{g^2}{4\pi}p^3\sqrt{s}$
$1^- \to 1^- 0^-$	$-ig\varepsilon_{\alpha\beta\gamma\delta}P_1^\alpha\varepsilon_1^{*\beta}P_d^\gamma\varepsilon_d^\delta(M)$	$\left(\dfrac{4\pi}{3}\right)^{\frac{1}{2}}\lambda_1 gp\sqrt{s}$	$\dfrac{1}{3}\dfrac{g^2}{4\pi}p^3\sqrt{s}$
$2^+ \to 1^- 0^-$	$-ig\varepsilon_{\alpha\beta\gamma\delta}P_1^\alpha\varepsilon_1^{*\beta}P_d^\gamma\varepsilon_d^{\delta\mu}(M)P_{1\mu}$	$\left(\dfrac{4\pi}{10}\right)^{\frac{1}{2}}\lambda_1 gp^2\sqrt{s}$	$\dfrac{1}{10}\dfrac{g^2}{4\pi}p^5\sqrt{s}$

Thus, if particle 2 is a photon, one has

$$m\Gamma(0 \to 1 + \gamma) = \frac{1}{8s}(s - m_1^2)^3 (g^2/4\pi + h^2/4\pi) \tag{3}$$

not assuming parity conservation. The parity-conserving matrix element of $1^+ \to 1^- 0^-$ decay is obtained from (1) by crossing symmetry (replacing P_2 by $-P_d$ and ε_2^* by ε_d):

$$T(1^+ \to 1^- 0^-) = g'\varepsilon_d\varepsilon_1^* + h(\varepsilon_d P_1)(\varepsilon_1^* P_d) ,$$
$$T_1(0) = -\left(\frac{4\pi}{3}\right)^{\frac{1}{2}} g_L s^{\frac{1}{2}}p/m_1 , \qquad T_1(\pm 1) = i\left(\frac{4\pi}{3}\right)^{\frac{1}{2}} g_T p^2 s/m^2 ; \tag{4}$$

$$m\Gamma = \tfrac{1}{6}p^5 s^{\frac{1}{2}}[g_L^2/(4\pi m_1^2) + 2g_T^2 s/(4\pi m^4)] ,$$
$$h = g_L + g_T P_1 P_d/m^2 , \qquad g' = -g_T[(P_1 P_d)^2 - sm_1^2]/m^2 . \tag{5}$$

g_L and g_T are the longitudinal and transverse coupling constants [Gilman 68].

For *decays into two spin-$\frac{1}{2}$ particles*, we shall only need electromagnetic and weak Born terms, which are of the particular form of Eq. (2) in 1.1.4. For $0 \to \frac{1}{2}\frac{1}{2}$ decays, the matrix element can be decomposed into

$$T = g_+ T_+ + g_- T_- , \tag{6}$$

$$T_\pm(0 \to \tfrac{1}{2}\tfrac{1}{2}) = (P_1 + P_2)^\sigma \bar{u}_1(P_1\lambda_1)\gamma_\sigma(1 \pm \gamma_5)u_2(-P_2\lambda_2) = 2im\delta_{\lambda_1\lambda_2}T_\pm(\lambda_1) , \tag{7}$$

$$T_\pm(\tfrac{1}{2}) = [(E_1 \pm p)(E_2 \mp p)]^{\frac{1}{2}} , \qquad T_\mp(-\tfrac{1}{2}) = T_\pm(\tfrac{1}{2}) . \tag{8}$$

For the weak decays, one has $m_2 = 0$ for the neutrino, $g_+ = 0$ and

$$T_+(\tfrac{1}{2}) = 0 , \qquad T_+(-\tfrac{1}{2}) = [2p(E_1 - p)]^{\frac{1}{2}} = m_1(1 - m_1^2/m^2)^{\frac{1}{2}} , \qquad m\Gamma = 4\frac{g_-^2}{4\pi}m_1^2 p^2 . \tag{9}$$

Pilkuhn

For decays in to $e\bar{e}$ or $\mu\bar{\mu}$, one has

$$m_1 = m_2, \quad g_+ = g_- \text{ (parity conservation)}, \quad T_\pm(\pm\tfrac{1}{2}) = m_1, \tag{10}$$

The important thing to notice is the factor m_1 both in (9) and (10), which suppresses $e\bar{\nu}$ and $e\bar{e}$ pairs relative to $\mu\bar{\nu}$ and $\mu\bar{\mu}$ pairs. (The decomposition (6) also applies to $\pi^0 \to e^+ e^-$ decay which proceeds via two-photon intermediate states.)

For decays into $e\bar{e}$ and $\mu\bar{\mu}$ via a one-photon virtual state (Dalitz pairs), it is convenient to first consider the decay of a virtual photon of invariant mass $m_\gamma = s^{\frac{1}{2}}$:

$$T(\gamma \to e\bar{e}) = e\varepsilon^\nu(\lambda_\gamma)\,\bar{u}(P_1 \lambda_1)\,\gamma_\nu u(-P_2 \lambda_2),$$
$$T_1(\lambda_1, \lambda_2) = 2e(4\pi/3)^{\frac{1}{2}}(m_e \delta_{\lambda_1 \lambda_2} + 2^{-\frac{1}{2}} m_\gamma \delta_{\lambda_1, -\lambda_2}), \tag{11}$$

$$m\Gamma(\gamma \to e\bar{e}) = \frac{\alpha}{3}(1 - 4m_e^2/m_\gamma^2)^{\frac{1}{2}}(m_\gamma^2 + 2m_e^2). \tag{12}$$

Then, from Eq. (2) in 1.1.3, one obtains

$$m_V \Gamma(V \to e\bar{e}) = \frac{4\pi}{f^2}\frac{\alpha^2}{3}(1 - 4m_e^2/m_V^2)^{\frac{1}{2}}(m_V^2 + 2m_e^2) \approx \frac{4\pi}{f^2}\frac{\alpha^2}{3}m_V^2 \tag{13}$$

for $V = \varrho, \omega, \phi$.

For other decays involving Dalitz pairs, see 1.2.4. The angular distributions for the decays $1^- \to 1^- 0^-$ and $1^- \to \frac{1}{2}\frac{1}{2}$ in reactions $ab \to cd$ are given by

$$W(\vartheta, \varphi) = \frac{1}{4\pi}(1 + A/2) \quad \text{and} \quad \frac{1}{4\pi}[1 + A(m_V^2 - 4m_e^2)/(m_V^2 + 2m_e^2)] \tag{14}$$

respectively, with A given by Eq. (7) in 1.2.1.

1.2.3 Two-body baryon decays

Matrix elements and widths for mesonic baryon decays are given in Tab. 1. In the first place, these expressions apply to the *weak mesonic decays* of the Λ, Σ, Ξ and Ω^- states. For spin-$\frac{1}{2}$ decays, a special nomenclature has been developed:

$$S = A, \quad P = [\lambda(m^2, m_1^2, m_2^2)]^{\frac{1}{2}} B, \tag{1}$$

$$\alpha = \frac{2\mathrm{Re}S^*P}{|S|^2 + |P|^2} = \frac{2|SP|\cos\Delta}{|S| + |P|^2}, \quad \beta = \frac{2\mathrm{Im}S^*P}{|S|^2 + |P|^2} = \frac{-2|SP|\sin\Delta}{|S|^2 + |P|^2} = (1 - \alpha^2)^{\frac{1}{2}}\sin\phi, \tag{2}$$

$$\gamma = \frac{|S|^2 - |P|^2}{|S|^2 + |P|^2} = (1 - \alpha^2)^{\frac{1}{2}}\cos\phi, \quad \alpha^2 + \beta^2 + \gamma^2 = 1, \quad \begin{matrix} -\pi/2 \\ \pi/2 \end{matrix} < \phi < \begin{matrix} \pi/2 \\ 3\pi/2 \end{matrix} \quad \text{for} \quad \begin{matrix} \gamma > 0 \\ \gamma < 0 \end{matrix}. \tag{3}$$

With this notation, the decays distribution (Eq. (3) in 1.2.1) is

$$W(\vartheta, \varphi) = \frac{1}{4\pi}[1 - 2\alpha \sin\vartheta \sin\varphi\,\mathrm{Im}\varrho_{\frac{1}{2}-\frac{1}{2}}] = \frac{1}{4\pi}(1 + \alpha \vec{P}\hat{p}_1), \tag{4}$$

where $\vec{P}$ is the polarization of the decaying baryon d and $\hat{p}_1$ a unit vector along the momentum $\vec{p}_1$ of the final baryon. The polarization of the final baryon is given by

$$\vec{P}_1 = [(\alpha + \vec{P}\hat{p}_1)\hat{p}_1 + \beta(\vec{P} \times \hat{p}_1) + \gamma\hat{p}_1 \times (\vec{P} \times \hat{p}_1)](1 + \alpha\vec{P}\hat{p}_1)^{-1}, \tag{5}$$

where $\vec{P}_1$ is defined in that rest frame of particle 1 obtained by a rotation-free Lorentz transformation from d to 1 along $\hat{p}_1$. The decay matrix elements of Tab. 1 also apply to the strong decays $\Lambda(1405) \to \pi\Sigma$ (with $T_p = 0$) and to the strong decays of the baryon decuplet (with $T_d = 0$). In this case T_s and T_{p3} are real.

For parity conserving decays into vector mesons, one uses $T = V^\mu \cdot \varepsilon_\mu(\lambda)$, where

$$V^\mu(S_d = \tfrac{1}{2}) = \bar{u}_1(\lambda_1)\left[G^V\gamma^\mu + G^T\tfrac{1}{2}(\gamma^\mu\gamma^\nu - \gamma^\nu\gamma^\mu)P_{V\nu}(m + m_1)^{-1} - G^S P_V^\mu(m + m_1)^{-1}\right]u(M)$$
$$= \bar{u}_1(\lambda_1)\left[(G^V + G^T)\gamma^\mu - G^T(P_d + P_1)^\mu(m + m_1)^{-1} - G^S P_V^\mu(m + m_1)^{-1}\right]u(M). \tag{6}$$

$$V^\mu(S_d = \tfrac{3}{2}) = \bar{u}_1(\lambda_1)\gamma_5\left[G_1 g^{\mu\nu} - G'\frac{\gamma^\mu P_V^\nu}{m + m_1} + G_2\frac{P_d^\mu P_V^\nu - P_d P_V g^{\mu\nu}}{(m + m_1)^2}\right]u_\nu(M). \tag{7}$$

When gauge invariance is imposed, (6) reduces to Eqs. (4) or (5) in 1.1.3, and $G' = G_1$ in (7). Finally, the parity-conserving $\frac{1}{2}^+ \to \frac{1}{2}^+ 2^+$ decay is given by

$$T = \bar{u}_1(\lambda_1)\left[G_1[(P_d + P_1)_\mu \gamma_\nu + (P_d + P_1)_\nu \gamma_\mu] + G_2(P_d + P_1)_\mu (P_d + P_1)_\nu\right]u(M)\,\varepsilon^{\mu\nu*}(\lambda_t). \tag{8}$$

Pilkuhn

Tab. 1. Decay matrix elements for baryon decays of both parities

Decay	T	T_S	$m\Gamma$
$\frac{1}{2}\to\frac{1}{2}0$	$\bar{u}_1(\lambda_1)\,(A - iB\gamma_5)\,u(M)$	$(2\pi)^{\frac{1}{2}}\,(T_s + 2i\lambda_1 T_p)$	$\dfrac{p}{8\pi s^{\frac{1}{2}}}\,(|T_s|^2 + |T_p|^2)$
$\frac{3}{2}\to\frac{1}{2}0$	$\bar{u}_1(\lambda_1)\,(A_3 - iB_3\gamma_5)\,P_2^\mu u_\mu(M)$	$\left(\dfrac{2\pi}{3}\right)^{\frac{1}{2}}(T_{p3} + 2i\lambda_1 T_d)$	$\dfrac{p}{24\pi s^{\frac{1}{2}}}\,(|T_{p3}|^2 + |T_d|^2)$
$\frac{3}{2}\to\frac{3}{2}0$	$-\bar{u}_{1\mu}(\lambda_1)\,(A - iB\gamma_5)\,u^\mu(M)$	$\dfrac{2}{3}\pi^{\frac{1}{2}}\Big\{|\lambda_1|\,(T_s + i(-1)^{\lambda_1+\frac{1}{2}}T_p)$ $+\dfrac{E_1}{2m_1}(\tfrac{9}{4} - \lambda_1^2)(T_s + i(-1)^{\lambda_1-\frac{1}{2}}T_p)\Big\}$	$\dfrac{p}{16\pi s^{\frac{1}{2}}}\Big\{|T_s|^2\Big(1 + \tfrac{1}{9}\Big(\dfrac{2E_1}{m_1}+1\Big)^2\Big)$ $+|T_p|^2\Big(1 + \tfrac{1}{9}\Big(\dfrac{2E_1}{m_1}-1\Big)^2\Big)\Big\}$

$$T_s = k_+ A, \qquad T_p = k_- B, \qquad T_{p3} = pk_+ A_3, \qquad T_{d3} = pk_- B_3, \qquad k_\pm = [(m\pm m_1)^2 - m_2^2]^{\frac{1}{2}} = [2m(E_1\pm m_1)]^{\frac{1}{2}}$$

1.2.4 Decay into 3 particles and sequential decays

For 3-particle-decays, the decay width Γ can be written in two different ways

$$m\Gamma(\mathrm{d}\to 123) = (64\pi^3)^{-1} \int \mathrm{d}E_1\,\mathrm{d}E_2 \sum_{\lambda_1\lambda_2\lambda_3} |T(\mathrm{d}\to 123)|^2 \tag{1}$$

$$= (2\pi)^{-4}\int \mathrm{d}\Omega_1\,\mathrm{d}s_{12}p_3 p_{12}\tfrac{1}{16}(ss_{12})^{-\frac{1}{2}}\sum_{\lambda_1\lambda_2\lambda_3} |T(\mathrm{d}\to 123)|^2\,, \tag{2}$$

$$p_3 = [\lambda(s, m_3^2, s_{12})/4s]^{\frac{1}{2}}, \qquad p_{12} = [\lambda(s_{12}, m_1^2, m_2^2)/4s_{12}]^{\frac{1}{2}}, \qquad s_{12} = s + m_3^2 - 2E_3 s^{\frac{1}{2}}\,. \tag{3}$$

$E_1, E_2,$ and $E_3 = s^{\frac{1}{2}} - E_1 - E_2$ are the energies in the rest frame of particle d. The allowed region of these energies is called the Dalitz plot, it is limited by collinear decays, which satisfy

$$(E_1^2 - m_1^2)^{\frac{1}{2}} \pm (E_2^2 - m_2^2)^{\frac{1}{2}} \pm (E_3^2 - m_3^2)^{\frac{1}{2}} = 0\,. \tag{4}$$

Eq. (1) is suited for proper 3-particle decays such as $\omega\to 3\pi$, $\mathrm{K}\to 3\pi$, $\eta\to 3\pi$ or $\mathrm{X}\to\eta\pi\pi$. If the Dalitz plot is small enough, T is essentially a constant times certain threshold factors analogous to 2-particle decays, and (1) can be integrated approximately analytically (see [Källen 64, 65, ref. in 1.1.6] for various examples). If two of the three particles (particles 1 and 2, say) are leptons or come from the decay of a resonance of mass m_R (sequential decay, for example $\mathrm{X}\to\varrho\gamma\to\pi^+\pi^-\gamma$), the second form is more appropriate. Here p_{12} is the momentum of particles 1 and 2 in their own cms, and p_3 is the decay momentum of $\mathrm{d}\to 3(12)$. Ω_1 specifies the momentum direction of particle 1 in the (1,2) cms. For a sequential decay, one has

$$\Gamma(\mathrm{d}\to 123) = \frac{1}{\pi}\int \mathrm{d}s_{12}\Gamma(\mathrm{d}\to r3)\,m_r\Gamma(r-12)\,[(m_r^2 - s_{12})^2 + m_r^2\Gamma_r^2]^{-1}\,. \tag{5}$$

For small Γ_r, and m_r^2 well inside the integration interval, one gets

$$\Gamma(\mathrm{d}\to 123) \approx \Gamma(\mathrm{d}\to r3)\frac{\Gamma(r\to 12)}{\Gamma_r}\left[1 - \frac{m_r\Gamma_r}{\pi}\frac{(m - m_3)^2 - (m_1 + m_2)^2}{[(m - m_3)^2 - m_r^2]\,[m_r^2 - (m_1 + m_2)^2]}\right]$$
$$+ \frac{1}{\pi}\frac{\mathrm{d}}{\mathrm{d}s}[\Gamma(\mathrm{d}\to r3)\,m_r\Gamma(r\to 12)]_{s=m_r^2}\ln\frac{(m - m_3)^2 - m_r^2}{m_r^2 - (m_1 + m_2)^2}\,. \tag{6}$$

Another important application of (5) is the internal conversion of γ-rays ("Dalitz pairs"), with $m_r = \Gamma_r = 0$ and $m\Gamma(r\to e\bar{e})$ given by Eq. (12) in 1.2.2. In particular, for $0^-\to 1^-\gamma$ and $1^-\to 0^-\gamma$ decays, one gets to zero[th] order in the electron mass [Pilkuhn 71]

$$\frac{\Gamma(\mathrm{d}\to e\bar{e}3)}{\Gamma(\mathrm{d}\to\gamma3)} = \frac{2\alpha}{3\pi}\left[\ln\frac{m(1 - c)}{m_e} - \frac{7}{4} - \frac{c}{(1 - c)^2}\left(1 + \frac{c}{2}\frac{3 - c}{1 - c}\ln c\right)\right], \qquad c = m_3^2 m^{-2}\,. \tag{7}$$

Decays such as $\omega\to\varrho\pi\to 3\pi$ or $\mathrm{A}_1\to\varrho\pi\to 3\pi$ cannot be handled by (5), because in this case ϱ resonances in the various pion combinations interfere with each other.

Pilkuhn

The angular distribution of $\omega \to \pi^+ \pi^- \pi^0$ decay is given by Eq. (7) in 1.2.1, with ϑ and φ replaced by the angles ϑ_n and φ_n of the normal to the decay plane. The distribution of $2^+ \to 1^- 0^- \to 3$ mesons is not given by Eq. (8) in 1.2.1, but depends both on ϑ_n, φ_n and ϑ, φ:

$$\tfrac{16}{15} \pi W = \tfrac{1}{3} \varrho_{00} f_0^2 + \varrho_{11} [\sin^2 \vartheta_n \cos^2 \vartheta + \cos^2 \vartheta_n \sin^2 \vartheta + \tfrac{1}{2} \sin 2\vartheta_n \sin 2\vartheta \cos(\varphi - \varphi_n)]$$

$$+ \varrho_{22} \sin^2 \vartheta_n \sin^2 \vartheta - \text{Re} \varrho_{21} [\sin^2 \vartheta_n \sin 2\vartheta \cos\varphi + \sin 2\vartheta_n \sin^2 \vartheta \cos\varphi_n]$$

$$- (\tfrac{8}{3})^{\frac{1}{2}} f_0 [\text{Re} \varrho_{10} (\sin \vartheta_n \cos \vartheta \cos\varphi_n + \cos \vartheta_n \sin \vartheta \cos\varphi) - \text{Re} \varrho_{20} \sin \vartheta_n \sin \vartheta \cos(\varphi_n + \varphi)] \tag{8}$$

$$+ \text{Re} \varrho_{2-1} [\sin^2 \vartheta_n \sin 2\vartheta \cos(2\varphi_n + \varphi) + \sin 2\vartheta_n \sin^2 \vartheta \cos(\varphi_n + 2\varphi)] + \varrho_{2-2} \sin^2 \vartheta_n \sin^2 \vartheta \cos(\varphi_n + \varphi)$$

$$- \varrho_{1-1} [\sin^2 \vartheta_n \cos^2 \vartheta \cos 2\varphi_n + \cos^2 \vartheta_n \sin^2 \vartheta \cos 2\varphi + \tfrac{1}{2} \sin 2\vartheta_n \sin 2\vartheta \cos(\varphi_n + \varphi)] ,$$

$$f_0 = 2 \cos \vartheta_n \cos \vartheta - \sin \vartheta_n \sin \vartheta \cos(\varphi_n - \varphi) .$$

1.2.5 References for 1.2

For journal abbreviation see p. 2

Dalitz   66   Proc. Int. School of Physics Enrico Fermi, Course 33 (New York: Academic Press):
                            R. H. Dalitz.
Gilman   68   PR 165, 1803: F. J. Gilman, H. Harari.
von Hippel   72   PR D 5, 624: F. von Hippel, C. Quigg.
Pilkuhn   71   NP, B 29, 462: H. Pilkuhn.
Rose   57   Elementary theory of angular momentum (New York: Wiley): M. E. Rose.

1.3 General formulas for two-particle reactions

1.3.1 Kinematics and partial-wave decomposition

For a general two-particle reaction $ab \to cd$, the Mandelstam variables are defined as follows

$$s = (P_a + P_b)^2 = (P_c + P_d)^2 = (\text{total c.m. energy})^2 ,$$

$$t = (P_a - P_c)^2 = (P_b - P_d)^2 = (\text{4-momentum transfer})^2 , \tag{1}$$

$$u = (P_a - P_d)^2 = (P_b - P_c)^2 = m_a^2 + m_b^2 + m_c^2 + m_d^2 - s - t .$$

When the target particle b is at rest in the laboratory frame $s = m_a^2 + m_b^2 + 2 E^{lab} m_b$. The initial and final c.m. momenta are denoted by q and q' and can be expressed in terms of the function λ (Eq. (12) in 1.1.1):

$$4q^2 s = (2 p_a^{lab} m_b)^2 = \lambda(s, m_a^2, m_b^2) . \tag{2}$$

The differential cross section for unpolarized particles is expressed in terms of the T-matrix elements

$$\frac{d\sigma}{dt} = \frac{\pi}{qq'} \frac{d\sigma}{d\Omega} = [64\pi q^2 s (2S_a + 1)(2S_b + 1)]^{-1} \sum_{\lambda_a \lambda_b \lambda_c \lambda_d} |T_{\lambda_a \lambda_b \lambda_c \lambda_d}(s, t)|^2 \tag{3}$$

where each particle's helicity λ_i varies between $+S_i$ and $-S_i$, and $d\Omega = d\phi \, d\cos\theta$ denotes the c.m. differential solid angle. For polarized target and beam, one has

$$\frac{d\sigma}{dt} = [64\pi q^2 s]^{-1} \sum_{\lambda_a \ldots \lambda_d} \varrho_{\lambda_a \lambda_a'}^{S_a} \varrho_{\lambda_b \lambda_b'}^{S_b} T_{\lambda_a \lambda_b \lambda_c \lambda_d} T_{\lambda_a' \lambda_b' \lambda_c \lambda_d}^* , \tag{4}$$

where ϱ^S is the polarization density matrix. Unpolarized particles have $\varrho_{\lambda \lambda'}^S = \delta_{\lambda \lambda'}(2S + 1)^{-1}$, which leads to (3). The expansion of the helicity amplitudes $T_{\lambda_a \ldots \lambda_d}$ in partial-wave amplitudes of total angular momentum J reads (compare with Eq. (2) of 1.2.1):

$$T_{\lambda_a \ldots \lambda_d}(s, t) = 8\pi \sqrt{s} \, e^{i(\lambda - \lambda')\phi} \sum_J (2J + 1) \, d_{\lambda \lambda'}^J(\theta) \, T_{\lambda_a \ldots \lambda_d}^J(s) , \qquad \lambda \equiv \lambda_b - \lambda_a , \quad \lambda' \equiv \lambda_d - \lambda_c . \tag{5}$$

For fixed J, the $T_{\lambda_a \ldots \lambda_d}^J$ form the matrix $T^J(ab \to cd)$ of the reaction $ab \to cd$, which consists of $(2S_a + 1)(2S_b + 1) \times (2S_c + 1)(2S_d + 1)$ elements. We now introduce a large matrix T^J which comprises the partial-wave T-matrices of all communicating open channels. The S-matrix is then defined by

$$S^J = 1 + 2i\sqrt{Q} \, T^J \sqrt{Q} , \tag{6}$$

where $Q = $ diagonal matrix of the q's of all open channels. Time reversal invariance is assumed; with a suitable choice of phases, $T^J = T_{\text{transposed}}^J$. Unitarity requires

$$S^{J+} S^J = 1, \qquad \text{Im} \, T^J = T^{J+} Q T^J . \tag{7}$$

Parity conservation requires (η_i = intrinsic parity of particle i)

$$T^J_{\lambda_a \lambda_b \lambda_c \lambda_d} = \eta_a \eta_b \eta_c \eta_d (-1)^{S_a + S_b - S_c - S_d}\, T_{-\lambda_a -\lambda_b -\lambda_c -\lambda_d}\,. \tag{8}$$

This allows one to split T^J into two matrices, each of which fulfills (7) separately:

$$T^{J\pm}_{\lambda_a \lambda_b \lambda_c \lambda_d} = T^J_{\lambda_a \lambda_b \lambda_c \lambda_d} \pm T^J_{\lambda_a \lambda_b -\lambda_c -\lambda_d}\,. \tag{9}$$

Meson-meson scattering: For $\pi\pi$ scattering, $\pi\pi \leftrightarrow K\bar{K}$ etc., where no spins are involved, one has $T^{J-} = 0$, $T^{J+} = T_l$, where l is the orbital angular momentum. The diagonal elements of T_l are denoted by $f_{l,c}$ (c = channel $= \pi\pi$ or $K\bar{K}$) and are parametrized by the phase shift δ_c and elasticity η_c

$$f_{lc} = \frac{1}{2iq_c}\,[\eta_{lc}\exp(2i\delta_{lc}) - 1] \qquad 0 < \eta_{lc} \leqq 1\,. \tag{10}$$

When only one channel is open, it has $\eta_l = 1$ by unitarity (7), and for small q^2, δ_l can be parametrized by the effective range approximation (ERA)

$$q^{2l+1}\cot\delta_l = a_l^{-1} + \tfrac{1}{2}q^2 r_l\,, \tag{11}$$

where a_l and r_l are the (real) scattering length and effective range.

Meson-baryon scattering: When particles a and c carry spin-parity 0^- and b and d carry $\tfrac{1}{2}^+$, the splitting (9) introduces the orbital angular momentum as follows:

$$T^{J+}_{\frac{1}{2}\frac{1}{2}} = T_{l+}\,, \qquad T^{J-}_{\frac{1}{2}\frac{1}{2}} = T_{(l+1)-}\,, \qquad (l)\pm \text{ means } J = l \pm \tfrac{1}{2}\,. \tag{12}$$

Now (10) and (11) hold for the amplitudes $f_{l\pm,c}$, phase shifts $\delta_{l\pm,c}$ and elasticities $\eta_{l\pm,c}$. Also needed are the non-spin flip amplitude f, the spin-flip amplitude g, and the amplitudes f_1, f_2:

$$-f_2 \sin\theta = g = \sum_{l=1}^{\infty} [f_{l+} - f_{l-}]\, P_l'(\cos\theta)\sin\theta\,,$$
$$f_1 + f_2\cos\theta = f = \sum_{l=0}^{\infty} [(l+1)f_{l+} + l f_{l-}]\, P_l(\cos\theta)\,, \tag{13}$$

for which the differential cross section and polarization P_d are ($\vec{P}$ is defined following Eq. (5) in 1.2.1; for meson-baryon scattering $P_x = P_z = 0$, $P_y = -2\,\mathrm{Im}\,\varrho_{\frac{1}{2}-\frac{1}{2}}$)

$$\frac{d\sigma}{d\Omega} = |f|^2 + |g|^2\,, \qquad \frac{d\sigma}{d\Omega}\,P = 2\,\mathrm{Im}\,f^* g\,. \tag{14}$$

Baryon-baryon scattering: For NN scattering and the $\Lambda N + \Sigma N$ system we introduce for (9) the notation

$$T^{J\pm}_{\sigma,\sigma'}\,, \qquad \sigma = 4\lambda_a\lambda_b\,, \qquad \sigma' = 4\lambda_c\lambda_d\,, \qquad \lambda_a \equiv \tfrac{1}{2}\,. \tag{15}$$

Here l is diagonal only for T^{J-}, namely, $l = J$; $\sigma = 1(-1)$ refers to the spin-singlet states $^1S_0$, $^1P_1$, $^1D_2\ldots$ (triplet states $-\,^3P_1$, $-\,^3D_2\ldots$). T^{J+} connects triplet states with $l = J \pm 1$:

$$\sigma = \pm 1: \quad \left(\frac{J}{2J+1}\right)^{\frac{1}{2}}|l = J \mp 1\rangle \mp \left(\frac{J+1}{2J+1}\right)^{\frac{1}{2}}|l = J \pm 1\rangle, \qquad T^{0+}_{1,1} = T(^3P_0)\,. \tag{16}$$

$$T^{1+}_{1,1} = \frac{1}{3}T(^3S_1) + \frac{2}{3}T(^3D_1) - \frac{\sqrt{2}}{3}[T(^3D_1 \to {}^3S_1) + T(^3S_1 \to {}^3D_1)]\,,$$

$$T^{1+}_{-1,-1} = \frac{2}{3}T(^3S_1) + \frac{1}{3}T(^3D_1) + \frac{\sqrt{2}}{3}[T(^3D_1 \to {}^3S_1) + T(^3S_1 \to {}^3D_1)]\,,$$

$$T^{1+}_{1,-1} = \frac{1}{3}\sqrt{2}[T(^3S_1) - T(^3D_1)] - \frac{2}{3}T(^3S_1 \to {}^3D_1) + \frac{1}{3}T(^3D_1 \to {}^3S_1)\,,$$

$$T^{1+}_{-1,1} = \frac{1}{3}\sqrt{2}[T(^3S_1) - T(^3D_1)] + \frac{1}{3}T(^3S_1 \to {}^3D_1) - \frac{2}{3}T(^3D_1 \to {}^3S_1)\,. \tag{17}$$

For elastic scattering, $T(S \to D) = T(D \to S)$ and $T^{J+}_{1,-1} = T^{J+}_{-1,1}$.

Photoproduction of pseudoscalar mesons: The partial wave helicity amplitudes $T^J_{\lambda_\gamma \lambda \lambda'}$ (where λ_γ, λ and λ' are the helicities of the photon and initial and final nucleon, respectively) are related to the electric and magnetic multipoles $E_{l\pm}$ and $M_{l\pm}$ by

$$T^J_{1,\frac{1}{2},\pm\frac{1}{2}} = \frac{1}{\sqrt{2}}\{lM_{l+} + (l+2)E_{l+} \pm (l+2)M_{(l+1)-} \mp lE_{(l+1)-}\},$$

$$T^J_{-1,\frac{1}{2},\pm\frac{1}{2}} = \frac{1}{\sqrt{2}}\sqrt{l(l+2)}\{M_{l+} - E_{l+} \pm M_{(l+1)-} \mp E_{(l+1)-}\}, \tag{18}$$

where $l\pm$ have the same meaning as in meson-baryon scattering. At threshold the unpolarized differential cross section reduces to

$$\left[\frac{k}{q}\frac{d\sigma}{d\Omega}\right]_{q=0} = |E_{0+}|^2, \tag{19}$$

where k and q are the photon and meson c.m. momenta.

1.3.2 Threshold behaviour and Coulomb corrections

It is useful to introduce the matrix K^J defined by

$$S^J = [1 + i\sqrt{Q}K^J\sqrt{Q}][1 - i\sqrt{Q}K^J\sqrt{Q}]^{-1}, \tag{1}$$

where K is real and symmetric due to unitarity and time-reversal invariance. The threshold behaviour is governed by the orbital angular momenta l_i and l_f in initial and final states. When the K-matrix is transformed from the helicity basis to the l-basis, one can make the effective range approximation (ERA), which reads, in the absence of the Coulomb interaction,

$$Q^l(K^J)^{-1}Q^l = (A^J)^{-1} + \tfrac{1}{2}(Q^2 - Q_0^2)^{\frac{1}{2}} R^J(Q^2 - Q_0^2)^{\frac{1}{2}}, \tag{2}$$

where A^J and R^J are the scattering length and effective range matrices. They are real and symmetric. Q_0 is the matrix Q at the energy at which the ERA is made (normally the threshold of one of the channels). In the one-dimensional case and for $q_0 = 0$, (2) reduces to Eq. (11) in 1.3.1. Q^l is the matrix with diagonal elements q_c^l. Occasionally as in Eq. (17) in 1.3.1, several l-values exist for given J and channel c.

Coulomb corrections: Let m denote the reduced mass and e_a and e_b the charges of particles a and b ($e^2/4\pi = \alpha$). Define

$$\eta = \frac{m}{q}\frac{e_a e_b}{4\pi}, \qquad \sigma_l = \arg\Gamma(l+1+i\eta) = \sigma_0 + \sum_{n=1}^{l}\arctan\frac{\eta}{n}, \qquad \sigma_0 = -0.5772\eta. \tag{3}$$

The amplitude for pure Coulomb scattering must be added to the elastic non-flip amplitude. The non-relativistic form is

$$f_c(\theta) = -\eta[2q\sin^2\tfrac{1}{2}\theta]^{-1}\exp\{2i\sigma_0 - 2i\eta\ln\sin\tfrac{1}{2}\theta\}. \tag{4}$$

In addition, the matrix T^J is replaced by

$$T'^J = e^{i\sigma_l}T^J e^{i\sigma_l} \tag{5}$$

where T^J still contains Coulomb corrections, and $e^{i\sigma_l}$ is a diagonal channel matrix analogous to Q^l. For uncoupled channel S-wave scattering, the effective range formula becomes

$$C_0^2 q\cot\delta_0 + 2q\eta h(\eta) = \frac{1}{a} + \frac{1}{2}rq^2,$$

where

$$C_0^2 = |\Gamma(1+i\eta)e^{-\frac{1}{2}\pi\eta}|^2 = \frac{2\pi\eta}{e^{2\pi\eta}-1}, \tag{6}$$

and

$$h(\eta) = \mathrm{Re}\,\frac{\Gamma'(-i\eta)}{\Gamma(-i\eta)} - \ln\eta = \eta^2\sum_{n=1}^{\infty}\frac{1}{n(n^2+\eta^2)} - \ln\eta - 0.5772\ldots.$$

For inner Coulomb corrections to the a's in NN-scattering, see 1.5.3. For the inclusion of vacuum polarization, see [Heller 60].

1.3.3 Born terms and Lagrangians

Born terms represent the contribution of 1-particle intermediate states (see 1.1.5) in the s-, t- or u-channel, with the vertex functions V constructed according to the Born rules of 1.2.2 for decay matrix elements. More precisely, for an intermediate particle e of mass m_e, the s- and t-channel Born terms are

$$T_s^{Born} = \sum_{\lambda_e} V_{e,cd}\frac{1}{m_e^2 - s}V_{ab,e} \qquad T_t^{Born} = \sum_{\lambda_e}V_{a,ce}\frac{1}{m_e^2-t}V_{be,d}, \tag{1}$$

Pilkuhn

where the comma separates incoming and outgoing particles. In T_s^{Born}, $V_{ab,e}$ corresponds to the "formation" of a resonance e, and $V_{e,cd}$ to its "decay" (for a true resonance $m_e^2 - s$ is replaced by $m_e^2 - im_e\Gamma_e - s$). In T_t^{Born}, e can be defined as outgoing at the a, ce vertex. At the be, d vertex, it can then be treated either as an incoming particle e of 4-momentum P_e, or as an outgoing antiparticle $\bar{e}$ of 4-momentum $-P_e$. The only new point is the continuation of the spin functions off the mass shell. For the spin $\frac{1}{2}$, 1, $\frac{3}{2}$ and 2 functions defined in 1.1.1, one uses

$$\sum_{\lambda_e = -\frac{1}{2}}^{+\frac{1}{2}} u_e(\lambda_e)\,\bar{u}_e(\lambda_e) = P_\mu\gamma^\mu + m_e \,, \tag{2}$$

$$\sum_{\lambda_e = -1}^{1} \varepsilon_\mu(\lambda_e)\,\varepsilon_\nu^*(\lambda_e) = -g_{\mu\nu} + P_\mu P_\nu/m_e^2 \equiv P_{\mu\nu} \,, \tag{3}$$

$$\sum_{\lambda_e = -\frac{3}{2}}^{\frac{3}{2}} u_\mu(\lambda_e)\,\bar{u}_\nu(\lambda_e) = \tfrac{1}{3}(P\gamma + m_e)\left[\frac{2}{m_e^2}P_\mu P_\nu - g_{\mu\nu} - \gamma_\nu\gamma_\mu + \frac{1}{2m_e}(\gamma_\mu P_\nu - \gamma_\nu P_\mu)\right], \tag{4}$$

$$\sum_{\lambda_e = -2}^{2} \varepsilon_{\mu\nu}(\lambda_e)\,\varepsilon_{\varrho\sigma}^*(\lambda_e) = \tfrac{1}{2}(P_{\mu\varrho}P_{\nu\sigma} + P_{\mu\sigma}P_{\nu\varrho}) - \tfrac{1}{3}P_{\mu\nu}P_{\varrho\sigma} \,, \tag{5}$$

where $P \equiv P_e$ ($= P_a + P_b$ for s-channel Born terms and $P_a - P_c$ for t-channel Born terms).

For pseudoscalar meson-baryon scattering (πN scattering, $\bar{K}$N scattering etc.), one can either use the pseudoscalar (ps) coupling (see Tab. 1 in 1.2.3) or the pseudovector (pv) coupling:

$$V_{\lambda_b, \lambda_c}^{ps} = iG_{ab,e}\bar{u}_e(\lambda_e)\gamma_5 u_b(\lambda_b), \qquad V_{\lambda_b, \lambda_c}^{pv} = -i\sqrt{4\pi}\,\frac{f_{ab,e}}{m_{\pi^+}}\,\bar{u}_e(\lambda_e)\gamma_5 P_{a\mu}\gamma^\mu u_b(\lambda_b)\,. \tag{6}$$

The latter coupling is not strictly a Born term since it contains an unnecessary 4-momentum P_a. Consequently the residues are identical for

$$G_{ab,e} = (4\pi)^{\frac{1}{2}} f_{ab,e}(m_b + m_e)/m_{\pi^+}\,. \tag{7}$$

However, if chiral symmetry is imposed [Weinberg 66], V must vanish identically for $P_a = 0$, in which case V^{pv} really contains only necessary 4-momenta, and V^{ps} is ruled out. The factor $m_{\pi^+}^{-1}$ is added for dimensional reasons and not for dynamical ones. Therefore, in SU(2) or SU(3)-comparisons it must not be exchanged against m_{π^0} for $a = \pi^0$, m_K for $a = K$ etc., unless a dynamical motivation is given.

For vector and tensor exchange in meson-baryon scattering, the Born approximation to the meson vertex follows from Tab. 2 in 1.2.1 and the replacement $P_1 \to -P_a$

$$V_{a,cv}(\lambda_v) = G_{a,cv}(P_a + P_c)_\mu \varepsilon^{*\mu}(\lambda_v), \qquad V_{a,ct}(\lambda_t) = G_{a,ct}(P_a + P_c)_\mu (P_a + P_c)_\nu \varepsilon^{*\mu\nu}(\lambda_t)\,. \tag{8}$$

Similarly, the Born approximation at the baryon vertex is given by Eqs. (6 ... 8) in 1.2.3. However, instead of Eq. (7) in 1.2.3 one normally needs the Born term for an outgoing spin $-\frac{3}{2}$ particle:

$$V_{b,d\bar{v}}^\mu = \bar{u}_\nu(\lambda_d)\left[G_1(b, d\bar{v})\,g^{\mu\nu} - G'(b, d\bar{v})\,\frac{\gamma^\mu P_e^\nu}{m_b + m_d} + G_2(b, d\bar{v})\,\frac{P_b^\mu P_e^\nu - P_b P_e g^{\mu\nu}}{(m_b + m_d)^2}\right]\gamma_5 u_b(\lambda_b)\,. \tag{9}$$

Again, from the point of view of SU(3)-symmetry, the definitions should be changed such that the mass denominators do not vary within the SU(3)-multiplet. See also 1.6.6.

For particles of relatively low spin and isospin, it is convenient (but not necessary) to derive Born terms from Lagrangians $\mathscr{L}(x)$ by $V_{ab,e} = \langle P_e \lambda_e | \mathscr{L}(0) | P_a \lambda_a, P_b \lambda_b \rangle$. The Lagrangians for (6) for the special case of πNN are

$$\mathscr{L}_{\pi NN}^{ps} = iG\gamma_5\vec{\tau}N\vec{\pi}, \qquad \mathscr{L}_{\pi NN}^{pv} = \sqrt{4\pi}fm_{\pi^+}^{-1}\bar{N}\gamma_\mu\gamma_5\vec{\tau}N\partial^\mu\vec{\pi}\,. \tag{10}$$

The Lagrangians for decays into spinless particles (see Tab. 2 in 1.2.1) are, for the special cases of $\varepsilon \to \pi\pi$, $\varrho \to \pi\pi$ and $f \to \pi\pi$ decays,

$$\mathscr{L}_{\varepsilon\pi\pi} = \tfrac{1}{2}G_{\varepsilon\pi\pi}m_\varepsilon\vec{\pi}\vec{\pi}\varepsilon, \qquad \mathscr{L}_{\varrho\pi\pi} = G_{\varrho\pi\pi}\vec{\varrho}_\mu(\vec{\pi}\times\partial^\mu\vec{\pi}), \qquad \mathscr{L}_{f\pi\pi} = 2\frac{G_{f\pi\pi}}{m_f}\partial_\mu\vec{\pi}\,\partial_\nu\vec{\pi}f^{\mu\nu}\,. \tag{11}$$

The εNN, ϱNN and fNN Lagrangians are (see Tab. 1 in 1.2.3, Eq. (5) in 1.1.3 and Eq. (8) in 1.2.3)

$$\mathscr{L}_{\varepsilon NN} = G_{\varepsilon NN}\bar{N}N\varepsilon, \qquad \mathscr{L}_{\varrho NN} = G_{\varrho NN}^V\bar{N}\gamma_\mu\vec{\tau}N\varrho^\mu - i\frac{G_{\varrho NN}^T}{4m_p}\bar{N}\tfrac{1}{2}[\gamma_\mu,\gamma_\nu]\vec{\tau}N(\partial^\mu\varrho^\nu - \partial^\nu\varrho^\mu)$$

$$\mathscr{L}_{fNN} = 2i\frac{G_{fNN}^{(1)}}{m_p}\bar{N}(\gamma_\mu\partial_\nu + \gamma_\nu\partial_\mu)Nf^{\mu\nu} + 4\frac{G_{fNN}^{(2)}}{m_p^2}\partial_\mu\bar{N}\,\partial_\nu Nf^{\mu\nu}\,. \tag{12}$$

Some authors use a factor $\frac{1}{2}$ in $\mathscr{L}_{\varrho NN}$ ("universality convention" of [Sakurai 60]), in which case $G_{\varrho NN}^2$ is larger by a factor 4.

Pilkuhn

1.3.4 Isospin and SU(3) properties of coupling constants

The phase conventions of Condon and Shortley are used for Clebsch-Gordan coefficients, and in meson-baryon product states the baryon comes first (compare [Particle 71]). For example, the $|I, I_3\rangle$ pion-nucleon states are

$$\left|\frac{1}{2}, \frac{1}{2}\right\rangle = \frac{1}{\sqrt{3}} |p\pi^0\rangle - \sqrt{\frac{2}{3}} |n\pi^+\rangle, \quad \left|\frac{1}{2}, -\frac{1}{2}\right\rangle = \sqrt{\frac{2}{3}} |p\pi^-\rangle - \frac{1}{\sqrt{3}} |n\pi^0\rangle. \tag{1}$$

a) Pion-baryon coupling constants

$$G \equiv G(p, p\pi^0) = -G(n, n\pi^0) = -\frac{1}{\sqrt{2}} G(p, n\pi^+) = \frac{1}{\sqrt{2}} G(n, p\pi^-), \tag{2}$$

$$G_{\Lambda\Sigma\pi} \equiv G(\Sigma^+, \Lambda\pi^+) = G(\Sigma^0, \Lambda\pi^0) = G(\Sigma^-, \Lambda\pi^-)$$
$$= -G(\Lambda, \Sigma^+\pi^-) = G(\Lambda, \Sigma^0\pi^0) = -G(\Lambda, \Sigma^-\pi^+). \tag{3}$$

The second row of (3) follows from the first by replacing a final state pion by an initial state antipion and observing the signs from charge conjugation,

$$C|\pi^\pm\rangle = -|\pi^\mp\rangle, \quad C|\pi^0\rangle = |\pi^0\rangle, \tag{4}$$

which follow from the Condon-Shortley phase convention in Eq. (1). Isospin invariance for Σ and Ξ couplings:

$$G_{\Sigma\Sigma\pi} \equiv G(\Sigma^+, \Sigma^+\pi^0) = -G(\Sigma^+, \Sigma^0\pi^+) = G(\Sigma^-, \Sigma^0\pi^-) = -G(\Sigma^-, \Sigma^-\pi^0),$$
$$= G(\Sigma^0, \Sigma^+\pi^-) = -G(\Sigma^0, \Sigma^-\pi^+), \quad G(\Sigma^0, \Sigma^0\pi^0) = 0, \tag{5}$$

$$G_{\Xi\Xi\pi} \equiv G(\Xi^0, \Xi^0\pi^0) = -G(\Xi^-, \Xi^-\pi^0) = -\frac{1}{\sqrt{2}} G(\Xi^0, \Xi^-\pi^+) = \frac{1}{\sqrt{2}} G(\Xi^-, \Xi^0\pi^-). \tag{6}$$

The SU(3) values of these coupling constants are expressed in terms of G and $\alpha = D/(D + F)$ as follows:

$$\alpha \equiv \sqrt{\tfrac{3}{4}} G_{\Lambda\Sigma\pi}/G, \quad G_{\Sigma\Sigma\pi} = 2(1 - \alpha)G, \quad G_{\Xi\Xi\pi} = -(2\alpha - 1)G. \tag{7}$$

For the decays $\Delta \to N\pi$, we have

$$G_\Delta = G(\Delta^{++}, p\pi^+) = G(\Delta^-, n\pi^-) = 3^{\frac{1}{2}} G(\Delta^+, \pi^+ n) = 3^{\frac{1}{2}} G(\Delta^0, \pi^- p) = (\tfrac{3}{2})^{\frac{1}{2}} G(\Delta^+, \pi^+ n) = (\tfrac{3}{2})^{\frac{1}{2}} G(\Delta^0, \pi^0 n). \tag{8}$$

The isospin relations for the coupling constants of the N*, Λ*, Σ*, Ξ* resonances are identical to those of N, Λ and Ξ*. The SU(3)-values of the decuplet (Δ, Υ, Ξ*, Ω) coupling constants are

$$G_\Delta = -6^{\frac{1}{2}} G_{\Upsilon\Sigma\pi} = 2^{\frac{1}{2}} G_{\Upsilon\Lambda\pi} = (\tfrac{3}{2})^{\frac{1}{2}} G_{\Xi^*\Xi\pi}. \tag{9}$$

b) Kaon-baryon coupling constants

$$G_{N\Lambda K} \equiv G(p, \Lambda K^+) = G(n, \Lambda K^0) = -G(\Lambda, pK^-) = G(\Lambda, n\overline{K}^0), \tag{10}$$

$$G_{N\Sigma K} \equiv -G(p, \Sigma^0 K^+) = G(n, \Sigma^0 K^0) = G(\Sigma^0, pK^-) = G(\Sigma^0, n\overline{K}^0)$$

$$= \frac{1}{\sqrt{2}} G(p, \Sigma^+ K^0) = -\frac{1}{\sqrt{2}} G(n, \Sigma^- K^+) = \frac{1}{\sqrt{2}} G(\Sigma^-, nK^-) = \frac{1}{\sqrt{2}} G(\Sigma^+, p\overline{K}^0). \tag{11}$$

The signs of the $\overline{K}$ coupling constants in (10) and (11) follow from

$$C|K^\pm\rangle = -|K^\mp\rangle, \quad C|K^0\rangle = |\overline{K}^0\rangle. \tag{12}$$

The SU(3) values of $G_{N\Lambda K}$ and $G_{N\Sigma K}$ are

$$G_{N\Lambda K} = -\frac{1}{\sqrt{3}} (3 - 2\alpha)G, \quad G_{N\Sigma K} = -(2\alpha - 1)G = G_{\Xi\Xi\pi}. \tag{13}$$

For $\Delta K\Sigma$ couplings, the SU(2) relations are the same as for $\Delta N\pi$ in Eq. (8). SU(3)-symmetry says

$$G_\Delta = -G_{\Delta\Sigma K} = 3^{\frac{1}{2}} G_{\Upsilon N\overline{K}} = -3^{\frac{1}{2}} G_{\Upsilon\Xi K}. \tag{14}$$

c) η-baryon and X-baryon coupling constants

Here all Clebsch-Gordan coefficients are $+1$, and the definitions are obvious. The SU(3)-values of the octet component η_8 defined in Eq. (2) in 1.1.2 are

Pilkuhn

$$G_{NN\eta 8} = \frac{1}{\sqrt{3}}(3-4\alpha)\,G, \qquad G_{\Sigma\Sigma\eta 8} = -G_{\Lambda\Lambda\eta 8} = G_{\Lambda\Sigma\pi} = \sqrt{\frac{4}{3}}\,\alpha G\,, \tag{15}$$

$$G_{\Upsilon\Sigma\eta} = G_{\Xi^*\Xi\eta} = -2^{-\frac{1}{2}}G_\Delta\,. \tag{16}$$

d) Vector and tensor meson couplings

The SU(2) and SU(3)-properties of the BBV (= baryon-baryon-vector) coupling constants in Eqs. (6), (7) in 1.2.3 and Eq. (9) in 1.3.3 follow from those of the BBP coupling constants (P = pseudoscalar meson) by replacing $\pi \to \varrho$, $K \to K^*$, $\eta_8 \to \phi_8$ and $X_1 \to \omega_1$. Similarly, for the BBT coupling constants in Eq. (8) in 1.2.3, one replaces $\pi \to A_2$, $K \to K_N$, $\eta_8 \to f_8$ and $X_1 \to f_1$. For the vector mesons, the coupling constants G^V have $\alpha^V = 0$ (but $\alpha^T \neq 0$) in the universality model (see also the remark following Eq. (8) in 1.1.4). Universality requires

$$2G^V_{NN\varrho} = G^V_{\Sigma\Sigma\varrho} = G^V_{\Xi\Xi\varrho} = G_{\pi\pi\varrho} = G_{KK\varrho} = 2G_{PPV}(1+\alpha_1)\,, \tag{17}$$

where α_1 is an SU(3)-breaking parameter [Diu 65]:

$$G_{KK\phi 8} = \sqrt{3}\,G_{PPV}(1-\alpha_1), \qquad G_{K\pi K^*} = G_{PPV}(1-\alpha_1/2) = -G_{K\eta K^*}/\sqrt{3}\,. \tag{18}$$

By charge conjugation, PPV coupling constants have no symmetric part ($\alpha_{PPV} = 0$), and the coupling constants of $P_1 P_8 V_8$ and $P_8 P_8 V_1$ are zero. (This follows from G-parity conservation (Eq. (7) in 1.1.2) in the coupling of the hyperneutral multiplet members.) The same is true for the PVT couplings:

$$G_{A_2\varrho\pi} = (\tfrac{2}{3})^{\frac{1}{2}}G_F, \qquad G_{K_N K^*\pi} = G_{K_N K\varrho} = -G_{K_N K\omega}/\sin\theta_v = \tfrac{1}{2}G_F, \qquad G_{f'K^*\overline{K}} = G_F\cos\theta_t\,. \tag{19}$$

The PPT couplings have no antisymmetric part ($\alpha_{PPT} = 1$):

$$G_{A_2\pi\eta} = (\tfrac{2}{5})^{\frac{1}{2}}G_8\cos\theta_p - G_1\sin\theta_p, \qquad G_{A_2\pi X} = (\tfrac{2}{5})^{\frac{1}{2}}G_8\sin\theta_p + G_1\cos\theta_p, \qquad G_{A_2 K\overline{K}} = -(\tfrac{3}{5})^{\frac{1}{2}}G_8\,,$$

$$G_{K_N K\pi} = (\tfrac{9}{10})^{\frac{1}{2}}G_8, \qquad G_{K_N K\eta} = (\tfrac{1}{10})^{\frac{1}{2}}G_8\cos\theta_p, \qquad G_{f\pi\pi} = -(\tfrac{3}{5})^{\frac{1}{2}}G_8\sin\theta_t + (\tfrac{3}{8})^{\frac{1}{2}}G_1\cos\theta_t\,, \tag{20}$$

$$G_{fK\overline{K}} = 5^{-\frac{1}{2}}G_8\sin\theta_t + 2^{-\frac{1}{2}}G_1\cos\theta_t, \qquad G_{f\eta\eta} = -(5^{-\frac{1}{2}}G_8\sin\theta_t + 8^{-\frac{1}{2}}G_1\cos\theta_t)\cos^2\theta_p\,.$$

The coupling constants for f′meson follow from those for f meson by replacing $\cos\theta_t$ by $-\sin\theta_t$ and $\sin\theta_t$ by $\cos\theta_t$. SU(3)-invariance for photon couplings is formulated in 1.6.6.

In the quark model, one puts $\cos\theta_t = \cos\theta_v = (\tfrac{2}{3})^{\frac{1}{2}}$ ($\theta = 35.3°$, "ideal mixing angle") and puts the coupling constants between ϕ and f and combinations of ϱ, π, ω, f and N equal to zero. This rule is quite accurate in the cases where it can be tested ($G_{\phi\varrho\pi} = G_{\phi NN} = G_{f'\pi\pi} = 0$).

1.3.5 Dispersion relations

For reactions ab→cd where the imaginary part of the scattering amplitude is known for all energies at fixed t and the real part is known at a small number of points (depending on the number of subtractions), coupling constants and scattering lengths can be determined from dispersion relations. These relations are most conveniently written down for amplitudes that are free from kinematical singularities (see [A. Martin 70], ref. in 1.1.6). For pseudoscalar meson-baryon scattering, these are the amplitudes A and B defined by

$$T(\lambda, \lambda') = \overline{u}_d(\lambda')\left[A + \tfrac{1}{2}(P_a + P_c)_\mu \gamma^\mu B\right] u_b(\lambda)\,. \tag{1}$$

They are related to f_1 and f_2 of Eq. (13) in 1.3.1 by

$$f_1 = \frac{1}{8\pi\sqrt{s}}\sqrt{(E_b + m_b)(E_d + m_d)}\left[A + B(s^{\frac{1}{2}} - \tfrac{1}{2}m_b - \tfrac{1}{2}m_d)\right]$$

$$\tag{2}$$

$$f_2 = \frac{1}{8\pi\sqrt{s}}\sqrt{(E_b - m_b)(E_d - m_d)}\left[A + B(s^{\frac{1}{2}} + \tfrac{1}{2}m_b - \tfrac{1}{2}m_d)\right]\,.$$

The forward scattering amplitude in the laboratory system for elastic scattering ($m = m_a = m_c$, $M = m_b = m_d$ in Eq. (1) and (2) in 1.3.1) is

$$\frac{1}{8\pi M}T_{\frac{1}{2}\frac{1}{2}}(\omega, 0) \equiv f(\omega) = \frac{1}{4\pi}[A(\omega,0) + \omega B(\omega,0)]\,, \qquad \omega = E_a^{lab} = \frac{1}{2M}(s - M^2 - m^2)\,, \tag{3}$$

for which the imaginary part above threshold is given by the optical theorem ($k = p_a^{lab}$ in Eq. (2) in 1.3.1):

$$\mathrm{Im}\,f(\omega) = k\sigma(\omega)/4\pi\,. \tag{4}$$

Pilkuhn

2*

The resulting dispersion relations for $K^{\pm}N$ and $\pi^{\pm}N$ scattering are [B. Martin 70]

$$\mathrm{Re}\,f_{\pm}(\omega) = \frac{P}{4\pi^2}\int_m^\infty k'\,d\omega'\left[\frac{\sigma_+(\omega')}{\omega'\mp\omega} + \frac{\sigma_-(\omega')}{\omega'\pm\omega}\right] + \sum_i \frac{R_i}{\omega_i\pm\omega} + \frac{1}{\pi}\int_{\bar\omega}^m d\omega'\,\frac{\mathrm{Im}\,f_-(\omega')}{\omega'\pm\omega}\,,\tag{5}$$

$$R_i = \frac{G_i^2}{4\pi}\frac{(M_i-M)^2-m^2}{4M^2}\,,\qquad \omega_i = \frac{1}{2M}(M_i^2-M^2-m^2)\,,\tag{6}$$

where P denotes the principle value and $\bar\omega$ the threshold of the lightest intermediate state ($\pi^0\Lambda$ for K^-p scattering, $\pi^0 n$ for π^-p scattering). These relations still need a subtraction. If the Pomeranchuk theorem holds, the dispersion relation for $f_- - f_+$ converges: $F^{(-)}\equiv\frac{1}{2}(f_- - f_+)$

$$\mathrm{Re}\,F^{(-)}(\omega) = \omega\sum_i \frac{R_i^2}{\omega_i^2-\omega^2} + \frac{\omega}{4\pi^2}P\int_m^\infty k'\,d\omega'\,\frac{\sigma_-(\omega')-\sigma_+(\omega')}{\omega'^2-\omega^2} + \frac{\omega}{\pi}\int_{\bar\omega}^m d\omega'\,\frac{\mathrm{Im}\,f_-(\omega')}{\omega'^2-\omega^2}\,.\tag{7}$$

For $\pi^{\pm}N$-scattering, the last integral is zero or negligible; $\omega_N = -m^2/2M$, and $R_N = -2\frac{G^2}{4\pi}\omega_N^2$ (the factor 2 comes from Eq. (2) in 1.3.4). Subtracted dispersion relations (subtraction energy ω_0) are

$$\mathrm{Re}\,f_{\pm}(\omega) - \mathrm{Re}\,f_{\pm}(\omega_0) = \pm\frac{\omega-\omega_0}{4\pi^2}P\int_m^\infty k'\,d\omega'\left[\frac{\sigma_+(\omega')}{(\omega'\mp\omega)(\omega'\mp\omega_0)} - \frac{\sigma_-(\omega')}{(\omega'\pm\omega)(\omega'\pm\omega_0)}\right]$$
$$\mp\sum_i\frac{R_i(\omega-\omega_0)}{(\omega_i\pm\omega)(\omega_i\pm\omega_0)} \mp \frac{\omega-\omega_0}{\pi}\int_{\bar\omega}^m\frac{\mathrm{Im}\,f_-(\omega')\,d\omega'}{(\omega'\pm\omega)(\omega'\pm\omega_0)}\,,\tag{8}$$

where one may chose $\omega_0=0$ or m. In the latter case one introduces the s-wave scattering length (Eq. (1) in 1.3.2) through

$$\mathrm{Re}\,f_{\pm}(m) = \left(1+\frac{m}{M}\right)a_{0\pm}\,.\tag{9}$$

These "fixed-t" relations allow the determination of coupling constants of s- and u-channel Born terms. They can be derived in some cases from axiomatic field theory (see the review by [Sommer 70]). To isolate t-channel singularities (meson exchange), one needs dispersion relations for backward scattering [Atkinson 62]. For elastic meson-baryon scattering ($m_a = m_c = 1$, $m_b = m_d = M$, $t = -4q^2$), these can be written [Engels 70]

$$\mathrm{Re}\,X_i(q^2) = X_{Bi}(q^2) + \frac{P}{\pi}\int_0^\infty \frac{\mathrm{Im}\,X_i(q'^2)\,dq'^2}{q'^2-q^2} + \frac{P}{\pi}\int_{-\infty}^1 \frac{\mathrm{Im}\,X_i(q'^2)\,dq'^2}{q'^2-q^2}\,,\tag{10}$$

$$X_1(q^2) = \frac{1}{4\pi}A(s,-4q^2) + \frac{M}{4\pi}\left(\frac{q^2+1}{q^2+M^2}\right)^{\frac{1}{2}}B(s,-4q^2)\,,\qquad X_{B1} = \frac{f^2/M}{q_B^2-q^2}\,,\tag{11a}$$

$$X_2 = \frac{M}{4\pi}\frac{B(s,-4q^2)}{[(q^2+M^2)(q^2+1)]^{\frac{1}{2}}}\,,\qquad X_{B2} = \frac{4Mf^2}{q_B^2-q^2}\,,\tag{11b}$$

$$X_3 = \frac{M}{4\pi}\left(\frac{q^2+1}{q^2+M^2}\right)^{\frac{1}{2}}B(s,-4q^2)\,,\qquad X_{B3} = \frac{f^2/M}{q_B^2-q^2}\,,\qquad q_B^2 \equiv \frac{1}{4M^2}-1\,.\tag{11c}$$

For the extension to the case of different meson masses, see [Hite 72].

1.3.6 Resonance formation and background

The term "resonance formation" is used for s-channel resonances $ab\to e\to cd$. The resonating partial wave amplitude is parametrized as follows:

$$f_l = (f_{l,back}+f_{l,res})e^{2i\xi}\,,\qquad f_{l,back} = \frac{1}{2i\sqrt{q_i q_f}}(1-e^{-2i\xi})\,,\tag{1}$$

$$f_{l,res} = \pm(m^2-s-im\Gamma)^{-1}\left[\frac{m}{q'}\Gamma_{cd}\frac{m}{q}\Gamma_{ab}\right]^{\frac{1}{2}}\,.\tag{2}$$

The sign is determined by the relative sign of the coupling constants and can be inferred from interference with other resonating partial wave amplitudes (see [Levi-Setti 69]).

In the elastic channel $ab\to ab$, the second factor reduces to $m\Gamma_{ab}/q$. Useful definitions are

$$x = \Gamma_{ab}/\Gamma\ (\text{"elasticity"})\,,\qquad t_{cd} = \Gamma_{cd}^{\frac{1}{2}}\Gamma_{ab}^{\frac{1}{2}}\Gamma^{-1}\times\mathrm{sign}(G_{cd}\cdot G_{ab})\,.\tag{3}$$

Pilkuhn

For a purely elastic amplitude ($\eta_l = 1$ in Eq. (9) in 1.3.1), one can write

$$\cot\delta = \frac{m^2 - s}{m\Gamma} = 2(m - \sqrt{s})/\Gamma', \qquad \Gamma' = \Gamma\frac{2m}{m + \sqrt{s}}, \tag{4}$$

$$\frac{1}{m\Gamma} = -\frac{\mathrm{d}}{\mathrm{d}s}(\cot\delta)_{s=m^2} = -\left[\frac{E_b(m^2)}{2m^2}\frac{\mathrm{d}}{\mathrm{d}E_a}(\cot\delta)\right]_{E_a = (2m)^{-1}(m^2 + m_a^2 - m_b^2)}. \tag{5}$$

1.3.7 References for 1.3

For journal abbreviations see p. 2

Atkinson	62	PR 128, 1908: D. Atkinson.
Engels	70	NP B 25, 141: J. Engels.
Heller	60	PR 120, 627: L. Heller.
Hite	72	PR D 5, 422: G. E. Hite, R. Jacob.
Levi Setti	69	Proc. Lund Conf. on Elem. Particles, (ed. G. von Dardel), p. 349.
Martin, B.	70	STMP 55, 73: B. R. Martin.
Particle	72	RMP 43, S 1 and PL 39 B, 1: Particle data group.
Sakurai	60	AP 11, 1: J. J. Sakurai.
Sommer	70	FP 18, 577: G. Sommer.
Weinberg	66	PRL 16, 879: S. Weinberg.

1.4 Tables of particle properties and decay coupling constants

1.4.1 Fundamental constants, leptons and absorption lengths

Tab. a). *Fundamental constants*

Symbol and name	Value [1,2]
N (Avogadro's number)	$6.022169(40) \cdot 10^{23}$ mole^{-1} (based on $A(^{12}C) = 12$)
c (velocity of light)	$2.9979250(10) \cdot 10^{10}$ cm sec^{-1}
e (elementary charge)	$4.803250(21) \cdot 10^{-10}$ esu $= 1.6021917(70) \cdot 10^{-19}$ Coulomb
α (fine-structure constant)	$e^2/\hbar c = 1/137.03602(21)$ ($= e^2/4\pi$ in the formulas)
MeV (Mega electron Volt)	$1.6021917(70) \cdot 10^{-6}$ erg
$\hbar$ ($1/2\pi$ times Planck's constant)	$6.582183(22) \cdot 10^{-22}$ MeV sec $= 1.0545919(80) \cdot 10^{-27}$ erg sec
$\hbar c$	$1.9732891(66) \cdot 10^{-11}$ MeV cm $= 197.32891(66)$ MeV fm
k (Boltzmann constant)	$1.380622(59) \cdot 10^{-16}$ erg/°K
1 eV per particle	$11604.85(49)$ °K
r_e^2 ($r_e =$ electron radius)	$e^4 m_e^{-2} c^{-4} = (2.817939(13)$ fm$)^2 = 79.40780(73)$ mb
μ_B (Bohr magneton)	$e\hbar/2m_e c = 0.5788381(18) \cdot 10^{-14}$ MeV/Gauss
μ_N (nuclear magneton)	$e\hbar/2m_p c = 3.152526(21) \cdot 10^{-18}$ MeV/Gauss
pc (magnetic rigidity)	$300\, H\varrho Z^{-1}$ (pc in MeV, H in Gauss, ϱ in cm)

Tab. b). *Leptons*

Symbol	m [GeV]	m^2 [GeV^2]	Γ^{-1} [sec]	Magnetic moment
$e^\mp$	0.0005110041(16)	$0.2611252 \cdot 10^{-6}$	stable	1.001159658(4)
$\mu^\mp$	0.1056594(4)	0.01116391	$2.2002(8) \cdot 10^{-6}$ [3]	1.00116616(31)
$\nu_e, \bar{\nu}_e$	$< 6 \cdot 10^{-8}$ [4]	≈ 0	$\left.\begin{array}{l} \Gamma_{lab}^{-1} = \infty \\ \Gamma^{-1} \text{ is undetermined} \end{array}\right.$	0
$\nu_\mu, \bar{\nu}_\mu$	< 0.0006 [5]	≈ 0		0

[1] From [Particle 71, 72].

[2] See 1.1.1 for the presentation of errors.

[3] $\Gamma = 2.9916(11) \cdot 10^{-10}$ eV, g (in Eq. (1) of 1.1.4) $= \dfrac{[192\pi^3 \Gamma m^{-5}]^{\frac{1}{2}}}{1 - 8m_e^2 m_\mu^{-2}}\left[1 + \dfrac{\alpha}{4\pi}(\pi^2 - \tfrac{25}{4})\right] = 1.1655(2) \cdot 10^{-5}$ GeV^{-2}

$= 1.4348(3) \cdot 10^{-49}$ erg cm^3 [Roos 71]. Experimental numbers from [Williams 72].

[4] From [Bergkvist 72].

[5] From [Backenstoss 71].

Tab. c). *Radiation lengths and ionization loss parameters for* e^- *and* e^+

Absorber, Z		ϱ	t_0		E_c	a	b		γ	
		$\mathrm{g\,cm}^{-3}$	$\mathrm{g\,cm}^{-2}$	cm	MeV		e^-	e^+	e^-	e^+
Water	4.69	1.00	35.7	35.7	62	141.49	35.64	25.22	0.834	0.764
Air	7.31	0.0012	36.4	30400	76	147.55	72.13	67.24	0.348	0.328
Al	13	2.70	23.9	8.86	34	66.84	17.13	12.63	0.768	0.704
Ilford G5	21.4	3.83	11.0	2.88	14	28.65	9.66	7.96	0.593	0.545
Fe	26	7.86	13.7	1.74	18	—	—	—	—	—
Cu	29	8.94	12.7	1.42	16	31.64	9.53	7.58	0.574	0.516
Pb	82	11.35	5.58	0.513	6.4	12.50	5.21	4.45	0.567	0.528

This table is from [Messel 70]. The radiation length t_0 is defined through the energy loss of high-energy electrons (due to bremsstrahlung) along their path x, $dE/dx = -E/t_0$. It is computed as $t_0^{-1} = 4\alpha r_e^2 nZ(Z+\xi)\ln(183Z^{-\frac{1}{3}})$, $\xi = \ln(1440Z^{-\frac{1}{3}}/183)$. For α and r_e see Tab. 1.4.1 a, for n see 1.4.1 d. The absorption length for photons due to pair creation is approximately $t_0 \cdot \frac{9}{7}$. E_c = critical energy (= natural energy unit for electromagnetic cascade). The ionization loss is parametrized as $-dE/dt = a - bE^{-\gamma}$, in units of mc^2 per radiation length. For air, the following volume percentages are used: N 78.03%, O 20.99%, Ar 0.94% and CO_2 0.03%. For other atomic and nuclear properties of materials see [Particle 71 or 72].

Tab. d). *High-energy absorption cross sections of hadrons on nuclei*

At high energies, it is convenient to define $\sigma_{abs} = \sigma_{tot} - \sigma_{el}$, where σ_{el} comprises besides truly elastic scattering also elastic scattering with target excitation but negligible momentum loss. For nuclear mass number $A > 5$, one can parametrize $\sigma_{abs} = \sigma_0 A^\alpha$. Values for σ_0 and α are given in the following table. In addition, [Gorin 71] give $\alpha = 0.76$ for π^-, 0.76 for K^- and 0.66 for $\bar{p}$ at 13.3 GeV/c (they also report $\bar{d}$ absorption cross sections). [Allaby 70] give for ^{4}He $\sigma_{abs} = 74(3)$ mb for π^- (30…60 GeV/c), 65(3) mb for K^- (30…40 GeV/c) and 118(5) mb for $\bar{p}$ (30…40 GeV/c). Total neutron cross sections at 10 GeV/c can be represented as $\sigma_{tot} = 2\pi(1.3A^{\frac{1}{3}} - 0.6)\,\mathrm{fm}^2$ for all nuclei [Engler 68]. Neutron and proton cross sections on nuclei are equal above 1 GeV/c. Proton absorption cross sections on Carbon have been parametrized by [Grigorov 70] as $216(7)\,\mathrm{mb}(1 + 0.068(12)\ln(E/20\,\mathrm{GeV}))$ for $E > 20$ GeV in a Cosmic-Ray experiment. For pp-collisions, accelerator experiments give $\sigma_{abs} = 30$ mb at 26 GeV and 31 mb at 500 GeV ([Holder 71]; experimental numbers revised 1972). This constancy with energy makes the energy dependence of [Grigorov 70] somewhat unlikely. See also [Trefil 71].

σ_0 and α are inversely correlated. From $\sigma(\pi^-p) < \sigma(pp) < \sigma(\bar{p}p)$ one may conclude $\alpha(\pi^-) > \alpha(p) > \alpha(\bar{p})$. The value $\alpha(p) = 0.7$ for $p > 20$ GeV/c is estimated on this basis. The absorption cross sections in air are computed for A(air) = 14.4.

Hadron	p [GeV/c]	σ_0 [mb]	α	σ_{abs}(air) [mb]	t_{abs}(air) $\left[\dfrac{\mathrm{g}}{\mathrm{cm}^2}\right]$ [1]	Reference
	1.1	≈44.0	≈0.69	≈277	≈ 86	Igo 67
π^-	20…60	29.0(7)	0.750(5)	205(3)	117(2)	Allaby 70
K^-	20…40	26.0(7)	0.760(7)	186(3)	128(2)	Allaby 70
$\bar{p}$	20	59.1(2.6)	0.648(10)	326(18)	74(4)	Allaby 70
	30	50.9(2.4)	0.674(9)	292(8)	82(2)	Allaby 70
	40	49.9(2.4)	0.674(10)	291(7)	83(2)	Allaby 70
K^+	3.2	21(2)	0.80(1)	177(18)	135(16)	Abrams 71
K^-	3.2	29(3)	0.76(1)	220(21)	109(11)	Abrams 71
$\bar{p}$	3.2	68(7)	0.65(1)	385(39)	62(6)	Abrams 71
p, n	> 20	≈38	≈0.70	≈245	≈ 98	Grigorov 70
	1.7	42	0.67	251	95	Igo 67

[1] The absorption length t_{abs} is defined as the intensity decrease per path element dx, $dI/dx = -I/t_{abs}$. It is related to σ_{abs} through $t_{abs}^{-1} = n\sigma_{abs}$, where n is the particle density, $n = NA^{-1}\varrho$, N is Avogadro's number and A the atomic weight. ϱ is the mass density in gcm^{-3} and is omitted when t_{abs} is given in gcm^{-2}.

Pilkuhn

1.4.2 The nonet of pseudoscalar mesons (0^- states)

Tab. 1. Masses, lifetimes and decay channels [1]

Name	m [GeV] m^2 [GeV²]	Γ^{-1} [sec] Γ [eV]	Decay channel [2]	see	Fraction	$\Gamma_{partial}$ [eV]	p or p_{max} [3] [MeV]
$\pi^\pm$	0.139576(8)	$2.6024(24)\cdot10^{-8}$	$\mu\nu$ [5]	d	0.99975	$2.5286\cdot10^{-8}$	29.798
	0.0194815	$2.5293(23)\cdot10^{-8}$	$e\nu$	d	0.000124(3)	$3.14\cdot10^{-12}$	69.789
	$m_{\pi^\pm}-m_{\pi^0}=4.604(4)$ MeV		$\mu\nu\gamma$		0.000124(25)	$3.14\cdot10^{-12}$	29.798
π^0	0.134972(12)	$0.56\cdot10^{-16}$	$\gamma\gamma$	a	0.9883(4)	11.56	67.486
	0.0182174	$11.7(1.2)$ [4]	$\gamma e\bar{e}$	a	0.0117(4)	0.137	67.482
	$\theta_{\pi\eta}=-0.106(5)$ from Eq.(6) in 1.1.2						
$K^\pm$	0.49384(11)	$1.235(4)\cdot10^{-8}$	$\mu\nu$ [5]	d	0.6377(29)	$3.399\cdot10^{-8}$	235.6
	0.24388	$5.330(18)\cdot10^{-8}$	$\pi^\pm\pi^0$	g	0.2093(30)	$1.116\cdot10^{-8}$	205.2
			$\pi^\pm\pi^\pm\pi^\mp$	h	0.0557(4)	$2.969\cdot10^{-9}$	125.6
			$\pi^\pm\pi^0\pi^0$	h	0.0170(5)	$0.906\cdot10^{-9}$	133.1 [6]
			$\pi^0\mu\nu$	e	0.0318(11)	$1.695\cdot10^{-9}$	215.2
			$\pi^0 e\nu$	e	0.0485(7)	$2.585\cdot10^{-9}$	228.5
			$\pi^0 e\nu\gamma$		0.0004(1)	$2.1\cdot10^{-11}$	228.5
	$m_{K^\pm}-m_{K^0}=-3.95(13)$ MeV		$\pi^\pm\pi^0\gamma$		0.0002(1)	$1\cdot10^{-11}$	205.2
K^0_S	0.49779(15)	$0.862(6)\cdot10^{-10}$	$\pi^\pm\pi^-$	g	0.687(5)	$524.9\cdot10^{-8}$	206.1
	0.24780	$7.64(6)\cdot10^{-6}$	$\pi^0\pi^0$	g	0.313(5)	$239.1\cdot10^{-8}$	209.1
			$\pi^+\pi^-\gamma$		0.0023(8)	$1.76\cdot10^{-8}$	206.1
K^0_L	0.49779(15)	$5.17(4)\cdot10^{-8}$	$\pi^0\pi^0\pi^0$	h	0.214(7)	$2.72\cdot10^{-9}$	139.3
	0.24780	$1.273\cdot10^{-8}$	$\pi^+\pi^-\pi^0$	h	0.126(3)	$1.60\cdot10^{-9}$	132.9 [6]
			$\pi\mu\nu$	f	0.268(7)	$3.41\cdot10^{-9}$	216.1
			$\pi e\nu$	f	0.389(8)	$4.95\cdot10^{-9}$	229.3
			$\pi e\nu\gamma$		0.013(8)	$1.7\cdot10^{-10}$	229.3
			$\pi^+\pi^-$	g	0.00157(5)	$2.00\cdot10^{-11}$	206.1
			$\pi^0\pi^0$	g	0.00094(19)	$1.20\cdot10^{-11}$	209.1
	$m_L-m_S=0.460(6)\,\Gamma_S$ [7]		$\gamma\gamma$	i	0.00050(5)	$0.64\cdot10^{-11}$	248.9
η [8]	0.5488(6)	$2.4\cdot10^{-19}$	$\gamma\gamma$	a	0.380(11)	1010(230)	274.4
	0.30118	2700(670)	$\pi^0\pi^0\pi^0$	c	0.300(11)	810	179.6
			$\pi^0\gamma\gamma$	c	0.031(11)	127	257.8
		charged	$\pi^+\pi^-\pi^0$	c	0.240(6)	624	174.6 [6]
		channels 27.8%	$\pi^+\pi^-\gamma$	b	0.049(2)	127	236.2
			$\pi^+\pi^-e\bar{e}$	b	0.0003(1)	0.83	235.9
X or η' [8]	0.9571(5)	$>3.3\cdot10^{-22}$	$\eta\pi^+\pi^-$	c	$\left.\begin{array}{l}\;\;0.440(\text{th})\\0.68(2)\;\;0.240(\text{th})\end{array}\right.$	for $\Gamma=2$ MeV $\left\{\begin{array}{l}8.8\cdot10^5\\4.8\cdot10^5\end{array}\right.$	230.9 [10]
	0.9160	$<2\cdot10^6$ [9]	$\eta\pi^0\pi^0$	c			237.7 [10]
			$\varrho^0\gamma$	b	0.29(3)	$5.8\cdot10^5$	164.5
	$\theta_\kappa=-10.4(2)^0$ from Eq. (2), (3)		$\gamma\gamma$	a	0.017(4)	$0.34\cdot10^5$	478.9
	in 1.2.2 [11]		$\pi^+\pi^-e\bar{e}$	b	0.0020(th)	400	457.9

[1]) The experimental numbers are from [Particle 72], unless stated otherwise.
[2]) Particles and antiparticles have identical total and partial widths. A complication arises in case f.
[3]) For decay into >2 particles, p_{max} gives the maximum momentum of the heaviest particle.
[4]) From [Bellettini 70].
[5]) $\mu\nu$ and $e\nu$ stand for $\mu^+\nu_\mu$ and $e^+\nu_e$ in π^+ and K^+ decays and for $\mu^-\bar{\nu}_\mu$ and $e^-\bar{\nu}_e$ in π^- and K^- decays.
[6]) The maximum momentum of π^0 is smaller by 0.74 MeV.
[7]) From [Carnegie 71]. [8]) See Ann. Rev. Nucl. Sci. 21 (71) for an introduction. [9]) From [Binnie 72].
[10]) The maximum pion momentum is 196.8 for $\pi^\pm$ and 203.1 for π^0.
[11]) The sign comes from [Bloodworth 72] and agrees with the quark model.

Fractions below 10^{-4}: $\pi^\pm\to\pi^0 e\nu\;1.02(7)\cdot10^{-8}$, $\pi^\pm\to e\nu\gamma\;3.0(5)\cdot10^{-8}$, $K^\pm\to\pi^\pm\pi^\mp e\nu\;3.7(2)\cdot10^{-5}$, $K^\pm\to\pi^0\pi^0 e\nu\;2\cdot10^{-5}$, $K^\pm\to\pi^\pm\pi^\mp\mu\nu\;0.9(4)\cdot10^{-5}$, $K^\pm\to e\nu\;1.3(2)\cdot10^{-5}$, $K^\pm\to\pi^\pm\pi^\pm\pi^\mp\gamma\;10(4)\cdot10^{-5}$. See 1.4.2$a$ for further decay modes below 10^{-4}.

a) Two-photon decays of π^0, η and X. Reviews [Paty 70, Mopurgo 71]

	$g_{\gamma\gamma}$ [GeV^{-1}][1]	$\dfrac{\Gamma(\gamma e\bar{e})}{\Gamma(\gamma\gamma)}$ [2]	$\dfrac{\Gamma(\gamma\mu\bar{\mu})}{\Gamma(\gamma\gamma)}$ [2]	$\dfrac{\Gamma(e\bar{e}e\bar{e})}{\Gamma(\gamma\gamma)}$ [3]	$\dfrac{\Gamma(e\bar{e})}{\Gamma(\gamma\gamma)}$ [4]	$\dfrac{\Gamma(\mu\bar{\mu})}{\Gamma(\gamma\gamma)}$ [4,5]
π^0	0.337(17)	0.0119	0	$3.47\cdot 10^{-5}$	$\approx 6\cdot 10^{-8}$	0
η	0.382(44)	0.0162	0.00055	$6.6\cdot 10^{-5}$	$>4.5\cdot 10^{-9}$	$(1-3)\cdot 10^{-5}$
X	<1.1[6]	0.0170	—	—	—	—

[1] $g = e^2 g_{\gamma\gamma}$ in Tab. 1 in 1.2.2 and $\Gamma_{\gamma\gamma} = \dfrac{\pi}{4}\alpha^2 m^3 g_{\gamma\gamma}^2$ (extra factor $\frac{1}{2}$ for identical particles).

[2] $c=0$ and an extra factor 2 in Eq. (7) in 1.2.4. For $\gamma\mu\bar{\mu}$, change the bracket to $(1 - \frac{9}{8}\xi^2 + \frac{1}{8}\xi^3)\ln(\xi^{-\frac{1}{2}} + (\xi^{-1}-1)^{\frac{1}{2}}) - (\frac{7}{4} - \frac{13}{8}\xi - \frac{1}{8}\xi^2)(1-\xi)^{\frac{1}{2}}$, $\xi = 4m_\mu^2/m^2$.

[3] This is roughly $(\Gamma(\gamma e\bar{e})/2\Gamma(\gamma\gamma))^2$.

[4] The matrix element is of the form of Eqs. (6), (7), and (10) in 1.2.2. Taking Im T from $\Gamma(\gamma\gamma)$ and Re $T=0$ leads to $\Gamma(l\bar{l})/\Gamma(\gamma\gamma) > \alpha^4 \dfrac{1}{pm} m_1^2 \ln^2((m+2p)/2m_1)$. See [Quigg 68, Litskevich 70].

[5] A large experimental value $5.9(2.2)\cdot 10^{-5}$ is given by [Hyams 69] for η decay.

[6] For $\Gamma(X\to\gamma\gamma < 40$ keV). From [Basile 71]. The most recent $X\to\gamma\gamma$ fraction is 0.029(9) [Apel 72].

b) The $\pi^+\pi^-\gamma$ decays of η and X

Both decays are dominated by the ϱ^0-meson and can be computed according to Eq. (5) in 1.2.4. For the X decay, approximation Eq. (6) in 1.2.4 can be used (the finite-width correction is about 10%). Putting $g = \sqrt{\alpha}g_{X\varrho\gamma}$ in Tab. 1 in 1.2.2 one gets $g_{X\varrho\gamma}^2/4\pi < 16$ GeV^{-2} for $\Gamma_X < 2$ MeV. The shape of the $\eta\to\pi^+\pi^-\gamma$ spectrum is calculated by [Cnops 68]. The decays of η and X into $\pi^+\pi^- e^+e^-$ are calculated according to Eq. (7) in 1.2.4.

c) The decays $\eta\to\pi\pi\pi$, $X\to\eta\pi\pi$ and $\eta\to\pi^0\gamma\gamma$

All information is contained in the Dalitz plot (1.2.4). Taking as particle 3 the π^0 in the $\pi^0\pi^0\pi^0$ and $\pi^+\pi^-\pi^0$ states and the η in the $\eta\pi^0\pi^0$ and $\eta\pi^+\pi^-$ states, the matrix elements must be symmetric in E_1 and E_2,

$$T(E_1, E_2, E_3) = T(E_2, E_1, E_3).$$

This is trivial for $1=2$ as in the $\pi^0\pi^0\pi^0$ or $\eta\pi^0\pi^0$ states, and follows from CP conservation for the $\pi^+\pi^-\pi^0$ and $\pi^+\pi^-\eta$ states. The initial state has $CP = -1$, the π^0 or η mesons have $CP = -(-1)^{l_3}$, where l_3 is the angular momentum of particle 3 around the $\pi^+\pi^-$ pair, and the $\pi^+\pi^-$ pair has $CP = +1$ and a spin $l_{+-} = l_3$, since the initial state has spin 0. Consequently l_{+-} must be even. The isospin of the (1, 2)-pair can be 0 or 2, and the $\pi^+\pi^-\pi^0$ and $\pi^0\pi^0\pi^0$ states have components only in $I=1$ and $I=3$ states.

The Dalitz plot of these decays is so small that only a linear expansion of T in E_i can be determined. Due to the symmetry in E_1 and E_2, only E_3 enters:

$$T = \text{const}(1 + bY^0), \qquad Y^0 = \frac{3T^0}{Q} - 1 = 3\,\frac{E_3 - m_3}{m - 2m_1 - m_3} - 1,$$

$b = -0.550(12)$ for $\eta\to\pi^+\pi^-\pi^0$ decay [Cnops 68a, Danburg 70] and $b=0$ for $\eta\to\pi^0\pi^0\pi^0$ decay. When the isospin-3 state is absent (see the end of 1.1.3), one gets $\Gamma(\eta\to\pi^0\pi^0\pi^0)/\Gamma(\eta\to\pi^+\pi^-\pi^0) \cong \frac{3}{2}\cdot 1.13\cdot(1+b^2/4)^{-1} = 1.58$. The $\pi^0\gamma\gamma$ channel is difficult to separate from the $\pi^0\pi^0\pi^0$ channel. It could be up to 10% of $\eta\to$ neutral decays. Theoretical considerations indicate hardly more than 1%. In that case, the fraction $\eta\to\pi^0\pi^0\pi^0$ would be increased to 0.331, which gives 1.4 for the $\pi^0\pi^0\pi^0/\pi^+\pi^-\pi^0$ ratio. For $X\to\eta\pi\pi$ decays the factor 3 in the definition of Y^0 is replaced by $2 + m_\eta/m_\pi$, and $b = -0.28(6)$ for $X\to\eta\pi^+\pi^-$ [Dufey 69].

d) The $\mu\bar{\nu}$ and $e\bar{\nu}$ decays of $\pi^\pm$ and $K^\pm$ have $V^\mu = 0$ in Eq. (2) in 1.1.4 and Γ given by Eq. (9) in 1.2.2, which leads to $g_\pi = 1.0578(5)\cdot 10^{-6}$ GeV^{-1} and $g_K = 0.2917\cdot 10^{-6}$ GeV^{-1}. The theoretical $e\bar{\nu}/\mu\bar{\nu}$ ratios are $1.228\cdot 10^{-4}$ for π (radiative corrections included, see [Marshak 69, ref. in 1.1.6] and $0.258\cdot 10^{-4}$ for K. The "reduced" coupling constants f are (see Eq. (6) in 1.1.4)

$$f = 2^{\frac{1}{2}}g_\pi/g_V^{\Delta Y=0} = 130.15\text{ MeV} = 0.932 m_{\pi^+} \quad \text{and} \quad f_K = 2^{\frac{1}{2}}g_K[g\sin\theta_A]^{-1} = 123.04\text{ MeV} = 0.8815 m_{\pi^+}.$$

e) The $\pi^0 l\bar{\nu}$ decays of $\pi^\pm$ and $K^\pm$ have $A^\mu = 0$ in Eq. (2) in 1.1.4 and $V^\mu = f_+(P_\pm + P_0)^\mu + f_-(P_\pm - P_0)^\mu$. (Reviews: [Gaillard 70], [Jones 70], [Haidt 71].) For the π decay, CVC (see 1.1.4) says $f_- = 0$, $f_+ = 2^{\frac{1}{2}}g_V^{\Delta Y=0}$, and the width is approximately [Källen 64, 65, ref. in 1.1.6]

$$\Gamma = (30\pi^3)^{-1} g_V^2 \Delta^5\cdot R(m_e^2/\Delta^2), \qquad \Delta \equiv m_{\pi^+} - m_{\pi^-} = 4.604\text{ MeV},$$

$$R(\varepsilon) = (1-\varepsilon)^{\frac{1}{2}}(1 - \tfrac{9}{2}\varepsilon + 4\varepsilon^2) + \tfrac{15}{2}\varepsilon^2 \ln(1 + (1+\varepsilon^{-1})^{\frac{1}{2}}).$$

Pilkuhn

For the K-decays, $f_\pm$ must be expanded in terms of $t = (P_\pm - P_0)^2 = (P_l + P_{\bar\nu})^2$: $f_\pm(t) = f_\pm(0)(1 - \lambda_\pm t/m_\pi^2)$. The Cabibbo theory says $f_+(K) = \frac{1}{2}f_+(\pi)\tan\theta_V = 2^{-\frac{1}{2}}g\sin\theta_V$. The widths are given by ($\xi \equiv f_-(0)/f_+(0)$):

$$\Gamma_{e3} \equiv \Gamma(K^+ \to \pi^0 e^+ \nu) = g^2\sin^2\theta_V(768\pi^3)^{-1}m^5|f_+(0)|^2(0.573 + 0.138\lambda_+ m_K^2/m_\pi^2)$$

$$\frac{\Gamma_{\mu3}}{\Gamma_{e3}} = \frac{\Gamma(K^+ \to \pi^0\mu^+\nu)}{\Gamma(K^+ \to \pi^0 e^+\nu)} = \frac{0.6457 + 3.8008\lambda_+ + 6.8120\lambda_+^2 + 0.1264\xi + 0.4757\xi\lambda_+ + 0.0192\xi^2}{1.0000 + 3.6995\lambda_+ + 5.4777\lambda_+^2}$$

Analyzed	Information	Result	Best fit
Dalitz plot in K_{13}	λ_+	0.030(7)	
Dalitz plot in $K_{\mu3}$	ξ, Λ, λ_+	$\xi = -1.00(40)$ $\lambda_+ = 0.043(17)$	$\lambda_- = 0.05(10)$
μ^+ polarization in $K_{\mu3}$	$\xi, \Lambda \equiv \xi(\lambda_- - \lambda_+)$	$\xi = -1.45(70)$ $\Lambda = 0.11(15)$	for $\lambda_- = 0$: $\xi = -0.85(20)$
$\Gamma_{\mu3}/\Gamma_{e3} = 0.626(19)$	relation above		$\lambda_+ = 0.045(12)$

The K* dominance model gives $\lambda_+ = 0.023$.

f) The decays of K_L^0 into $\pi^+ e\bar\nu_e$, $\pi^- \bar{e}\nu_e$, $\pi^+\mu\bar\nu_\mu$ and $\pi^-\bar\mu\nu_\mu$

The time development ($\tau =$ proper time) of a general state ψ_K is found from its expansion in terms of K_S and K_L defined in Eq. (4) in 1.1.2:

$$\psi_K(0) \equiv a_S K_S + a_L K_L; \qquad \psi_K(\tau) = a_S e^{-iM_S\tau}K_S + a_L e^{-iM_L\tau}K_L$$

with $M_i \equiv m_i - i\Gamma_i/2$. (Example: for $K^-p \to K^0 n$, one has $a_S \approx a_L \approx 2^{-\frac{1}{2}}$.) The time distribution of $\pi^\mp e^\pm\bar\nu$ decays $N^\pm(\tau)$ is ($a_S/a_L = \varrho e^{i\phi_\varrho}$):

$$N^\pm(\tau) = |1 + X|^2\varrho^2 e^{-\Gamma_S\tau} + |1 - X|^2 e^{-\Gamma_L\tau} \pm 2\,\mathrm{Re}\,\varepsilon(1 - |X|^2)(e^{-\Gamma_L\tau} + \varrho^2 c^{-\Gamma_S\tau})$$
$$+ \varrho e^{-\tau(\Gamma_S + \Gamma_L)\,2}[\pm 2(1 - |X|^2)\cos(\Delta m\tau + \phi_\varrho) - 4\,\mathrm{Im}\,X\sin(\Delta m\tau + \phi_\varrho)]$$

$X \equiv g/f = f(\bar{K} \to \pi^- l\nu)/f(K \to \pi^- l\nu)$. The $\Delta Y = \Delta Q$ rule says $X = 0$. Experimentally, $X = 0.05^{+0.025}_{-0.035} + 0.01(2)i$ in $K_L^0 \to \pi e\nu$ decays. The $K_L^0 \to \pi\mu\nu$ decay has $\lambda_+ = 0.08(1)$ [Chien 71]; λ_- and ξ are strongly correlated. For $\lambda_- = 0$, one gets $\xi = -0.68^{+0.12}_{-0.20}$. Due to the CP-violating parameter ε in Eq. (4) in 1.1.2, the $\pi^- e^+\nu$ and $\pi^+ e\bar\nu$ widths of K_L are slightly different: For a pure K_L state ($\varrho = 0$), the charge asymmetry $(N^+ - N^-)/(N^+ + N^-)$ is 0.00285(28) in K_{e3} [Marx 70] and 0.00405(135) in $K_{\mu3}$ [Dorfan 67]. Assuming $X = 0$, this is just $2\,\mathrm{Re}\,\varepsilon$, or $\mathrm{Re}\,\varepsilon = 0.00149(14)$.

g) $K \to \pi\pi$ decays (review: [Steinberger 70])

The matrix element for K decay into a $\pi\pi$-system of isospin I is (see 1.2.1)

$$T_I = T(K \to (\pi\pi)_I) = G_0^I e^{i\delta_I} = 8\pi m A_I e^{i\delta_I}, \qquad T(\bar{K} \to (\pi\pi)_I) = 8\pi m A_I^* e^{i\delta_I},$$

where $I = 0$ or 2 and δ_I is the $\pi\pi$-phase shift. Since the decays can violate time-reversal invariance, A_I may still be complex. In the following, the phase of $\bar{K}^0$ relative to K^0 is defined such that A_0 is real and positive. Defining

$$\varepsilon' \equiv i2^{-\frac{1}{2}}e^{i(\delta_2 - \delta_0)}\mathrm{Im}\,A_2/A_0, \qquad \omega \equiv 2^{-\frac{1}{2}}e^{i(\delta_2 - \delta_0)}\mathrm{Re}\,A_2/A_0$$

and
$$\Gamma(K_S \to \pi^0\pi^0)/\Gamma(K_S \to \pi^+\pi^-) = p(\pi^0\pi^0)/2p(\pi^+\pi^-) - 3\,\mathrm{Re}\,\omega \quad \text{yields} \quad \mathrm{Re}\,\omega = 0.0174\,.$$

Since $\delta_2 - \delta_0 \approx 45°$, one has $\omega \ll 1$.

Selection rules: $\Delta I = \frac{1}{2}$ would require $T_+ \equiv T(K^+ \to \pi^+\pi^0) = 0$, $A_2 = 0$ ($\omega = \varepsilon' = 0$). The weaker rule $\Delta I < \frac{5}{2}$ requires

$$T_2 = \sqrt{\tfrac{2}{3}}T_+ \quad \text{or} \quad 2|A_0|^2(|\omega|^2 + |\varepsilon'|^2) = \tfrac{2}{3}|A_+|^2 \quad \text{or} \quad |\omega|^2 + |\varepsilon'|^2 = \Gamma_+/3\Gamma_S$$

Decay	$\eta = T(K_L)/T(K_S)$	(Theory)	with $\omega = 0$
$K \to \pi^+\pi^-$	$0.00190(5)e^{i40(6)°}$	$\varepsilon + \varepsilon'/(1 + \omega)$	$\varepsilon = 0.00200(19)e^{i42°}$
$K \to \pi^0\pi^0$	$0.00217(5)e^{i48(10)°}$	$\varepsilon - 2\varepsilon'/(1 - 2\omega)$	$\lbrace\ |\varepsilon'| = 0.06(17)$

h) $K \to 3\pi$ decays

The Dalitz plot is treated as in Section c. Particle 3 is the π^- in the $K^+ \to \pi^+\pi^+\pi^-$ decay and the π^+ in the $K^+ \to \pi^0\pi^0\pi^+$ decay. For the $K_L \to 3\pi$ decays, all arguments run through as in Section c, if the minute $CP = +1$

Pilkuhn

admixture is neglected. The expansion uses a parameter which is closely related to the relativistic invariants:

$$T = A + B\hat{s}, \qquad \hat{s} = (2s_{12} - s_{13} - s_{23})/3m_{\pi^+}^2 = (2m^2 + 2m_3^2 - 6mE_3 - m_1^2 - m_2^2)/3m_{\pi^+}^2$$

$$|T|^2 = |A|^2(1 + 2\hat{s}\,\mathrm{Re}\,B/A) = |T(\hat{s}=0)|^2(1 + g\hat{s}) .$$

For its relation to the expansion in Y^0 used in Section c, see [Particle 71], p. 41, with $a_y \equiv 2b$. Isospin decomposition [Zemach 64]

$I(\pi\pi\pi)$	$T(\pi^+\pi^+\pi^-)$	$T(\pi^0\pi^0\pi^+)$	$T(\pi^+\pi^-\pi^0)$	$T(\pi^0\pi^0\pi^0)$
1	$2a_{\frac{1}{2}} + 2a_{\frac{3}{2}} - b_{\frac{1}{2}}\hat{s} - b_{\frac{3}{2}}\hat{s}$	$a_{\frac{1}{2}} + a_{\frac{3}{2}} + b_{\frac{1}{2}}\hat{s} + b_{\frac{3}{2}}\hat{s}$	$-a_{\frac{1}{2}} + 2a_{\frac{3}{2}} - b_{\frac{1}{2}}\hat{s} + 2b_{\frac{3}{2}}\hat{s}$	$-3a_{\frac{1}{2}} + 6a_{\frac{3}{2}}$
2	$(C_{\frac{3}{2}} + C_{\frac{5}{2}})\hat{s}$	$(C_{\frac{3}{2}} + C_{\frac{5}{2}})\hat{s}$	0	0
3	$d_{\frac{5}{2}} + d_{\frac{7}{2}}$	$-2d_{\frac{5}{2}} - 2d_{\frac{7}{2}}$	$-\frac{3}{2}d_{\frac{5}{2}} + 2d_{\frac{7}{2}}$	$3d_{\frac{5}{2}} - 4d_{\frac{7}{2}}$

The $\Delta I = \frac{1}{2}$ rule predicts the ratios 4/1/2/3 for the reduced widths $|A|^2$ (weights $\frac{1}{2} : \frac{1}{2} : 1 : 3$ from the identity of particles).

Decay	$\Gamma\,[10^{-10}\,\mathrm{eV}]$	g	Relative Dalitz plot areas [1]		
			U	NU	NUC
$K^+ \to \pi^+\pi^+\pi^-$	29.7	$-0.217(3)$ [2]	1.000	1.000	1.000
$K^+ \to \pi^0\pi^0\pi^+$	9.06	$0.511(18)$	1.247	1.184	1.155
$K_L \to \pi^+\pi^-\pi^0$	15.4	$0.400(33)$	1.219	1.268	1.268
$K_L \to \pi^0\pi^0\pi^0$	26.2	0	1.487	1.487	1.451

[1]) From [Mast 69]. U = uniform Dalitz plot, NU = nonuniform, includes the slopes g (with $g(++-) = -0.197$), NUC including final state Coulomb interactions.

[2]) From [Ford 72].

i) The decays $K_L^0 \to \gamma\gamma$ and $K_L^0 \to \mu\bar{\mu}$

The decay $K_L^0 \to \gamma\gamma$ should provide a reliable value for the imaginary part of the $K_L^0 \to \mu\bar{\mu}$ amplitude. Neglecting the real part, one thus finds a lower limit for $K_L^0 \to \mu\bar{\mu}$ decays which is 3 times larger than the present experimental upper limit $\Gamma(K_L^0 \to \mu\bar{\mu}) < 1.8 \cdot 10^{-9}\,\Gamma(K_L^0)$ [Clark 71]. See also footnote [4]) to Tab. *a*.

1.4.3 The vector meson nonet (1^- states)

Tab. 1. Masses, total and partial width: see p. 27.

a) Decays of ϱ, ω and ϕ mesons into $e\bar{e}$ and $\mu\bar{\mu}$ pairs and the pion electric form factor in the timelike region

V	$\Gamma(V \to e\bar{e})\,[\mathrm{keV}]$ [1]	Fraction $[10^{-5}]$	$\dfrac{\Gamma(V \to \mu\bar{\mu})}{\Gamma(V \to e\bar{e})}$	$f^2/4\pi$
ϱ^0	6.1(5)	4.0(4)	0.9986	2.56(22)
ω	0.76(8)	8.2(1.0)	0.9987	18.4(20)
ϕ	1.41(12)	34.5(2.7)	0.9995	11.00(9)

[1]) From [Bizot 70, Lefrancois 71].

The $V \to e\bar{e}$ decay width is given in Eq. (13) in 1.2.2. Its cleanest determination comes from the e^+e^- colliding beam experiments of the type $e\bar{e} \to V \to 2$ or 3 mesons. For $V \to M\bar{M}$ decays (M = pseudoscalar meson), the differential and total cross sections are (β = meson velocity)

$$\frac{d\sigma}{d\Omega} = \frac{\alpha^2\beta^3}{8s}\sin^2\theta|F_M(s)|^2, \qquad \sigma = \frac{\pi\alpha^2\beta^3}{3s}|F_M(s)|^2, \qquad s = 4E_e^2, \qquad \beta = p_M/E_e$$

where $F_M(s)$ is the electric form factor of the meson M (see Eq. (6) in 1.1.3). For an isolated resonance, $|F_M(s)|^2$ is of the form given by Eq. (5) in 1.1.5. For $s = m_V^2$, one has

$$\sigma(e\bar{e} \to V \to M\bar{M}) = 12\pi\Gamma(V \to e\bar{e})\,\Gamma(V \to M\bar{M})\,(m_V\Gamma_V)^{-2} .$$

Pilkuhn

The pion form factor $F_\pi(s)$ in $e\bar{e} \to V \to \pi\pi$ decay is complicated by the large width of the ϱ resonance and by the decay $\omega \to \pi\pi$ ($\varrho - \omega$ interference, see also [Marshall 71]):

$$F_\pi(s) = \frac{m_\varrho^2(1 + d\Gamma_\varrho/m_\varrho)}{m_\varrho^2 - s - im_\varrho\Gamma_\varrho(p/p_\varrho)^3 m_\varrho s^{-\frac{1}{2}}} + \frac{\xi_\omega e^{i\alpha_\omega} m_\omega^2}{m_\omega^2 - s - im_\omega\Gamma_\omega},$$

where d is a model-dependent correction [Gounaris 68]

$$d = \frac{3}{\pi}\frac{m_\pi^2}{p_\varrho^2}\log\left(\frac{m_\varrho + 2p_\varrho}{2m_\pi}\right) + \frac{m_\varrho}{2\pi p_\varrho} - \frac{m_\pi^2 m_\varrho}{\pi p_\varrho^3} = 0.48 \quad (d = 0.66(37) \text{ from the fit to } F_\pi),$$

$$\xi_\omega^2 = \Gamma(\omega \to \pi\pi)\,\Gamma(\omega \to e\bar{e}) \cdot \left(\frac{E_\pi}{p_\pi}\right)^3 \cdot 36a^{-2}m_\omega^{-2}, \quad \alpha_\omega = (87.5 \pm 14)^\circ,$$

$$\Gamma^{\frac{1}{2}}(\omega \to \pi\pi) = 0.20(5)\,\Gamma_\omega^{\frac{1}{2}} \quad \text{[Lefrancois 71]}.$$

In ϱ and ω production in strong interactions, the phase α depends also on the production amplitudes [Roos 70].

Tab. 1. Masses, total and partial widths

Name	m[GeV] m^2[GeV^2]	Γ[MeV] $m\Gamma$[1000 MeV^2]	Decay channel	see	Fraction	$\Gamma_{partial}$ [MeV]	$\dfrac{G^2}{4\pi}$ [1]	p or p_{max} [MeV]
$\varrho^\pm$	0.765(10) 0.585	125(20) 95.6	$\pi^\pm\pi^0$ $\pi^\pm\gamma$	b c	0.999 0.0009(th)	125 0.12(th)	$= \dfrac{1}{4\pi}G^2(\varrho^0)(\text{th})$ 0.076 GeV^{-2}	356.9 369.7
ϱ^0 [2]	0.7754(73) 0.6012	149(23) 115.5	$\pi^+\pi^-$ $\pi^0\gamma$	b c	0.999 0.0008(th)	149 0.12(th)	2.56(25) 0.076 GeV^{-2}	361.7 375.9
ω	0.7839(3) 0.6145	10.0(6) 7.84	$\pi^+\pi^-\pi^0$ $\pi^0\gamma$ [3] $\pi^+\pi^-$ $\pi^0e\bar{e}$	c d c	0.897(40) 0.090(10) 0.012(3) 0.00077(th)	8.97 0.90 0.12 0.0077	0.545 GeV^{-2}	327.7 [4] 380.2 366.1 380.2
ϕ	1.0195(6) 1.0394 $\theta_v = 39.6(7)^\circ$ from Eqs.(2) and (3) in 1.1.2	4.3(3) 4.38	K^+K^- K_LK_S $\pi^+\pi^-\pi^0(\varrho\pi)$ $\eta\gamma$ $\pi^0\gamma$	b b c c	0.49(2) 0.31(2) 0.17(2) 0.026(12) 0.0025(9) [5]	2.11 1.32 0.73 0.11 0.011	$\Big\}$1.69(18) 0.077 GeV^{-2} 0.0028 GeV^{-2}	125.0 108.3 461.3 [4] 361.6 500.8
$K^{*\pm}$	0.8917(5) 0.7951	50.3(9) 44.9	$K^\pm\pi^0$ $K^0\pi^+, \bar{K}^0\pi^-$ $K^\pm\gamma$	b b c	0.342 0.657 0.0014(th)	17.2 33.0 0.07(th)	0.84(7) $2\cdot0.84$ 0.076 GeV^{-2}	289.5 285.6 309.2
K^{*0} $\bar{K}^{*0}$ [6]	0.8979(8) 0.8062 $\|m(K^{*0}) - m(K^{*-})\| = 6.2(1.5)$ MeV	52.2(1.2) 47.0	$K^\pm\pi^\mp$ [7] $K^0\pi^0, \bar{K}^0\pi^0$ $K^0\gamma, \bar{K}^0\gamma$	b b c	0.666 0.328 0.0054(th)	34.8 17.1 0.28(th)	$2\cdot0.84$ 0.84 $\dfrac{4}{4\pi}G^2(K^+\gamma)$	292.4 291.4 311.0

See subsection 1.4.3a, p. 26, for decays into $e\bar{e}$ and $\mu\bar{\mu}$ pairs.

[1]) The coupling constants are calculated assuming $R = 0$ (see 1.2.1). So far this assumption is barely tested in $\phi \to K\bar{K}$ decays, where the variation with p^3 predicts a $K^+K^-/K_L^0K_S^0$ ratio of 1.54.

[2]) From [Benaksas 72].

[3]) This decay mode could include $20\%\,\pi^0\pi^0\gamma$ states [Dakin 71].

[4]) See footnote [3]), Tab. 1 in 1.4.2.

[5]) From [Lefrancois 71].

[6]) From [Aguilar 71]. The P-wave Breit-Wigner form is used (see 1.4.3b).

[7]) Hypercharge conservation is always understood. Thus K^{*0} decays into $K^+\pi^-$ or $K^0\pi^0$ and $\bar{K}^{*0}$ decays into $K^-\pi^+$ or $\bar{K}^0\pi^0$.

b) $\Gamma = \frac{2}{3}p^3 m^{-2}G^2/4\pi$ according to Tab. 2 in 1.2.1. The ϱ^0 and ϕ widths come from e^+e^--collisions [Bizot 70, Gourdin 70, Lefrancois 71], $\varrho^\pm$ come from strong production. For the K* width, the experimental fractions are about $\frac{1}{3}$ for $K\pi^0$ decays and $\frac{2}{3}$ for $K\pi^\pm$ decays, but the values in the table include deviations due to the variation of p^3. From universality (Eq. (17) in 1.3.4), broken SU(3) symmetry (Eq. (18) in 1.3.4), and $G_{\varrho^0\pi\pi}^2/4\pi = 2.56$, $G_{K^*K\pi}^2/4\pi = 0.84$, one obtains $G_{PPV}^2/4\pi = 0.771$, $\alpha_1 = -0.09$ and from Eq. (2) in 1.1.2 $G_{\phi K\bar{K}}^2/4\pi = 1.63$ and $G_{\omega K\bar{K}}^2/4\pi = 1.11$.

Pilkuhn

c) $\Gamma = e^2 p^3 G_\gamma^2/12\pi = \alpha p^3 G_\gamma^2/3$, with $G_\gamma = g/e$ according to Tab. 1 in 1.2.2. The theoretical values (marked "th") are obtained by assuming vector meson dominance, namely $G_{\omega\pi\varrho} = G_{\omega\pi\gamma} \cdot f_\varrho = G_{\varrho\pi\gamma} \cdot f_\omega$, $G_{\phi\pi\varrho} = G_{\phi\pi\gamma} \cdot f_\varrho$ and for the K* from SU(3): $g(K^{*+} \to K^+\gamma) = -\frac{1}{2}g(K^{*0} \to K^0\gamma) = g\,(\varrho \to \pi\gamma)$. See also [Esaybeg 71], the review by [Mopurgo 71], and 1.6.6. The $\eta\gamma$ and $\pi^0\gamma$ decays of ϕ are from [Lefrancois 71]. The decays with $e\bar{e}$ pairs instead of a photon are calculated according to Eq. (7) in 1.2.4.

d) The decays $\omega \to \pi^+\pi^-$ and $\phi \to \pi^+\pi^-$ are isospin-forbidden. For $\varrho-\omega$ interference see section a.

1.4.4 The tensor meson nonet (2^+ states)

Name	m[GeV] m^2[GeV²]	Γ[MeV] $m\Gamma$[1000 MeV²]	Decay channel	Fraction[2]	$\Gamma_{partial}$ [MeV] exp.	SU(3)[3]	p or p_{max} [MeV]
$A_2^\pm$	1.309(7)	100(20)	$\varrho^\pm\pi^0$	0.391 ⎱0.782	39	⎱ 71	416.6
			$\varrho^0\pi^\pm$	0.391 ⎰(22)	39	⎰	410.1
	1.713[1]	130.9	$\eta\pi^\pm$	0.179(18)	18	14.0	528.7
			$K^+\bar{K}^0, K^-K^0$ [5]	0.027(7)	3	7.3	427.1
			$X\pi^\pm$	0.012(11)	1	0	278.6
A_2^0	1.309(7)	100(20)	$\varrho^+\pi^-$	0.391 ⎱0.782	39	⎱ 71	415.6
			$\varrho^-\pi^+$	0.391 ⎰(22)	39	⎰	415.6
	1.713	130.9	$\eta\pi^0$	0.179(18)	18	14.0	529.4
			K^+K^- [6]	0.0135 ⎱0.027	1	⎱ 7.3	429.4
			$K_SK_S + K_LK_L$ [6]	0.0135 ⎰(7)	1	⎰	
			$X\pi^0$	0.012(11)	1	0	424.8
f	1.266(10)	157(25)	$\pi^+\pi^-$	0.54 ⎱0.81	84	⎱ 126	609.6
			$\pi^0\pi^0$	0.27 ⎰	42	⎰	618.5
	1.603	198.8	K^+K^- [6]	≈ 0.03	4	⎱ 7.0	396.1
	$\theta_t = 32.0(2.5)$ from Eqs. (2) and (3) in 1.1.2		$K_SK_S + K_LK_L$ [6]	≈ 0.03	4	⎰	391.1
			$\pi^+\pi^+\pi^-\pi^-$	0.06(2)	9		
f'	1.514(5)	73(23)	K^+K^-	0.36 ⎱0.72(12)	23	⎱ 45.0	573.7
			$K_SK_S + K_LK_L$	0.36 ⎰	23	⎰	570.3
	2.292	110.5	$K^{*+}K^- + K^{*-}K^+$	0.05 ⎱0.1(1)	4	⎱ 9	293.3
			$K^{*0}\bar{K}^0 + \bar{K}^{*0}K^0$	0.05 ⎰	4	⎰	280.7
			$\pi^+\pi^-$	0 ⎱< 0.14	0	⎱ 2.5	744.0
			$\pi^0\pi^0$	0 ⎰	0	⎰	744.8
			$\eta\pi^+\pi^-$	0.12 ⎱0.18(10)	7		623.6
			$\eta\pi^0\pi^0$	0.06 ⎰	4		625.8
			$\eta\eta$	0.0 < 0.40	0	12	521.4
$K_N^\pm$ and K_N^0 $\bar{K}_N^0$	1.420(3) 2.016 1.419(4) 2.014[4]	107(15) 150.7	$\underline{K_N^\pm}$ / $\underline{K_N^0, \bar{K}_N^0}$ [5]				for $m^+ = m^0$ = 1.419
			$K^\pm\pi^0$ / $K^0\pi^0, \bar{K}^0\pi^0$	0.189 ⎱0.569	20	⎱ 64.0	616
			$K^0\pi^+, \bar{K}^0\pi^-$ / $K^\pm\pi^\mp$	0.380 ⎰(40)	41	⎰	
			$K^{*\pm}\pi^0$ / $K^{*0}\pi^0, \bar{K}^{*0}\pi^0$	0.091 ⎱0.274	10	⎱ 24	415
			$K^{*0}\pi^+, \bar{K}^{*0}\pi^-$ / $K^{*\pm}\pi^\mp$	0.183 ⎰(32)	20	⎰	
			$K^\pm\varrho^0$ / $K^0\varrho^0, \bar{K}^0\varrho^0$	0.030 ⎱0.092	3	⎱ 7	324
			$K^0\varrho^+, \bar{K}^0\varrho^-$ / $K^\pm\varrho^\mp$	0.062 ⎰(35)	7	⎰	
			$K^\pm\omega$ / $K^0\omega, \bar{K}^0\omega$	0.045(18)	5	5	304
			$K^\pm\eta$ / $K^0\eta, \bar{K}^0\eta$	0.020(18)	2	2.0	482

[1]) A_2 mass and width from [Bowen 71, Crennel 71, Foley 71, Grayer 71]. Earlier experiments indicated a "split" resonance.

[2]) The fractions are taken from [Particle 71]. The charge decomposition uses only Clebsch-Gordon coefficients.

[3]) The SU(3) values of the partial widths are from [Aguilar 71], who take the A_1 widths from [Alston 71]. Eqs. (19) and (20) of 1.3.4 are used, with $\theta_p = -10°$ and $\theta_t = 30°$.

[4]) From [Aguilar 71]. Resonance curve with constant width.

[5]) Hypercharge conservation is understood. Compare Footnote [7]) of Tab. 1 in 1.4.3.

[6]) Strong f-A_2^0 interference may occur [Biswas 72], which is however of opposite sign for charged and neutral kaons.

1.4.5 Other multi-meson states

Name	$I(S^P)$	m [GeV]	Γ [MeV]	Decay mode
A_1	$1(1^+)$	1.070	50—200	$\pi\pi\pi\,(\varrho\pi)$
B	$1(1^+)$	1.233	≈ 100	$\omega\pi$
Q	$\frac{1}{2}(1^+)$	1.24—1.40	100—300	$K\pi\pi(K^*\pi + K\varrho)$
ε	$0(0^+)$	≈ 0.75	100—300	$\pi\pi$
S^*	$0(0^+)$	≈ 1.0	150—300	$K\bar{K}$

The A_1 and Q states appear as broad bumps, the A_1 160 MeV above the $\varrho\pi$ threshold and the Q 270 MeV above the $K^*\pi$ threshold, respectively. There is no evidence that they are resonances with factorizing residues. Alternative explanations are "diffraction dissociated" π and K states. For the A_1 resonance [Ballam 70] finds $|g_1/g_0| = 0.48 \pm 0.02$ $(= |g_T/g_L\sqrt{2}|$ in the notation of Eq. (4) in 1.2.2). See [Particle 71, 72] for further states.

1.4.6 The baryon octet and Ω^-

Tab. 1. *Masses, lifetimes, decay channels and magnetic moments*[1])
These states have spin-parity $\frac{1}{2}^+$ except Ω^- which has $\frac{3}{2}^+$

Name	m [GeV] m^2 [GeV^2]	Γ^{-1} [sec] Γ [eV]	Decay channel	see	Fraction	$\Gamma_{partial}$ [eV]	Magnetic moment $[e\hbar/2m_p c]$	p or p_{max} [MeV]
p	0.9382592(52) 0.8803303	$>2\cdot10^{28}$ 0					2.792782(17)	
n	0.9395527(52) 0.8827593	918(14) $7.17\cdot10^{-19}$	$pe\bar{\nu}$	a	1	$7.17\cdot10^{-19}$	$-1.91315(7)$	1.2
Λ	1.11559(6) 1.24454 $\theta_{\Lambda\Sigma} = -0.0135(7)$ from Eq. (6) in 1.1.2	$2.52(2)\cdot10^{-10}$ $2.61\cdot10^{-6}$	$p\pi^-$ [2]) $n\pi^0$ $pe\bar{\nu}$ $p\mu\bar{\nu}$	c c b b	0.644(7) 0.355(7) 0.00080(6) 0.00014(6)	$1.681\cdot10^{-6}$ $0.927\cdot10^{-6}$ $2.1\ \cdot10^{-9}$ $0.37\cdot10^{-9}$	$-0.73(7)$[5])	100.5 103.9 163.2 130.9
Σ^+	1.18942(11) 1.41467	$0.800(6)\cdot10^{-10}$ $8.23\cdot10^{-6}$	$p\pi^0$ $n\pi^+$ $p\gamma$ $n\pi^+\gamma$	c c d d	0.517(8) 0.483(8) 0.00124(18) 0.00013(3)	$4.25\cdot10^{-6}$ $3.97\cdot10^{-6}$ $1.02\cdot10^{-8}$ $0.11\cdot10^{-8}$	2.6(5)[3])	189.0 185.1 224.6 185.1
Σ^0	1.19251(10) 1.42208	$1.41\cdot10^{-19}$(th) $<10^{-14}$(exp) 4670(th)	$\Lambda\gamma$ $\Lambda e\bar{e}$	d d	0.995 0.00544 (th)	4647 25.4	0.99[4])	74.44 74.43
Σ^-	1.19737(7) 1.43369	$1.49(3)\cdot10^{-10}$ $4.42\cdot10^{-6}$	$n\pi^-$ $ne\bar{\nu}$ $n\mu\bar{\nu}$	c b b	0.998 0.00109(5) 0.00045(4)	$4.41\cdot10^{-6}$ $4.82\cdot10^{-9}$ $1.99\cdot10^{-9}$	-0.63[4])	193.0 230.1 209.6
Ξ^0	1.3147(7) 1.7284	$3.03(18)\cdot10^{-10}$ $2.17\cdot10^{-6}$	$\Lambda\pi^0$	c	1	$2.17\cdot10^{-6}$	-1.71[4])	135.1
Ξ^-	1.32131(18) 1.74586	$1.66(4)\cdot10^{-10}$ $3.97\cdot10^{-6}$	$\Lambda\pi^-$ $\Lambda e\bar{\nu}$	c b	1 0.00067(23)	$3.97\cdot10^{-6}$ $2.7\ \cdot10^{-9}$	-0.39[4])	139.1 189.7
Ω^-	1.6725(5) 2.7973	$1.3(4)\cdot10^{-10}$ 0.81(th)$\cdot10^{-10}$ $5.1(1.6)\cdot10^{-6}$	$\Xi^0\pi^-$ $\Xi^-\pi^0$ ΛK^-	e e e	0.17(th) 0.08(th) 0.72(th)	$0.87\cdot10^{-6}$ $0.41\cdot10^{-6}$ $3.67\cdot10^{-6}$		293.9 289.9 211.2

[1]) The experimental numbers are from [Particle 72], unless stated otherwise.   [2]) Results of [Baltay 71] included.
[3]) From [Alley 71].   [4]) See 1.6.6.   [5]) From [Dahl 71].

Fractions $<10^{-4}$: $\Sigma^- \to \Lambda e\bar{\nu}$ $5.90(57)\cdot10^{-5}$, $\Sigma^+ \to \Lambda\bar{e}\nu$ $1.86(42)\cdot10^{-5}$. See also notes b, d and c.

Pilkuhn

a) Nuclear and neutron β-decay [Blin-Stoyle 69]

For allowed transitions, g_T, g_S, g_P and g_3 in Eqs. (3) and (4) in 1.1.4 can be neglected and g_V and g_A are taken at $t = 0$. Radiative corrections, on the other hand, are important [Källen 67, 68]. The momentum spectrum of an allowed transition is

$$d\Gamma = [g_V'^2 |M_V|^2 + g_A'^2 |M_A|^2] \, F(Z, E) (E_{max} - E)^2 p^2 \, dp (2\pi^3)^{-1} \left[1 + \frac{\alpha}{2\pi} g(E, E_{max})\right].$$

p, E, m denote the electron's momentum, total energy and mass, F is the Fermi function, g is the universal function that determines the radiative correction δ_R', and g_V', g_A' are effective coupling constants including model-dependent radiative corrections C and D:

$$g_V' = g_V(1 + \alpha C/4\pi), \qquad g_A' = g_A(1 + \alpha D/4\pi).$$

$$ft(1 + \delta_R') = K[g_V'^2 |M_V|^2 + g_A'^2 |M_A|^2]^{-1}, \qquad K \equiv 2\pi^3 m^5 \ln 2 = 1.23063 \cdot 10^{-94} \quad \text{in cgs units,}$$

$$f = \int_1^{E_{max}} pE(E_{max} - E)^2 F(Z, E) \, dE,$$

and

$$\delta_R' = \frac{\alpha}{2\pi} [\int (E_{max} - E)^2 p^2 \cdot g \, dp][\int (E_{max} - E)^2 p^2 \, dp]^{-1} = \frac{\alpha}{2\pi} \left[3 \ln \frac{m_p}{m} - \tfrac{3}{4} - 2h(E_{max} - 1)\right].$$

The function $h(\varepsilon)$ is tabulated by [Källen 67, 68]. CVC implies

$$|M_V|^2 = [I(I + 1) + I_{3,b}(I_{3,b} - 1)] \, (1 - \delta_c)$$

for transitions between states of isospin $I_b = I_d = I$ (δ_c corrects for isospin impurities; $\delta_c = 0$ for ^{26m}Al and ≈ 0.007 for ^{14}O).

Decay	t [sec] [1]	δ_R'	ft [sec]	$|M_V|^2$	$|M_A|^2$	$g_{V,A}'$ $[10^{-49}$ erg cm$^3]$ $[10^{-5}$ GeV$^{-2}]$	
^{14}O → ^{14}N	70.58(4)	1.3(5)%	3081(11) [2]	1.986(8)	0	$g_V' = 1.4150(11)$	1.1149(9) [3]
^{26m}Al → ^{26}Mg	6.346(5)	1.12%	3073(5) [3]	2.000(?)	0	$= 1.41494(9)$	1.1146(9) [4]
n → p	637(10)	1.5%	1100(16) [5]	1	3	$g_A' = 1.25(2) \, g_V'$ [8]	
^{3}H → ^{3}He	3.8695(13) $\cdot 10^8$	1.5%	1159(11) [6]	1	2.84(6)	$= 1.437(22) \cdot 10^{-5}$ GeV^{-2}	
			1148(3) [7]		2.91(5)		

[1] t is the half-life t_0 defined in Eq. (3) of 1.1.5.   [2] [G. Clark 71].   [3] [Blin-Stoyle 70].

[4] [Freeman]. The Cabibbo angle (Eq. (5) in 1.1.4) from this value of g_V' and from g (footnote [3] of Tab. b in 1.4.1) is $\theta_V = 0.186(4)$.

[5] [Christensen 72].   [6] [Salgo 69].   [7] [Bergkvist 72].

[8] [Paul 70, Roos 71a]. [Christensen 69] find $g_A = 1.26(2) \, g_V$, neglecting radiative corrections, from electron asymmetry in neutron decay.

b) Semileptonic hyperon decays and the Cabibbo angle

Decay	g_V/g_V (np)	g_A/g_A (np)	g_A/g_V Exp.	fit	Fraction $[10^{-4}]$	g_T
$\Sigma^- \to \Lambda e\bar{\nu}$	0	$(\tfrac{2}{3})^{\frac{1}{2}} \alpha$	$(-0.24 \pm 0.18)^{-1}$	∞	0.63	2.35
$\Sigma^+ \to \Lambda \bar{e}\nu$	0	$(\tfrac{2}{3})^{\frac{1}{2}} \alpha$	—	∞	0.20	2.35
$\Lambda \to pe\bar{\nu}$	$(\tfrac{3}{2})^{\frac{1}{2}} \tan\theta_V$	$6^{-\frac{1}{2}}(3 - 2\alpha) \tan\theta_A$	$0.74^{+0.12}_{-0.09}$ [1]	0.71	8.46	-0.57
$\Lambda \to p\mu\bar{\nu}$	$(\tfrac{3}{2})^{\frac{1}{2}} \tan\theta_V$	$6^{-\frac{1}{2}}(3 - 2\alpha) \tan\theta_A$	—	0.71	1.39	-0.57
$\Sigma^- \to n\mu\bar{\nu}$	$-\tan\theta_V$	$(2\alpha - 1) \tan\theta_A$	$\pm 0.21(12)$	-0.33	10.6	0.54
$\Sigma^- \to ne\bar{\nu}$	$-\tan\theta_V$	$(2\alpha - 1) \tan\theta_A$	$0.4^{+1.5}_{-0.5}$ [2]	-0.33	5.0	0.54
$\Xi^- \to \Lambda e\bar{\nu}$	$-(\tfrac{3}{2})^{\frac{1}{2}} \tan\theta_V$	$-6^{-\frac{1}{2}}(3 - 4\alpha) \tan\theta_A$	—	0.19	5.2	-0.04
$\Xi^- \to \Sigma^0 e\bar{\nu}$	$2^{-\frac{1}{2}} \tan\theta_V$	$2^{-\frac{1}{2}} \tan\theta_A$	—	1.23	0.8	0.65

[1] [Baggett 72].   [2] [Ellis 72].

Pilkuhn

The Cabibbo fit (see 1.1.4) contains 3 parameters: α (which is mainly determined by the $\Sigma \to \Lambda e\nu$ fractions) and θ_V and θ_A. g_T is taken from SU(3)-estimates and g_P is neglected (g_S and g_3 are put equal to zero in Eqs. (3) and (4) in 1.1.4, and the contribution from $F_3(\Sigma\Lambda)$ (analogous to the F_3 in Eq. (4) in 1.1.3 is also neglected). The fit reproduced in the table [Ebenhöh 71] has $\alpha_{weak} = 0.633(19)$, $\theta_A = \theta_V = \theta = 0.239(5)$ and $\tan\theta = 0.2437$. The two-angle fit gives $\theta_V = 0.250(10)$ and $\theta_A = 0.227(12)$, which is not much better. For the PCAC relation, given by Eq. (7) in 1.1.4, and not assuming SU(3)-invariance, one has to use $\alpha_{weak} = 0.6124$, which comes from $\Sigma \to \Lambda e\nu$ decays alone ($g_{A,\Sigma\Lambda} = 0.5005 \, g_{A,np} = 0.719 \cdot 10^{-5} \, \mathrm{GeV}^{-2}$). See also [Roos 71a]. For an independent determination of θ_V see footnote [4]) in 1.4.6a.

c) Pionic hyperon decays

The nomenclature is explained in 1.2.3. For time-reversal invariant decays, $\Delta = \delta_S - \delta_P$ (see the remark following Eq.(5) in 1.2.1). α and ϕ are the measured quantities, γ and Δ are derived

Decay	α	ϕ	γ	Δ	$A \cdot 10^7$	$B \cdot 10^7$
$\Lambda \to \pi^- p$	0.645(16)	$-(6.3 \pm 3.5)^\circ$	0.76	$(7.5 \pm 3.9)^\circ$	3.30(4)	22.67(71)
$\Lambda \to \pi^0 n$	0.649(46)			$(6.8 \pm 2.0)^\circ$		
$\Sigma^- \to \pi^- n$	$-0.069(8)$	$(10 \pm 15)^\circ$	0.98	$(249^{+12}_{-115})^\circ$	4.06(7)	$-1.19(93)$
$\Sigma^+ \to \pi^+ n$	0.066(16)	$(167 \pm 20)^\circ$	-0.97	$(-73^{+133}_{-10})^\circ$	0.04(9)	41.43(76)
$\Sigma^+ \to \pi^0 p$	$-0.991(19)$	$(22 \pm 90)^\circ$	0.12	$(183^{+11}_{-12})^\circ$	3.32(30)	$-25.02(40)$
$\Xi^- \to \pi^- \Lambda$	$-0.41(4)$	$(-3 \pm 9)^\circ$	0.90	$(-10 \pm 18)^\circ$	4.05(7)	$-16.1(1.4)$
$\Xi^0 \to \pi^0 \Lambda$	$-0.35(8)$	$(25 \pm 21)^\circ$	0.85	$(48^{+16}_{-37})^\circ$	3.32(11)	$-9.9(2.2)$

The columns for $\alpha, \phi, \gamma, \Delta$ are taken from [Particle 71], those for A and B (where one has put $\cos\Delta = \pm 1$) from [Filthuth 69]. Our A and B are dimensionless. For dimension $10^5/\mathrm{sec}$, Γ in Tab. 1 in 1.2.3 contains an additional factor of $m_{\pi^+}^{-1}$. The conversion factor to the dimensionless quantities A and B is $2.1715 \cdot 10^{-7}$ sec.

The $\Delta I = \tfrac{1}{2}$ rule says $T(\Lambda \to \pi^0 n) = -2^{-\frac{1}{2}} T(\Lambda \to \pi^- p)$. The phase Δ for this decay is obtained from the πN phase-shift analysis and the $\Delta I = \tfrac{1}{2}$ rule. The other $\Delta I = \tfrac{1}{2}$ constraints are $T(\Xi^0 \to \pi^0 \Lambda) = -2^{-\frac{1}{2}} T(\Xi^- \to \pi^- \Lambda)$ and $T(\Sigma^+ \to \pi^+ n) - T(\Sigma^- \to \pi^- n) = 2^{\frac{1}{2}} T(\Sigma^+ \to \pi^0 p)$. They are all satisfied, for both the A and B amplitudes. Also the Lee-Sugawara relation $2T(\Xi^- \to \pi^- \Lambda) + T(\Lambda \to \pi^- p) = 3^{\frac{1}{2}} T(\Sigma^+ \to \pi^0 p)$ holds for both A and B [Marshak 69, ref. in 1.1.6].

d) Electromagnetic Σ decays

The $\Sigma^0 \to \Lambda\gamma$ decay is given by Eq. (4) in 1.1.3, with $t = 0$ and $F_2(0) = 1$:

$$T_\mu(\Sigma^0 \to \Lambda) = -\frac{1}{2} \frac{\kappa^{\Lambda\Sigma}}{m_\Sigma + m_\Lambda} \bar{u}_\Lambda(\gamma_\mu\gamma_\nu - \gamma_\nu\gamma_\mu) u_\Sigma p_\gamma^\nu.$$

The width is given by Eq. (5) in 1.2.1 with

$$T_{\frac{1}{2}}(\lambda_1 \lambda_\gamma) = -(4\pi)^{\frac{1}{2}} e\kappa^{\Sigma\Lambda}(m_\Sigma - m_\Lambda)\left[\delta_{\lambda_1, -\frac{1}{2}}\delta_{1, \lambda_\gamma} + \delta_{\lambda_1, \frac{1}{2}}\delta_{-1, \lambda_\gamma}\right],$$

$$\Gamma = \frac{\alpha}{2}(\kappa^{\Sigma\Lambda})^2 (m_\Sigma - m_\Lambda)^3 \frac{m_\Sigma + m_\Lambda}{m_\Sigma^3}$$

and $\kappa^{\Sigma\Lambda} = \sqrt{3}\,\kappa^\Lambda \approx 1.44$ according to SU(3). The experimental width is not yet known. The general formulas for $\Sigma \to \Lambda e\bar{e}$ decay are given by [Alff 65]. For small F_3 in Eq. (4) in 1.1.3, the branching ratio is given by Eq. (7) in 1.2.4 plus a small correction. The form factor of the $\Sigma^+ \to p\gamma$ decay differs from $T_\mu(\Sigma^0 \to \Lambda)$ by a factor $1 + \varepsilon\gamma_5$. In baryon pole dominance models [Graham 65], $\varepsilon \ll 1$. For further radiative hyperon decays see [Gaillard 71].

e) The Ω^- particle

Due to its hypercharge -2, this particle is difficult to produce. It should belong to the baryon decuplet, particularly because of the equal mass spacing between Δ, Υ, $\Xi(1530)$ and Ω. Only $28\,\Omega$ events have been identified so far. The theoretical decay fractions are from [Ram Mohan 70], who also estimates the following decay fractions: $\Xi^0(1530)\,\pi^-$ 0.0092, $\Xi^-(1530)\,\pi^0$ 0.0046, $\Xi^0 e\bar{\nu}$ 0.006, $\Xi^0\mu\bar{\nu}$ 0.004.

Pilkuhn

1.4.7 The baryon decuplet except Ω^- ($\frac{3}{2}^+$-states)

Name	m [GeV] m^2 [GeV^2]	Γ [MeV] $m\Gamma$ [1000 MeV^2]	Decay channel	see	Fraction	$\Gamma_{partial}$ [MeV]	p [MeV]
Δ^{++}	1.2300(18) 1.513	111.1(2.4) 136.7	$p\pi^+$	a	1	111.1	225.4
Δ^+	1.2315(th) 1.517[1])	115.7 142.5	$p\pi^0$ $n\pi^+$ $p\gamma$	a a b	0.672 0.321 0.0062	78 37.0 0.7	228.9 225.6 258.3
Δ^0	1.2329(23) 1.520	117.6(2.6) 145.0	$n\pi^0$ $p\pi^-$ $n\gamma$	a a b	0.665 0.328 0.0062	78.2 38.6 0.69	229.0 228.0 258.4
Δ^-	1.2344(8) 1.5237[1])	117.0 144.4	$n\pi^-$	a	1	117.0	228.2
Υ^+ [2])	1.3826(10) 1.9116	35.9(2.6) 49.6	$\Lambda\pi^+$ $\Sigma^+\pi^0$ $\Sigma^0\pi^+$ $\Sigma^+\gamma$	c c c b	0.90(3) 0.055 } 0.045 } 0.10(3) 0.0057(th)	32.3 2.0 1.6 0.20	205.3 128.4 120.0 179.7
Υ^0	1.3843(th) 1.9163[1])	37.7(th) 52.2	$\Lambda\pi^0$ $\Sigma^+\pi^-$ $\Sigma^-\pi^+$ $\Lambda\gamma$ $\Sigma^0\gamma$	c c c b b	0.91 0.05 0.04 0.0070(th) 0.0014(th)	34.3(th) 1.9 1.5 0.249 0.049	209.5 126.3 115.8 242.6 178.5
Υ^-	1.3859(20) 1.9207	36.3(6.3) 50.3	$\Lambda\pi^-$ $\Sigma^-\pi^0$ $\Sigma^0\pi^-$	c c c	0.90(3) 0.049 0.051	32.7 1.8 1.9	208.6 122.5 124.3
Ξ^0(1530)	1.5323(7) [3]) 2.3479	11.0(1.8) 16.9	$\Xi^0\pi^0$ $\Xi^-\pi^+$ $\Xi^0\gamma$	c c b	0.376 0.597 0.027(th)	3.93 6.57 0.27	158.4 147.2 202.1
Ξ^-(1530)	1.5362(16) [3]) 2.3599	16.2(4.6) 24.9	$\Xi^-\pi^0$ $\Xi^0\pi^-$	c c	0.316 0.684	3.66(th) 7.90(th)	155.4 159.4

---

[1]) The masses of Δ^+, Δ^-, and Υ^0 have been determined from the equal-spacing rule for SU(2)-multiplets (compare the Σ masses in 1.4.6). SU(2)-symmetry implies the mass differences $\Xi^- - \Xi^0 = \Upsilon^- - \Upsilon^0 - \Delta^- - \Delta^+$ and $\Upsilon^0 - \Upsilon^+ = \Delta^0 - \Delta^+$.   [2]) Υ denotes the resonance $\Sigma(1385)$.   [3]) [Kirsch 72].

a) The Δ (or N (1236)) resonances*

The positions and widths of the decays $\Delta^{++} \to p\pi^+$ and $\Delta^0 \to p\pi^-$ are taken from [Carter 71], who uses the Layson formula (Eq. (4) in 1.3.6), with $m\Gamma$ taken from Tab. 1 in 1.2.3, $T_{d3} = 0$, $A_3 = G_\Delta/m_{\pi^+}$, and with a suppression factor N taken from Tab. 1 in 1.2.1 with $R \approx 0.92$ fm. However, a satisfactory description of $\cot\delta$ must include a nonresonant phase ξ according to Eq. (1) in 1.3.6, with $\tan\xi = 0.01 \, (q/m_\pi)^3$ [Particle 71]. One then gets $m(\Delta^{++}) = 1.234$ GeV, $\Gamma(\Delta^{++}) = 120$ MeV and $R = 0.75$ fm. The isospin relations for the decay coupling constant G_Δ are given in Eq. (8) in 1.3.4. The partial and total widths of Δ^+ and Δ^- and the fractions of Δ^0 have been computed from the isospin invariance of G_Δ and the decay momentum dependence $\Gamma \approx p^3$ (i.e. $R \approx 0$), which dependence is supported by the ratio $\Gamma(\Delta^0)/\Gamma(\Delta^{++})$.

At the resonance position m, one has for the decay $\Delta^{++} \to p\pi^+$

$$m\Gamma = \frac{G_\Delta^2}{4\pi} \cdot \frac{2}{3} p^3 \frac{(m+m_p)^2 - m_\pi^2}{4mm_\pi^2(1 + R^2 p^2)} \qquad \frac{G_\Delta^2}{4\pi} = \begin{matrix} 0.69 & \text{for} & R = 0.75 \text{ fm} \\ 0.40 & \text{for} & R = 0 \quad \text{fm} . \end{matrix}$$

Pilkuhn

For $\xi = 0$, $G_A^2/4\pi$ is 0.83 for $R = 1$ and 0.36 for $R = 0$. An independent determination of G_A comes from the use of Δ-exchange Born terms in πN-scattering. A determination from the amplitudes in the region $|v| < v_1 = m_\pi + t/4m_p$, $|t| < 4m_\pi^2$ gives $G_A^2/4\pi = 0.26$ (see [Ebel 71]). Since Born terms have $R = 0$, this value must be compared with the above values for $R = 0$. Finally, [Ball 72] determine the position of the complex Δ pole as $m_\Delta - i\frac{1}{2}\Gamma_\Delta = (1211 - 50i)$ MeV, using a 2-channel, 4-parameter ND^{-1} model.

b) Radiative decays of decuplet states

$\Gamma(\Delta^+ \rightarrow p\gamma)$ is taken from [Pfeil 70]. The other widths are calculated using the threshold behaviour $\Gamma \approx p^3$ and the isospin- and SU(3)-relations $G(\Delta^+ p) = G(\Delta^0 n) = G(\Upsilon^+ \Sigma^+) = G(\Xi^0(1530)\Xi^0)$, $G(\Upsilon^0\Lambda) = -(\frac{3}{4})^{\frac{1}{2}} G(\Delta^0 n)$, $G(\Upsilon^0\Sigma^0) = \frac{1}{2}G(\Delta^0 n)$, $G(\Upsilon^-\Sigma^-) = G(\Xi^-(1530)\Xi^-) = 0$.

c) The resonances Υ and $\Xi(1530)$

As Υ and $\Xi(1530)$ have $I = 1$ and $I = \frac{1}{2}$, the isospin relations between their decay coupling constants have the forms given by Eq. (5) in 1.3.4 for $\Sigma\Sigma\pi$ and Eq. (6) in 1.3.4 for $\Xi\Xi\pi$ vertices, respectively. The fractions of Υ^+, Υ^- and $\Xi^0(1530)$ as well as the partial widths of Υ^0 and $\Xi^-(1530)$ have been computed from the isospin invariance of G and the threshold behaviour $\Gamma \approx p^3$. However, in contrast to the procedure used for the Δ resonance, it appears that $\Gamma(s)$ has been taken as a constant in the resonance formula, in which case $\Gamma_{\Upsilon\Lambda\pi}(m^2)$ is smaller by about 1 MeV and $\Gamma_{\Upsilon\Sigma\pi}(m^2)$ by some 20% (see the remarks at the end of 1.1.5). The $\Sigma\pi/\Lambda\pi$ ratio increases by a factor 2 across the resonance [Pilkuhn 69]. See also the SU(3) comparison in 1.5.7.

1.4.8 The nonets $\frac{1}{2}^-$ and $\frac{3}{2}^-$

a) $\frac{1}{2}^-$ nonet (S-wave resonances)

Resonance	m [GeV]	Γ [MeV]	Decay channel	Fraction	$\Gamma_{partial}$ [MeV]	t(from Eq.(3) in 1.3.6) exp.	theor.
N(1535)	1.55(5)	105(55)	Nη	0.35	37	0.45(5)	0.22
			Nπ	0.55	58	0.34(5)	0.22
			N$\pi\pi^0$	0.10	10		
Λ(1405) mainly singlet, $\theta = 20°$	1.405(5)	40(10)	$\Sigma\pi$	1	40(10)		
Λ(1670) mainly octet	1.67	26(12)	$\Sigma\pi$	0.45	12	$-0.28(5)$	-0.37
			$\Lambda\eta$	0.35	9	0.26(5)	0.22
			N$\bar{K}$	0.20	5	0.17(5)	0.14
Σ(1750)	1.75	65(15)	N$\bar{K}$	≈ 0.15	≈ 9.8	$-0.25(5)$	1.61
			$\Lambda\pi$		3.2(th)		-0.27
			$\Sigma\eta$		125(th)		-0.09
			$\Sigma\pi$		0(th)		1.59
Ξ(1825)	? [1]	?	$\Lambda\bar{K}$		80(th)		
			$\Sigma\bar{K}$		17(th)		
			$\Xi\pi$		62(th)		

[1] The masses 1.825 and 1.812 GeV are SU(3)-predictions. Ξ resonances beyond the $\Xi(1530)$ are not firmly established. One resonance is reported at 1.606 GeV ($\Gamma = 21(7)$ MeV) by [Ross 72].

Pilkuhn

b) $\frac{3}{2}^-$ nonet (D-wave resonances)

Resonance	m [GeV]	Γ [MeV]	Decay channel	Fraction	$\Gamma_{partial}$ [MeV]	t (from Eq.(3) in 1.3.6) exp.	theor.
N(1520)	1.525(15)	125(25)	Nπ	0.5	60	0.52(5)	0.54
			$\Delta\pi$ [1]	0.5	60		
			Nη	0.006	7		0.02
Λ(1520) From N$\overline{K}$ and $\Sigma\pi$ decays: mainly octet, $\theta = 25(5)^0$ From $\Upsilon\pi$ decays: $\theta = 73(10)^0$ [2]	1.518(2)	16(2)	N$\overline{K}$	0.46(1)	7.20	0.45(5)	0.44
			$\Sigma\pi$	0.41(1)	6.72	0.43(5)	0.43
			$\Lambda\pi\pi$ [2]	0.02(1) $\}$0.11(1)	0.32		
			$\Upsilon\pi$	0.09(1)	1.44		
			$\Sigma\pi\pi$	0.010(1)	0.16		
			$\Lambda\gamma$	0.008(2)	0.13		
Λ(1690)	1.69	56(29)	N$\overline{K}$	0.20	11	0.18(5)	0.18
			$\Sigma\pi$	0.60	34	$-0.36(5)$	-0.34
			$\Lambda\pi\pi$	0.02	1		
			$\Sigma\pi\pi$	$\leqq 0.18$	$\leqq 10$		
Σ(1670)	1.67	50	N$\overline{K}$	≈ 0.08 [4]		0.08(5)	0.07
			$\Sigma\pi$		32(th)	0.20(5)	0.21
			$\Lambda\pi$		1.9(th)	0.10(5)	0.05
			Λ(1405)π				
Ξ(1812)	? [3]	?	$\Lambda\overline{K}$		7(th)		
			$\Sigma\overline{K}$		5(th)		
			$\Xi\pi$		4(th)		
			Ξ(1530)π		3.4(th)		

[1] Uncorrelated N$\pi\pi$ states are included but are not dominant.
[2] From [Mast 72].
[3] See footnote[1] on page 34.
[4] Mass, width and N$\overline{K}$ fraction are estimated from a partial-wave analysis of K^-p-scattering. There is disagreement between formation and production experiments.

The masses and total widths of these resonances are taken from [Particle 71]. The amplitudes t measured at the resonance position and the theoretical predictions for decays into the baryon octet (N, Λ, Σ, Ξ) are from [Plane 70]. The decays into the baryon decuplet (Δ, Υ, Ξ(1530)) are from [Burkhardt 71]. See also [Deans 71, ref. in 1.5.9], [Wagner 71, Flaminio 70] and the review of [Samios 70]. The mixing angle of Λ(1520) is taken from its N$\overline{K}$ and $\Sigma\pi$ widths as $25(5)^0$ [Burkhardt 71], in contradiction with the value $73(10)^0$ obtained from recent $\Upsilon\pi$ widths [Mast 72].

    The decays of the $\frac{1}{2}^-$ resonances into baryon and meson states are described according to Tab. 1 in 1.2.3 with $A = G$, $B = 0$: $G^2/4\pi = 2m^2\Gamma_{partial}\,p^{-1}k_+^{-2}$. For the Λ(1405), $G^2/4\pi = 0.18\,\Gamma_{tot}/m_\pi = 0.052$.

1.4.9 The octets $\frac{5}{2}^-$ and $\frac{5}{2}^+$ and further pion-nucleon resonances

Name	(J^P)	m [GeV]	Γ [MeV]	Decay channel	Fraction
N(1670)	$(\frac{5}{2}^-)$, D-wave	1.638(13)	140(35)	$N\pi$	0.40
				$\Delta\pi$	0.44
				$N\pi\pi$	0.16
Λ(1830)	$(\frac{5}{2}^-)$, D-wave	1.835	112(38)	$\Sigma\pi$	0.30
				$N\overline{K}$	0.10
				$\Lambda\pi\pi$	0.11
Σ(1760) [1])	$(\frac{5}{2}^-)$, D-wave	1.759(4)	107(11)	$N\overline{K}$	0.42
				$\Lambda\pi$	0.15
				$\Lambda(1520)\pi$	0.14
				$Y\pi$	0.04
				$\Sigma\pi$	0.01
N(1688)	$(\frac{5}{2}^+)$, F-wave	1.686(6)	140(40)	$N\pi$	0.60
				$\Delta\pi$	0.26
				$N\pi\pi$	0.14
Λ(1815)	$(\frac{5}{2}^+)$, F-wave	1.820(5)	82(18)	$N\overline{K}$	0.62
				$Y\pi$	0.17
				$\Sigma\pi$	0.11
Σ(1915)	$(\frac{5}{2}^+)$, F-wave	1.910	70	$N\overline{K}$	0.11
				$\Lambda\pi$	0.07
				$\Sigma\pi$	0.04
N(1470)	$(\frac{1}{2}^+)$, P-wave	1.470(35)	280(120)	$N\pi$	0.60
				$N\pi\pi$	0.40
Δ(1650)	$(\frac{1}{2}^-)$, S-wave	1.655(40)	165(35)	$N\pi$	0.28
				$N\pi\pi$	0.72
Δ(1670)	$(\frac{3}{2}^-)$, D-wave	1.685(35)	240(65)	$N\pi$	0.15
				$N\pi\pi$	
N(1700)	$(\frac{1}{2}^-)$, S-wave	1.715(50)	250(150)	$N\pi$	0.65
				ΛK	0.05
N(1780)	$(\frac{1}{2}^+)$, P-wave	1.755(105)	250(200)	$N\pi$	0.30
				ΛK	0.07
				$N\eta$	0.10

[1]) From [Barletta 72].

The resonance name indicates its isospin: N has $I = \frac{1}{2}$, Δ has $I = \frac{3}{2}$, Λ has $I = 0$ and Σ has $I = 1$. Numerical values are from [Particle 71]. The list of N and Δ resonances (including those of 1.4.7 and 1.4.8) is complete up to 1800 MeV. Further SU(3)-classifications are given by [Plane 70] and [Flaminio 70].

1.4.10 References for 1.4

For journal abbreviation see p. 2.

Abrams	71	BNL report 16092: R. J. Abrams, R. L. Cool, G. Giacomelli, T. F. Kycia, B. A. Leontic, K. K. Li, A. Lundby, D. N. Michael, J. Teiger.
Aguilar	71	PR D 4, 2583: M. Aguilar-Benitez, R. L. Eisner, J. B. Kinson. See also PRL 25, 1362 and PL 26, 446.
Alff	65	PR 137 B, 1105: C. Alff, N. Gelfand, U. Nauenberg, M. Nussbaum, J. Schultz, J. Steinberger, H. Brugger, L. Kirsch, R. Plano, D. Berley, A. Prodell.

Allaby	71	SJNP 13, 295: J. V. Allaby, Yu. B. Bushnin, S. P. Denisov, A. N. Diddens, R. W. Dobinson, S. V. Donskov, G. Giacomelli, Yu. P. Gorin, A. Klovning, A. I. Petrukhin, Yu. D. Prokoshkin, R. S. Shuvalov, C. A. Stahbrandt, D. A. Stoyanova.
Alley	71	PR D 3, 75: P. W. Alley, J. R. Benbrook, V. Cook, G. Glass, K. Green, J. F. Hague, R. W. Williams.
Alston	71	PL 34 B, 156: M. Alston-Garnjost, A. Barbaro-Galtieri, W. F. Buhl, S. E. Derenzo, L. D. Epperson, S. M. Flatte, J. H. Friedman, G. R. Lynch, R. L. Ott, S. D. Protopopescu, M. S. Rabin, F. T. Solmitz.
Apel	72	PL 40B, 680: W. D. Apel *et al.*
Backenstoß	71	PL 36 B, 403: G. Backenstoß, H. Daniel, H. Koch, U. Lynen, Ch. von der Malsburg, G. Poelz, H. P. Povel, H. Schmitt, K. Springer, L. Tauscher.
Baggett	72	ZP 249, 279: M. Baggett, N. Baggett, F. Eisele, H. Filthuth, H. Frehse, V. Hepp, R. Howard, E. Leitner, G. Zech.
Ball	72	PRL 28, 1143: J.S. Ball *et al.*
Ballam	70	PR D 1, 94: J. Ballam, A. D. Brody, G. B. Chadwick, Z. G. T. Guiragossian, W. B. Johnson, R. R. Larsen, D. W. G. S. Leith, K. Moriyasu.
Baltay	71	PR D 4, 670: C. Baltay, A. Bridgewater, W. A. Cooper, M. Habibi, N. Yeh.
Barletta	72	NP B 40, 45: W.A. Barletta.
Basile	71	NP B 33, 29: M. Basile, D. Bollini, P. Dalpiaz, P. L. Frabetti, T. Massam, F. Navach, F. L. Navarria, M. A. Schneegans, A. Zichichi.
Bellettini	70	NC 66 A, 243: G. Bellettini *et al.*
Benaksas	72	PL 39 B, 289: D. Benaksas *et al.*
Bergkvist	72	NP B 39, 317: K.E. Bergkvist.
Binnie	72	PL 39 B, 275: D. M. Binnie *et al.*
Biswas	72	PR D5, 1564: N.N. Biswas *et al.*
Bizot	70	PL 32 B, 416: J. C. Bizot, J. Buon, Y. Chatelus, J. Jeanjean, D. Lalanne, H. Nguyen Ngoc, J. P. Perez-Y. Jorba, P. Petroff, F. Richard, F. Rumpf, D. Treille.
Blin-Stoyle	69	Isospin in nuclear physics (North-Holland, ed. D. H. Wilkinson): R. J. Blin-Stoyle.
Blin-Stoyle	70	NP A 150, 369: R. J. Blin-Stoyle, J. M. Freeman.
Bloodworth	72	NP B 39, 525: I. J. Bloodworth *et al.*
Bowen	71	PRL 26, 1663: D. Bowen, D. Earles, W. Faissler, M. Gettner, M. Glaubmann, B. Gottschalk, G. Lutz, J. Moromisato, E. I. Shibata, Y. W. Tang, E. von Goeler, H. R. Blieden, G. Finocchiaro, J. Kirz, R. Thun.
Burkhardt	71	NP B 27, 64: E. Burkhardt, H. Filthuth, E. Kluge, H. Oberlack, R. Armenteros, M. Ferro-Luzzi, D. W. G. S. Leith, R. Levi-Setti, J. Meyer, A. Minten, R. Barloutaud, P. Granet, J. P. Porte.
Carnegie	71	PR D 4, 1: R. K. Carnegie, R. Cester, V. L. Fitch, M. Strovink, L. R. Sulak.
Chien	71	UCLA−1054: C.-Y. Chien *et al.*
Christensen	69	PL 28 B, 411: C. J. Christensen, V. E. Krohn, C. R. Ringo.
Christensen	72	PR D5, 1628: C. J. Christensen, A. Bahnsen, W. K. Brown, B. M. Rustad.
A. Clark	71	PRL 26, 1667: A. R. Clark *et al.*
G. Clark	71	PL 35 B, 503: G. J. Clark *et al.*
Cnops	68	PL 26 B, 398: A. M. Cnops, G. Finocchiaro, P. Mittner, P. Zanella, J. P. Dufey, B. Gobbi, M. A. Pouchon, A. Müller.
Cnops	68a	PL 27 B, 113: A. M. Cnops, G. Finocchiaro, P. Mittner, J. P. Dufey, B. Gobbi, M. A. Pouchon, A. Müller.
Crennel	71	PL 35 B, 185: D. J. Crennel, H. A. Gordon, Kwan-Wu Lai, J. M. Scarr.
Dahl	71	NC 3 A, 1: E. Dahl-Jensen, N. Doble, D. Evans, A. J. Herz, U. Liebermeister, Ph. Rosselet, C. Busi, G. Önengüt, P. Tolun, M. Gailloud, R. Weill, G. Hansl, G. Romano, V. Rossi.
Dakin	71	PL 36 B, 607: J. T. Dakin, M. G. Hauser, M. N. Kreisler, R. E. Mischke, J. J. Ritsko.
Danburg	70	PR D 2, 2564: J. S. Danburg *et al.*
Dorfan	67	PRL 19, 987: D. Dorfan, J. Enstrom, D. Raymond, M. Schwartz, S. Wojcicki, D. H. Miller, M. Paciotti.
Dufey	69	PL 29 B, 605: J. P. Dufey *et al.*
Ebenhöh	71	ZP 241, 473: H. Ebenhöh, F. Eisele, H. Filthuth, W. Föhlisch, V. Hepp, E. Leitner, W. Presser, H. Schneider, T. Thouw, G. Zech.
Ellis	72	NP B 39, 76: R. J. Ellis *et al.*

Pilkuhn

Engler	68	PL 28 B, 64: J. Engler, K. Horn, J. König, F. Mönnig, P. Schludecker, H. Schopper, P. Sievers, H. Ullrich.
Esaybegyan	71	SJNP 12, 317: S. V. Esaybegyan, DzL. Chkareuli.
Filthuth	69	Top. Conf. on Weak Interactions, CERN report 69-7: H. Filthuth.
Flaminio	70	BNL report 14572: E. Flaminio, W. Metzger, N. P. Samios, M. Goldberg.
Foley	71	PRL 26, 413: K. J. Foley, W. A. Love, S. Ozaki, E. D. Platner, A. C. Saulys, E. H. Willen, S. J. Lindenbaum.
Ford	72	PL 38 B, 335: W. T. Ford, P. A. Piroué, R. S. Remmel, A. J. S. Smith, P. A. Souder.
Freeman	71	Harwell report AERE-R-7083: J. M. Freeman *et al.*
Gaillard	70	CERN report 70–14: M. K. Gaillard, L. M. Chounet.
Gaillard	71	NC 6 A, 559: M. Gaillard.
Gorin	71	SJNP 13, 192: Yu. P. Gorin, S. P. Denisov, S. V. Donskov, V. A. Kachanov, V. M. Kutin, A. I. Petrukhin, Yu. D. Prokoshkin, E. A. Razuvaev, D. A. Stoyanova, R. S. Shuvalov.
Gounaris	68	PRL 21, 244: G. Gounaris, J. J. Sakurai.
Gourdin	70	STMP 55, 191: M. Gourdin.
Gourdin	71	Int. Conf. Meson Reson. and rel. Electromagnetic Phenomena, Bologna: M. Gourdin.
Graham	65	PR 140 B, 1144: R. H. Graham, S. Pakvasa.
Grayer	71	PL 34 B, 333: G. Grayer, H. Dietl, W. Koch, H. Lippmann, E. Lorenz, G. Lütjend, W. Männer, J. Meissburger, U. Stierlin, P. Weilhammer.
Grigorov	70	SJNP 11, 455: N. L. Grigorov, V. E. Nesterov, I. D. Rapoport, I. A. Savenko, G. A. Skuridin.
Haidt	71	PR D 3, 10: D. Haidt, J. Stein, S. Natali, G. Piscitelli, F. Romano, E. Fett, J. Lemonne, T. I. Pedersen, S. N. Tovey, V. Brisson, P. Petiau, C. D. Esveld, J. J. M. Timmermans, B. Aubert, L. M. Chounet, Le Dong, F. Bobisut, H. Huzita, F. Sconza, A. Marzari-Chiesa, A. E. Werbrouk.
Holder	71	PL 35 B, 361: experimental numbers revised 1972: M. Holder *et al.*
Hyams	69	PL 29 B, 128: B. D. Hyams, W. Koch, D. C. Potter, L. von Lindern, E. Lorenz, G. Lütjens, U. Stierlin, P. Weilhammer.
Igo	67	NP B 3, 181: G. J. Igo, J. L. Friedes, H. Palevsky, R. Sutter, G. Bennett, W. D. Simpson, D. M. Corley, R. L. Stearns.
Jones	70	Progr. Nucl. Phys. 12, 1: P. B. Jones.
Källén	67	NP B 1, 225: G. Källén.
Källén	68	STMP 46: G. Källén.
Kirsch	72	NP B 40, 349: L. Kirsch *et al.*
Lefrancois	71	Int. Symp. on Electron and Photon Interactions (Cornell): J. Lefrancois.
Litskevich	70	SJNP 11, 599: I. K. Litskevich.
Marshall	71	Int. Conf. Meson Reson. and rel. Electromagnetic Phenomena, Bologna: R. Marshall.
Marx	70	PL 32 B 219: J. Marx, D. Nygren, J. Peoples, T. Kirk, J. Steinberger.
Mast	69	PR 183, 1200: T. S. Mast, L. K. Gershwin, M. Alston-Garnjost, R. O. Bangerter, A. Barbaro-Galtieri, J. J. Murray, F. T. Solmitz, R. D. Tripp.
Mast	72	PRL 28, 1220: T. S. Mast *et al.*
Messel	70	Electron-photon shower distribution function tables for lead, copper and air absorbers. (Oxford: Pergamon Press): H. Messel, D. F. Crawford.
Mopurgo	71	Lectures at the E. Majorana School, Erice: G. Mopurgo.
Particle	70	PL 33 B, 1: Particle data group.
Particle	71, 72	RMP, S 1 and PL 39 B, 1: Particle data group.
Paty	70	Herceg Novi Lectures: M. Paty.
Paul	70	NP 154, 160: H. Paul.
Pfeil	70	STMP 55, 213: W. Pfeil, D. Schwela.
Pilkuhn	69	NCL 1, 854: H. Pilkuhn, A. Swoboda.
Plane	70	NP B 22, 93: D. E. Plane, P. Baillon, C. Bricman, M. Ferro-Luzzi, J. Meyer, E. Pagiola, N. Schmitz, E. Burkhardt, H. Filthuth, E. Kluge, H. Oberlack, R. Barloutaud, P. Granet, J. P. Porte, J. Prevost.
Quigg	68	UCRL-18487: C. Quigg, J. D. Jackson.
Ram Moham	70	PR D 1, 266: L. R. Ram Moham.
Roos	70	Proc. Daresbury Study Weckend No. 1, DNPL/R 7, p. 173: M. Roos.
Roos	71	NP B 29, 296: M. Roos, A. Sirlin.

Pilkuhn

Roos	71a	PL 36 B, 130: M. Roos.
Ross	72	PL 38 B, 177: R. T. Ross et al.
Salgo	69	NP A 138, 417: R. C. Salgo, H. H. Staub.
Samios	70	BNL report 15284 (Dubna Conf.): N. P. Samios.
Steinberger	70	CERN report 70-1: J. Steinberger.
Trefil	71	NP B 34, 109: J. S. Trefil.
Wagner	71	NP B 25, 411: F. Wagner, C. Lovelace.
Williams	72	PR D6, 737: R. Williams, D. Williams.
Zemach	64	PR 133 B, 1200: C. Zemach.

1.5 Effective range parameters and coupling constants of stable vertices

1.5.1 πN and KN scattering and the ηn, Xn and KΛ channels

	πN				KN	
	in units of m_π^{-2l-1}		in units of fm^{2l+1}		in units of fm^{2l+1}	
	$I-\frac{3}{2}$	$I=\frac{1}{2}$	$I=\frac{3}{2}$	$I=\frac{1}{2}$	$I=1$	$I=0$
a_s	-0.103	0.185	$-0.146(10)$	$0.262(30)$	$-0.30(2)$	$-0.10(5)$
r_s	≈ 13	≈ 2	≈ 19	≈ 3	$0.40(15)$	—
$a_{\mathrm{p}\frac{1}{2}}$	-0.040	-0.085	-0.113	-0.240	$-0.035(6)$	$0.11(4)$
$a_{\mathrm{p}\frac{3}{2}}$	0.219	-0.024	0.619	-0.068	$0.007(8)$	—

The πN scattering lengths are the recommended values of [Ebel 71] S-wave effective ranges are computed from the parameters b of the expansion (not valid for $q^2 < 0$)

$$\mathrm{Re}\, f_{0+}^{I}(q^2) = a_{0+}^{I} + b_{0+}^{I} q^2 , \qquad r_{0+}^{I} = -2[a_{0+}^{I} + b_{0+}^{I}(a_{0+}^{I})^{-2}]$$

$$b^{\frac{1}{2}} - b^{\frac{3}{2}} = 0.027(15)\, m_\pi^{-3} , \qquad b^{\frac{1}{2}} + 2b^{\frac{3}{2}} = -0.18(6)\, m_\pi^{-3}$$

(see [Achuthan 71] and [Höhler 72]). Actually, the effective range analysis is of very restricted use in πN scattering, as is indicated by the large values of r_s. The relations between isospin and charge amplitudes are

$$f^{\frac{3}{2}} = f(\pi^+ p) , \qquad f^{\frac{1}{2}} = \tfrac{3}{2} f(\pi^- p) - \tfrac{1}{2} f(\pi^+ p) ,$$

where $a(\pi^+ p)$ is defined by Eq. (6) in 1.3.2 and $a(\pi^- p)$ by a more complicated formula (see [Oades 71]). In practice, however, one uses the more accurate data at higher energies and extrapolates to $q = 0$ using fixed-t dispersion relations. Reviews: [Hamilton 63 and 67], [Moorhouse 69]. For the determination of other expansion parameters, see [Höhler 72].

The KN effective range parameters are taken from [A. Martin 69], [B. Martin 69] and [Cutkosky 69] for $I = 1$ and from [Stenger 64] and [Chand 67] for $I = 0$. The relations between isospin and charge amplitudes are

$$f^1 = f(\mathrm{K}^+ p) , \qquad f^0 = f(\mathrm{K}^+ p) - 2 f(\mathrm{K}^+ n) , \qquad f(\mathrm{K}^+ n) = f(\mathrm{K}^0 n) ,$$

where $f(\mathrm{K}^0 n)$ enters $f(\mathrm{K}_\mathrm{L}^0 p)$ through Eq. (4) in 1.1.2:

$$f(\mathrm{K}_\mathrm{L} p) = \tfrac{1}{2}(1 - 2\varepsilon) f(\mathrm{K}^0 p) + \tfrac{1}{2}(1 + 2\varepsilon) f(\overline{\mathrm{K}}^0 p) .$$

Reviews: [Bransden 69], [B. Martin 70].

The reactions $\pi^- p \to \eta n$ and $\pi^- p \to \mathrm{K}\Lambda$ below 2 GeV can be described in terms of resonance decays alone [Deans 71]. For the reaction $\pi^- p \to Xn$, thus far only the cross section has been measured [Basile 71].

Pilkuhn

1.5.2 The $\bar{\mathrm{K}}\mathrm{N}$, $\pi\Lambda$ and $\pi\Sigma$ channels

	1	2	$I=1$ 3	4	5	6	7	$I=0$ 8	9
	$\bar{\mathrm{K}}\mathrm{N}\to\bar{\mathrm{K}}\mathrm{N}$	$\bar{\mathrm{K}}\mathrm{N}\to\pi\Sigma$	$\bar{\mathrm{K}}\mathrm{N}\to\pi\Lambda$	$\pi\Sigma\to\pi\Sigma$	$\pi\Sigma\to\pi\Lambda$	$\pi\Lambda\to\pi\Lambda$	$\bar{\mathrm{K}}\mathrm{N}\to\bar{\mathrm{K}}\mathrm{N}$	$\bar{\mathrm{K}}\mathrm{N}\to\pi\Sigma$	$\pi\Sigma\to\pi\Sigma$
K_s	-0.01	-0.71	-0.38	0.34	-0.21	0.17	-2.40	-1.21	-1.05
A_s	$-0.05(4)$ $+0.63(6)\mathrm{i}$	—	—	$-0.31(30)$ $+0.24(31)\mathrm{i}$	—	$-0.45(61)$	$-1.74(4)$ $+0.70(1)\mathrm{i}$	—	$-0.14(20)$

Recent effective-range K-matrix fits (Eq. (2) in 1.3.2) to $\mathrm{K}^-\mathrm{p}$ and $\mathrm{K}^0_\mathrm{L}\mathrm{p}$ reactions are given by [Berley 70], [Thompson 70] and [Kim 71]. Berley and Thompson assume diagonal R^J-matrices. The resulting matrices are still very unreliable, especially in the $(\pi\Lambda, \pi\Sigma)\to(\pi\Lambda, \pi\Sigma)$ part. The pure s-wave zero-range fit $(R=0)$ to the $\bar{\mathrm{K}}\mathrm{N}$ data below 280 MeV/c which is given in the table is due to [A. Martin 70]. It is invariant against simultaneous sign changes of elements in columns 2, 3, 7 or 2, 5, 7 of the table. The scattering lengths $A_\mathrm{s}(c)$ are defined as $f^I_{0+}(q_c=0)$ for each channel c. They are well determined only for $c=\bar{\mathrm{K}}\mathrm{N}$. Attempts to work below the $\bar{\mathrm{K}}\mathrm{N}$ threshold employ the impulse approximation in $\mathrm{K}^-\mathrm{d}$ reactions and yield for the two real scattering lengths

$$A_\mathrm{s}(\pi\Lambda) = -0.25(10), \qquad A^0_\mathrm{s}(\pi\Sigma) = -0.5(2) \qquad [\text{Cline 71}].$$

All numbers are in units of fm. For $\pi\Lambda$ scattering, $\delta_\mathrm{s}-\delta_\mathrm{p}$ can be determined from Ξ decays (1.4.6 c). Both the $\mathrm{K}^-\mathrm{p}$ Coulomb interaction and the $\mathrm{K}^- - \mathrm{K}^0$ mass difference give important deviations from charge independence and are included according to the method of [Dalitz 60]. See also [A. Martin 70], [B. Martin 70] and [Oades 72].

1.5.3 NN, YN, nd and $\bar{\mathrm{N}}\mathrm{N}$ scattering

	pp [1]	np [2]	nn [3]	Λp [4]	Λn [4]	Σ^+p [4]
a_s	7.786(8)	23.719(13)	16.7(6)	1.93	2.45	2.40
r_s	2.840(9)	2.714(87)	3.2(1.6)	2.14	2.11	2.80
a_t	—	$-5.414(4)$	—	1.34	1.06	-0.71
r_t	—	1.763(5)	—	2.25	2.47	1.39

[1]) From [Slobodrian 68]; vacuum polarization is included. The pp scattering length a_N in the absence of the Coulomb potential is approximately related to a_pp and r_pp by

$$a_\mathrm{N}^{-1} = a_\mathrm{pp}^{-1} + 2q\eta(\ln 2q\eta r + 0.330) = a_\mathrm{pp}^{-1} - 0.0690.$$

This yields $a_\mathrm{N} = 16.8$ which agrees with a_nn but not with $a_\mathrm{s}(\mathrm{np})$, thus providing evidence for isospin breaking.

[2]) From [Houk 71].

[3]) From [Verondini 71]. See also [Noyes 71] and [Slaus 71].

[4]) From [de Swart 71]. No errors are given. The large deviation from isospin invariance is due to the $\Lambda\Lambda\pi$ and $\Lambda\Lambda\varrho$ coupling constants (see 1.5.5). For $\Sigma^-\mathrm{p}$ data, a zero range expansion is given by [Gell 70]. See also [de Swart 71, Letessier 71, Eisele 71].

In this table, s stands for the singlet s-wave $(J=0)$ and t for the triplet s-wave $(J=1)$. The effective range parameters a and r are given in units of fm. For pp and Σ^+p scattering, Eq. (6) in 1.3.2 is applicable whereas for the other channels Eq. (11) in 1.3.1 is used. Notice the unusual sign convention!

nd-*scattering:* Let a_2 and a_4 denote the doublet and quartet scattering lengths. [Dilg 71] finds
$$a_{coh} = a_4 + \tfrac{1}{2}a_2 = -6.672(7)\,\text{fm}, \qquad a_2 = -0.65(4)\,\text{fm}, \qquad a_4 = -6.35(2)\,\text{fm}.$$

$\bar{\mathrm{N}}\mathrm{N}$-*scattering:* The scattering amplitudes are related to the isospin amplitudes $T(I)$ as follows:

$$T(\bar{\mathrm{p}}\mathrm{p}\to\bar{\mathrm{p}}\mathrm{p}) = \tfrac{1}{2}T(0)+\tfrac{1}{2}T(1), \qquad T(\bar{\mathrm{p}}\mathrm{p}\to\bar{\mathrm{n}}\mathrm{n}) = \tfrac{1}{2}T(0)-\tfrac{1}{2}T(1), \qquad T(\bar{\mathrm{p}}\mathrm{n}\to\bar{\mathrm{p}}\mathrm{n}) = T(1).$$

The scattering lengths are given by [Bryan 68] as

$$a(1) = -0.82 + 0.87\,\mathrm{i}, \qquad a(0) = -1.01 + 0.65\,\mathrm{i}$$

for triplet s-waves and

$$a(1) = -1.00 + 0.60\,\mathrm{i}, \qquad a(0) = -0.62 + 1.33\,\mathrm{i}$$

for singlet s-waves, all expressed in units of fm.

Pilkuhn

1.5.4 Meson-meson scattering

	$\pi\pi$ scattering			$K\pi$ scattering		$K\bar{K}$ scattering
	$I=0$	$I=1$	$I=2$	$I=\frac{1}{2}$	$I=\frac{3}{2}$	$I=0$
a_s	0.17(2)	—	$-0.050(5)$	0.16(2)	$-0.078(8)$	$1.3+i\ 0.35$
r_s	$-7.3(7)$	—	6.0(6)	$-5.0(5)$	17(2)	1.4
a_p	—	0.033(3)	—	0.014(1)	0	—
Reference	[Weinberg 66]			[Ebel 71]	[1])	[Beusch 70]

[1]) Based on [Griffith 68].

The above table gives the effective range parameters according to soft-meson theory (with $f_\pi = f_K = 126(6)$ MeV, compare 1.4.2 d). The units are m_π^{-2I-1}, except for $K\bar{K}$ scattering which is given in units of fm. Experimental information comes from the rare K_{e4}-decay ($K^\pm \to \pi\bar{\pi}ev$): $a_s^0 = 0.52(12)$, $r_s^0 \approx 0$ [Basile 71a] and from the extrapolation of high-energy peripheral $\pi N \to \pi\pi N$ reactions $a_s \approx 0.2$, $r_s^0 = -0.6$, $a_p^1 \approx 0.04$, $a_s^2/a_s^0 \approx -3$ [Scharenguivel 70, Cline 70]. [Botke 70] finds $a_s^0 = 0.2(1)$, $r_s^0 = -5.2^{+1.5}_{-2.0}$, [Maung 70] finds $a_s^0 = 0.28(21)$ for $r_s^0 \approx 1$. For dispersion relation fits see [Morgan 70a]. See also the reviews by [Morgan 70, Baltay 70] and especially [Petersen 71].

1.5.5 Pion-baryon coupling constants

	$pn\pi^+\ (a)$	$pp\pi^0\ (b)$	$\Lambda\Sigma\pi\ (c)$	$\Sigma\Sigma\pi\ (d)$	$\Lambda\Lambda\pi\ (e)$	$\Xi\Xi\pi\ (f)$
f^2 [1])	0.081(3)	0.0774(44)	0.0405(49)	0.0418(69)		0.0092
$G^2/4\pi$	14.64(54)	14.0(8)	11.1(1.2)	12.2(2.0)	0.023	3.4

[1]) The connection between G and f is given in Eq. (7) in 1.3.3.

a) $G^2/4\pi = 180.770\ f^2$. The value $(81^{+3}_{-4}) \cdot 10^{-3}$ is recommended by [Ebel 71]. It comes from fixed-t dispersion relations for $\pi^\pm p$ scattering (1.3.5). The most accurate determination gives 81.5 ± 1.6 [Samaranayake 69]. See also b). The true $\pi^\pm$ coupling constants are larger by a factor $2^{\frac{1}{2}}$ (see Eq. (2) in 1.3.4).

b) This value comes from a multi-energy phase-parameters search of pp scattering [Breit 71], who also find $G^2/4\pi = 14.7(9)$ for np scattering, which confirmes a). A supposed deviation of G from its isospin invariant value is not yet established.

c) This value comes from the Goldberger-Treiman relation, Eq. (7) in 1.1.4. For $NN\pi$, the left hand side is $1.0578 \cdot 10^{-6}\ \mathrm{GeV}^{-1} \cdot 0.4052 = 0.426 \cdot 10^{-6}\ \mathrm{GeV}^{-1}$, and the right hand side is $0.02784 \cdot \mathrm{GeV} \cdot 1.437 \cdot 10^{-5} \cdot \mathrm{GeV}^{-2} = 0.400 \cdot 10^{-6}\ \mathrm{GeV}^{-1}$, which is only 6% smaller, the difference being probably due to the π baryon form factor. The value $g_A(\Lambda\Sigma) = 0.719 \cdot 10^{-5}\ \mathrm{GeV}^{-2}$ (1.4.6b) leads to $f(\Lambda\Sigma\pi) = 0.1893$, or $f(\Lambda\Sigma\pi) = 0.2013$ after adding 6% for the form factor. The resulting value of $G^2/4\pi$ is 11.08, 11.05 or 11.12 for the respective choices of $m_{\Sigma^0}, m_{\Sigma^+}$ or m_{Σ^-} inserted in Eq. (7) in 1.3.3. The theoretical basis of the Goldberger-Treiman relation is less well founded than the dispersion relations. However, attempts to extract $f_{\Lambda\Sigma\pi}$ from $\bar{K}N$ scattering [Chan 70] or hypernuclear physics (see [de Swart 71]) are premature. Other methods [Engels 71] are also disputable.

d) From superconvergence relations for the amplitude B (Eq. (1) in 1.3.5) of the $\pi^+\Sigma^- \to \pi^-\Sigma^+$ reaction using resonance saturation [Engels 71].

e) The difference between the binding energies of $^4_\Lambda H$ and $^4_\Lambda He$ is explained by $\Lambda\Lambda\pi$ and $\Lambda\Lambda\varrho$ couplings. $G_{\Lambda\Lambda} = -0.046\ G_{\Lambda\Sigma}$ both for π and the tensor coupling of ϱ.

f) From a superconvergence relation with $\Xi(1530)$ saturation [Engels 71]. This is an order of magnitude estimate only, using $\Gamma(\Xi(1530)) = 11$ MeV.

The $\Delta\Delta\pi$ coupling constant occurs indirectly in some models. See [Sutherland 67] for a review.

Pilkuhn

1.5.6 Other meson-baryon coupling constants

a) Kaons. The $N\Lambda K$ and $N\Sigma K$ coupling constants are determined from dispersion relations for forward $K^{\pm}p$ scattering (Eq. (7) in 1.3.5). Placing the Σ pole at the Λ mass yields $G_Y^2 \equiv (G_{N\Lambda K}^2 + 0.84\, G_{N\Sigma K}^2)/4\pi$. [A. Martin 70a] finds $G_Y^2 = 8.8 \pm 3.0$. The conformal mapping method [Chao 71] gives 14_{-3}^{+4}. [Cutkosky 70] finds 9 ± 3 from K^+p phase shifts. $G_{N\Sigma K}^2/4\pi$ appears to be < 1.0 [A. Martin 70, Queen 69a, B. Martin 69], which implies $G_{N\Lambda K}^2/4\pi \approx 8 \pm 3$. See also the reviews by [Queen 69, B. Martin 70, O'Brien 71].

From the p-wave component in the $\overline{K}N \to \pi\Sigma$ relation, [Tripp 68] finds $\frac{3}{2}\, G_{\overline{K}N Y}^2/G_{\pi\Lambda Y}^2 = 1.5$, whereas [Bunnel 70] finds this ratio to be ≈ 1 for $K^-d \to \Lambda\pi^-p$ production. SU(3)-symmetry conditions (Eqs. (9) and (14) in 1.3.4) predict unity for this expression. The $\overline{K}N\Lambda(1405)$ coupling constant is given by $G^2/4\pi = (2\pi k_+^2)^{-1}\, 4m^2 \int \mathrm{Im} f_s^0(\overline{K}N)\cdot d\omega$ (compare 1.4.8 b). the integration extending over the resonance. The solution of [B. Martin 69] gives 0.19 ± 0.03. [Tripp 68] finds 0.28. SU(3)-symmetry results without mixing yield $G^2(\overline{K}N\Lambda(1405))/4\pi = \frac{2}{3} G^2(\pi\Sigma\Lambda(1405))/4\pi = 0.035$ (compare 1.4.8).

The ratio $G_{\Lambda\Sigma^+ K^+}^2/G_{\Lambda\pi^+p}^2 = 1.0 \pm 0.3$ is determined by [Barloutaud 71] from a Regge pole model of baryon exchange in backward $\overline{K}N \to \pi\Sigma$ production. These authors also find $G_{N\Lambda K}^2/G^2 = 1.0 \pm 0.3$. Additional information about kaon coupling constants comes from the Cabibbo theory (see 1.5.7).

b) η- and X-mesons. From a comparison of backward peaks in $\pi^-p \to \eta n$ and $\pi^-p \to \pi^0 n$ production, [Chase 69] finds 0.18 ± 0.06 for $G_{NN\eta}^2/G_{NN\pi}^2$. [Boright 70] on the other hand, finds 0.45 ± 0.11 for this ratio. Analyses of NN scattering give $G_{NN\eta}^2/4\pi \approx 3$ [Green 68, Gersten 71], but here the η and X mesons are not separated. Setting $G_{NN\eta} = 0$, one finds $G_{NNX}^2/4\pi = 13.8$ [Bryan 71].

c) Vector mesons. The ϱNN and ωNN coupling constants can be obtained from NN scattering (see [Ebel 71]). However, it is difficult to determine G^V and G^T individually. For the ϱ meson, the G^T/G^V ratio is better determined from πN scattering, which yields ratio values ranging between 3 to 4. [Baacke 68, Engels 70, Strauß 70]. Values of this magnitude agree with the nucleon electromagnetic form factors, analyzed in terms of a vector-dominance model [Wataghin 69]. For the ωNN couplings, the electromagnetic form factors imply $G^T/G^V = 0$. With these restrictions, one then obtains $G_{\omega NN}^{V\,2}/4\pi = 9 \pm 3$ and $G_{\varrho NN}^{T\,2}/4\pi = 16 + 3$. Recent determinations of $G_{\varrho NN}^T/G_{\varrho NN}^V$ from NN scattering give values closer to 5 [Schierholz 68, 72, Gersten 71, Bryan 71], without changing $G_{\varrho NN}^T$ much. Thus $G_{\varrho NN}^V$ appears to be known less accurately. The universality prediction of Sakurai (see in 1.3.7) is $G_{\varrho NN}^{V\,2}/4\pi = \frac{1}{4} g_{\varrho^0\pi\pi}^2/4\pi = 0.53$ (compare 1.4.3 b).

Both ϕNN coupling constants appear to be very small, $G_\phi^2/G_\omega^2 < 0.03$. This conclusion is based on the smallness of the ϕ/ω ratio in $\overline{p}p$ annihilations [Baltay 66] and in K^-N reactions [Hoogland 71], and is in agreement with the quark model [Kokkedee 69, ref. in 1.1.6].

d) Scalar and tensor mesons. The existence of scalar mesons as "good resonances" is as yet uncertain (see 1.4.5). The question whether the scattering amplitude has poles at $t = m^2 - im\Gamma$ corresponding to the "exchange of scalar mesons" is completely open, but in view of the difficulties encountered already with the "good" Δ resonance (see 1.4.7a) one may be pessimistic. The existing "determinations" of scalar meson exchange coupling constants are only convenient parametrizations. In particular, NN scattering analyses need the exchange of a low-lying scalar and isoscalar meson (σ or ε, mass between 400 and 700 MeV), which may represent the effect of uncorrelated $\pi\pi$ states in the t channel or $N\Delta$ states in the s channel. See [Ebel 71, Section 3.3 ii]. Determinations from πN scattering are given for $\varepsilon(m_\varepsilon \equiv 700$ MeV), S* and the tensor meson f (representing both f and f'). See [Ebel 71, Section 3.1 vii].

Coupling constants of vector and tensor mesons are frequently determined from Regge pole exchange models for high-energy peripheral interactions. See [Collins 71, Chapter 7] for a review. In particular, the ϱ- and A_2-exchange in the $\pi^-p \to \pi^0 n$ and $\pi^-p \to \eta n$ reactions, respectively, can be compared with the $\varrho + A_2$ exchange in the $K^-p \to \overline{K}^0 n$ or $K^+n \to K^0 p$ reactions. However, theoretical uncertainties about the extrapolation to the poles are so large that only coupling constant ratios can be determined. See also [Renner 71, Baacke 72].

1.5.7 SU(3)-comparison of meson-baryon coupling constants

a) PBB coupling constants

SU(3)-symmetry	$\Sigma\Sigma\pi$	$\Xi\Xi\pi$	$N\Lambda K$	$N\Sigma K$	$NN\eta_8$	$\Lambda\Lambda\eta_8$	$\Sigma\Sigma\eta_8$
pv-coupling⎱ f^2	0.0487	0.0041	0.0851	0.0041	0.0082	0.0405	0.0405
$\alpha_{pv} = 0.6124$⎰ $G^2/4\pi$	14.21	1.45	18.42	0.95	1.48	11.02	11.83
ps-coupling⎱	3.50	3.82	10.82	3.82	0.00	11.1	11.1
$\alpha_{ps} = 0.7554$⎰ $G^2/4\pi$							

Due to the mass splittings within SU(3)-multiplets, SU(3)-comparisons of coupling constants are model-dependent. For the pseudoscalar meson-baryon (PBB) coupling constants, one must choose between pseudovector (f) and pseudoscalar (G) coupling (see Eq. (6) in 1.3.3). In addition the parameter α Eq. (7) in 1.3.4 must be determined. Because of the large $\pi - K$ mass difference, it appears better not to use kaon coupling constants in the determination of α. In the above table, α is determined from the first and third columns of the table in 1.5.5.

For the $\Sigma\Sigma\pi$ coupling constant, pv- and ps-SU(3) symmetries lead to widely different results, the agreement with the values of the table in 1.5.5 being much better for pv-symmetry. For kaon couplings, G_Y^2 of subsection 1.5.6 comes out too large, especially so for pv-symmetry, while the ratio $G_{N\Sigma K}^2/G_{N\Lambda K}^2 \ll 1$ comes out slightly better for pv-symmetry. Independent support for pv-SU(3) symmetry of kaon couplings comes from the Cabibbo fit to weak decays (1.4.6b). With $\theta_A = \theta_V$ imposed, a fit to all axial decay constants with the single parameter $\alpha_{weak} \approx 0.63$ is possible. Then, if kaon PCAC is fulfilled, the strong kaon coupling constants must also obey pv-SU(3) symmetry. For possible complications due to form factors, see [Pilkuhn 70].

b) Decuplet decays

Decay	Γ_{exp}	Γ according to SU(3)-symmetry		
		$R = 0.75$ fm	$R = 0$	$R = 0, \Delta$ omitted
$\Upsilon^+ \to \Lambda\pi^+$	31.3	51.9	48.2	30.0
$\Upsilon^+ \to \Sigma^+\pi^0$	1.6	2.9	2.1	1.3
$\Xi^0(1530) \to \Xi^-\pi^+$	6.6	15.4	11.6	7.2

Here SU(3)-symmetry (Eq. (9) in 1.3.4) reduces the number of coupling constants to a single term, which is normally taken as G_Δ (1.4.7a). The widths which then result for Υ and Ξ(1530) are shown in the above table. However, an SU(3) comparison with $R = 0.75$ fm makes no sense, since $1 + R^2 p^2$ would be negative for the $\Upsilon N\overline{K}$ vertex, i.e. $G_{\Upsilon N\overline{K}}^2$ should be large and negative, which is certainly not the case (see 1.5.6). On the other hand for $R = 0$ the agreement with Γ(exp) is poor. Inasmuch as $R = 0$ cannot reproduce the shape of the Δ resonance, one could omit this resonance completely from the analysis, in which case the agreement is satisfactory (last column of the table). It may also be noted that the P-wave formula of Tab. 2 in 1.2.1 for decay into two spinless particles reproduces all decuplet decay widths, including that of Δ. The difference between this formula and the formula of 1.4.7a is a factor $k_+^2/4m_\pi^2$, the omission of which even improves the shape of the Δ resonance curve. (The resulting $G_{1\Delta}^2/4\pi$ is 20, see [Ebel 71].)

1.5.8 Nuclear coupling constants

The concept of scattering amplitudes as analytic functions with poles and cuts also applies to scattering on nuclei. This point of view may provide a better parametrization scheme than that based on the use of nuclear potentials. See [Ericson 70] for a review. The dispersion relation for forward scattering of a nucleon of kinetic energy $E = \omega - m$ and momentum $k = (\omega^2 - m^2)^{\frac{1}{2}}$ on a target of zero spin and isospin is

$$\mathrm{Re} f(E) = f(0) + \sum_\beta \frac{r_\beta E}{E_\beta(E_\beta - E)} + \frac{k^2}{2\pi^2} P \int_0^\infty \frac{\sigma(k')\,dk'}{k'^2 - k^2} + \frac{E}{\pi} \int_{-2m}^{E_b} \frac{\mathrm{Im} f(E')\,dE'}{(E' - E)\,E'}.$$

This relation has a subtraction at $\omega_0 = m$ and corresponds to relation (8) in 1.3.5 with the antiparticle scattering σ_- neglected. In $n\alpha$ scattering, the ^{3}He pole occurs at $E_0 = -15.4$ MeV with residue $r_0 = 3.2(4)$ [Locher 72]. In nd scattering, the positions and residues of the proton and triton poles are [Locher 70]

$$E_p = -(1 - m_n/m_d) B_d = -1.113 \text{ MeV}, \qquad r_{pnd} = 0.081(2),$$

$$E_t = -(1 + m_n/m_d) B_{nt} = -9.93 \text{ MeV}, \qquad r_{dnt} = 0.382(40),$$

where B_d is the deuteron binding energy and B_{nt} the binding energy of the last neutron in the triton. In potential theory, the residues are related to the normalization factor N of the asymptotic wave function of the bound neutron by

$$\psi(r \to \infty) = N Y_l^m(\hat{r})\, r^{-1}\, e^{-\alpha r}, \qquad r_{pnd} = N^2/m_n, \qquad r_{dnt} = \tfrac{3}{8} N^2 \frac{m_n + m_d}{m_n m_d}.$$

For the deuteron this is equivalent to $r_{pnd} = \alpha m_n^{-1}(1 - r_t\alpha)$, where r_t is the triplet np scattering length and $\alpha^2 = m_n B_d$. For π-nucleus scattering, see [Ericson 70] or [Locher 71].

Pilkuhn

1.5.9 References for 1.5

For journal abbreviation see p. 2.

Achutan	71	ZP 242, 167: P. Achutan, G. E. Hite, G. Höhler.
Baacke	68	ZP 214, 381: J. Baacke, G. Höhler, F. Steiner.
Baacke	72	NP B: J. Baacke, T. H. Chang, H. Kleinert.
Baltay	66	PR 142, 932: C. Baltay, J. Lach, J. Sandweiss, H. D. Taft, N. Yeh, D. L. Stonehill, R. Stump.
Baltay	70	Experimental meson spectroscopy. Columbia University Press: C. Baltay, A. H. Rosenfeld.
Barloutaud	71	NP B 26, 557: SABRE collaboration.
Basile	71	NC 3 A, 371: M. Basile, D. Bollini, P. Dalpiaz, P. L. Frabetti, T. Massam, F. Navach, F. L. Navarria, M. A. Schneegans, A. Zichichi.
Basile	71a	PL 36 B, 619: P. Basile, S. Brékin, A. Diamant-Berger, P. Kunz *et al.*
Berley	70	PR D 1, 1996: D. Berley, P. Yamin, R. Kofler, A. Mann, G. Meissner, S. Yamanoto, J. Thompson, W. Willis. Err.PR D 3, 2297.
Beusch	70	in Baltay 70, p. 185: W. Beusch.
Boright	70	PL 33 B, 615: J. P. Boright, D. R. Bowen, D. E. Groom, J. Orear, D. P. Owen, A. J. Pawlicki, D. H. White.
Botke	70	NP B 23, 253: J. C. Botke.
Bransden	69	in High En. Physics, Vol. 3 (ed. EHS Burhop, Academic Press).
Breit	71	Buffalo Univ. preprint: G. Breit, M. Tischler, S. Mukherjee, J. Lucas.
Bryan	68	NP B 5, 201: R. A. Bryan, R. J. N. Phillips.
Bryan	71	Nijmegen preprint: R. A. Bryan, A. Gersten.
Bunnel	70	NCL 3, 224: K. O. Bunnel, D. Cline, R. Laumann, J. Mapp, J. L. Uretsky.
Chan	70	PR D 2, 2635: C. H. Chan, L. L. Smalley.
Chand	67	AP 42, 81: R. Chand.
Chao	71	PRL 25, 1463: Y-A. Chao, E. Pietarinen.
Chase	69	PL 30 B, 659: R. C. Chase, E. Coleman, H. W. Courant, E. Marquit, E. W. Petraske, H. F. Romer, K. Ruddick.
Cline	70	NP B 18, 77: D. Cline, K. J. Braun, V. R. Scherer.
Cline	71	PRL 26, 1194, PL 35 B, 606: D. Cline, R. Laumann, J. Mapp.
Collins	71	PRep 1, 103: P. D. B. Collins.
Cutkosky	70	PR D 1, 2547: R. E. Cutkosky, B. B. Deo.
Dalitz	60	AP 10, 307: R. H. Dalitz, S. F. Tuan.
Deans	71	Tampa + Huntsville preprint: S. R. Deans, J. E. Rush.
Dilg	71	PL 36 B, 208: W. Dilg, L. Koester, W. Nistler.
Ebel	71	NP B 33, 317: G. Ebel, A. Müllensiefen, H. Pilkuhn, F. Steiner, D. Wegener, M. Gourdin, C. Michael, J. L. Petersen, M. Roos, B. R. Martin, G. Oades, J. J. de Swart.
Eisele	71	PL 37 B, 204: F. Eisele, H. Filthuth, W. Föhlisch, V. Hepp, G. Zech.
Engels	70	NP B 15, 365: J. Engels, G. Höhler, B. Petersson.
Engels	71	NP B 31, 531: J. Engels, H. Pilkuhn.
Ericson	70	NP A 148, 1: T. E. O. Ericson, M. P. Locher.
Gersten	71	PR D 3, 2076: A. Gersten, R. H. Thompson, A. E. S. Green.
Green	71	PR 174, 1304: A. E. S. Green, T. Ueda.
Griffith	68	PR 176, 1705: R. W. Griffith.
Hamilton	63	RMP 35, 737: J. Hamilton, W. S. Woolcock.
Hamilton	67	in High En. Physics, Vol. 1 (ed. EHS Burhop, Academic Press).
Höhler	72	NP B 39, 237: G. Höhler, H. P. Jacob, R. Strauß.
Hoogland	70	NP B 21, 381: W. Hoogland, J. C. Kluyver, G. G. G. Massaro, A. G. Tenner, A. Minguzzi-Ranzi, S. Focardi, D. Merril, G. Lamidey, U. Karshon, G. Yekutieli.
Houk	71	PR C 3, 1886: T. L. Houk.
Kim	71	PRL 27, 356: J. K. Kim.
Koester	71	PRL 27, 956: L. Koester, W. Nistler.
Letessier	71	NC 5 A, 82: J. Letessier, A. Tounsi.
Locher	70	NP B 23, 116: M. P. Locher.
Locher	72	NP B 36, 634: M. P. Locher.
Locher	71	CERN report 71–14: Proc. Spring School on pion interactions at low and medium energies.

Pilkuhn

A. Martin   70   NP B 16, 479: A. D. Martin, G. G. Ross.
A. Martin   70a   NP B 20, 287: A. D. Martin, R. Perrin.
B. Martin   69   PR 183, 1352 + 1345: B. R. Martin, M. Sakitt.
B. Martin   70   STMP 55, 73: B. R. Martin.
Maung   70   PL 33 B, 521: T. Maung, G. E. Masek, E. Miller, H. Ruderman, W. Vernon, K. M.
                              Crowe, N. T. Dairiki.
Moorhouse   69   Ann. Rev. Nucl. Sci. 19, 301: R. G. Moorhouse.
Morgan   70   STMP 55, 1: D. Morgan, J. Pisut.
Morgan   70a   PR D 2, 520: D. Morgan, G. Shaw.
Noyes   71   PR C 4, 995: H. P. Noyes, H. M. Lipinski.
Oades   71   Helv. Physics Acta 44, 5 + 160: G. C. Oades, G. Rasche.
Oades   72   PR D 4, 2153: G. C. Oades, G. Rasche.
O'Brien   71   NP, B: D. M. O'Brien, W. S. Woolcock.
Pilkuhn   70   STMP 55, 168: H. Pilkuhn.
Petersen   71   PRep 2, 155: J. L. Petersen.
Queen   69   FP 17, 467: N. M. Queen, M. Restignoli, G. Violini.
Queen   69a   NP B 11, 115: N. M. Queen, S. Leeman, F. E. Yeomans.
Renner   71   NP B 35, 397: B. Renner, P. Zerwas.
Samaranayake   69   Lund Conf.: V. K. Samaranayake, W. S. Woolcock.
Scharenguievel   70   NP B 22, 16: J. H. Scharenguievel, L. J. Gutay, D. H. Miller, F. T. Meiere, S. Marateck.
Schierholz   68   NP B 7, 483: G. Schierholz.
Schierholz   72   NP B 40, 335: G. Schierholz.
Slaus   71   PRL 26, 789: I. Slaus, J. W. Sunier, G. Thompson, J. C. Young, J. W. Verba, O. J.
                              Margoziotis, P. Doherty, R. T. Cahill.
Stenger   64   PR 134 B, 1111: V. J. Stenger, W. E. Slater, D. H. Stork, H. K. Ticho, G. Goldhaber,
                              S. Goldhaber.
Strauß   70   Karlsruhe Thesis: R. Strauß.
Sutherland   67   NC 48, 188: D. G. Sutherland.
de Swart   71   STMP 60, 138: J. J. de Swart, M. M. Nagles, T. A. Rijken, P. A. Verhoeven.
Thompson   70   Proc. Duke Conf. on Hyperon Resonances (ed. E. C. Fowler): J. Thompson.
Tripp   68   PRL 21, 1721: R. D. Tripp, R. O. Bangerter, A. Barbaro-Galtieri, T. S. Mast.
Verondini   71   Rivista NC 1, 33: E. Verondini.
Wataghin   69   NP B 10, 107: V. Wataghin.
Weinberg   66   PRL 17, 616: S. Weinberg.

1.6 Electromagnetic form factors and SU(3)-comparison

1.6.1 Electromagnetic form factors of nucleons

Instead of F_1 and F_2 defined in Eq. (5) in 1.1.3, one normally parametrizes the magnetic and electric form factors according to

$$G_M = F_1 + \kappa F_2 , \qquad G_E = F_1 + \kappa F_2 t/4m^2 , \qquad G_M(0) = 2m\mu .$$

These obey the scaling law

$$G_{E\mathrm{p}}(t) \cong \mu_\mathrm{p}^{-1} G_{M\mathrm{p}}(t) \cong \mu_\mathrm{n}^{-1} G_{M\mathrm{n}} \equiv G(t) .$$

See the reviews of [Rutherglen 69] and [Wilson 70]. A deviation from the scaling law is reported by [Berger 71]:

$$\mu_\mathrm{p} G_{E\mathrm{p}}/G_{M\mathrm{p}} = 1 + dt , \qquad d = 0.051\,(18)\,\mathrm{GeV}^{-2} = 0.0020\,(7)\,\mathrm{fm}^2 .$$

$G_{M\mathrm{n}}$ is known less accurately [Bartel 69, Berger 71]. A simple parametrization of $G(t)$ is the dipole fit (t measured in GeV^2):

$$G(t) \approx G_D(t) \equiv [1 - t/0.71]^{-2} .$$

Small deviations from the dipole fit ($G_{E\mathrm{p}}/G_D < 1$ for $0.4 < -t < 3\,\mathrm{GeV}^2$, $\mu_\mathrm{p}^{-1} G_{M\mathrm{p}}/G_D < 1$ for $t > -0.8\,\mathrm{GeV}^2$ and > 1 for $t < -0.8\,\mathrm{GeV}^2$) are reported by [Albrecht 66, Bartel 67 and 70, Berger 68 and 69, Coward 68, Goitein 70, Price 71]. $G_{E\mathrm{n}}$ can be obtained from elastic $e - \mathrm{d}$ scattering; the result is model-dependent. [Galster 71] finds

$$G_{E\mathrm{n}}(t) = \mu_\mathrm{n} t (4M^2)^{-1} [1 - pt(4M^2)^{-1}]^{-1} G_{E\mathrm{p}} \quad \text{for} \quad -1\,\mathrm{GeV}^2 < t ,$$

with $p = 10.7$ for the Hamada-Johnston wave function and $p = 5.6$ for the Feshbach-Lomon wave function. See also [Bumiller 70]. The slope at $t = 0$ can be measured by scattering thermal neutrons on atoms [Krohn 66]:

$$[-\,\mathrm{d}\,G_{En}(t)/\mathrm{d}t]_{t=0} = 0.496(10)\,\mathrm{GeV}^{-2}\,.$$

1.6.2 Electromagnetic form factors of ^{3}He and d

The electric form factor for different nuclei has been reviewed by [Hofstadter 67]. In the meantime, and especially for the light nuclei, form factors have been extended to large momentum transfers and the "elementary particle" description has been used as the only method of analysis reliable for large t values. The nuclei ^{3}He and ^{3}H both have spin $\frac{1}{2}$. Therefore the electromagnetic vertex T_μ is again of the form (5) in 1.1.3. The electric and magnetic form factors (defined in 1.6.1) of ^{3}He are parametrized as [McCarthy 70]

$$G(t) = \mathrm{e}^{a^2 t} + b^2\,t\,\mathrm{e}^{c^2 t} + d\,\exp[-(\sqrt{-t} - q_0)^2\,p^{-2}]$$

	a [fm]	b [fm]	c [fm]	d	q_0 [fm^{-1}]	p [fm^{-1}]
G_E	0.675(8)	0.366(25)	0.836(32)	$-0.00678(83)$	3.98(9)	0.90(16)
G_M	0.654(24)	0.456(29)	0.821(53)	0	—	—

G_E and G_M *can* be Fourier-transformed. The electric and magnetic mean square radii are 1.88(5) and 1.95(11) fm. See [Yang 71] for attempts to explain these results in potential theory.

The electromagnetic deuteron vertex is characterized by 3 vertex functions [Glaser 57]

$$T_\mu = \varepsilon_\varrho^{*\prime}\left[F_0\,g^{\varrho\sigma} + \frac{F_2}{2M_\mathrm{d}^2}\left((P-P')^\varrho(P-P')^\sigma + \frac{t}{3}\,g^{\varrho\sigma}\right)\right](P+P')^\mu\,\varepsilon_\sigma$$

$$+\,\varepsilon_\varrho^{*\prime}\,G_1(g^{\mu\varrho}\,g^{\nu\sigma} - g^{\mu\sigma}\,g^{\nu\varrho})\,(P-P')^\nu\,\varepsilon_\sigma\,.$$

These are related to the more familiar charge (c), magnetic (M) and quadrupole (Q) form factors by [Gourdin 63]

$$F_c = F_0 + \tfrac{2}{3}\eta\,[F_0 + F_2(1+\eta) - G_1]\,, \qquad F_c(0) = 1\,, \qquad \eta \equiv -t/4m_\mathrm{d}^2\,,$$

$$F_Q = F_0 - G_1 + F_2(1+\eta)\,, \qquad F_Q(0) = m_\mathrm{d}^2\,Q_\mathrm{d}\,, \qquad F_M = G_1(1+\eta)^{\frac{1}{2}}\,.$$

$$Q_d = 0.287(2)\,fm^2 \qquad [\text{Reid } 72]$$

The differential cross section is

$$\frac{\mathrm{d}\sigma}{\mathrm{d}\Omega} = \frac{\mathrm{d}\sigma}{\mathrm{d}\Omega}_{Mott}\left[A + B\,\mathrm{tg}^2\left(\frac{\theta}{2}\right)\right]\,,$$

$$A = F_c^2 + \tfrac{8}{9}\eta^2 F_Q^2 + \tfrac{2}{3}\eta(\eta+1)\,F_M^2\,, \qquad B = \tfrac{4}{3}\eta(1+\eta)^2\,F_M^2\,.$$

A fit to all data including [Galster 71] by K. H. Schmidt (private communication) gives

$$A = (1 - t/0.138\,\mathrm{GeV}^2)^{-4.66}\,.$$

Recent form factor measurements for ^{4}He have been reported by [Frosch 67] and for ^{6}Li and ^{7}Li by [Suelzle 67].

1.6.3 Electric form factors of pions

The pion form factor $F_\pi(t)$ is defined in Eq. (6) in 1.1.3. Its parametrization in the timelike region $(t \to s > 0)$ is given in 1.4.3a. In the spacelike region $(t < 0)$ one finds from π^+ electroproduction

$$F_\pi = \left[1 - \frac{t}{(0.56 \pm 0.08)^2\,\mathrm{GeV}^2}\right]^{-1} \qquad \text{for} \quad -0.4\,\mathrm{GeV}^2 < t < 0\,,$$

using the dispersion theory of Zagury [Mistretta 68].

Beyond the resonance region [Driver 71], the form factor is given by

$$F_\pi = \left[1 - \frac{t}{m_\varrho^2}\right]^{-1}\left[1 - \frac{0.85\,t}{m_\varrho^2}\left(1 + \frac{0.65\,t}{m_\varrho^2}\right)\right] \qquad \text{for} \quad -0.85\,\mathrm{GeV}^2 < t < 0.18\,\mathrm{GeV}^2\,,$$

using an electric Born term model with absorption corrections [Schmidt 71].

Pilkuhn

1.6.4 N—Δ and N—N* transition form factors

Reviews: [Clegg 69, Wilson 70, Gayler 71]. The slope at $t = 0$ of transition form factor for each of the resonances Δ, $N(1520)$, $N(1688)$ is greater than the slope of the nucleon form factors. The ratio of longitudinal to transversal excitation cross section is smaller than 0.2 [Brasse 71, Bartel 71]. The M_{1+} multipole dominates the excitation of the Δ resonance [Albrecht 70, 71, Galster 71, Hellings 71] for $-t < 1$ $(\text{GeV/c})^2$. The transition form factor for the excitation of the $\Delta(1236)$ resonance may be parametrized by [Bartel 69]:

$$G_M^*(t) = G_M^*(0) \left\{ G_M^V(t) \frac{0.85}{1-t/2.9} + G_{Ep}(t) \frac{0.15}{1-t/0.97} \right\} \left(\frac{E_2^*(t) + m_p}{E_2^*(0) + m_p} \right)^{\frac{1}{2}}$$

where G_M^V is the magnetic isovector form factor, E_2^* is the energy of the target proton in the hadron cms, $G_M^*(0) = (m_p/m_\Delta)^{\frac{1}{2}} \mu^*$, and $\mu^* = (8\pi q_\Delta m_\Delta \Gamma_\Delta/3m_p k_\Delta^2)^{\frac{1}{2}} |M_{1+}^I(m_\Delta)|$; M_{1+}^I is defined in Eq. (18) in 1.3.1, and k_Δ, q_Δ are the photon and pion momenta in the hadron cms. [Pfeil 70, ref. in 1.4.10] gives $\mu^* = 1.24(2)\mu_p$. See also [Ebel 71, ref. in 1.1.6]. G_M^* can also be described by a simple dipole fit [private communication by D. Wegener]:

$$G_M^*(t) = 1.18(3)\mu_p \left[1 - t/0.56(1) \right]^{-2} .$$

1.6.5 Deep inelastic electron scattering

Inelastic reactions of the type $ep \to e\ldots$ are measured by summing over all elastic and inelastic states (reviews: [Taylor 70, Drees 71, Kendall 71]). The differential cross section is of the form

$$\frac{d^2\sigma}{d\Omega'_{lab} dE'_{lab}} = 4\alpha^2 E'^2_{lab} t^{-2} [2W_1(v, t) \sin^2(\theta_{lab}/2) + W_2(v, t) \cos^2(\theta_{lab}/2)] ,$$

where E'_{lab}, Ω'_{lab} and θ_{lab} are the energy, solid and polar angles of the outgoing electron and v is the energy of the virtual photon, all quantities being defined in the laboratory system. The variables t and v are defined by

$$t = -4E_{lab} E'_{lab} \sin^2(\theta_{lab}/2) , \quad v = E_{lab} - E'_{lab} = (s - m_p^2 - t)/2m_p , \quad s = (P_p + P_e - P'_e)^2 .$$

The functions W_i are related to the transverse and longitudinal cross sections by

$$W_1 = \frac{v + t/2m_p}{4\pi^2\alpha} \sigma_T , \quad W_2 = \frac{v + t/2m_p}{4\pi^2\alpha} \frac{-t}{v^2 - t} (\sigma_T + \sigma_L) .$$

Separation of σ_L and σ_T is possible for $-11\text{ GeV}^2 < t < -1.5\text{ GeV}^2$ and $4\text{ GeV}^2 < s < 16\text{ GeV}^2$. So far, the ratio σ_L/σ_T could be constant $(= 0.18(10))$ [Miller 72], but also $-0.035t$ or $-tv^{-2}$. W_2 apparently depends only on the combination $x = -t/2Mv$ ("scaling behaviour") [Friedman 71]:

$$vW_2(t, v) = (1 - x)^3 [1.274 + 0.5989(1 - x) - 1.675(1 - x)^2] .$$

W_1 shows scaling behaviour too. For neutrons, W_2 exhibits scaling behaviour and becomes considerably smaller than $W_2(p)$ as $-t$ increases [Bloom 70].

1.6.6 SU(3)-comparison of electromagnetic coupling constants

a) The $p\gamma\gamma$ and $pv\gamma$ coupling constants [Gourdin 71], [Mopurgo 71, ref. in 1.4.10]

Let G_{88}, G_{18} and G_{81} be the amplitudes G_{jk} between vector mesons of multiplet j and pseudoscalar mesons of multiplet k. They can be computed from the G_γ-values of 1.4.3, using the signs which agree best with the vector meson dominance (VMD) model (G in units GeV^{-1}):

$$G_{\omega\pi\gamma} = 2.6(3), \quad G_{\phi\pi\gamma} = -0.19(5), \quad G_{\phi\eta\gamma} = -0.98(12)$$

$$G_{88} = 2.1(2), \quad G_{18} = 2.5(1.2), \quad G_{81}^{quad} = 15(4), \quad G_{81}^{lin} = 5.7(1.5) .$$

The last value corresponds to linear mass mixing in the pseudoscalar multiplet, $\theta_p^{lin} = -23°$. The quark model predicts $G_{81} = G_{18} = 2^{\frac{1}{2}} G_{88}$, and VMD predicts $G_{81}/G_{88} \approx 1.3$–$2.8$. The main objection against quadratic mass mixing in the pseudoscalar multiplet is the resulting large $X \to \varrho\gamma$ width. The following table is from [Gourdin 71], with $G_{88} = 2.0(2)$, $G_{18} = 2.8(1.0)$, $G_{81}^{quad} = 7.9$ and $G_{81}^{lin} = 3.2$. Using a fraction of 0.3 for $X \to \pi^+ \pi^- \gamma$ decays, the predicted X width is 5.8 MeV (440 keV) and the fraction of $X \to \gamma\gamma$ decays is 87 keV = 1.5% (9.2 keV = 2.1%) (experimental value (1.7 ± 0.4)%. Numbers for linear mass mixing in parentheses). Using only $g_{\gamma\gamma}$ from 1.4.2a, one gets $g_8 = 0.389(19)\,\text{GeV}^{-1}$, $g_1 = 1.08(27)\,(0.51(12))\,\text{GeV}^{-1}$, and $\Gamma(X \to \gamma\gamma) = 39^{+23}_{-18}\,(5.6^{+3.5}_{-2.7})\,\text{keV}$ [1].

Decay	G_{88}	G_{18}	G_{81}	Γ^{quad} [keV]	Γ^{lin} [keV]
$\varrho^0 \to \pi\gamma$	1	0	0	126	114
$\varrho^0 \to \eta\gamma$	$3^{\frac{1}{2}} \cos\theta_p$	0	$-3^{\frac{1}{2}} \sin\theta_p$	141	108
$\omega \to \pi\gamma$	$3^{\frac{1}{2}} \sin\theta_v$	$3^{\frac{1}{2}} \cos\theta_v$	0	1120	1120
$\omega \to \eta\gamma$	$-\sin\theta_v \cos\theta_p$	$\cos\theta_v \cos\theta_p$	$-\sin\theta_v \sin\theta_p$	12.7	13.7
$\phi \to \pi\gamma$	$3^{\frac{1}{2}} \cos\theta_v$	$-3^{\frac{1}{2}} \sin\theta_v$	0	7.6	7.6
$\phi \to \eta\gamma$	$-\cos\theta_v \cos\theta_p$	$-\sin\theta_v \cos\theta_p$	$-\cos\theta_v \sin\theta_p$	122	93
$\phi \to X\gamma$	$-\cos\theta_v \sin\theta_p$	$-\sin\theta_v \sin\theta_p$	$\cos\theta_v \cos\theta_p$	5.4	1.5
$X \to \omega\gamma$	$-\sin\theta_v \sin\theta_p$	$\cos\theta_v \sin\theta_p$	$\sin\theta_v \cos\theta_p$	175	12.3
$X \to \varrho\gamma$	$3^{\frac{1}{2}} \sin\theta_p$	0	$3^{\frac{1}{2}} \cos\theta_p$	1490	111

[1] $g_{\pi\gamma\gamma} = \left(\tfrac{3}{4}\right)^{\frac{1}{2}} g_8$, $g_{\eta\gamma\gamma} = \tfrac{1}{2}\cos\theta_p g_8 - \sin\theta_p g_1$, $g_{X\gamma\gamma} = \tfrac{1}{2}\sin\theta_p g_8 + \cos\theta_p g_1$.

b) The $\phi\gamma$ and $\omega\gamma$ coupling constants

The experimental $f^2/4\pi$, are given in 1.4.3a. SU(3) predictions imply that the photon couples to the state $\sqrt{3}\,\varrho^0 + \phi_8$, in which event $f^{-1}\begin{pmatrix}\phi\\\omega\end{pmatrix} = \begin{pmatrix}\cos\\\sin\end{pmatrix}\theta_v f_p^{-1} 3^{-\frac{1}{2}}$ according to Eq. (2) in 1.1.2. This establishes $\theta_v = 33.3°$ and 40.2° for the ϕ and ω resonances, respectively, in agreement with $\theta_v = 39.6(7)°$ from the mass formula (Eq. (3) in 1.1.2). The current mixing model (Eq. (2) in 1.1.3) has extra mass factors and does not work as well.

c) Magnetic moments of baryons

As emphasized in 1.3.3, the definition of κ in Eq. (5) in 1.1.3 contains some unmotivated SU(3) breaking through the mass denominator. Define $\kappa = (m/m_p)\,\hat{\kappa}$, where m_p is the proton mass. Then $\hat{\kappa} F_2(t)$ may obey SU(3) symmetry at large t. Deviations for small t due to vector meson states have been calculated by [Divakaran 70], who finds $\mu(\Lambda) = -0.64$, $\mu(\Sigma^+) = 2.61$ as well as the other magnetic moments given in 1.4.6.

1.6.7 References for 1.6

For journal abbreviation see p. 2

Albrecht	66	PRL 17, 1192: W. Albrecht, H. J. Behrend, F. W. Brasse, W. Flauger, H. Hultschig, K. G. Steffen.
Albrecht	70	NP B 25, 1: W. Albrecht, F. W. Brasse, H. Dorner, W. Flauger, K. H. Frank, J. Gayler, V. Krobel, J. May, P. D. Zimmermann, A. Caurau, A. Diaczek, J. C. Dumas, G. Tristram, J. Valentin, A. Aubret, E. Charelas, E. Ganssauge.
Albrecht	71	NP B 27, 615: W. Albrecht, F. W. Brasse, H. Dorner, W. Flauger, K. H. Frank, J. Gayler, V. Korbel, P. D. Zimmermann, A. Caurau, A. Diaczek, J. C. Dumas, G. Tristram, J. Valentin, A. Aubret, E. Charelas, E. Ganssauge.
Bartel	67	PL 25 B, 236: W. Bartel, B. Dudelzak, H. Krehbiel, J. M. McElroy, U. Meyer-Berkhout, R. J. Morrison, H. Nguyen Ngoc, W. Schmidt, G. Weber.
Bartel	69	PL 30 B, 285: W. Bartel, F. W. Büsser, W. R. Dix, R. Felst, D. Harms, H. Krehbiel, P. E. Kuhlmann, W. Schmidt, W. Walter, G. Weber.
Bartel	69a	DESY report F 22, 69–3: W. Bartel.
Bartel	70	PL 33 B, 245: W. Bartel, F. W. Büssner, W. R. Dix, R. Felst, D. Harms, H. Krehbiel, P. E. Kuhlmann, J. McElroy, G. Weber.
Bartel	71	PL 35 B, 181: W. Bartel, F. W. Büsser, W. R. Dix, R. Felst, D. Harms, H. Krehbiel, P. E. Kuhlmann, J. McElroy, J. Meyer, G. Weber.

Pilkuhn

Berger	68	PL 28 B, 276: Ch. Berger, E. Gersing, G. Knop, B. Langenbeck, K. Rith, F. Schumacher.
Berger	69	Univ. Bonn preprint: Ch. Berger, V. Burkert, G. Knop, B. Langenbeck, K. Rith.
Berger	71	PL 35 B, 87: Ch. Berger, V. Burkert, G. Knop, B. Langenbeck, K. Rith.
Bloom	70	SLAC-PUB 796: E. D. Bloom, G. Buschhorn, R. L. Cothell, D. H. Coward, H. de Staebler, J. Dress, C. L. Jordan, G. Miller, L. W. Mo, H. Tiel, R. E. Taylor, M. Breidenbach, W. R. Ditzler, J. I. Friedman, G. C. Hartmann, H. W. Kendall, J. S. Poucher.
Brasse	71	DESY 71–2: F. W. Brasse, W. Fehrenbach, W. Flauger, K. H. Frank, J. Gayler, V. Korbel, J. May, P. D. Zimmermann.
Bumiller	70	PRL 25, 1774: F. A. Bumiller, F. R. Buskirk, J. W. Stewart, E. B. Dally.
Clegg	69	Liverpool Conf. (ed. D. W. Graben, N. P. Daresbury, Lab.): A. B. Clegg.
Coward	68	PRL 20, 292: D. H. Coward, H. de Staebler, R. A. Early, J. Litt, A. Minten, L. W. Mo, W. K. H. Panofsky, R. E. Taylor, M. Breidenbach, J. I. Friedman, H. W. Kendall, P. N. Kirk, B. C. Barish, J. Mar, J. Pine.
Divakaran	70	NP B 15, 601: P. P. Divakaran.
Drees	71	STMP 60, 107: J. Drees.
Driver	71	PL 35 B, 81: C. Driver, K. Heinboth, K. Höhne, G. Hofmann, P. Karow, D. Schmidt, G. Specht, J. Rathje, J. I. Friedmann.
Friedman	71	Trieste Conf.: J. I. Friedman.
Friedman	71 a	SLAC-PUB 907: J. I. Friedman, H. W. Kendall, E. D. Bloom, D. H. Coward, H. de Staebler, J. Drees, C. L. Jordan, G. Miller, R. E. Taylor.
Frosch	67	PR 160, 874: R. F. Frosch, J. S. McCarthy, R. E. Rand, M. R. Yearian.
Galster	71	NP B 32, 221: S. Galster, H. Klein, J. Moritz, K. H. Schmidt, D. Wegener, J. Bleckwenn.
Gayler	71	Daresbury Study Weekend: J. Gayler.
Glaser	57	NC 5, 1197: V. Glaser, B. Jakšić.
Goitein	70	PR 1 D, 2449: M. Goitein, R. J. Budnitz, L. Carroll, J. R. Chen, J. R. Dunning, K. Hanson, D. C. Imrie, C. Mistretta, R. Wilson.
Gourdin	63	NC 28, 533: M. Gourdin.
Gourdin	71	Int. Conf. Meson Reson. and rel. Electromagn. Phenomena, Bologna: M. Gourdin.
Hellings	71	Daresbury NPL-P 65: R. D. Hellings, J. Allison, A. B. Clegg, F. Foster, G. Hughes, P. Kummer, R. Siddle, B. Dickinson, M. Ibbotson, R. Wawson, H. R. Montgomery, W. J. Shuttleworth, A. Sofair, J. Farman.
Hofstädter	67	Landolt-Börnstein, New Series I/2 (ed. H. Schopper): R. Hofstädter, H. R. Collard.
Kendall	71	Int. Symp. on Electron and Photon Interactions (Cornell): H. W. Kendall.
Krohn	66	PR 148, 1303: V. E. Krohn, G. R. Ringo.
McCarthy	70	PRL 25, 884: J. S. McCarthy, I. Sick, R. R. Whitney, M. R. Yearian.
Miller	72	PR D 5, 528: G. Miller, E. D. Bloom, G. Buschhorn, D. H. Coward, H. de Staebler, J. Drees, C. L. Jordan, L. W. Mo, R. E. Taylor, J. I. Friedman, G. C. Hartmann, H. W. Kendall, R. Verdier.
Mistretta	68	PRL 20, 1523: C. Mistretta, D. Imrie, J. A. Appel, R. Bunditz, L. Carrol, M. Gotein, K. Hanson, R. Wilson.
Price	71	PR D 4, 45: L. E. Price, J. R. Dunning jr., M. Goitein, K. Hanson, T. Kirk, R. Wilson.
Reid	72	PRL 29, 494: R. V. Reid, M. L. Vaida.
Rutherglen	69	Liverpool Conf.: J. G. Rutherglen.
Schmidt	71	DESY 71-22: W. Schmidt.
Suelzle	67	PR 162, 992: L. R. Suelzle, M. R. Yearian, H. Crannell.
Taylor	70	Kiev Conf.: R. E. Taylor.
Wilson	70	Kiev Conf.: R. Wilson.
Yang	71	PL 36 B, 1: S. N. Yang, A. D. Jackson.

Pilkuhn

2 Notations, constants and general relations

The symbols most frequently encountered in the following tables are defined in this section. Additional quantities will be defined where it is necessary. For convenience some general relations are also summarized.

2.1 Notation and relations

All data in this volume refer to reactions with two particles in the initial and final state, respectively:

$$1 + 2 \rightarrow 3 + 4 \, .$$

Here particle 1 is moving and particle 2 (target) is stationary in the laboratory system. The kinematical variables of the various particles are, if necessary, distinguished by indices which indicate the number of the particle (1 to 4). In elastic scattering, particle 1 and particle 3 are the same particle and particle 4 is the recoiling target. (Indices referring to particle 3 are occassionally suppressed when no confusion exists, e.g. $\theta = \theta_3$ or $\Omega = \Omega_3$.)

Quantities defined in the center of mass system (c.m.) are denoted by an asterisk, whereas quantities in the laboratory system (lab.) are asterisk free.

The symbols commonly used in this compilation are listed in the following table.

The $c = \hbar = h/2\pi = 1$ system of units is used.

2.1.1 Variables of particle i

	Center of mass system	Laboratory system
momentum (3-dimensional)	$\vec{p}_i^{\,*}$	$\vec{p}_i$
kinetic energy	T_i^*	T_i
total energy	E_i^*	E_i
4-momentum	$P_i^* = (E_i^*, \vec{p}_i^{\,*})$	$P_i = (E_i, \vec{p}_i)$
scattering angle	θ_i^*	θ_i
rest mass of particle	m_i	m_i
solid angle	Ω^*	Ω
velocity/velocity of light	β_i^*	β_i
transverse momentum	p_T	p_T

The following relations hold (in c.m. or lab.):

$$P^2 = E^2 - p^2 \, ,$$
$$E^2 = p^2 + m^2 \, ,$$
$$T = E - m \, .$$

2.1.2 Cross sections

σ_{tot} total cross section for $1 + 2 \rightarrow$ anything

σ_{int} total ($=$ integrated) cross section for $1 + 2 \rightarrow 3 + 4$

σ_{el} total ($=$ integrated) cross section for $1 + 2 \rightarrow 1 + 2$

σ_{ch} total ($=$ integrated) charge exchange cross section

$\dfrac{d\sigma}{d\Omega}$ differential cross section in lab. system

$\dfrac{d\sigma}{d\Omega^*}$ differential cross section in c.m. system

$\dfrac{d\sigma}{dt}$ invariant differential cross section.

2.1.3 Polarization

For the definition of the polarization parameters P, A and R see Vol. I/7.

Carlson et al.

2.1.4 Relativistic invariants

The invariants s, t and u are defined and expressed in c.m. and lab. variables. The target particle m_2 is assumed to be at rest in the lab. system in deriving these expressions.

total c.m. energy squared

$$s = (P_1 + P_2)^2 = (E_1^* + E_2^*)^2 = m_1^2 + m_2^2 + 2E_1 m_2$$

4-momentum transfer 1–3 squared

$$t = (P_1 - P_3)^2 = m_1^2 + m_3^2 - 2E_1^* E_3^* + 2p_1^* p_3^* \cos\theta_3^* = m_1^2 + m_3^2 - 2E_1 E_3 + 2p_1 p_3 \cos\theta_3$$
$$= (P_2 - P_4)^2 = m_2^2 + m_4^2 - 2E_2^* E_4^* + 2p_2^* p_4^* \cos\theta_4^* = (m_2 - m_4)^2 - 2m_2(E_4 - m_4)$$

4-momentum transfer 1–4 squared

$$u = (P_1 - P_4)^2 = m_1^2 + m_4^2 - 2E_1^* E_4^* + 2p_1^* p_4^* \cos\theta_4^* = m_1^2 + m_4^2 - 2E_1 E_4 + 2p_1 p_4 \cos\theta_4$$
$$= (P_2 - P_3)^2 = m_2^2 + m_3^2 - 2E_2^* E_3^* + 2p_2^* p_3^* \cos\theta_3^* = (m_2 - m_3)^2 - 2m_2(E_3 - m_3) .$$

These 3 invariants satisfy the relation

$$s + t + u = \sum_{i=1}^{4} m_i^2 .$$

For elastic scattering $m_1 = m_3$ and $m_2 = m_4$, so that

$$p_1^* = p_2^* = p_3^* = p_4^* \equiv p^*$$

and

$$E_1^* = E_3^* ; \qquad E_2^* = E_4^* .$$

In this case the expressions for t and u may be simplified:

$$t = -4p^{*2} \sin^2 \tfrac{1}{2}\theta_3^* = -2m_2(E_4 - m_4) = -2m_2(E_1 - E_3)$$
$$u = -2p^{*2}(1 + \cos\theta_3^*) + \frac{(m_1^2 - m_2^2)^2}{s} .$$

2.1.5 Lorentz transformations

The target m_2 is assumed to be at rest.
The c.m. velocity $\beta_{c.m.}$ and the Lorentz factor $\gamma_{c.m.} = 1/\sqrt{1 - \beta_{c.m.}^2}$.

$$\beta_{c.m.} = \frac{p_1}{E_1 + m_2} \qquad \gamma_{c.m.} = \frac{E_1 + m_2}{\sqrt{s}}$$

where, as before

$$s = m_1^2 + m_2^2 + 2E_1 m_2 = (m_1 + m_2)^2 + 2T_1 m_2 .$$

The Lorentz transformations for momentum and energy are then written as

$$p_T = p_i \sin\theta_i = p_i^* \sin\theta_i^*$$

$$\begin{cases} p_i \cos\theta_i = \gamma_{c.m.}(p_i^* \cos\theta_i^* + \beta_{c.m.} E_i^*) \\ E_i = \gamma_{c.m.}(E_i^* + \beta_{c.m.} p_i^* \cos\theta_i^*) \end{cases} \quad \text{or} \quad \begin{cases} p_i^* \cos\theta_i^* = \gamma_{c.m.}(p_i \cos\theta_i - \beta_{c.m.} E_i) \\ E_i^* = \gamma_{c.m.}(E_i - \beta_{c.m.} P_i \cos\theta_i) . \end{cases}$$

The following relation is often useful

$$p_1^* = p_2^* = \frac{p_1 m_2}{\sqrt{s}} .$$

For $i = 3, 4$, the relation between the angles θ_i and θ_i^* is given by

$$\operatorname{tg}\theta_i = \frac{1}{\gamma_{c.m.}} \operatorname{tg}\tfrac{1}{2}\theta^* \cdot \frac{1 + \cos\theta_i^*}{X_i^* + \cos\theta_i^*} = \frac{1}{\gamma_{c.m.}} \frac{\sin\theta_i^*}{X_i^* + \cos\theta_i^*}$$

with

$$X_i^* = \frac{\beta_{c.m.}}{\beta_i^*} .$$

Carlson et al.

For i $= 2 = 4$, in case of elastic scattering, $\beta_4^* = \beta_2^* = \beta_{\text{c.m.}}$, thus $X_4^* = 1$ and

$$X_3^* = \frac{E_1 m_2 + m_1^2}{E_1 m_2 + m_2^2},$$

so that the angle transformation reduces to

$$\operatorname{tg}\theta_i = \frac{1}{\gamma_{\text{c.m.}}}\operatorname{tg}\tfrac{1}{2}\theta_i^* \quad (i = 2 = 4).$$

In the high energy limit this result is also valid for particle $i = 1 = 3$. The differential cross sections $\dfrac{d\sigma}{dt}$, $\dfrac{d\sigma}{d\Omega^*}$ and $\dfrac{d\sigma}{d\Omega}$ are related by

$$\frac{d\sigma}{dt_i} = \frac{\pi}{p_1^* p_i^*} \cdot \frac{d\sigma}{d\Omega_i^*}$$

$$\frac{d\sigma}{d\Omega_i^*} = \left(\frac{p_i^*}{p_i}\right)^2 \sqrt{1 - \left(\frac{p_2^* m_i}{p_i^* m_2}\sin\theta_i\right)^2} \cdot \frac{d\sigma}{d\Omega_i}.$$

In the case of elastic scattering this reduces to

$$\frac{d\sigma}{dt_i} = \frac{\pi}{p_i^{*2}} \cdot \frac{d\sigma}{d\Omega_i^*} = \frac{\pi}{p_i^2}\sqrt{1 - \left(\frac{m_i}{m_2}\sin\theta_i\right)^2} \cdot \frac{d\sigma}{d\Omega_i}.$$

2.1.6 Diffraction scattering

The angular distribution of many 2-body reactions exhibits a pronounced forward diffraction peak. In this case, the diffractional cross section is often parametrized by assuming

$$\text{a linear } t\text{-dependence} \qquad \frac{d\sigma}{dt} = A \cdot e^{Bt}$$

or

$$\text{a quadratic } t\text{-dependence} \qquad \frac{d\sigma}{dt} = A' \cdot e^{B't + C't^2}.$$

The constants A, B and A', B' and C' are determined from a fit to the experimental data.

2.1.7 Optical theorem

When the spin dependence can be neglected, the imaginary part of the elastic forward scattering amplitude $f(E, t)$ is related to the total cross section by

$$\operatorname{Im} f(E, 0) = \frac{p^*}{4\pi}\sigma_{\text{tot}}.$$

Hence,

$$\left(\frac{d\sigma}{d\Omega}\right)_{\theta = 0} = |\operatorname{Re} f(E, 0)|^2 + \frac{p^{*2}}{16\pi^2}\sigma_{\text{tot}}^2$$

$$\left(\frac{d\sigma}{d|t|}\right)_{t = 0} = \frac{\pi}{p^{*2}}|\operatorname{Re} f(E, 0)|^2 + \frac{\sigma_{\text{tot}}^2}{16\pi}.$$

At high energies, $|\operatorname{Re} f / \operatorname{Im} f|^2$ is small and one obtains

$$\left(\frac{d\sigma}{d|t|}\right)_0 \approx \frac{5.095 \cdot 10^{-2}}{\text{mb} \cdot (\text{GeV/c})^2}\sigma_{\text{tot}}^2$$

to a good approximation.

Carlson et al.

2.2 Units and constants

2.2.1 Symbols

Symbol	Particle	Mass [GeV/c^2]	I	J^P
$\pi^\pm$	π-meson, pion	0.139576 ± 0.000011	1	0^-
π^0	π-meson, pion	0.134972 ± 0.000012	1	0^-
$K^\pm$	K-meson, kaon	0.49384 ± 0.00011	$\frac{1}{2}$	0^-
K^0	K-meson, kaon	0.49779 ± 0.00015	$\frac{1}{2}$	0^-
p	proton	0.9382592 ± 0.0000052	$\frac{1}{2}$	$\frac{1}{2}^+$
n	neutron	0.9395527 ± 0.0000052	$\frac{1}{2}$	$\frac{1}{2}^+$
d	deuteron	1.8755	0	1
N	nucleon			

2.2.2 Units

Quantity	Unit and conversion
cross section	barn (b), $\quad 1\,\mathrm{b} = 10^{-24}\,\mathrm{cm}^2$
mass	eV/c^2, $\quad 1\,\mathrm{eV/c}^2 = 1.781 \cdot 10^{-36}\,\mathrm{kg}$
momentum	eV/c, $\quad 1\,\mathrm{eV/c} = 5.341 \cdot 10^{-28}\,\mathrm{Nsec}^{-1}$
energy	eV, $\quad 1\,\mathrm{eV} = 1.602 \cdot 10^{-19}\,\mathrm{J}$
fermi (femtometer)	$1\,\mathrm{fm} = 10^{-13}\,\mathrm{cm}$
velocity of light	$c = 2.9979250 \cdot 10^{10}\,\mathrm{cm\ sec}^{-1}$
Planck's constant	$\hbar = h/2\pi = 6.582183 \cdot 10^{-22}\,\mathrm{MeV\ sec}$
	$\hbar c = 197.329\,\mathrm{MeV\ fermi}$

2.3 Abbreviations for experimental techniques

A	Activation methods
BC	Bubble chamber
C	Scintillation counter
CC, CLC	Cloud chamber
CTR	Counters
DBC	Deuterium bubble chamber
DCC	Diffusion cloud chamber
DSC	Double scattering
E, EM	Emulsions
EBC	Helium bubble chamber
FC	Fission counter
HBC	Hydrogen bubble chamber
IC	Ionisation chamber
LC	Luminescent chamber
MMS	Missing mass spectrometer
PBC	Propane bubble chamber
PC	Proportional counter
POLT	Polarized target
SC, SPC	Spark chamber
SSC	Solid-state counter
TSC	Triple scattering
XBC	Heavy liquid bubble chamber

3 Particle production in proton-proton interactions

3.1 Introduction

3.1.1 General remarks

This paper gives a compilation of cross sections for the *production of charged pions, kaons, and nucleons in proton-proton collisions*. The production process can formally be written as

$$p + p \rightarrow c + X, \tag{1}$$

where c is the detected particle and X denotes all non-detected reaction products. Processes of this type have recently been called inclusive reactions [69 F 1].

Experiments on particle production are usually among the first ones to be performed on a new accelerator for the very practical reason that information on particle fluxes is needed for the design of secondary particle beams. The absolute accuracy of these measurements is usually rather poor, because the number of interactions of an accelerator beam in the internal target is difficult to determine and the acceptance of the experimental set-up is often badly known. Moreover, mostly nuclear targets are used and elementary nucleon cross sections cannot be deduced unambiguously. Results on nuclear targets have been disregarded in the present compilation. Typical examples of data of this type can be found in papers by Baker *et al.* [61 B 1] on experiments at 30 GeV/c, Lundy *et al.* [65 L 1] at 12 GeV/c, Bushnin *et al.* [69 B 2] and Binon *et al.* [69 B 3] at 70 GeV/c, Allaby *et al.* [70 A 1] at 19.2 GeV/c, Cho *et al.* [71 C 8] at 12.4 GeV/c and Gorin *et al.* [71 G 3] at 70 GeV/c.

Data on *particle production with incident pions and kaons* is, in principle, available from many hydrogen bubble chamber experiments, and the inclusive type of analysis was recently given some attention: see, for instance, the following papers:

Elbert *et al.* [68 E 1, 70 E 1] on $\pi^- p \rightarrow \pi^- X$ at 25 GeV/c,
Piotrowska [70 P 1] on $\pi p \rightarrow \pi X$ or NX between 8 and 16 GeV/c,
Beaupré *et al.* [71 B 6] on $\pi^+ p \rightarrow \pi^\pm X$ at 8 and 16 GeV/c,
Foster *et al.* [71 F 1] on $K^- p \rightarrow \pi^\pm X$ at 9 GeV/c,
Ko and Lander [71 K 1] on $K^+ p \rightarrow \pi^- X$ at 11.8 GeV/c,
Biswas *et al.* [71 B 3] on $\pi^\pm p \rightarrow \pi^- X$ at 18.5 GeV/c,
Beaupré *et al.* [71 B 5] on $K^\pm p \rightarrow K^0 X$ between 5 and 10 GeV/c,
Stone *et al.* [71 S 3] on $K^+ p \rightarrow \pi^- X$ at 12.7 GeV/c and $\pi^+ p \rightarrow \pi^- X$ at 7 GeV/c,
Swanson *et al.* [71 S 1] on $\gamma p \rightarrow \pi^- X$ between 5 and 15 GeV/c, and
Chen *et al.* [71 C 5] where several reactions are compared.

Counter experiments with these projectiles are rare; the only example seems to be a study of the reaction $\pi^- p \rightarrow \pi^\pm X$ in the backward direction at 5 GeV/c [71 A 2].

The subject of this paper is restricted to reactions initiated by proton projectiles, because only for this case, i.e. reaction (1), does an abundance of data exist, both from counter experiments and from hydrogen bubble chamber exposures.

Review articles on particle production have been given by: Turkot [68 T 1], mainly concerned with proton-proton reactions; by Panvini [69 P 1], also with emphasis on proton-proton reactions; and by Erwin [70 E 2], mainly concerned with results in 25 GeV/c $\pi^- p$ reactions but nevertheless quite general. Review papers that treat inclusive reactions in general are for instance those of Ranft and Ranft [71 R 5], Quigg [71 Q 1], Van Hove [71 H 2], Deutschmann [71 D 3], Horn [71 H 3] and Berger [71 B 4]. Material on particular reaction channels produced in proton-proton interactions and their partial cross sections can be found in the tabulations of Hansen *et al.* [70 H 1] and Benary *et al.* [70 B 3].

The list of literature and the data collection cited in this paper were compiled up to the end of 1971.

3.1.2 Definition of kinematical variables

The notation of Chapter 2 of this volume will be used. Thus index 1 is assigned to the projectile, index 2 to the target, which is assumed to be at rest in the laboratory system, index 3 to the detected particle, and index 4 to the configuration X.

The direct result of a counter experiment is the double differential cross section $d^2\sigma/d\Omega_3 dp_3 \ (\theta_3, p_3)$, where $(d\Omega_3) = 2\pi \sin\theta_3 d\theta_3$ is the solid angle, and θ_3 and p_3 the angle and momentum in the laboratory system. The cross section is often also expressed in other variables such as, for example, the following.

i) p_3^*, θ_3^*. The Lorentz transformation of p_3, θ_3 into these c.m.s. variables is given in Chapter 2 of this volume.

ii) p_T, p_L, p_L^*. The transverse momentum p_T and the longitudinal momenta p_L and p_L^* are defined by:

$$p_T = p_3 \sin\theta_3 = p_3^* \sin\theta_3^*$$
$$p_L = p_3 \cos\theta_3 \tag{2}$$
$$p_L^* = p_3^* \cos\theta_3^* \, .$$

These variables are of interest, because the functional dependence of the cross sections on momentum and angle can be approximately factorized into a dependence on p_T and a dependence on p_L. A graphical representation in a (p_T, p_L^*) plane is called a Peyron plot.

iii) x, x^*. The reduced longitudinal moments x and x^* [69 F 1] are often used in comparing results at different incident momenta. Their definitions are

$$x^* = \frac{p_L^*}{p_{3,\max}^*}, \qquad x = \frac{p_L}{p_{3,\max}} \, . \tag{3}$$

Other definitions are $x'^* = p_L^*/p_1^*$, $x''^* = p_L^*/p_{L,\max}^*$ and $x'''^* = 2p_L^*/\sqrt{s}$. At asymptotic energies and $p_T \ll p_L^*$ these definitions lead to identical results; however, at 20 GeV and small p_T, the numerical values differ by 10% or more; $p_{3,\max}^*$ is given by

$$p_{3,\max}^* = \frac{1}{2\sqrt{s}} \sqrt{\{s - (m_3 + m_{4,\min})^2\}\{s - (m_3 - m_{4,\min})^2\}} \, , \tag{4}$$

where $m_{4,\min}$ is the minimum value of m_4, allowed by the conservation laws of baryon number, charge, and strangeness: for $m_3 = \pi^-, \pi^+, K^-, K^+, \bar{p}$ and p, one has $m_{4,\min} = 2p + \pi^+, p + n, 2p + K^+, \Lambda + p, 3p$ and p, respectively. With this definition, x^* is bounded by

$$-\sqrt{1 - \frac{p_T^2}{p_{3,\max}^{*\,2}}} \leqq x^* \leqq \sqrt{1 - \frac{p_T^2}{p_{3,\max}^{*\,2}}} \, . \tag{5}$$

iv) y, y^*. The rapidity y_3 for particle 3 is defined by

$$y_3 = \operatorname{arc\,sinh}\frac{p_L}{m_T} = \operatorname{arc\,cosh}\frac{E_3}{m_T} = \operatorname{arc\,tanh}\frac{p_L}{E_3} = \frac{1}{2}\ln\left(\frac{E_3 + p_L}{E_3 - p_L}\right) \tag{6}$$

and similarly for y_3^* in the c.m. system. The transverse mass is given by

$$m_T = \sqrt{m_3^2 + p_T^2} \, . \tag{7}$$

Analogous definitions hold for particles 1, 2 and 4.

Under Lorentz transformations y transforms in the following way:

$$y^* = y + \ln\sqrt{\frac{1 - \beta_{\text{c.m.}}}{1 + \beta_{\text{c.m.}}}} = y - \ln\{\gamma_{\text{c.m.}}(1 + \beta_{\text{c.m.}})\} \, . \tag{8}$$

Thus, $y_3 - y_1$ and $y_3 - y_2$ are Lorentz invariants.

At asymptotic energies y_3^* varies between the following limits:

$$y_{3,\min}^* \equiv -\frac{1}{2}\ln\left(\frac{s}{m_{T2}}\right) \leqq y_3^* \leqq +\frac{1}{2}\ln\left(\frac{s}{m_{T2}}\right) \equiv y_{3,\max}^* \, . \tag{9}$$

The rapidity magnifies the c.m. region around $x^* = 0$ and is therefore useful in the study of the production of low-energy pions. A rapidity spectrum is simply translated but not deformed by a Lorentz translation.

v) ξ_3. The reduced rapidity ξ_3 has been introduced by Van Hove [71 H 2] as:

$$\xi_3 = \frac{y_3 - y_2}{y_1 - y_2} \, . \tag{10}$$

ξ_3 is a Lorentz invariant. $\xi_3 = 1$ for $x^* = 1$, $\xi_3 = \frac{1}{2}$ for $x^* = 0$ and $\xi_3 = 0$ for $x^* = -1$ in the limit $s \to \infty$.

vi) t, m_4. The missing mass m_4 and the square of the four-momentum transfer t are defined as:

$$m_4^2 = E_4^{*\,2} - \vec{p}_4^{*\,2} = s - 2\sqrt{s}\,E_3^* + m_3^2 \, , \tag{11}$$

$$t = (P_1 - P_3)^2 = (E_1 - E_3)^2 - (\vec{p}_1 - \vec{p}_3)^2 \quad \text{(lab. or c.m.)}$$
$$= t_{\min} + t' \, , \tag{12}$$

Diddens/Schlüpmann

where

$$t_{\min} = m_1^2 + m_3^2 - 2(E_1 E_3 - p_1 p_3)$$

$$= m_1^2 + m_3^2 - \frac{(E_1^2 + p_1^2)\, m_3^2 + (E_3^2 + p_3^2)\, m_1^2}{E_1 E_3 + p_1 p_3} \qquad \text{(lab. or c.m.)}, \tag{13}$$

$$t' = -4 p_1 p_3 \sin^2 \tfrac{1}{2}\theta_3 \qquad \text{(lab. or c.m.)}. \tag{14}$$

A number of relations exist between these variables. An exact relation that can be derived from the four-vector relation $P_4^2 = (\sqrt{-t} + P_2)^2$ is

$$2 m_2 (E_1 - E_3) = m_4^2 - m_2^2 - t \qquad \text{(lab. only)}. \tag{15}$$

The relation between x'^* and y_3^* is given by

$$y_3^* = \operatorname{arc\,sin} h \, \frac{x'^* p_1^*}{m_{\mathrm{T}}} \xrightarrow{s \to \infty} \operatorname{arc\,sin} h \, \frac{x^* \sqrt{s}}{2 m_{\mathrm{T}}}. \tag{16}$$

For $s \to \infty$ this can be approximated by

$$x^* \approx +1, \qquad y_3^* \approx \ln\!\left(\frac{x^* \sqrt{s}}{m_{\mathrm{T}}}\right) = \frac{1}{2} \ln\!\left(\frac{s}{m_{\mathrm{T}^2}}\right) + \ln x^*, \tag{17}$$

$$x^* \sqrt{s} \approx 0, \qquad y_3^* \approx \frac{x^* \sqrt{s}}{2 m_{\mathrm{T}}}, \tag{18}$$

$$x^* \approx -1, \qquad y_3^* \approx -\ln\!\left(\frac{|x^*| \sqrt{s}}{m_{\mathrm{T}}}\right) = -\frac{1}{2} \ln\!\left(\frac{s}{m_{\mathrm{T}^2}}\right) - \ln|x^*|. \tag{19}$$

The relation between these variables is illustrated in Fig. 1 [71 D 2].

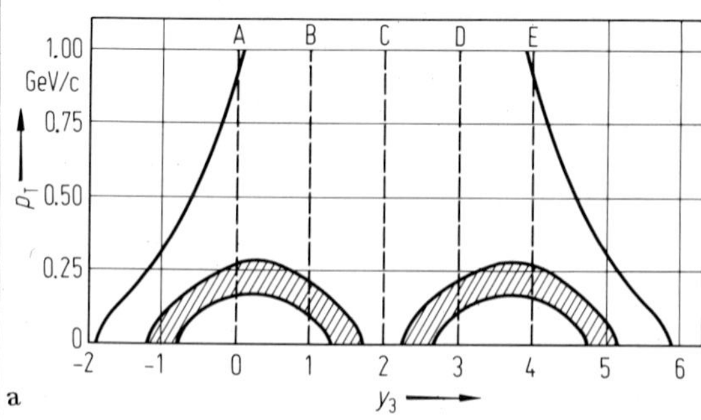

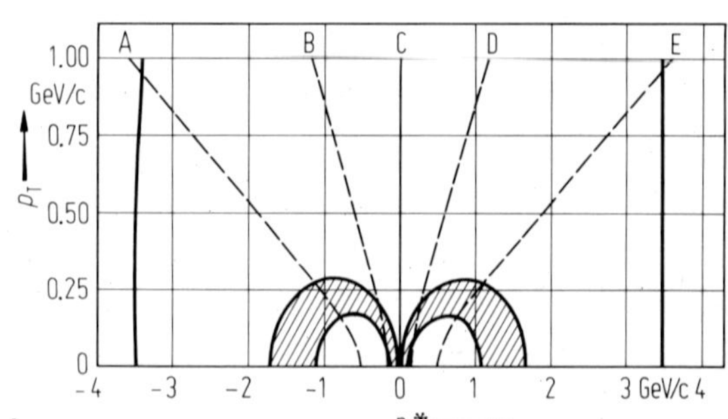

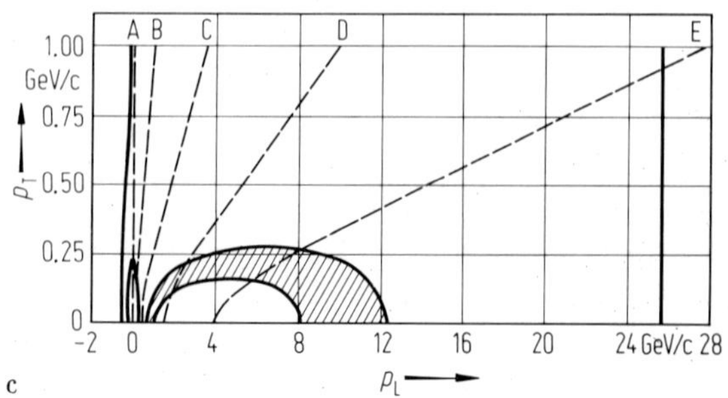

Fig. 1a⋯c. Comparison of the lab. rapidity plot (a), the c.m. Peyron plot (b) and the lab. frame (c). The phase-space boundary for the process $pp \to \pi X$ at 25.6 GeV/c is indicated in all three figures. The shaded band shows the region of phase space populated by pions that originate from Δ_{1236} decay, $pp \to \Delta\Delta$, $\Delta \to N\pi$. The Δ is produced in the near-forward direction and has a width of 120 MeV. The letters A to E denote lines of constant rapidity [71 D 2].

At high energies and small angles, the following simplifications can be made in the expressions for t and m_4:

$$E_1 \approx p_1, \qquad E_3 \approx p_3, \qquad 2 \sin\tfrac{1}{2}\theta_3 \approx \sin\theta_3, \tag{20}$$

$$m_4^2 \approx s(1 - |x^*|). \tag{21}$$

$$t_{\min} \approx m_1^2 (1 - |x^*|) - m_3^2 \left(\frac{1 - |x^*|}{|x^*|}\right), \tag{22}$$

$$t' \approx -\frac{p_{\mathrm{T}}^2}{|x^*|}. \tag{23}$$

Thus the part $t_{\min}$ of the four-momentum transfer squared is uniquely determined by the mass m_4 at some fixed momentum p_1; the angular dependence is all in t'.

The relation between these various variables, $p_\mathrm{T}, p_\mathrm{L}^*, x^*, y_3^*, t, t'$ and m_4 is sketched in Fig. 2.

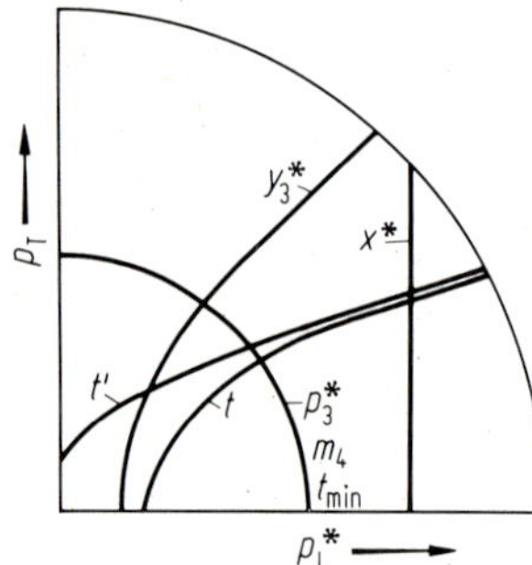

Fig. 2. The locus of various kinematical quantities on a Peyrou plot.

The Lorentz invariant phase space element $\mathrm{d}^3\vec{p}_3/E_3$ defines the transformation of the double differential cross sections [51 L 1]. Thus

$$E_3 \frac{\mathrm{d}^3\sigma}{\mathrm{d}^3\vec{p}_3} \equiv \frac{E_3}{p_3^2}\frac{\mathrm{d}^2\sigma}{\mathrm{d}\Omega_3\,\mathrm{d}p_3} = \frac{E_3}{\pi}\frac{\mathrm{d}^2\sigma}{\mathrm{d}p_\mathrm{T}^2\,\mathrm{d}p_\mathrm{L}} = \frac{1}{\pi}\frac{\mathrm{d}^2\sigma}{\mathrm{d}p_\mathrm{T}^2\,\mathrm{d}y_3} \tag{24}$$

are Lorentz invariant expressions.

Integration of the cross section over the momentum space gives

$$\iint \frac{\mathrm{d}^2\sigma}{\mathrm{d}\Omega_3\,\mathrm{d}p_3}\,\mathrm{d}\Omega_3\,\mathrm{d}p_3 = \langle n_3\rangle\,\sigma_{\mathrm{abs}}, \tag{25}$$

where $\langle n_3\rangle$ is the mean multiplicity of the produced particle and $\sigma_{\mathrm{abs}} = \sigma_{\mathrm{tot}} - \sigma_{\mathrm{el}}$ is the absorption cross section, which is the difference of the total cross section σ_{tot} and the elastic cross section σ_{el}. Summing over all produced charged particles, and integrating also the elastically scattered particles results in the definition of the mean charged particle multiplicity $\langle n_{\mathrm{ch}}\rangle$

$$\sum_{m_{3,\mathrm{ch}}} \iint \frac{\mathrm{d}^2\sigma}{\mathrm{d}\Omega_3\,\mathrm{d}p_3}\,\mathrm{d}\Omega_3\,\mathrm{d}p_3 = \langle n_{\mathrm{ch}}\rangle\,\sigma_{\mathrm{tot}}. \tag{26}$$

The invariant cross section expressed in t and m_4 is

$$\frac{\mathrm{d}^2\sigma}{\mathrm{d}t\,\mathrm{d}m_4^2} = \frac{\pi}{\lambda(s, m_1, m_2)}\left(\frac{E_3}{p_3^2}\frac{\mathrm{d}^2\sigma}{\mathrm{d}\Omega_3\,\mathrm{d}p_3}\right) \tag{27}$$

with

$$\lambda(s, m_1, m_2) = \sqrt{\{s - (m_1 + m_2)^2\}\{s - (m_1 - m_2)^2\}}$$
$$= 2p_1 m_2 = 2p_1^*\sqrt{s} \tag{28}$$
$$\underset{s\to\infty}{=} s.$$

A measurement of a high-energy lab. spectrum, corresponding to particles produced forward in the c.m. system, is simultaneously a measurement of a low-energy lab. spectrum, corresponding to particles emitted backward in the c.m. system, because of the symmetry of the proton-proton system around $\theta_3^* = 90°$.

3.1.3 Acknowledgements

We appreciate the readiness of Drs. H. Bøggild, R. M. Edelstein, D. B. Smith and Z. Ming Ma to contribute unpublished material for tabulation. We thank Dr. G. Cocconi for making some critical remarks on the manuscript. One of us (K. S.) thanks the Fachbereich Physik der F. U. Berlin for its hospitality.

3.2 Discussion of the data

3.2.1 $\pi^\pm$, $K^\pm$, $p^\pm$ data

The threshold momentum needed for pion production in pp collisions is 1.22 GeV/c for negative pions and 0.79 GeV/c for positive pions. Particle production in the threshold region has not been included in our tables. It has been reviewed, for example, by Gell-Mann [54 G 1] and by Marshak [52 M 1].

The proton exposures in hydrogen bubble chambers and the counter experiments that are of relevance for single particle spectra are summarized in 3.5, Tab. 2. In the last column are marked the experiments of which data have been included in 3.6. More general information on particular channels of particle production can be found in the tabulations of Hansen et al. [70 H 1] and Benary et al. [70 B 3].

Reproduced in 3.6 are the $\pi^\pm$ data of Melissinos et al. at 3.7 GeV/c [62 M 1], the K^+ data of Reed et al. at 3.7 and 3.2 GeV/c [68 R 1], the K^+ data of Hogan et al. at 3.3, 3.7 and 3.9 GeV/c [68 H 1] and the p data of Abolins et al. at 6 GeV/c [70 A 3]. [70 A 3] measured separately the spectra of protons produced together with pions only, and of those produced with kaons only; the data have been added by us. Six and higher prongs have been neglected; they should contribute not more than 5%. The above-mentioned data is all that is known to us below 10 GeV/c primary momentum. The kaon spectra are at momenta close to the threshold for kaon production (2.34 GeV/c for K^+, 3.30 GeV/c for K^-). They have been included in the table because of lack of other data below 10 GeV/c.

Akerlof et al. [71 A 1] measured an extensive set of $\pi^\pm$ and proton spectra at 12.5 GeV/c and also obtained data on $K^\pm$ and $\bar{p}$ production. Lamb et al. [66 L 1] measured spectra of $\pi^\pm$ emitted under 180°; however, the points near the kinematical limit seem to disagree with more recent results.

In the 20–30 GeV/c region one of the first experiments was made by Diddens et al. [64 D 1], who measured particle production at 19 and 24 GeV/c at a lab. angle of 115 mrad. The low-energy points seem to be incorrect. Dekkers et al. [65 D 1] published $\pi^\pm$, $K^\pm$ and $p^\pm$ spectra at 18.8 and 23 GeV/c, at lab. angles of 0 and 100 mrad; compared to more recent data these results seem to be too low by a factor ≈ 1.5. The results of these last three experiments ([66 L 1, 64 D 1, 65 D 1]) have not been included in 3.6. Anderson et al. [67 A 1] measured pion and proton spectra at a few angles, mostly at 30 GeV/c. A large set of data for secondary $\pi^\pm$, $K^\pm$, and $p^\pm$ was obtained by Allaby et al. at 19.2 GeV/c [70 A 1] and at 24.0 GeV/c [72 A 1], together with some measurements at 14.25 GeV/c. Contributions from bubble chambers in this energy range are from the experiments of Bøggild et al. at 19.2 GeV/c [71 B 1, 71 B 2], Smith et al. [69 S 1] at five momenta between 13 and 28 GeV/c, Sims et al. [71 S 2] at 28.5 GeV/c and Mück et al. [72 M 1] at 24.0 GeV/c. The first two experiments give only spectra on π^-, the last two also on positive particles.

Very recently, particle spectra have been measured at the CERN Intersecting Storage Rings (ISR) [71 R 1, 71 N 1, 72 B 3, 72 B 4] at equivalent lab. energies of about 1000 GeV. Many more data can be expected from the ISR in the near future and the interested reader is advised to consult conference proceedings, for instance [72 O 1].

The kinematical range of the three main counter experiments in the field, [71 A 1, 67 A 1, 70 A 1, 72 A 1] is illustrated in a p_T, p_L^* diagram in Fig. 3. Typical experimental results of Akerlof et al. [71 A 1] are shown in Fig. 4, those of Anderson et al. [67 A 1] in Fig. 5, those of Allaby et al. [70 A 1, 72 A 1] in Fig. 6, the ISR results of Ratner et al. [71 R 1] and Bertin et al. [72 B 3] in Fig. 7 and Fig. 8, respectively.

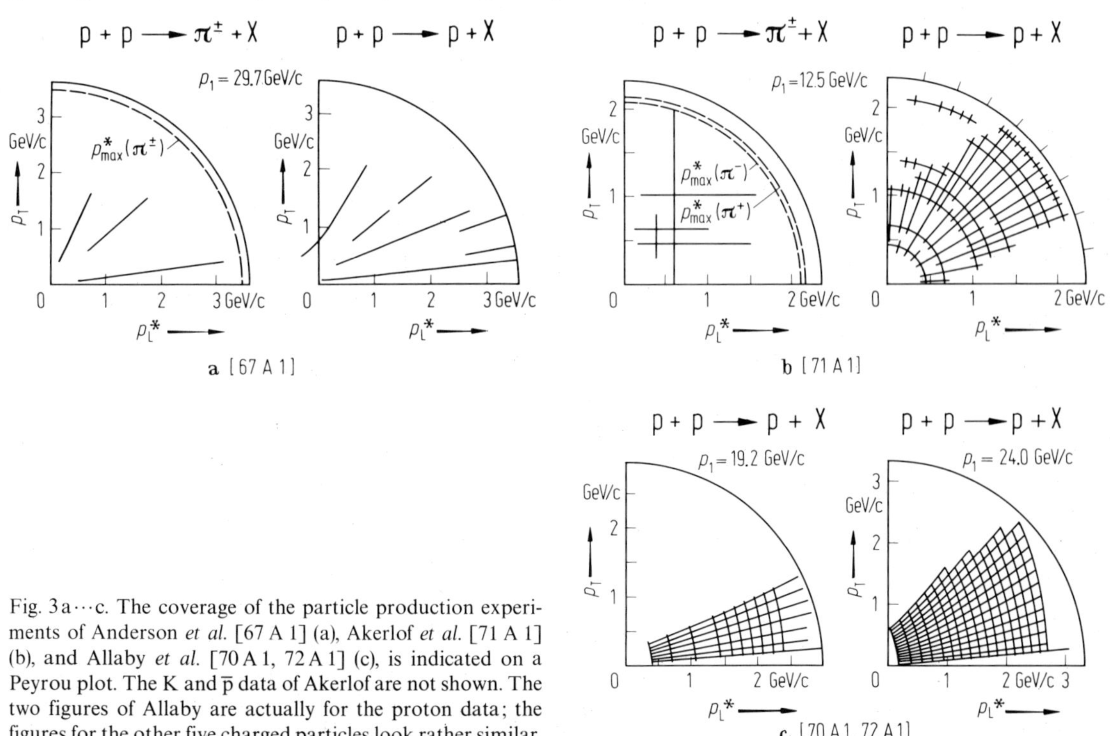

Fig. 3a···c. The coverage of the particle production experiments of Anderson et al. [67 A 1] (a), Akerlof et al. [71 A 1] (b), and Allaby et al. [70 A 1, 72 A 1] (c), is indicated on a Peyrou plot. The K and $\bar{p}$ data of Akerlof are not shown. The two figures of Allaby are actually for the proton data; the figures for the other five charged particles look rather similar.

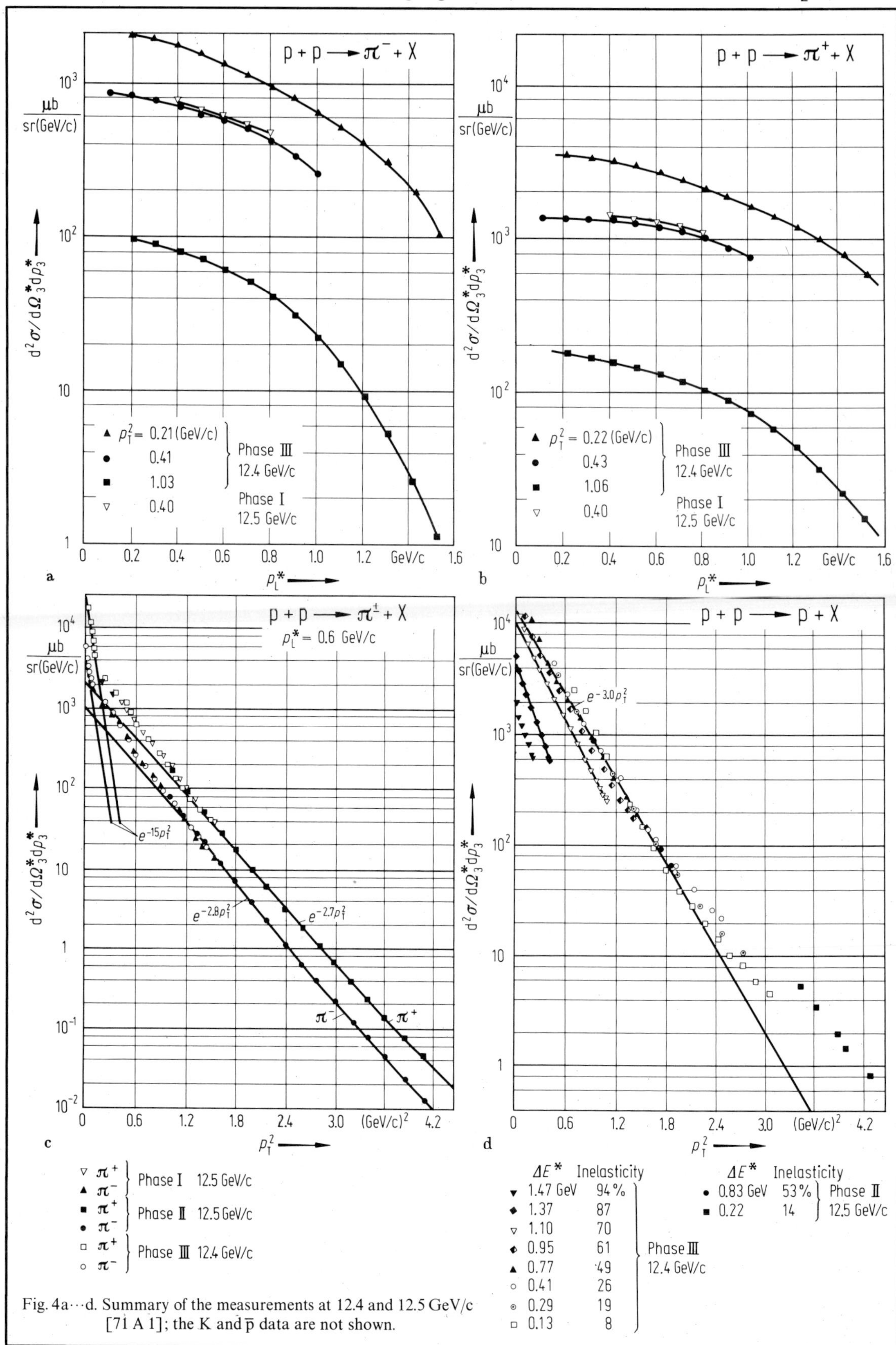

Fig. 4a···d. Summary of the measurements at 12.4 and 12.5 GeV/c
[71 A 1]; the K and p̄ data are not shown.

Diddens/Schlüpmann

It should be realized that in counter experiments spectra are normally measured with a momentum resolution that is typically around $\Delta p/p \approx \pm 1\%$. In bubble chamber experiments the data have usually been sampled in much larger bins.

The data are collected in 3.6; Tab. 2 in 3.5 gives some general information on the accepted experiments, especially the scale or systematic error to be applied. For all data in 3.6 lab. angles corresponding to $\theta_3^* > 90°$ have been converted to lab. angles corresponding to $180 - \theta_3^*$ in the c.m. system.

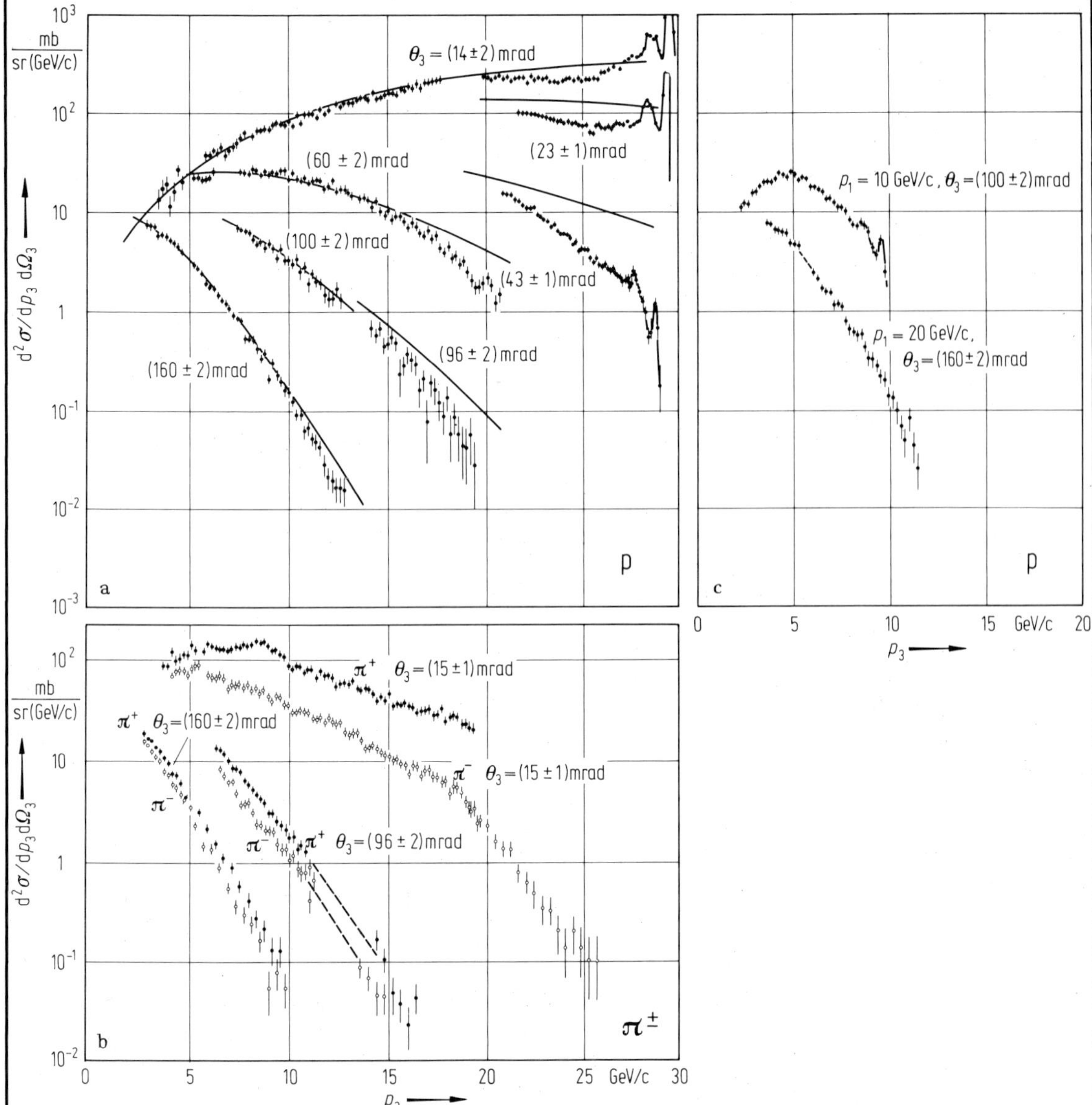

Fig. 5a···c. Summary of the lab. spectra of [67 A 1]. a) p spectra at $p_1 = 30$ GeV/c, b) $\pi^{\pm}$ spectra at 30 GeV/c, and c) p spectra at 20 and 10 GeV/c.

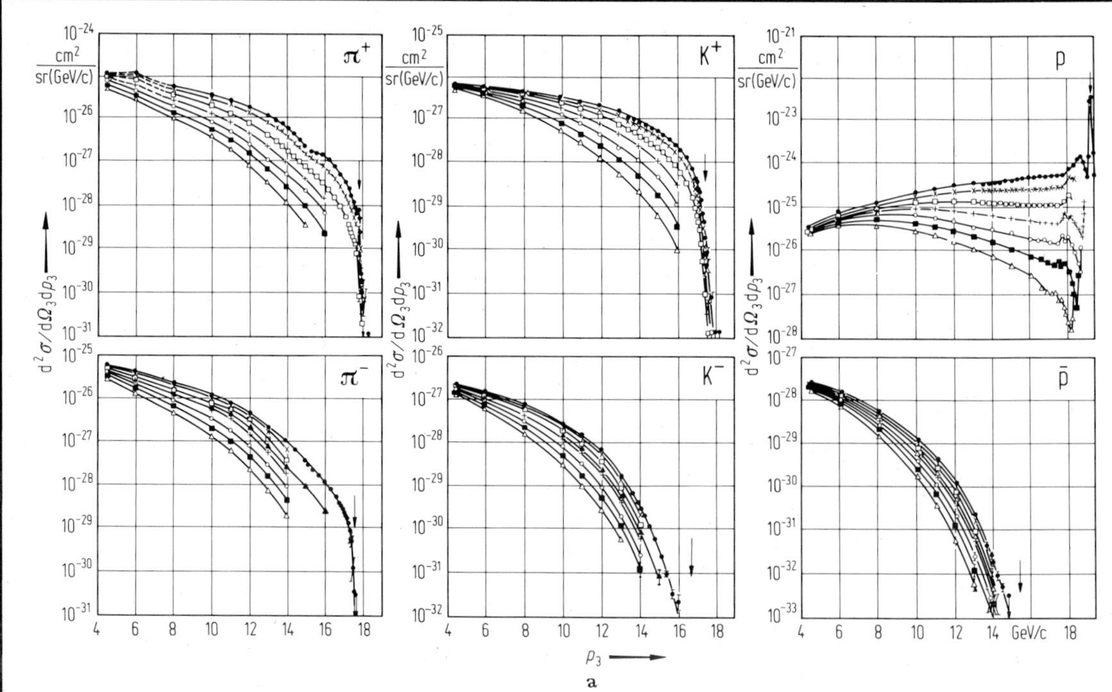

Fig. 6a. Summary of the lab. spectra at $p_1 = 19.2$ GeV/c [70 A 1]. The parameters for each curve are, from top to bottom,
$\theta_3 = 12, 20, 30, 40, 50, 60$ and 70 mrad, respectively.

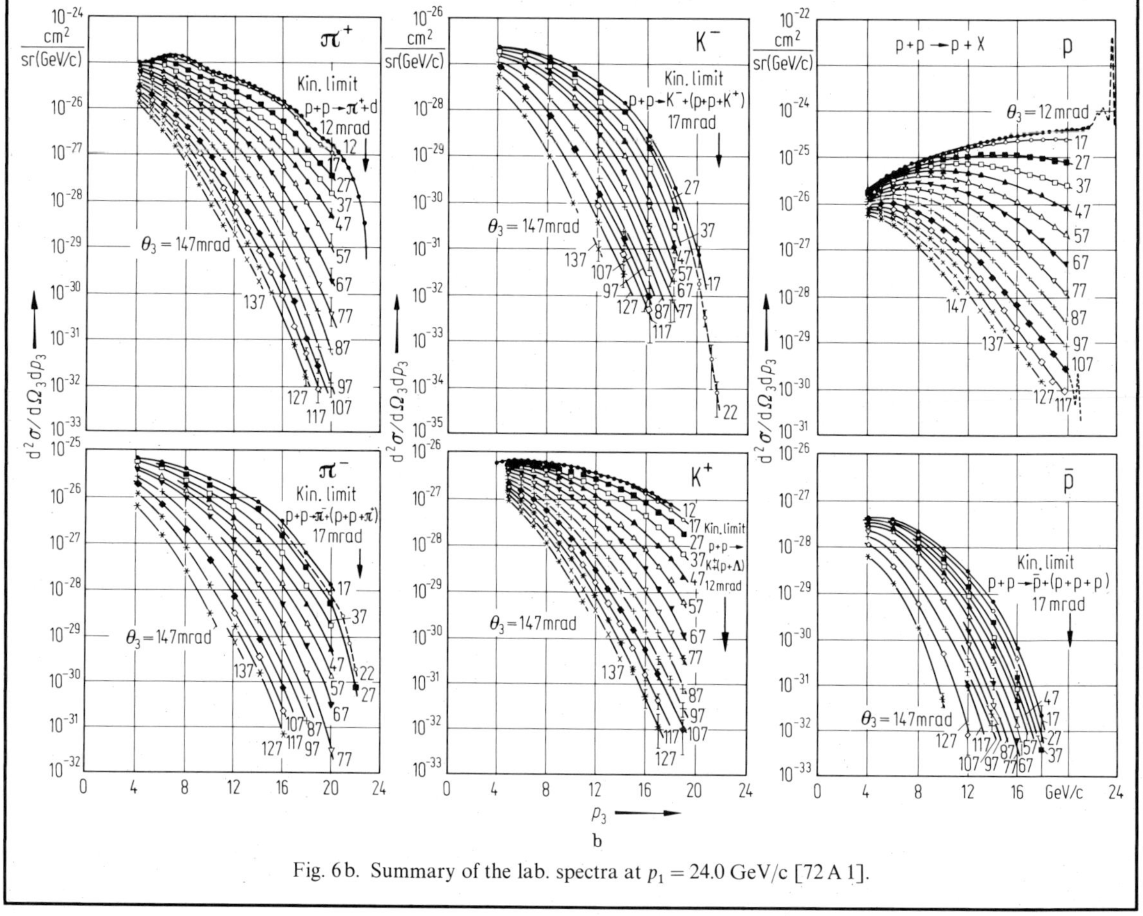

Fig. 6b. Summary of the lab. spectra at $p_1 = 24.0$ GeV/c [72 A 1].

Diddens/Schlüpmann

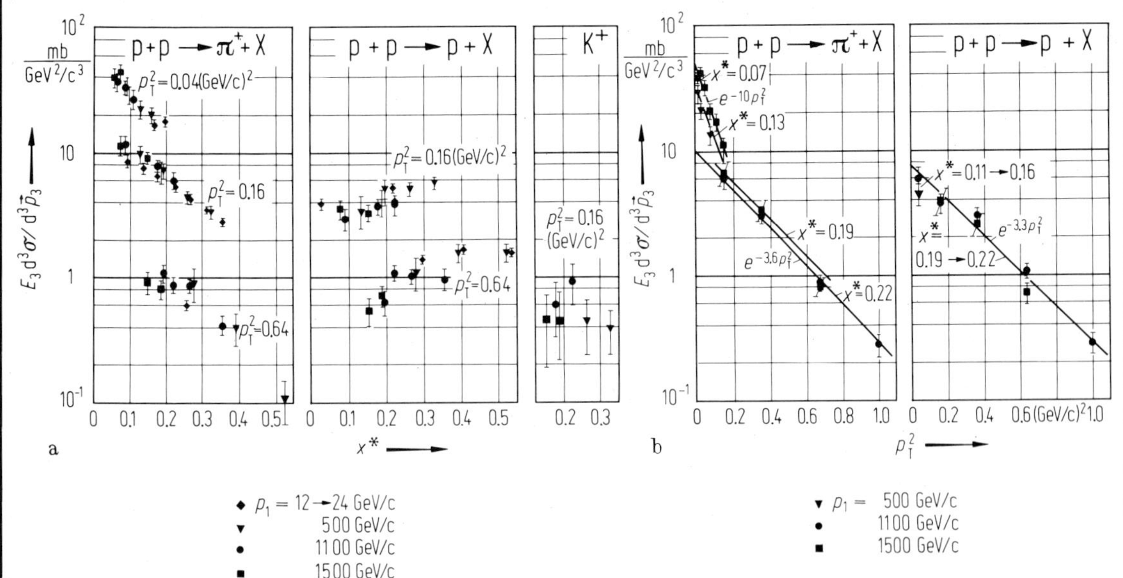

Fig. 7a, b. Positive particle spectra [71 R 1] measured at the CERN ISR. The figure compares the invariant cross section as a function of x^* (a) and p_T^2 (b) with data at much lower energies. Within the errors of about $\pm 15\%$ the cross section shows scaling behaviour.

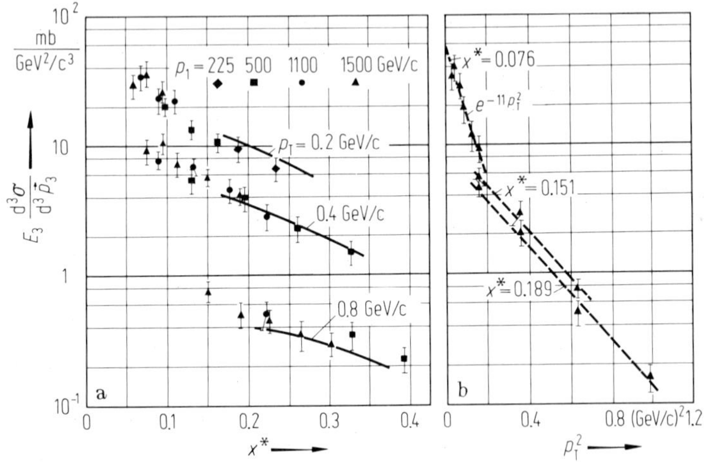

Fig. 8a, b. π^- spectra [72 B 3] measured at the CERN ISR. In (a) the invariant cross section as a function of x^* is compared with data at $p_1 \approx 20$ GeV/c shown by the full lines. In (b) the data as a function of p_T^2 are shown; the dashed lines are meant to guide the eye. The data show scaling behaviour, within the errors of $\pm 15\%$.

3.2.2 Production of other particles: n, γ, K_1^0, Λ, d, Σ, Ξ

The spectrum of neutrons produced in ≈ 25 GeV/c p-beryllium interactions has been observed recently [71 G 1 and 72 E 1]; results are plotted in Fig. 9, together with a p spectrum for comparison; the absolute normalization of the neutron spectrum is only approximate.

The spectrum of γ-rays has been measured at 23.1 GeV/c [62 F 1] and, recently, at the CERN ISR [71 N 1]; the spectra are reproduced in Figs. 10 and 11. γ-ray spectra yield information on the π^0 spectrum.

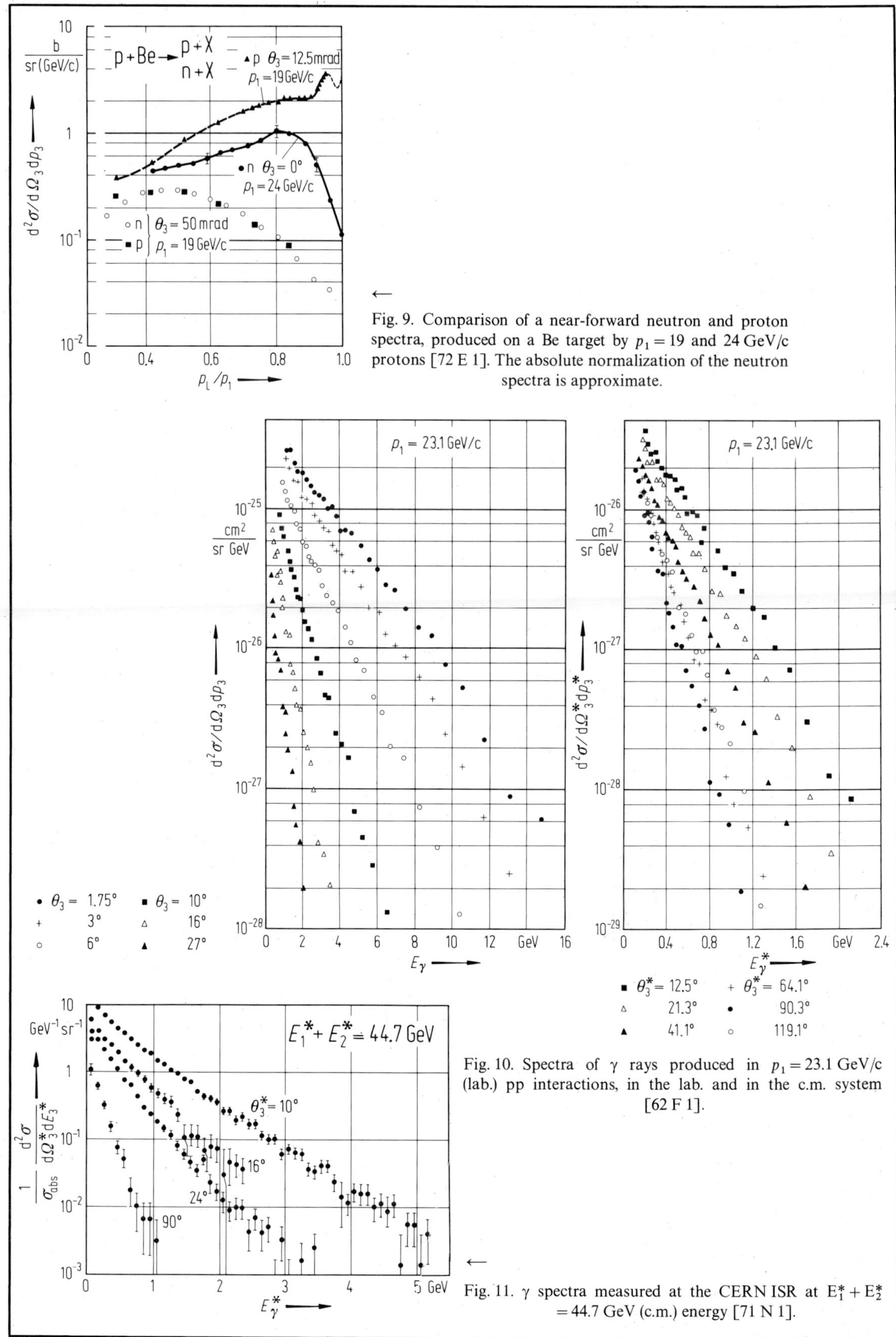

Fig. 9. Comparison of a near-forward neutron and proton spectra, produced on a Be target by $p_1 = 19$ and 24 GeV/c protons [72 E 1]. The absolute normalization of the neutron spectra is approximate.

Fig. 10. Spectra of γ rays produced in $p_1 = 23.1$ GeV/c (lab.) pp interactions, in the lab. and in the c.m. system [62 F 1].

Fig. 11. γ spectra measured at the CERN ISR at $E_1^* + E_2^* = 44.7$ GeV (c.m.) energy [71 N 1].

Diddens/Schlüpmann

A K_1^0 spectrum – usually assumed to be similar to the spectrum of the charged kaons – is shown in Fig. 12; it is due to Berger *et al.* [71 B 7].

The same authors have also measured a Λ spectrum, shown in Fig. 12 as well. Charged hyperon spectra produced in pp collisions are not available. Recently, momentum spectra of Σ^- and Ξ^- particles produced in p-tungsten collisions have been measured [72 B 2]. Normalized by means of a simultaneously measured π^- spectrum on tungsten, these data have been sketched into Fig. 13, together with the spectra of the other particles.

A measurement of deuteron production in 21.1 GeV/c pp collisions, at a lab. angle of 40 mrad, resulted in the momentum spectrum shown in Fig. 14 [69 A 1]; in this respect also see [64 D 1].

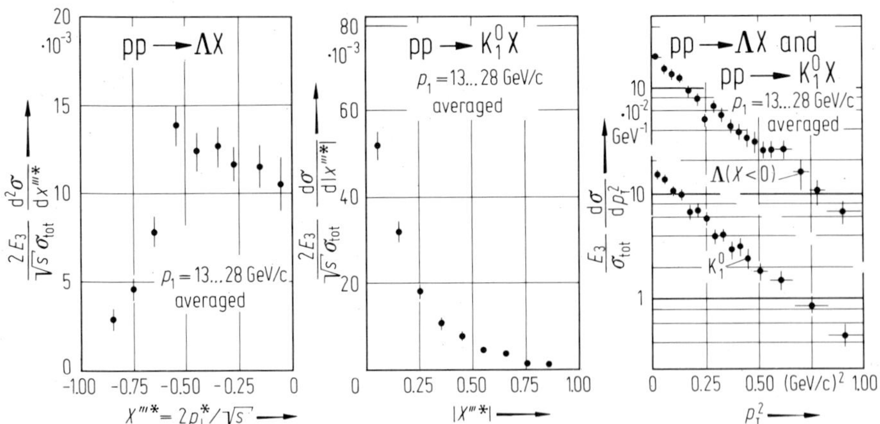

Fig. 12. K_1^0 and Λ spectra as a function of x^* and p_T^2. The spectra are from a proton exposure in a hydrogen bubble chamber at five energies between $p_1 = 13$ and 28 GeV/c [69 S 1]; the figure is adapted from [71 B 7].

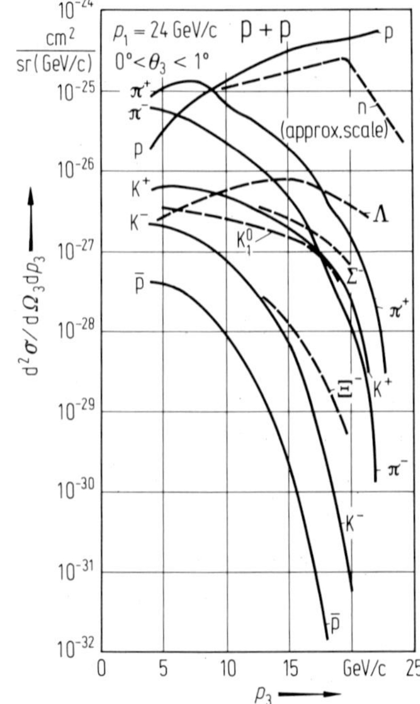

Fig. 13. Sketch of the particle spectra at $p_1 = 24$ GeV/c in the nearforward direction. The charged particle data are from [72 A 1]; the Λ and K_1^0 spectra are adapted from [71 B 7]; the Σ^- and Ξ^- spectra have been produced on a tungsten target and the cross section scale is derived from a normalization by means of a simultaneously measured π^- spectrum [72 B 2]; the neutron spectrum is produced on a beryllium target and its cross section scale is guessed [72 E 1].

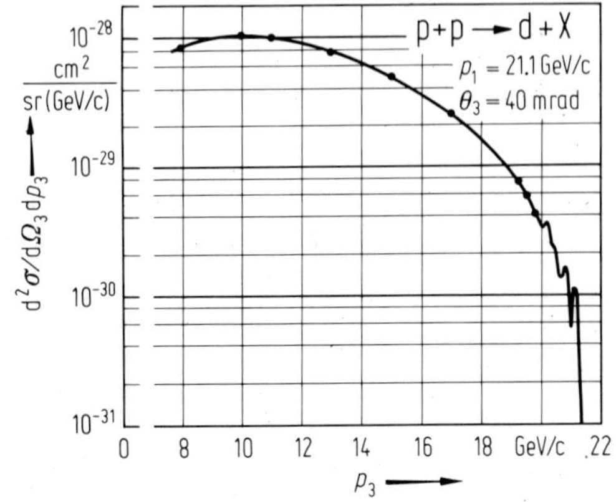

Fig. 14. Deuteron spectrum at $p_1 = 21$ GeV/c and $\theta_3 = 40$ mrad production angle [69 A 1].

3.3 Systematics of particle production

The following observations are at present regarded as the characteristic features of particle production:

i) The total cross section (σ_{tot}), the elastic cross section (σ_{el}) and consequently the absorption cross section ($\sigma_{abs} = \sigma_{tot} - \sigma_{el}$) are roughly constant above a few GeV, as shown in Fig. 15 [71 D 1]. A slow variation, for example, a logarithmic dependence on the beam momentum, can, however, not be ruled out. The cross section points $\sigma_{tot} = (40 \pm 2)$ mb and $\sigma_{el} = (7 \pm 1)$ mb at ≈ 500 GeV/c are from a recent experiment at the CERN ISR [71 H 1].

ii) The multiplicity of the outgoing particles is low, in the sense that most of the energy available in the c.m. is not spent in particle production, but instead is distributed as kinetic energy over the outgoing particles.

iii) The multiplicity increases slowly with energy. Fig. 16a displays the energy dependance of $\langle n_{ch} \rangle$, as derived from accelerator data, from the Echo-Lake experiment (cosmic ray particles incident on a hydrogen target surrounded by spark chamber detectors, [70 J 1]) and from preliminary ISR data. The data are compared to a logarithmic and to a powerlaw dependence on the incident energy. Other cosmic ray data, using emulsions as target material and as detector, are shown in Fig. 16b. For further discussions, see, for instance, [71 B 8 and 71 M 1].

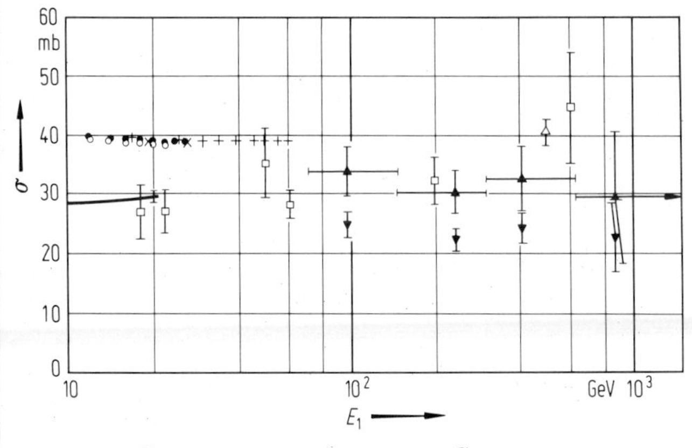

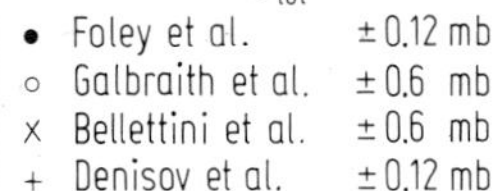

Fig. 15. The total cross section, σ_{tot}, and the absorption cross section, σ_{abs}, as a function of the lab. energy, E_1. The figure (from [71 D 1]) contains accelerator and cosmic-ray data; the point at 500 GeV from Holder *et al.* [71 H 1] has been measured at the CERN ISR

σ_{tot}		
• Foley et al.	± 0.12 mb	
○ Galbraith et al.	± 0.6 mb	
× Bellettini et al.	± 0.6 mb	
+ Denisov et al.	± 0.12 mb	
△ Holder et al.	± 2.0 mb	

σ_{abs}	
□ Grigorov et al.	
Echo lake experiment	
▲ With target wall corrections	
▼ With Monte Carlo corrections	
—— $\sigma_{tot} - \sigma_{el}$ from accelerators	

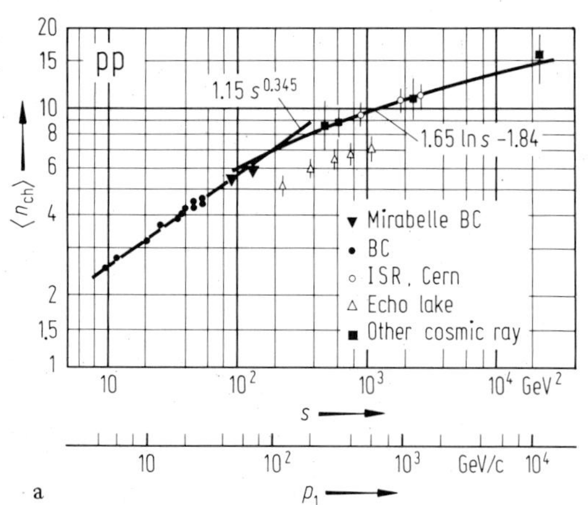

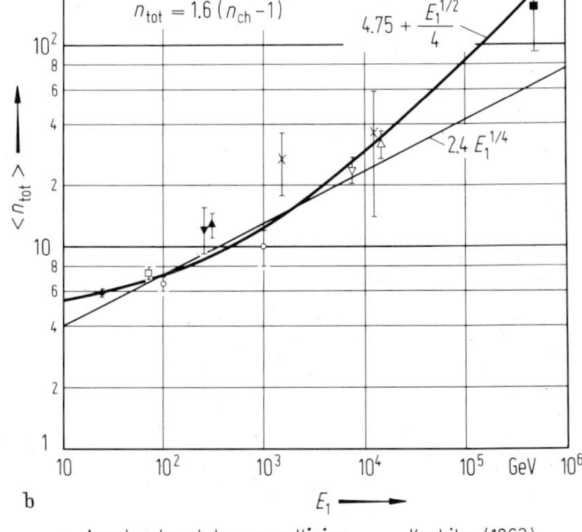

Fig. 16a. The average charged multiplicity, $\langle n_{ch} \rangle$, as a function of the square of the c.m. energy. The figure is from D. R. O. Morrison [72 O 1].

• Accelerator data pp collision	× Koshiba (1963)
□ Lal et al. (1962)	■ Perkins (1960)
○ Hansen and Fretter (1960)	▽ Malhotra (1963)
▼ Lohrmann et al. (1961)	△ Malhotra et al. (1965)
▲ Guseva et al. (1962)	

Fig. 16b. The total multiplicity, $\langle n_{tot} \rangle$, as a function of the lab. energy, E_1 measured in cosmic-ray experiments with emulsions as detector [66 P 1]. The total number of created particles is $n_{tot} = 1.6 \, (n_{ch-1})$.

iv) The multiplicity- or prong-number distribution can be approximated by a Poisson distribution [69 W 1]; however, it has not exactly this shape [70 C 2].

v) Most of the produced secondary particles are pions. Bøggild *et al.* [71 B 2] estimate the average numbers for particles produced in 19.2 GeV/c proton-proton interactions to be:

$$1.4\,p + 0.6\,n + 1.6\,\pi^+ + 1.4\,\pi^0 + 1.0\,\pi^- \,. \tag{29}$$

From the data of Allaby *et al.* [70 A 1] these numbers can be estimated for the less abundant charged particles to be approximately

$$0.04\,K^+ + 0.02\,K^- + 0.001\,\bar{p}\,. \tag{29a}$$

Berger *et al.* [71 B 7] derive the following numbers of strange particles (charged + neutrals):

$$0.07\,K + 0.04\,\bar{K} + 0.03\,\Lambda\,. \tag{29b}$$

At ISR energies the preliminary results indicate abundances of K^-, K^+ and $\bar{p}$ relative to pions of approximately 5 % [71 R 1, 72 B 3].

vi) The transverse momentum of the produced particles is low; for pions at accelerator energies $\langle p_T \rangle \approx 0.4\,\text{GeV}/c$. It is found that $\langle p_T \rangle$ is rather energy independent; cosmic-ray experiments using nuclear targets suggest a very slow increase with energy.

vii) For a description of the longitudinal momentum distribution, the concept of a "leading" particle has been introduced by cosmic-ray investigators. Usually two particles emerge from a collision with the same or similar quantum numbers as the incident particles and carrying a large fraction of the c.m. momentum of the two incident particles with them, typically about 50 %. They are called leading particles. The remaining particles cluster around $p_L^* = 0$. This effect is illustrated on a Peyrou plot in Fig. 17. The leading particles have a rather flat p_L^* distribution. Non-leading particles show a strong fall-off of the spectra with increasing p_L^*. Further illustration is provided in Fig. 18, where the x^* spectra at fixed-small p_T of Allaby [70 A 1, 72 A 1] are shown in one graph. It may be noted that the five non-leading particles still have characteristic differences in their behaviour: π^+ and K^+ are "more leading" than the remaining three negative particles. Fig. 13 shows that the Σ^- is even "more leading" then the π^+ [72 B 2]. An explanation of these characteristics is possibly the following: in an appreciable fraction of the collisions nucleon isobars are produced with leading particle properties and their high longitudinal velocity components are transferred to their decay products, $\pi^{\pm}$, K^+, Λ, and Σ^-. Particles that are not so easily produced this way (K^-, Ξ^-, and $\bar{p}$) have on the contrary a much steeper fall-off with momentum, more characteristic of a direct production process in the collision.

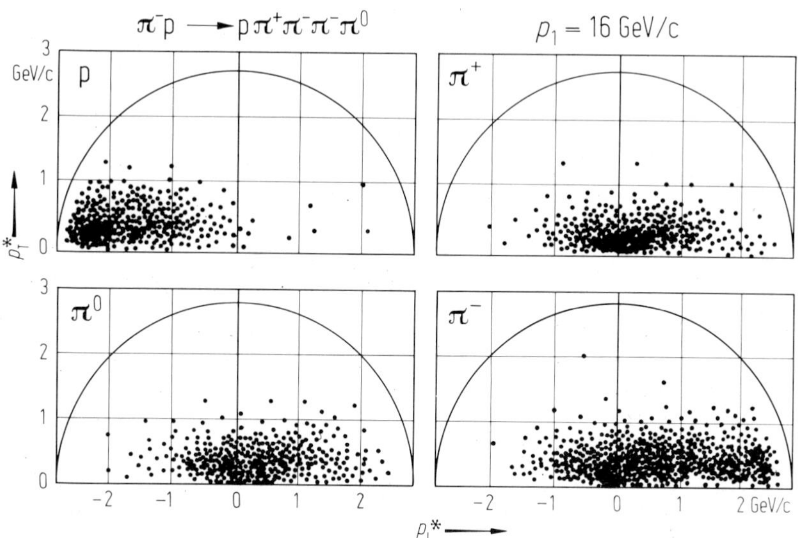

Fig. 17. Peyrou plot for the reaction $\pi^- p \to \pi^- \pi^- \pi^0 \pi^+ p$ at $p_1 = 16$ GeV/c. The protons and part of the π^- show leading particle behaviour, continuing in the direction of the incoming particles; the produced π^+, π^0, and part of the π^- show pionization behaviour, clustering around $p_L^* \approx 0$ [69 H 1].

Diddens/Schlüpmann

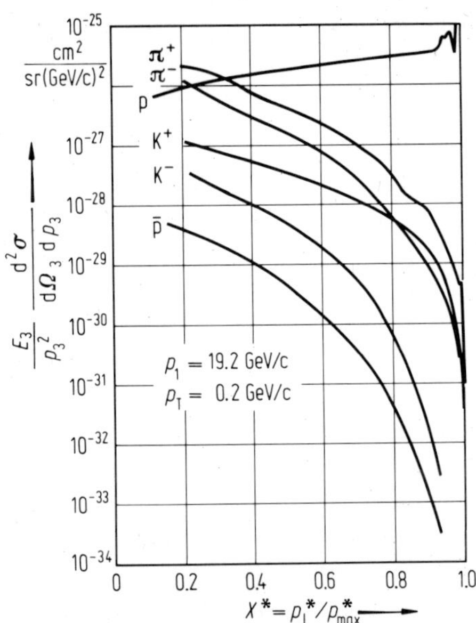

Fig. 18. Longitudinal momentum spectra of charged particle production in $p_1 = 19.2$ GeV/c pp interactions at a fixed $p_T = 0.2$ GeV/c [70 A 1 and 72 A 1].

    viii) The dependence of the cross section on p_T and p_L^* can be approximately factorized:

$$f(p_T, p_L^*) \approx f_1(p_T) \cdot f_2(p_L^*). \tag{30}$$

This is illustrated in Fig. 19, where the 19.2 and 24.0 GeV/c data of Allaby *et al.* [70 A 1, 72 A 1] are plotted against x^* for fixed p_T and against p_T for fixed x^* (these curves are produced by interpolation from the original data; experimental points are not shown, therefore, except to indicate a typical error). The similarity of the curves for a particular particle shows the extent to which the factorization in p_L^* and p_T is valid. Especially the proton data show deviations. Perhaps factorization is better satisfied in another set of variables, for instance p_T and y^*. The π^- data of Bøggild *et al.* [71 B 1, 72 B 1] at 19.2 GeV/c are shown in Fig. 20 plotted in these variables.

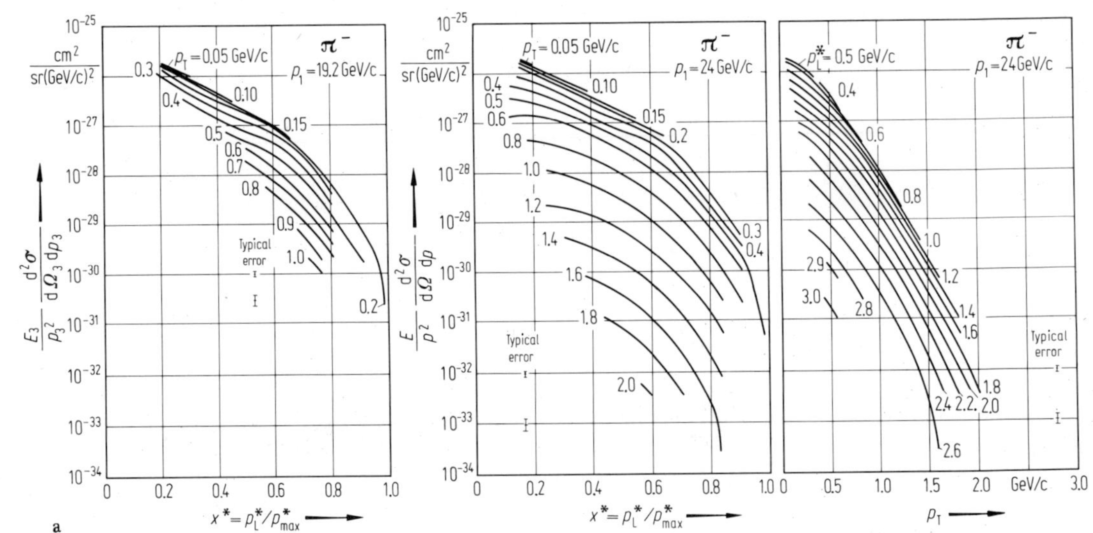

Fig. 19 a···d. Spectra as a function of x^* for fixed p_T and as a function of p_T for fixed x^*.

π^- (a), π^+ (b), K^- (c) and K^+ (d) production in $p_1 = 19.2$ GeV/c and 24 GeV/c pp interactions. The data are interpolated [70 A 1 and 72 A 1].

Diddens/Schlüpmann

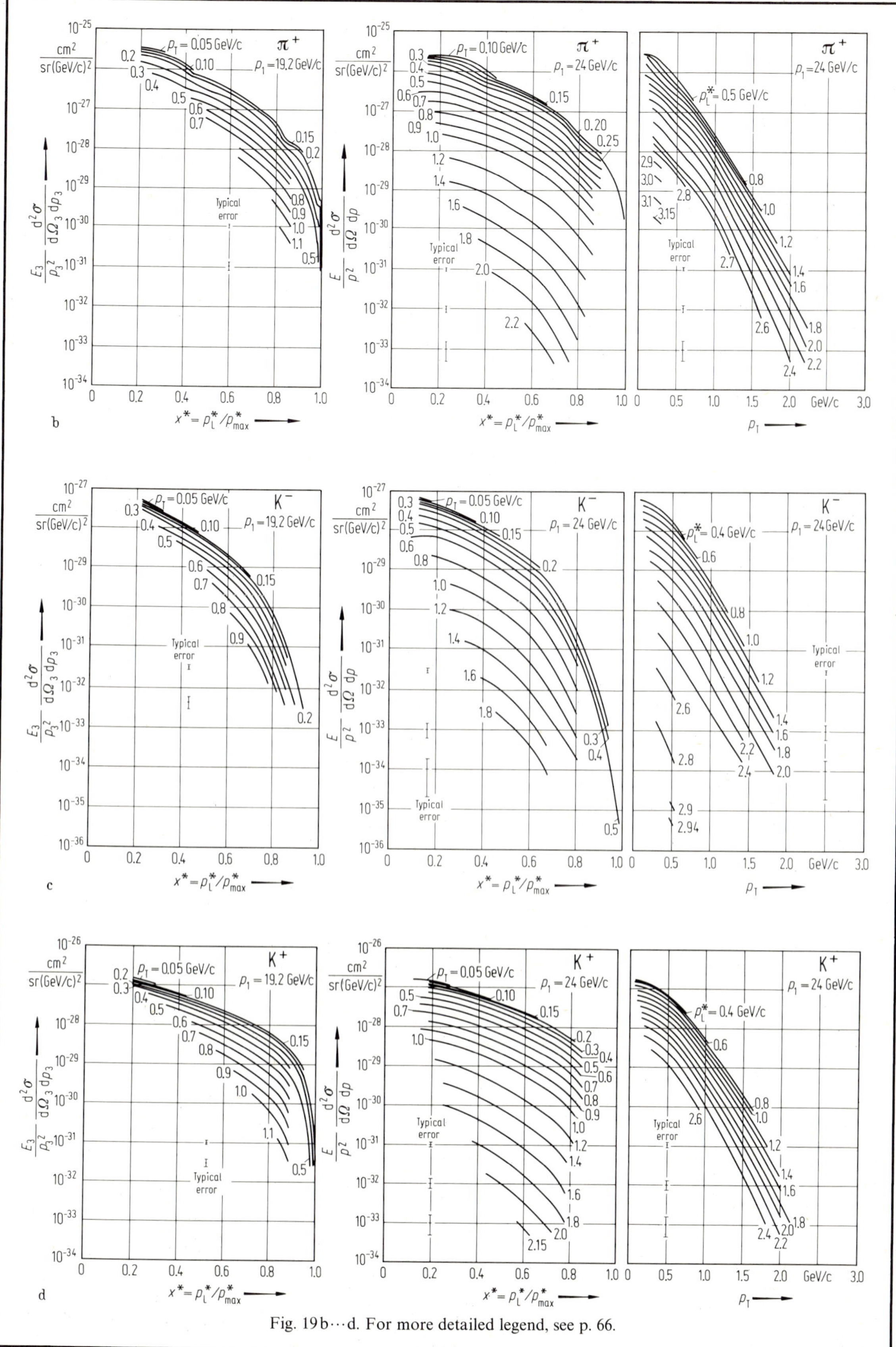

Fig. 19b···d. For more detailed legend, see p. 66.

Diddens/Schlüpmann

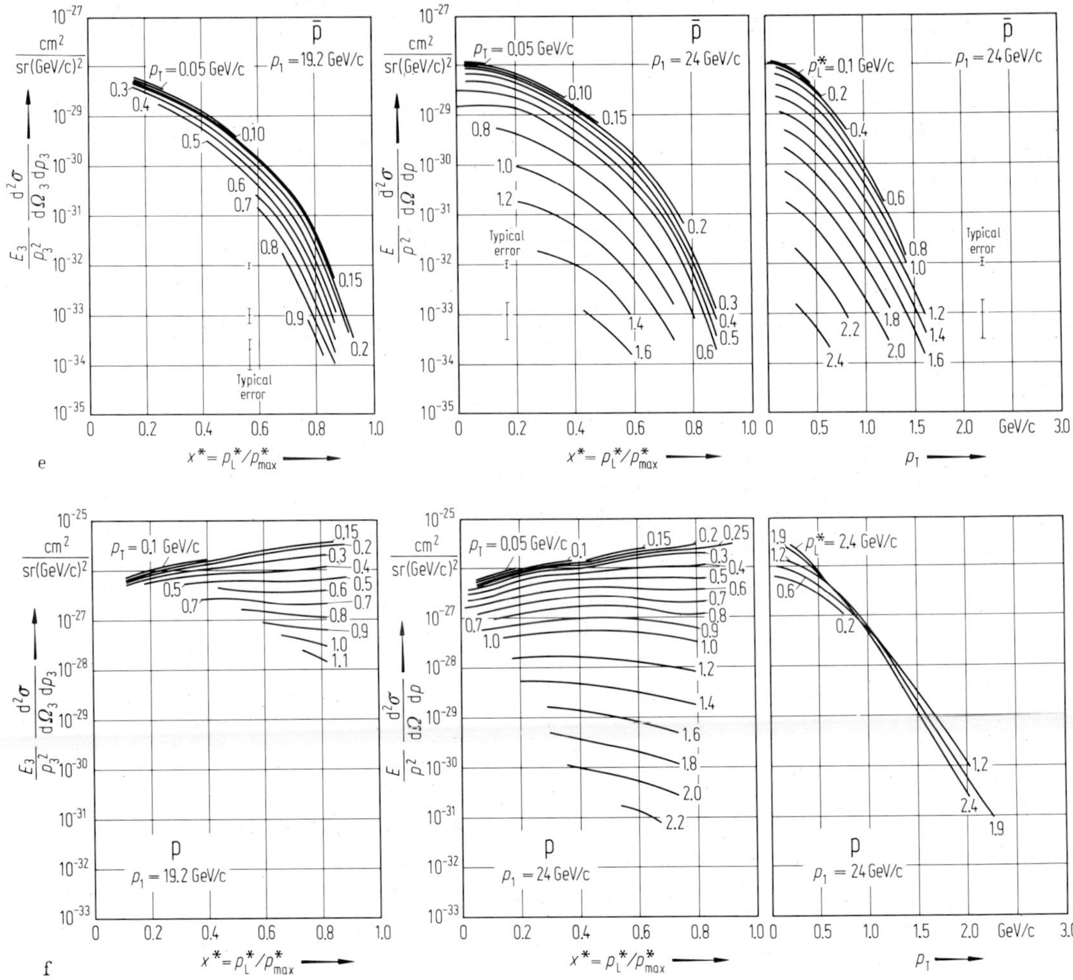

Fig. 19e, f. Spectra as a function of x^* for fixed p_T and as a function of p_T for fixed x^*. $\bar{p}$ (e) and p (f) production in $p_1 = 19.2$ GeV/c and 24 GeV/c pp interactions. The data are interpolated [70 A 1 and 72 A 1].

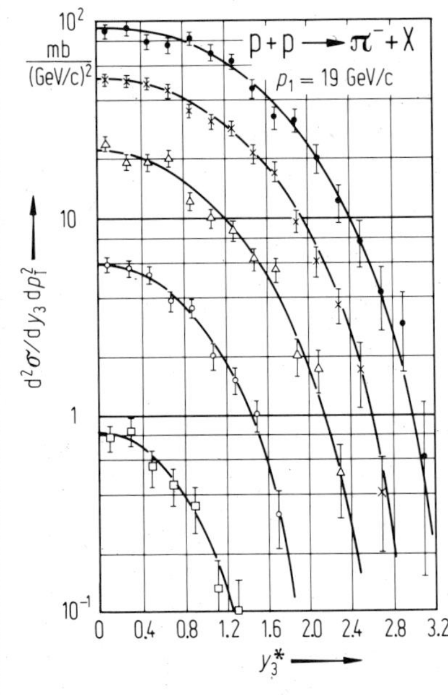

Fig. 20. Spectra of π^- produced in 19.2 GeV/c pp interactions, as a function of y_3^*, the c.m. rapidity [71 B 1, 71 B 2 and 72 B 1]

●   0 $< p_T <$ 0.16 GeV/c
×   0.16 $< p_T <$ 0.32
△   0.32 $< p_T <$ 0.48
○   0.48 $< p_T <$ 0.80
□   0.80 $< p_T <$ 1.12 GeV/c

ix) The spectra approximately show "scaling" behaviour in energy, i.e. at asymptotic energies the invariant cross section E_3 $(d^3\sigma/d^3\vec{p})$ is a homogeneous function of the reduced longitudinal momentum x^* and the transverse momentum p_T and has no further energy dependence. This scaling behaviour is demonstrated in Figs. 7, 8, 21, 22 and 23. It can be concluded from these figures that, for pions and protons, an energy-independent limit seems to be reached already approximately in the 20 GeV/c region[1]. It should be realized, however, that the present generation of experiments rarely reaches an absolute accuracy better than 10%. Much work remains to be done to clarify the accuracy of the scaling hypothesis and, if true, to discover the manner in which an energy-independent limit is approached. Some evidence against the scaling hypothesis has also been collected, from a comparison of low-energy spectra to cosmic-ray data ([71 L 1] $\gtrsim 10^4$ GeV). Scaling behaviour will be discussed further in the next section.

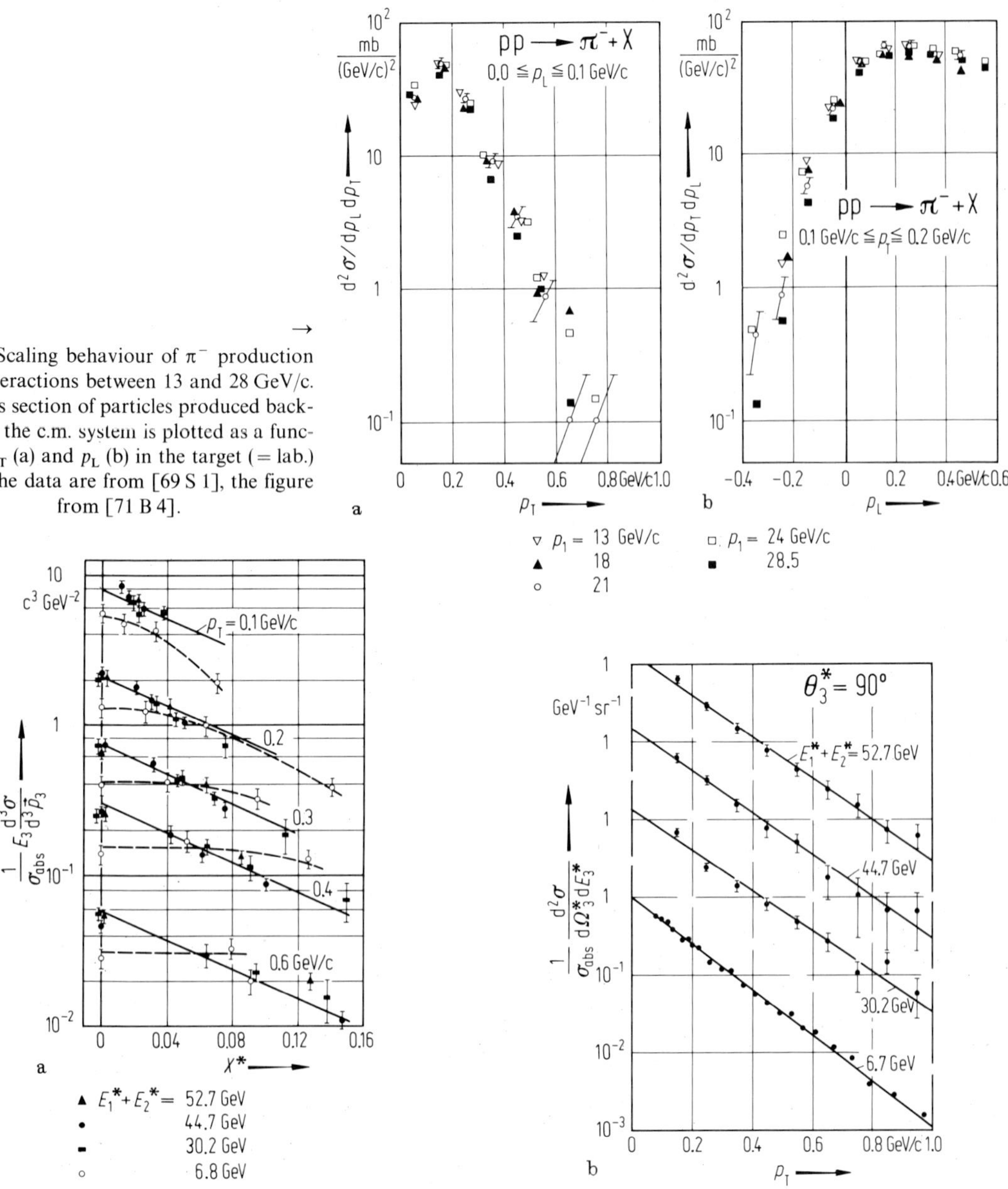

Fig. 21. Scaling behaviour of π^- production in pp interactions between 13 and 28 GeV/c. The cross section of particles produced backwards in the c.m. system is plotted as a function of p_T (a) and p_L (b) in the target (= lab.) frame. The data are from [69 S 1], the figure from [71 B 4].

Fig. 22a. Energy dependence of the invariant cross section for the production of γ rays in pp interactions at the c.m. energies noted in the figure [71 N 1].

Fig. 22b. Similarity of the transverse momentum dependence of γ-ray production in pp interactions at various energies in the c.m. system [71 N 1].

---

[1] When the data become more accurate it might be necessary to divide the invariant cross section by the total cross section σ_{tot}, in order to account for the weak energy dependence of the latter.

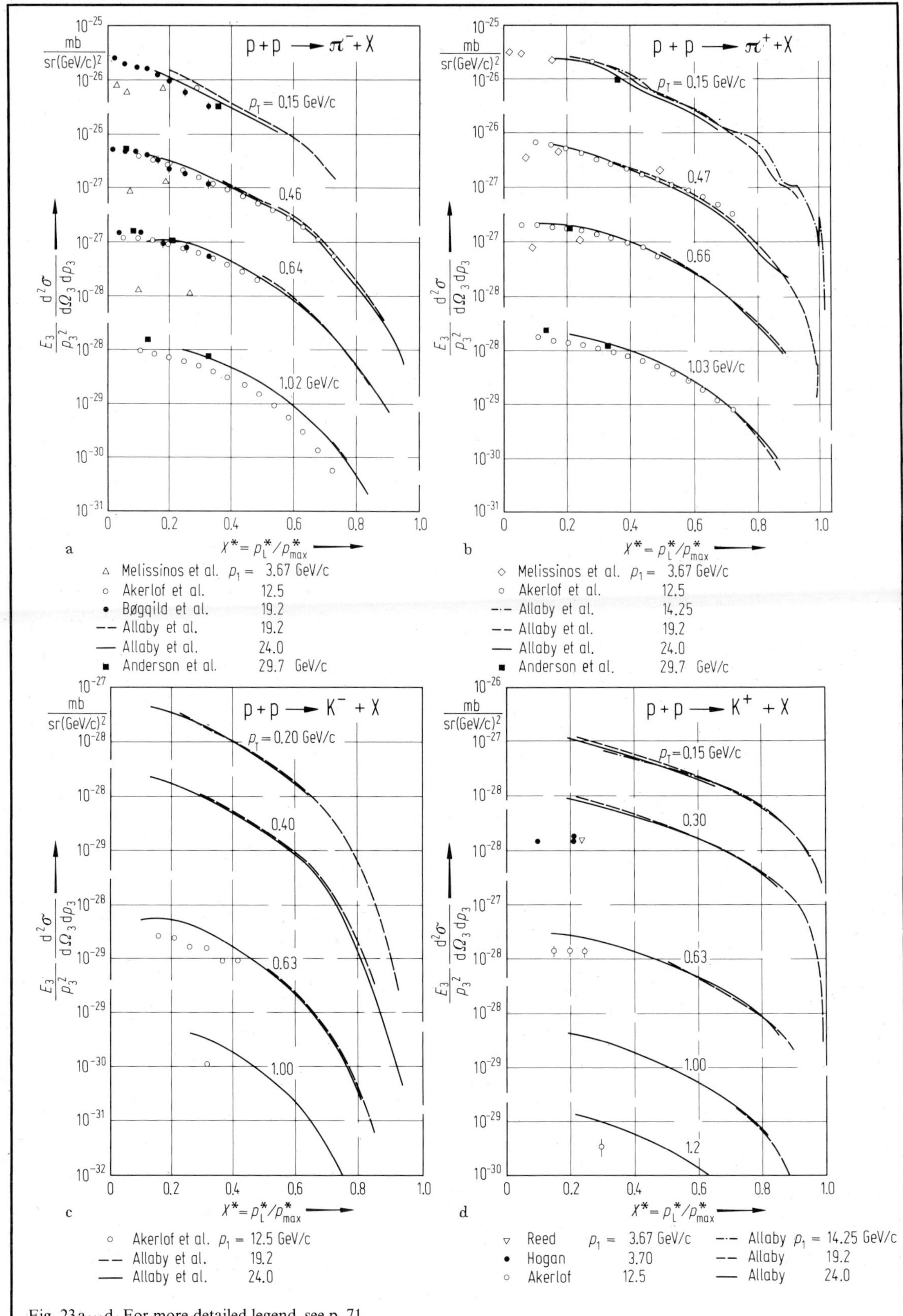

Fig. 23a···d. For more detailed legend, see p. 71.

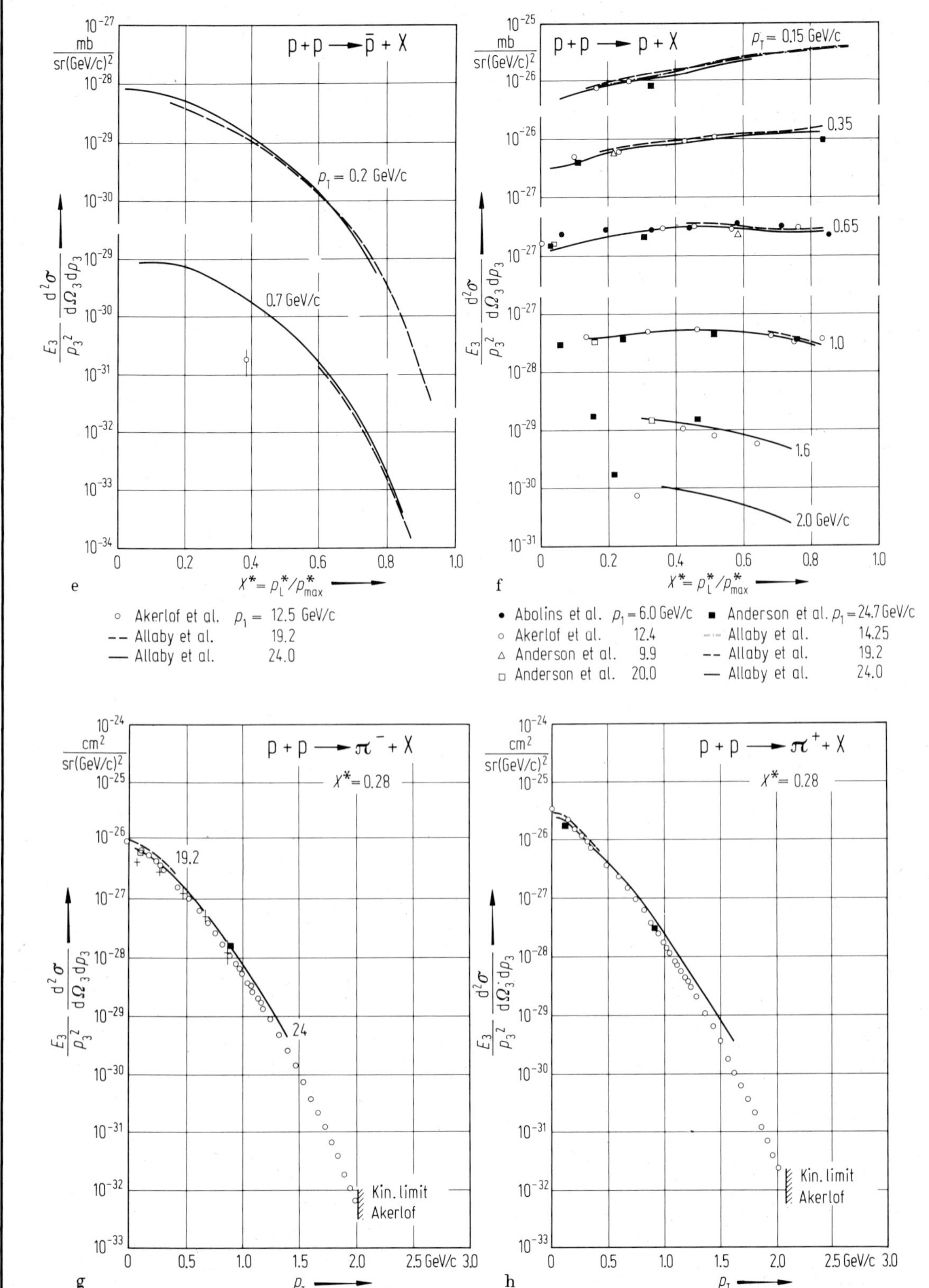

Fig. 23 a···h. Energy dependence of the invariant cross section for charged particle production at accelerator energies. The invariant cross section has been plotted as a function of x^* for π^- (a), π^+ (b), K^- (c), K^+ (d), $\bar{p}$ (e), p (f), and as a function of p_T for π^- (g), and π^+ (h) at various energies. The pion points in the 3 GeV/c region have not yet reached an energy-independent limit; nor have the 12 GeV/c kaon and antiproton cross sections.

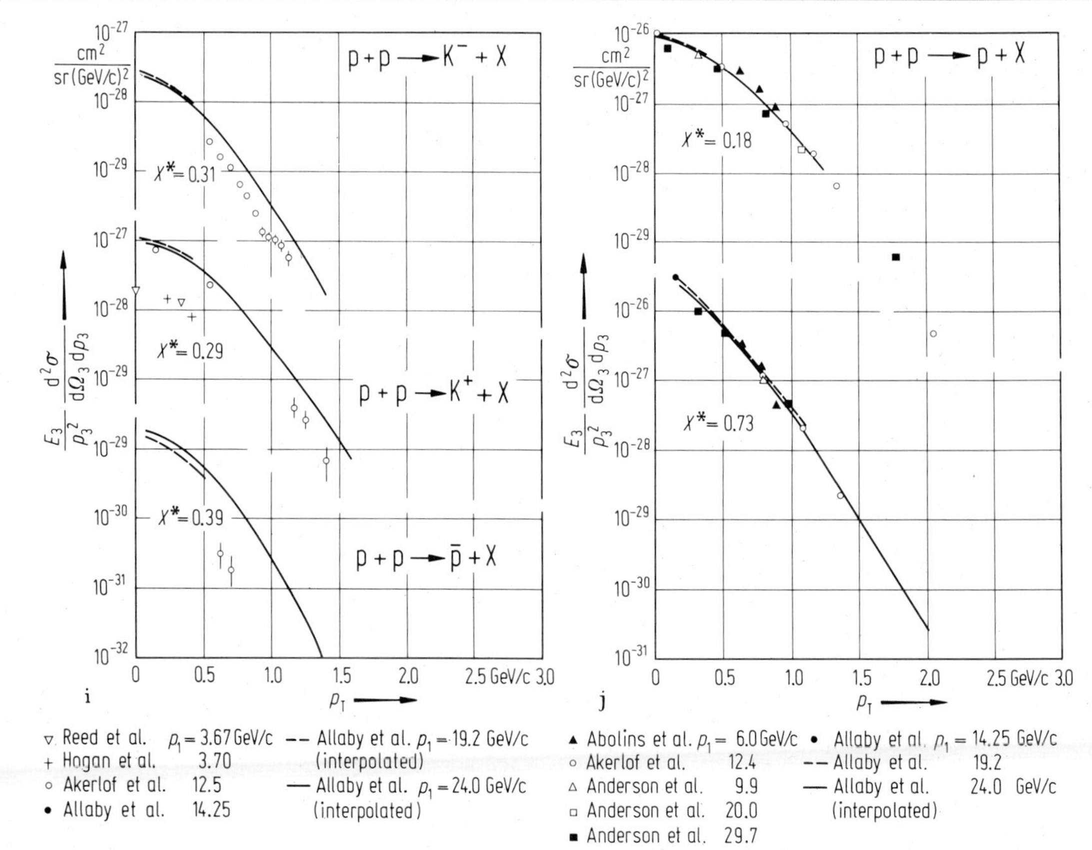

▽ Reed et al. $p_1 = 3.67$ GeV/c	–– Allaby et al. $p_1 = 19.2$ GeV/c	▲ Abolins et al. $p_1 = 6.0$ GeV/c
+ Hogan et al. 3.70	(interpolated)	○ Akerlof et al. 12.4
○ Akerlof et al. 12.5	—— Allaby et al. $p_1 = 24.0$ GeV/c	△ Anderson et al. 9.9
• Allaby et al. 14.25	(interpolated)	▫ Anderson et al. 20.0
		■ Anderson et al. 29.7

• Allaby et al. $p_1 = 14.25$ GeV/c
–– Allaby et al. 19.2
—— Allaby et al. 24.0 GeV/c
(interpolated)

Fig. 23 i, j. Energy dependence of the invariant cross section for charged particle production at accelerator energies. The invariant cross section has been plotted as a function of p_T for K⁻, K⁺, p̄ (i) and p (j) at various energies. The pion points in the 3 GeV/c region have not yet reached an energy-independent limit; nor have the 12 GeV/c kaon and antiproton cross sections

Some more detailed remarks can be made on the p_L^* and p_T distributions.

On theoretical grounds, two regions are often regarded separately with respect to the p_L^* dependence of the invariant cross section: a "pionization region" around $x^* = 0$ or $\xi^* = \frac{1}{2}$, and a "fragmentation region" $a \leqq |x^*| \leqq 1 - \varepsilon$, where a is a somewhat ambiguously defined number > 0.

The p_L^* distributions do not lead themselves to a description by a simple function of x^*. Over a limited range an exponential dependence is often adequate, but the leading particle is approximately described by a flat p_L^* distribution [67 A 1]. The π^+ longitudinal spectra show some structure at small p_T (see Fig. 19b); its origin is not understood in detail but is presumably related to the production of nucleon isobars.

Inspection of Fig. 24 shows that the p_T distributions are neither exponential in p_T, nor Gaussian. One might call the distribution either a flat-topped exponential or the sum of two (or more) Gaussians. In the first case the exponent varies between 3 and 9 GeV/c^{-1}; in the latter case of two Gaussians, exponents of 15 and 3 GeV^{-2} seem to be reasonable.

As to the deviations from a universal p_T distribution, $\langle p_T \rangle$ perhaps depends somewhat on the incoming energy [69 S 1], but the effect must be rather small (see Figs. 7, 8 and 25). Bøggild *et al.* [71 B 1] find $\langle p_T \rangle$ slowly increasing with p_L^* for negative pions produced at 19.2 GeV/c (see Fig. 26); for the leading protons the opposite is true (Fig. 19f); $\langle p_T \rangle$ depends on the particle produced, as shown in Figs. 24 and 27 [62 B 1]. If particular channels are chosen, then $\langle p_T \rangle$ also depends on the prong number, demonstrated, for instance, by Smith *et al.* [69 S 1] for π^- and π^+ distributions, produced in pp interactions between 13 and 28.5 GeV/c. These, and other authors fit their data to the expression

$$\frac{dN}{dp_T} = \frac{4N_T}{3\sqrt{\pi}}\, a_T^{\frac{5}{2}} p_T^{\frac{3}{2}} e^{-a_T p_T}$$

$$\frac{dN}{dp_L^*} = N_T a_L\, e^{-a_L p_L^*}$$

(31)

These formulae usually fit data with limited statistics reasonably well. The first expression is, however, somewhat misleading, since it suggests that the invariant cross section has a zero at $p_T = 0$, which is not the case in reality (see Fig. 4).

In the pionization region the low-energy pion spectrum has been fitted by a Bose-Einstein distribution function [71 B 3, 71 E 3] with a temperature $kT \approx$ pion mass, in agreement with the thermodynamical description of Hagedorn [65 H 1].

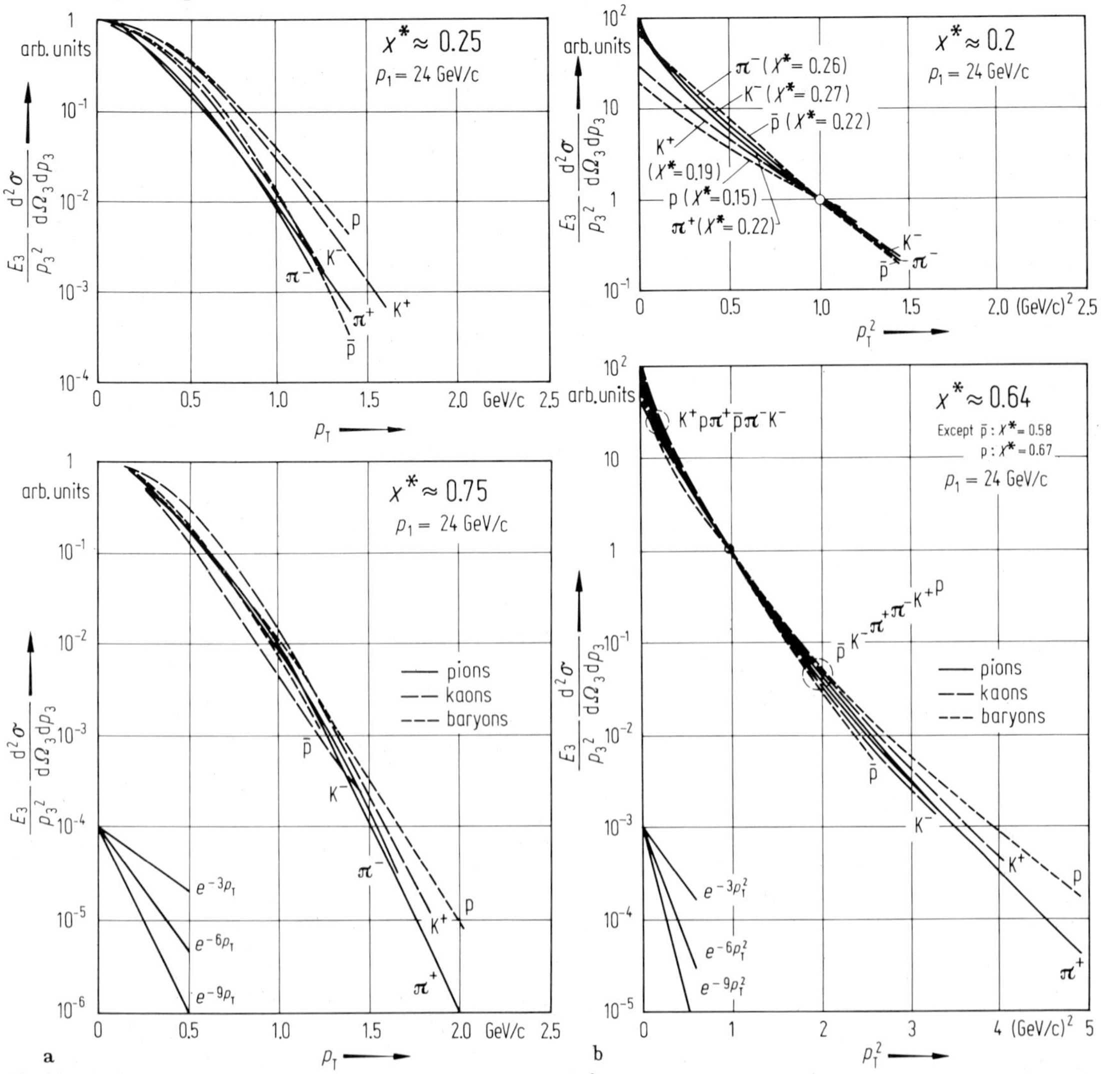

Fig. 24. The invariant cross section as a function of p_T (a) and p_T^2 (b) for charged particle production at $p_1 = 24$ GeV/c, in a region of low x^* and high x^* [72 A 1], respectively.

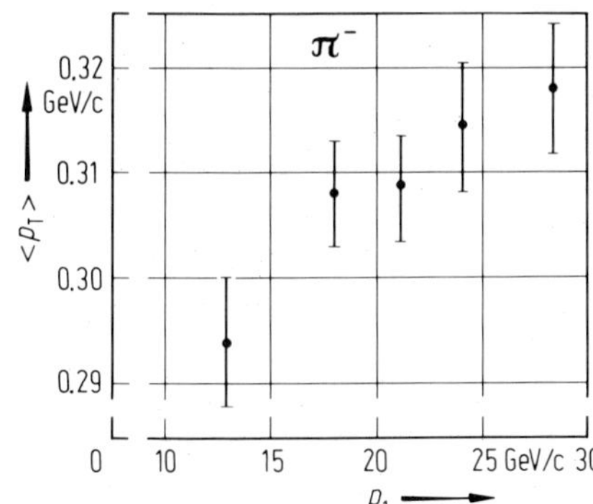

Fig. 25. Energy dependence of the average transverse momentum $\langle p_T \rangle$ in π^- production in pp interactions [69 S 1].

Diddens/Schlüpmann

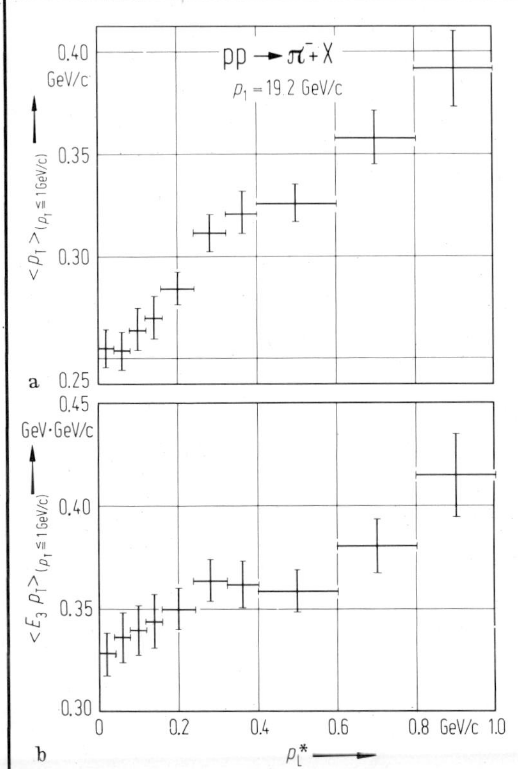

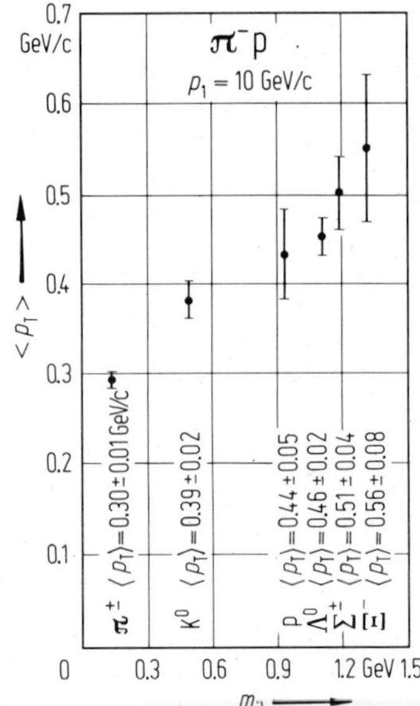

Fig. 26. Dependence of the average transverse momentum on the c.m. longitudinal momentum in π^- production in pp interactions at $p_1 = 19.2$ GeV/c [71 B 1]. In (a) deviations from factorization of the cross section $d^2\sigma/d^3\vec{p}_3$ are demonstrated; in (b) similar deviations for the invariant cross section are shown.

Fig. 27. Dependence of the average transverse momentum on the type of particle produced in $p_1 = 10$ GeV/c $\pi^- p$ interactions [62 B 1].

3.4 Theory and empirical formulae

3.4.1 Theory

The theory of particle production is still unsatisfactory. Early papers treated particle production in a manner analogous to electromagnetic bremsstrahlung [Heisenberg, 46 H 1] and thermo-radiation [51 F 1]. The bremsstrahlung ideas have been taken up more recently again by Kastrup [66 K 1], Anderson and Collins [67 A 2] and Mack [68 M 1]; actual data are fitted in the last two works. Also Feynman [69 F 1] revives it in his exposition on single-particle distributions.

Hagedorn, and Hagedorn and Ranft [65 H 1], postulated an equation of state for hadronic matter, and introduced the concept of a "continuous superposition of moving fireballs, each with another velocity with respect to the c.m. frame and decaying thermodynamically in its own rest frame." As a result, a semi-empirical formula was obtained, which represents a fit to the existing data. Other authors have introduced two temperatures in order to account for the observed asymmetry of the p_L, p_T distributions [68 W 1].

The multiperipheral model links single particle production to other strong interaction processes. Originally proposed by Amati, Bertocchi, Fubini, Stanghellini and Tonin [62 B 2] it was applied to inclusive reactions by Caneschi and Pignotti [69 C 1] and subsequently by Risk and Friedman [71 R 2], Peccei and Pignotti [71 P 1], Chen, Wang and Wong [71 C 7], Edelstein, Rithenberg and Rubinstein [71 E 2], and Risk [71 R 3]. The multiperipheral model incorporates features like the approximate factorization in longitudinal and transverse momentum dependence and a logarithmic increase of the multiplicity. Recent work has been stimulated by the suggestion of Mueller [70 M 1] that inclusive spectra can be related to a discontinuity in the imaginary part of the forward three-body scattering amplitude, and thereby become accessible to Regge theory in much the same way as total cross sections. Based on Mueller's work, Chan et al. [71 C 4] and Ellis et al. [71 E 1] have speculated about the rate of approach of the single particle distribution to the limiting distribution at infinite energies; they predict the $\pi^\pm$, $K^\pm$ and $\bar{p}$ production in pp interactions to reach their limits at relatively low energies; p-production will approach its limits as $s^{-\frac{1}{2}}$.

Feynman [69 F 1] and Benecke et al. [69 B 1] provided a stimulus by introducing the concepts of scaling and limiting distribution [see point (ix) in the previous section]. Feynman was guided by a parton model and postulated that the invariant cross section will reach asymptotically an energy-independent limit, depending only on the two variables x^* and p_T. In addition, Feynman expects the invariant cross section for pion production to be flat in the rapidity variable y_3^* near $x^* = 0$; thus

$$d\sigma \propto \frac{dp_L^*}{E_3^*},\tag{32}$$

i.e. the pion spectrum should have a bremsstrahlung shape for low energies. This feature makes the pion multiplicity increase logarithmically. The present data do not yet seem to verify this last point, Eq. (32) (see Fig. 20, for instance).

There is not much evidence yet that the nucleon multiplicity increases with energy, like the π multiplicity does. It might therefore be more sensible that for protons the expression

$$\frac{1}{\pi}\frac{d^2\sigma}{dp_T^2 dx^*} \approx \frac{p_1^*}{\pi}\frac{d^2\sigma}{dp_T^2 dp_L^*} = \frac{p_1^*}{E_3^*}\left(E_3\frac{d^3\sigma}{d^3\vec{p}_3}\right).\tag{33}$$

is energy-independent: integration of a universal x^*-dependence and a universal (steep) p_T dependence of this expression leads to a constant cross section. For $x^* \not\approx 0$ this equation can be written approximately as

$$\frac{1}{\pi}\frac{d^2\sigma}{dp_T^2 dx^*} \approx \frac{1}{x^*}\left(E_3\frac{d^3\sigma}{d^3\vec{p}_3}\right).\tag{33'}$$

Such a behaviour has been called "naïve scaling". The distinction between naïve scaling and Feynman scaling can only be made near $x^* = 0$ and the present data are not yet accurate enough for this.

Benecke et al. used the concept of fragmentation of the target particle and the projectile. They postulated that the fragment distributions will reach asymptotically a limit in the target frame and the projectile frame. It can be shown by a Lorentz transformation that this statement and the scaling hypothesis are identical (except near $x^* = 0$).

Scaling seems to be approximately satisfied already in the 20 GeV/c region for pions and protons as mentioned before. Data have been subjected to tests on scaling by Drell [70 D 1], Van der Velde [70 V 1], Bali et al. [70 B 1], Bøggild et al. [71 B 1], Chen et al. [71 C 3], Carazza et al. [71 C 1], Ratner et al. [71 R 1], Ranft [71 R 4], Michejda [71 M 1], Bertin et al. [72 B 3] and Panvini et al. [72 P 1].

The predictions of various models discussed above have been compared in a paper by De Tar [71 D 2].

Several authors [69 G 1, 70 C 1, 70 G 1, 70 B 2, 70 A 2, 71 C 2, 71 G 2] have tried to make a connection between inelastic proton scattering and inelastic electron scattering, in analogy with the ideas of Wu and Yang [65 W 1] on the connection between elastic proton-proton and elastic electron-proton scattering. At the present time the relation is at most qualitative. Because the most appropriate variables in such a comparison are t and m_4, the invariant cross section for $p + p \to p + X$ at 24 GeV/c [72 A 1] has been plotted in this manner in Fig. 28.

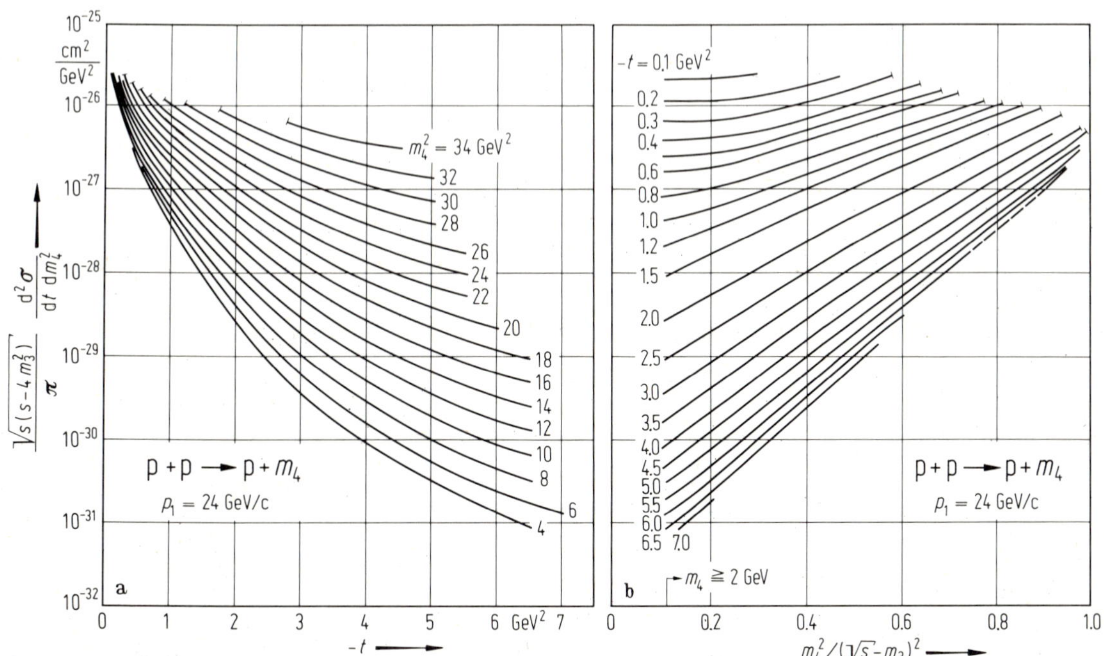

Fig. 28. Invariant cross section of proton production in $p_1 = 24$ GeV/c pp interactions, as a function of t (a) and $m_4^2/(\sqrt{s} - m_3)^2$ (b) [72 A 1]. The data are interpolated.

Inspired by a Regge-pole analysis, fits of the type

$$\frac{s}{\pi}\,\frac{d^2\sigma}{dt\,dm_4^2}\approx E_3\,\frac{d^3\sigma}{d^3\vec{p}_3}=\beta(t)\,(1-|x^*|)^{\gamma(t)} \tag{34}$$

have been tried, notably by Chen, Wang and Wong [71 C 7], Chou [71 C 6], Risk [71 R 3], Peccei and Pignotti [71 P 1] and Ko, Lander and Risk [71 K 2]. $\gamma(t)$ can be parametrized as

$$\gamma(t)=1-2\alpha(t)=1-2\alpha_0-2\alpha' t\,. \tag{35}$$

The values of α_0 and α', fitted to the 19.2 GeV/c data of Allaby *et al.* [70 A 1] by Chen, Wang and Wong, are given in 3.4.3, Tab. 1a and a graph due to Chou is shown in Fig. 29. x^* as used by Chen *et al.* is defined by the asymptotic formula (21): $1-|x^*|=m_4^2/s$; in the fitted region $p_{\rm T}\leqq 1$ GeV/c. The relation between the fitted values of α_0, α' and $\beta(t)$ and the known Regge trajectories and residue function is not always clear although in some cases success has been achieved [71 R 3, 71 K 2]; formula (34) has, however, at least some merit as a parametrization of a certain amount of data.

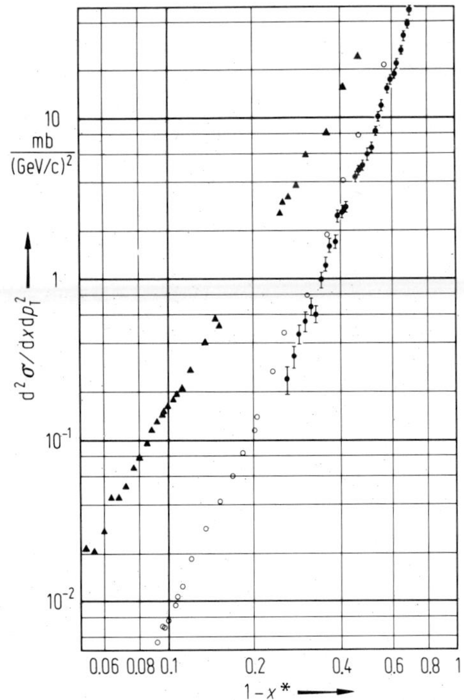

←

Fig. 29. The cross section for pion production at fixed $p_{\rm T}$ as a function of $1-x^*$ [71 C 6].

$$\left.\begin{array}{l}\pi^+\ \blacktriangle\\[2pt]\pi^-\ \circ\end{array}\right\}\ \text{from}\ p_1=19.2\ \text{GeV/c}\quad pp\ (p_{\rm T}=0)$$

$$\pi^-\ \text{⦙}\quad \text{from}\ p_1=30\ \text{GeV/c}\quad pp\ (p_{\rm T}=0.18\ \text{GeV/c})$$

3.4.2 Empirical and semi-empirical formulae

Several simple formulae have been proposed in order to give an approximate description of particle spectra.

Cocconi, Koester and Perkins (CKP) [61 C 1] introduced the concept of factorization in the longitudinal and the transverse momentum distribution and described the spectra by a formula of the type

$$\frac{d^2N}{d^3\vec{p}_3^*}=\frac{1}{\sigma_{\rm abs}}\,\frac{d^2\sigma}{d^3\vec{p}_3^*}=A\,e^{-p_{\rm L}^*/a}e^{-p_{\rm T}/b}\,. \tag{36}$$

The constants in the formula are determined by qualitative fits to actual data with utilization of the concepts of a constant $p_{\rm T}$ distribution, constant total cross section, and an increase of the multiplicity proportional to $E_1^{\frac{1}{4}}$. This equation gives an approximate description of the data and is handy for analytical manipulation because of its simplicity. From the c.m. equation an equally simple expression has been deduced for the forward lab. spectra. The constants for both cases are given in 3.4.3, Tab. 1 b; A is expressed in units of GeV$^{-3}\cdot$ interaction^{-1}. The CKP formulae do not reproduce the scaling in the Feynman sense.

Ranft [71 R 6] proposes two formulae, one for $\pi^\pm$ production, one for p production:

$$E_3 \frac{\mathrm{d}^2 N}{\mathrm{d}^3 \vec{p}_3} = U e^{-A x^{*2}} (e^{-B p_\perp^2} + C e^{-D p_\perp}) \qquad \text{for} \quad \pi^\pm , \tag{37}$$

$$\sqrt{s}\, \frac{\mathrm{d}^2 N^*}{\mathrm{d}^3 \vec{p}_3^*} = Q (1 + V x^* + W x^{*2})(e^{-B p_\perp^2} + C e^{-D p_\perp}) \quad \text{for} \quad \text{p}. \tag{38}$$

Here x^* is defined by $x^* = 2 p_L^* / \sqrt{s}$. The fitted constants are given in 3.4.3, Tab. 1 c. According to Ranft this formula gives a better description at large transverse momenta. The first formula fulfils Feynman scaling, the second naïve scaling. Very similar formulae have been proposed by Bali *et al.* [70 B 1], Bøggild *et al.* [71 B 1], and Sims *et al.* [71 S 2].

Other phenomenological formulae, not to be discussed here, have been suggested by Trilling [66 T 1], Sanford and Wang [67 S 1], and Liland and Pilkuhn [69 L 1].

3.4.3 Constants used in formulae

Tab. 1 a). Constants to be used in the formulae of Chen, Wang and Wong, Eqs. (34) and (35)

Particle	α_0	α' (GeV^{-2})
π^-	-2.0	0.78
π^+	-1.32	0.53
K^-	-3.55	0.70
K^+	-1.11	0.63
$\bar{p}$	-5.7	0.70
p	-0.1	0.43

Tab. 1 b). Constants to be used in the CKP formula, Eq. (36), for the evaluation of the proton, neutron and pion fluxes in inelastic proton-proton interactions

Particle	Lab. system			c.m. system		
	A (GeV^{-3}·inter-action^{-1})	a (GeV/c)	b (GeV/c)	A (GeV^{-3}·inter-action^{-1})	a (GeV/c)	b (GeV/c)
p or n	$1.7/p_1$	∞	0.22	$1.7/p_1^*$	∞	0.22
π^+, π^- or π^0	$14/(p_1^{\frac{1}{2}})$	$0.20(p_1^{\frac{3}{4}})$	0.22	5.0	$0.33(p_1^{*\frac{1}{2}})$	0.22

Tab. 1 c). Constants to be used in the Ranft formulae, Eqs. (37) and (38)

Particle	c.m. system							
	U (GeV^{-2}·inter-action^{-1})	Q (GeV^{-2}·inter-action^{-1})	A	V	W	B (GeV/c^{-2})	C	D (GeV/c^{-1})
p		1.39		0.43	-0.84	3.78	0.47	3.60
π^+	0.787		8.46			6.11	0.69	4.12
π^-	0.448		11.02			5.17	0.81	4.34

3.5 Survey on experiments

Tab. 2. Experiments on the production of a charged particle in pp interactions

p_1 (lab) (GeV/c)	Technique[a]	Detected particle	Systematic error (%)	Reference	Remarks	in 3.6
3.67	S	$\pi^\pm$	15	[62 M 1]		X
2.78 3.20 3.67	S	$\pi^\pm$	20	[66 R 1]	Graphs only	
3.20 3.67	S	K^+	25	[68 R 1]		X
3.35 3.70 3.86	S	K^+	16	[68 H 1]	Normalized to π^+ data of [62 M 1]	X
12.5	S	$\pi^\pm$	—	[66 L 1]	Backward	
12.5	S	$\pi^\pm$, $K^\pm$, $p^\pm$	10	[71 A 1]	Mostly $\pi^\pm$, p^+	X
19.0 24.0	S	$\pi^\pm$, $K^\pm$, $p^\pm$	—	[64 D 1]	Also d	
18.8 23.1	S	$\pi^\pm$, $K^\pm$, $p^\pm$	—	[65 D 1]	$\theta_3 = 0°$	
19.2	S	$\pi^\pm$, $K^\pm$, $p^\pm$	12	[70 A 1]	Also nuclei	X
14.25 19.2 24.0	S	$\pi^\pm$, $K^\pm$, $p^\pm$	12	[72 A 1]	Mostly at 24.0 GeV/c	X
9.9 20.0 29.7	S	$\pi^\pm$, p^+	20	[67 A 1]	Mostly at 29.7 GeV/c	X
500 1080 1500	S	π^+, p^+	15	[71 R 1]		
225 500 1080 1500	S	π^-, $\bar{p}$	15	[72 B 3]		
6.0	HBC	p^+	≈ 10	[70 A 3]		X
19.2	HBC	π^-	5	[71 B 1, 71 B 2]		X
12.9 18.0 21.1 24.1 28.4	HBC	π^-	5	[69 S 1]		X
12.0 24.0	HBC	$\pi^\pm$, p^+	≈ 5	[72 M 1]		
28.5	HBC	$\pi^\pm$	≈ 5	[71 S 2]		

[a]) Explanation of symbols: S = Scintillation counters, HBC = Hydrogen bubble chamber.

Diddens/Schlüpmann

3.6 Data tables: Cross sections for the production of charged particles in pp collisions as a function of various kinematical variables

The lab. cross sections are for particles produced forward in the c. m. system.
The tables have been ordered according to particle, in the sequence $\pi^-, \pi^+, K^-, K^+, \bar{p}.p.$ In the group for each particle the tables are roughly in order of increasing energy; data of one experiment remain consecutive.

3.6.1 pp→π⁻X

$p_1 = 3.670$ GeV/c; $p_1^* = 1.156$ GeV/c; $E_1^* + E_2^* = 2.978$ GeV; $s = 8.868$ GeV²; $\gamma_{c.m.} = 1.5871$; $\beta_{c.m.} = 0.77652$; $y_1^* = 1.0365$; $p_{3,max}^* = 0.798$ GeV/c; [62 M 1]

p_3	θ_3	$\dfrac{d^2\sigma}{d\Omega_3\,dp_3}$	Error	p_T	p_L^*	θ_3^*	$\dfrac{1}{p_3^{*2}}\cdot\dfrac{d^2\sigma}{d\Omega_3^*\,dp_3^*}$	$\dfrac{E_3}{p_3^2}\cdot\dfrac{d^2\sigma}{d\Omega_3\,dp_3}$	x^*	y_3^*	t	m_4	$\dfrac{d^2\sigma}{dt\,dm_4^2}$
GeV/c	mrad	cm² sr⁻¹ GeV⁻¹ c	%	GeV/c	GeV/c	deg	cm² sr⁻¹ GeV⁻³ c³	cm² sr⁻¹ GeV⁻² c³			GeV²	GeV	cm²/GeV⁴
0.437	0.0	3.896E−27	5	0.000	0.128	0.00	4.934E−26	9.355E−27	0.1608	0.8228	0.632	2.785	4.268E−27
0.604	0.0	5.140E−27	5	0.000	0.195	0.00	3.644E−26	8.730E−27	0.2441	1.1350	0.636	2.731	3.983E−27
0.759	0.0	6.183E−27	5	0.000	0.254	0.00	2.858E−26	8.277E−27	0.3180	1.3588	0.624	2.676	3.776E−27
0.910	0.0	6.972E−27	5	0.000	0.310	0.00	2.283E−26	7.753E−27	0.3881	1.5369	0.604	2.620	3.537E−27
1.057	0.0	6.917E−27	5	0.000	0.364	0.00	1.694E−26	6.599E−27	0.4559	1.6856	0.581	2.563	3.010E−27
1.203	0.0	6.022E−27	5	0.000	0.417	0.00	1.146E−26	5.039E−27	0.5225	1.8139	0.555	2.504	2.299E−27
1.348	0.0	4.572E−27	5	0.000	0.469	0.00	6.964E−27	3.410E−27	0.5882	1.9269	0.527	2.444	1.555E−27
1.492	0.0	2.850E−27	5	0.000	0.521	0.00	3.556E−27	1.918E−27	0.6533	2.0279	0.498	2.382	8.752E−28
1.636	0.0	2.276E−27	5	0.000	0.573	0.00	2.369E−27	1.397E−27	0.7180	2.1194	0.469	2.319	6.371E−28
1.779	0.0	1.421E−27	5	0.000	0.624	0.00	1.253E−27	8.014E−28	0.7824	2.2030	0.438	2.254	3.656E−28
1.922	0.0	8.514E−28	5	0.000	0.675	0.00	6.443E−28	4.443E−28	0.8465	2.2800	0.408	2.187	2.027E−28
2.064	0.0	2.835E−28	5	0.000	0.726	0.00	1.862E−28	1.377E−28	0.9104	2.3514	0.377	2.117	6.280E−29
0.384	246.8	2.111E−27	5	0.094	0.087	47.01	3.084E−26	5.848E−27	0.1096	0.4990	0.537	2.785	2.668E−27
0.526	274.3	3.783E−27	5	0.142	0.133	47.00	3.108E−26	7.446E−27	0.1665	0.6246	0.493	2.731	3.397E−27
0.658	285.8	4.314E−27	5	0.186	0.173	47.00	2.314E−26	6.700E−27	0.2169	0.6893	0.437	2.676	3.057E−27
0.787	291.9	3.879E−27	5	0.226	0.211	47.01	1.474E−26	5.007E−27	0.2646	0.7277	0.377	2.620	2.284E−27
0.913	295.5	2.979E−27	5	0.266	0.248	47.00	8.469E−27	3.299E−27	0.3110	0.7527	0.313	2.563	1.505E−27
1.039	297.9	2.346E−27	5	0.305	0.284	47.00	5.185E−27	2.279E−27	0.3563	0.7696	0.248	2.504	1.040E−27
1.163	299.5	1.846E−27	5	0.343	0.320	46.99	3.266E−27	1.599E−27	0.4012	0.7818	0.182	2.444	7.293E−28
1.287	300.7	1.595E−27	5	0.381	0.355	47.00	2.311E−27	1.247E−27	0.4456	0.7907	0.115	2.382	5.688E−28
1.410	301.6	1.102E−27	5	0.419	0.391	47.00	1.332E−27	7.853E−28	0.4897	0.7975	0.047	2.319	3.582E−28
1.533	302.3	7.339E−28	5	0.456	0.426	47.00	7.515E−28	4.806E−28	0.5336	0.8027	−0.021	2.254	2.193E−28
1.656	302.8	4.887E−28	5	0.494	0.461	47.00	4.295E−28	2.961E−28	0.5773	0.8070	−0.089	2.187	1.351E−28
1.779	303.3	1.709E−28	5	0.531	0.495	47.01	1.303E−28	9.638E−29	0.6208	0.8101	−0.157	2.117	4.397E−29
0.321	390.2	1.651E−27	5	0.122	0.040	71.96	2.957E−26	5.609E−27	0.0499	0.2128	0.427	2.785	2.559E−27
0.432	442.5	2.599E−27	5	0.185	0.060	72.01	2.641E−26	6.323E−27	0.0753	0.2564	0.326	2.732	2.884E−27
0.538	464.7	3.369E−27	5	0.241	0.078	72.01	2.235E−26	6.469E−27	0.0982	0.2775	0.219	2.677	2.951E−27
0.642	476.5	2.917E−27	5	0.294	0.096	71.99	1.369E−26	4.650E−27	0.1200	0.2896	−0.110	2.620	2.121E−27
0.744	483.7	2.352E−27	5	0.346	0.112	72.00	8.254E−27	3.216E−27	0.1409	0.2969	−0.001	2.563	1.467E−27

p_3	θ_3	$\dfrac{d^2\sigma}{d\Omega_3\,dp_3}$	Error	p_T	p_L^*	θ_3^*	$\dfrac{1}{p_3^{*2}}\cdot\dfrac{d^2\sigma}{d\Omega_3^*\,dp_3^*}$	$\dfrac{E_3}{p_3^2}\cdot\dfrac{d^2\sigma}{d\Omega_3\,dp_3}$	x^*	y_3^*	t	m_4	$\dfrac{d^2\sigma}{dt\,dm_4^2}$
GeV/c	mrad	cm² sr⁻¹ GeV⁻¹ c	%	GeV/c	GeV/c	deg	cm² sr⁻¹ GeV⁻³ c³	cm² sr⁻¹ GeV⁻² c³			GeV²	GeV	cm²/GeV⁴

$p_1 = 3.670$ GeV/c; $p_1^* = 1.156$ GeV/c; $E_1^* + E_2^* = 2.978$ GeV; $s = 8.868$ GeV²; $\gamma_{c.m.} = 1.5871$; $\beta_{c.m.} = 0.77652$; $y_1^* = 1.0365$; $p_{3,\,max}^* = 0.798$ GeV/c; [62 M 1] (cont.)

p_3	θ_3	$\dfrac{d^2\sigma}{d\Omega_3\,dp_3}$	Error	p_T	p_L^*	θ_3^*	$\dfrac{1}{p_3^{*2}}\cdot\dfrac{d^2\sigma}{d\Omega_3^*\,dp_3^*}$	$\dfrac{E_3}{p_3^2}\cdot\dfrac{d^2\sigma}{d\Omega_3\,dp_3}$	x^*	y_3^*	t	m_4	$\dfrac{d^2\sigma}{dt\,dm_4^2}$
0.845	488.4	1.799E-27	5	0.396	0.129	72.00	4.908E-27	2.158E-27	0.1615	0.3018	-0.112	2.504	9.844E-28
0.946	491.6	1.395E-27	5	0.447	0.145	72.00	3.043E-27	1.491E-27	0.1819	0.3054	-0.223	2.443	6.800E-28
1.046	493.9	9.942E-28	5	0.496	0.161	71.99	1.777E-27	9.589E-28	0.2020	0.3080	-0.335	2.382	4.375E-28
1.145	495.6	6.550E-28	5	0.545	0.177	71.99	9.779E-28	5.763E-28	0.2219	0.3099	-0.446	2.319	2.629E-28
1.245	497.0	3.965E-28	5	0.594	0.193	72.00	5.011E-28	3.205E-28	0.2418	0.3113	-0.559	2.254	1.462E-28
1.344	498.1	1.584E-28	5	0.642	0.209	72.00	1.719E-28	1.185E-28	0.2615	0.3123	-0.671	2.187	5.406E-29
1.444	498.9	3.957E-29	5	0.691	0.225	72.00	3.722E-29	2.753E-29	0.2814	0.3134	-0.784	2.117	1.256E-29
0.288	457.8	2.122E-27	5	0.127	0.016	83.01	4.317E-26	8.185E-27	0.0196	0.0825	0.371	2.785	3.734E-27
0.385	526.0	2.713E-27	5	0.193	0.024	83.01	3.130E-26	7.498E-27	0.0297	0.0993	0.241	2.731	3.421E-27
0.478	555.3	2.440E-27	5	0.252	0.031	83.01	1.838E-26	5.321E-27	0.0387	0.1071	0.109	2.676	2.428E-27
0.569	571.0	1.873E-27	5	0.307	0.038	83.01	9.990E-27	3.393E-27	0.0472	0.1114	-0.024	2.620	1.548E-27
0.658	580.5	1.416E-27	5	0.361	0.044	83.00	5.644E-27	2.199E-27	0.0555	0.1142	-0.158	2.563	1.003E-27
0.747	586.7	1.005E-27	5	0.414	0.051	83.00	3.112E-27	1.368E-27	0.0637	0.1161	-0.292	2.504	6.241E-28
0.836	591.0	7.379E-28	5	0.466	0.057	83.00	1.828E-27	8.950E-28	0.0717	0.1174	-0.426	2.444	4.083E-28
0.924	594.1	5.259E-28	5	0.517	0.064	83.00	1.067E-27	5.756E-28	0.0796	0.1183	-0.560	2.382	2.626E-28
1.012	596.5	3.501E-28	5	0.569	0.070	83.00	5.923E-28	3.492E-28	0.0875	0.1189	-0.695	2.319	1.593E-28
1.100	598.3	2.098E-28	5	0.620	0.076	83.00	3.006E-28	1.923E-28	0.0953	0.1194	-0.829	2.254	8.772E-29
1.188	599.7	8.733E-29	5	0.670	0.082	83.00	1.074E-28	7.404E-29	0.1031	0.1198	-0.964	2.186	3.378E-29

$p_1 = 12.500$ GeV/c; $p_1^* = 2.332$ GeV/c; $E_1^* + E_2^* = 5.028$ GeV; $s = 25.281$ GeV²; $\gamma_{c.m.} = 2.6796$; $\beta_{c.m.} = 0.92776$; $y_1^* = 1.6420$; $p_{3,\,max}^* = 2.107$ GeV/c; [71 A 1]

p_3	θ_3	$\dfrac{d^2\sigma}{d\Omega_3\,dp_3}$	Error	p_T	p_L^*	θ_3^*	$\dfrac{1}{p_3^{*2}}\cdot\dfrac{d^2\sigma}{d\Omega_3^*\,dp_3^*}$	$\dfrac{E_3}{p_3^2}\cdot\dfrac{d^2\sigma}{d\Omega_3\,dp_3}$	x^*	y_3^*	t	m_4	$\dfrac{d^2\sigma}{dt\,dm_4^2}$
3.139	0.0	3.209E-26	11	0.000	0.600	0.00	1.661E-26	1.023E-26	0.2847	2.1645	0.601	4.371	1.371E-27
3.182	44.5	2.311E-26	10	0.141	0.600	13.26	1.150E-26	7.270E-27	0.2847	1.8245	0.521	4.353	9.737E-28
3.224	62.1	1.872E-26	9	0.200	0.600	18.43	8.974E-27	5.812E-27	0.2847	1.6323	0.442	4.335	7.785E-28
3.285	80.6	1.556E-26	8	0.265	0.600	23.79	7.072E-27	4.741E-27	0.2847	1.4462	0.328	4.308	6.350E-28
3.325	90.3	1.325E-26	7	0.300	0.600	26.56	5.821E-27	3.988E-27	0.2848	1.3569	0.253	4.291	5.342E-28
3.363	98.8	1.096E-26	6	0.332	0.600	28.93	4.663E-27	3.262E-27	0.2847	1.2841	0.181	4.274	4.369E-28
3.546	129.6	5.966E-27	6	0.458	0.600	37.38	2.193E-27	1.684E-27	0.2847	1.0490	-0.162	4.193	2.255E-28
3.714	150.5	3.910E-27	5	0.557	0.600	42.87	1.269E-27	1.054E-27	0.2847	0.9129	-0.477	4.117	1.411E-28
3.885	167.6	2.541E-27	5	0.648	0.600	47.21	7.320E-28	6.545E-28	0.2847	0.8126	-0.797	4.038	8.766E-29
4.031	179.9	1.641E-27	5	0.721	0.600	50.24	4.294E-28	4.073E-28	0.2847	0.7457	-1.071	3.970	5.456E-29
4.168	190.0	1.109E-27	5	0.787	0.600	52.68	2.663E-28	2.662E-28	0.2848	0.6936	-1.327	3.905	3.566E-29
4.299	198.7	7.634E-28	5	0.849	0.600	54.74	1.694E-28	1.777E-28	0.2847	0.6507	-1.574	3.841	2.380E-29
4.424	206.1	5.048E-28	6	0.905	0.600	56.46	1.042E-28	1.142E-28	0.2848	0.6155	-1.807	3.780	1.529E-29
4.556	213.3	3.692E-28	6	0.964	0.600	58.11	7.084E-29	8.107E-29	0.2847	0.5822	-2.056	3.714	1.086E-29
4.637	217.4	3.074E-28	4	1.000	0.600	59.04	5.646E-29	6.632E-29	0.2847	0.5638	-2.208	3.673	8.883E-30
4.671	219.0	2.552E-28	6	1.015	0.600	59.40	4.604E-29	5.466E-29	0.2848	0.5566	-2.270	3.655	7.321E-30
4.782	224.2	1.852E-28	6	1.063	0.600	60.56	3.153E-29	3.875E-29	0.2847	0.5338	-2.480	3.598	5.190E-30
4.857	227.5	1.643E-28	4	1.095	0.600	61.29	2.693E-29	3.384E-29	0.2847	0.5195	-2.621	3.558	4.533E-30
4.889	228.8	1.260E-28	6	1.109	0.600	61.58	2.032E-29	2.578E-29	0.2848	0.5140	-2.679	3.542	3.453E-30
5.003	233.5	9.953E-29	6	1.158	0.600	62.61	1.518E-29	1.990E-29	0.2847	0.4941	-2.895	3.481	2.666E-30
5.064	235.8	8.825E-29	4	1.183	0.600	63.11	1.307E-29	1.743E-29	0.2848	0.4845	-3.008	3.448	2.335E-30

$p_1 = 12.500$ GeV/c; $p_1^* = 2.332$ GeV/c; $E_1^* + E_2^* = 5.028$ GeV; $s = 25.281$ GeV^2; $\gamma_{c.m.} = 2.6796$; $\beta_{c.m.} = 0.92776$; $y_1^* = 1.6420$; $p_{3,max}^* = 2.107$ GeV/c; [71 A 1] (cont.)

p_3	θ_3	$\dfrac{d^2\sigma}{d\Omega_3\,dp_3}$	Error	p_T	p_L^*	θ_3^*	$\dfrac{1}{p_3^{*2}}\cdot\dfrac{d^2\sigma}{d\Omega_3^*\,dp_3^*}$	$\dfrac{E_3}{p_3^2}\cdot\dfrac{d^2\sigma}{d\Omega_3\,dp_3}$	x^*	y_3^*	t	m_4	$\dfrac{d^2\sigma}{dt\,dm_4^2}$
GeV/c	mrad	cm^2 sr^{-1} GeV^{-1} c	%	GeV/c	GeV/c	deg	cm^2 sr^{-1} GeV^{-3} c^3	cm^2 sr^{-1} GeV^{-2} c^3			GeV^2	GeV	cm^2/GeV^4
5.104	237.3	6.882E-29	7	1.200	0.600	63.43	1.000E-29	1.349E-29	0.2848	0.4783	-3.083	3.426	1.807E-30
5.260	242.9	4.680E-29	4	1.265	0.600	64.63	6.325E-30	8.900E-30	0.2847	0.4554	-3.377	3.339	1.192E-30
5.446	248.9	2.597E-29	4	1.342	0.600	65.90	3.231E-30	4.770E-30	0.2848	0.4314	-3.724	3.233	6.389E-31
5.623	254.2	1.466E-29	5	1.414	0.600	67.01	1.691E-30	2.608E-30	0.2848	0.4106	-4.057	3.129	3.493E-31
5.794	258.9	8.212E-30	5	1.483	0.600	67.98	8.827E-31	1.418E-30	0.2847	0.3926	-4.377	3.025	1.899E-31
5.957	263.1	4.281E-30	5	1.549	0.600	68.84	4.312E-31	7.188E-31	0.2847	0.3766	-4.685	2.922	9.628E-32
6.115	266.8	2.296E-30	6	1.612	0.600	69.58	2.176E-31	3.756E-31	0.2848	0.3628	-4.979	2.819	5.030E-32
6.268	270.2	1.390E-30	6	1.673	0.600	70.27	1.244E-31	2.218E-31	0.2848	0.3503	-5.265	2.715	2.971E-32
6.416	273.3	7.862E-31	6	1.732	0.600	70.89	6.668E-32	1.226E-31	0.2848	0.3389	-5.543	2.611	1.642E-32
6.560	276.2	4.392E-31	7	1.789	0.600	71.46	3.539E-32	6.697E-32	0.2847	0.3284	-5.815	2.505	8.970E-33
6.700	278.8	2.694E-31	7	1.844	0.600	71.97	2.069E-32	4.022E-32	0.2848	0.3191	-6.076	2.398	5.387E-33
6.836	281.3	1.274E-31	9	1.898	0.600	72.46	9.343E-33	1.864E-32	0.2847	0.3102	-6.334	2.289	2.497E-33
6.968	283.5	7.841E-32	11	1.949	0.600	72.89	5.506E-33	1.126E-32	0.2848	0.3025	-6.579	2.178	1.508E-33
7.098	285.6	4.735E-32	12	2.000	0.600	73.29	3.189E-33	6.672E-33	0.2848	0.2951	-6.822	2.064	8.937E-34
2.407	131.8	1.441E-26	9	0.316	0.400	38.34	1.134E-26	5.997E-27	0.1898	0.9880	0.107	4.470	8.032E-28
2.641	170.1	7.493E-27	6	0.447	0.400	48.18	4.613E-27	2.841E-27	0.1898	0.7742	-0.331	4.371	3.805E-28
2.847	193.6	4.622E-27	6	0.548	0.400	53.85	2.347E-27	1.625E-27	0.1899	0.6591	-0.716	4.282	2.177E-28
3.031	210.2	3.017E-27	6	0.632	0.400	57.69	1.309E-27	9.964E-28	0.1898	0.5839	-1.062	4.201	1.335E-28
3.200	222.8	2.404E-27	12	0.707	0.400	60.51	9.123E-28	7.520E-28	0.1898	0.5298	-1.379	4.125	1.007E-28
3.357	232.8	1.734E-27	17	0.774	0.400	62.68	5.856E-28	5.170E-28	0.1898	0.4886	-1.673	4.053	6.924E-29
1.919	241.2	7.484E-27	11	0.458	0.210	65.38	7.474E-27	3.910E-27	0.0997	0.4254	-0.751	4.477	5.237E-28
2.295	201.0	7.773E-27	10	0.458	0.310	55.93	5.948E-27	3.393E-27	0.1471	0.6087	-0.523	4.423	4.545E-28
2.705	170.2	7.389E-27	9	0.458	0.410	48.18	4.338E-27	2.735E-27	0.1946	0.7758	-0.358	4.354	3.664E-28
3.140	146.5	6.723E-27	8	0.458	0.510	41.95	3.063E-27	2.143E-27	0.2420	0.9262	-0.240	4.274	2.871E-28
3.638	126.3	5.894E-27	7	0.458	0.620	36.47	2.069E-27	1.621E-27	0.2942	1.0749	-0.148	4.174	2.172E-28
4.105	111.9	5.114E-27	6	0.458	0.720	32.48	1.441E-27	1.247E-27	0.3417	1.1962	-0.090	4.075	1.670E-28
4.581	100.2	4.430E-27	6	0.458	0.820	29.20	1.019E-27	9.675E-28	0.3891	1.3067	-0.050	3.969	1.296E-28
5.065	90.6	3.718E-27	5	0.458	0.920	26.48	7.079E-28	7.343E-28	0.4366	1.4075	-0.024	3.856	9.836E-29
5.554	82.6	3.022E-27	5	0.458	1.020	24.19	4.830E-28	5.443E-28	0.4841	1.5000	-0.008	3.737	7.290E-29
6.047	75.9	2.509E-27	5	0.458	1.120	22.25	3.407E-28	4.150E-28	0.5315	1.5853	-0.001	3.613	5.559E-29
6.593	69.6	2.000E-27	5	0.458	1.230	20.43	2.299E-28	3.034E-28	0.5837	1.6720	-0.000	3.468	4.064E-29
7.093	64.7	1.489E-27	5	0.458	1.330	19.01	1.485E-28	2.100E-28	0.6312	1.7453	-0.004	3.329	2.812E-29
7.595	60.4	9.445E-28	5	0.458	1.430	17.77	8.248E-29	1.244E-28	0.6786	1.8138	-0.012	3.184	1.666E-29
8.099	56.6	4.881E-28	5	0.458	1.530	16.68	3.759E-29	6.028E-29	0.7261	1.8780	-0.024	3.029	8.073E-30
2.049	317.8	2.623E-27	11	0.640	0.110	80.26	1.931E-27	1.283E-27	0.0522	0.1669	-1.929	4.315	1.719E-28
2.362	274.5	2.844E-27	10	0.640	0.210	71.84	1.753E-27	1.206E-27	0.0997	0.3153	-1.581	4.287	1.616E-28
2.710	238.6	2.884E-27	9	0.640	0.310	64.18	1.470E-27	1.066E-27	0.1471	0.4568	-1.300	4.244	1.427E-28
3.088	208.9	2.834E-27	8	0.640	0.410	57.37	1.188E-27	9.187E-28	0.1946	0.5907	-1.075	4.186	1.230E-28
3.532	182.3	2.712E-27	7	0.640	0.520	50.92	9.185E-28	7.684E-28	0.2468	0.7276	-0.881	4.109	1.029E-28
3.956	162.6	2.532E-27	6	0.640	0.620	45.93	7.099E-28	6.404E-28	0.2942	0.8425	-0.745	4.028	8.578E-29

Diddens/Schlüpmann

p_3	θ_3	$\dfrac{d^2\sigma}{d\Omega_3\,dp_3}$	Error	p_T	p_L^*	θ_3^*	$\dfrac{1}{p_3^{*\,2}}\cdot\dfrac{d^2\sigma}{d\Omega_3^*\,dp_3^*}$	$\dfrac{E_3}{p_3^2}\cdot\dfrac{d^2\sigma}{d\Omega_3\,dp_3}$	x^*	y_3^*	t	m_4	$\dfrac{d^2\sigma}{dt\,dm_4^2}$
GeV/c	mrad	cm² sr⁻¹ GeV⁻¹ c	%	GeV/c	GeV/c	deg	cm² sr⁻¹ GeV⁻³ c³	cm² sr⁻¹ GeV⁻² c³			GeV²	GeV	cm²/GeV⁴

$p_1 = 12.500$ GeV/c; $p_1^* = 2.332$ GeV/c; $E_1^* + E_2^* = 5.028$ GeV; $s = 25.281$ GeV²; $\gamma_{c.m.} = 2.6796$; $\beta_{c.m.} = 0.92776$; $y_1^* = 1.6420$; $p_{3,\,max}^* = 2.107$ GeV/c; [71 A 1] (cont.)

p_3	θ_3	$\dfrac{d^2\sigma}{d\Omega_3\,dp_3}$	Error	p_T	p_L^*	θ_3^*	$\dfrac{1}{p_3^{*\,2}}\cdot\dfrac{d^2\sigma}{d\Omega_3^*\,dp_3^*}$	$\dfrac{E_3}{p_3^2}\cdot\dfrac{d^2\sigma}{d\Omega_3\,dp_3}$	x^*	y_3^*	t	m_4	$\dfrac{d^2\sigma}{dt\,dm_4^2}$
4.397	146.2	2.309E-27	6	0.641	0.720	41.66	5.396E-28	5.254E-28	0.3417	0.9492	-0.638	3.938	7.037E-29
4.849	132.4	1.913E-27	5	0.640	0.820	37.98	3.760E-28	3.947E-28	0.3891	1.0486	-0.553	3.840	5.286E-29
5.312	120.8	1.604E-27	5	0.640	0.920	34.83	2.674E-28	3.021E-28	0.4366	1.1405	-0.488	3.734	4.046E-29
5.783	111.0	1.223E-27	5	0.641	1.020	32.13	1.745E-28	2.115E-28	0.4840	1.2253	-0.439	3.620	2.833E-29
2.272	292.8	1.979E-27	13	0.656	0.170	75.46	1.262E-27	8.727E-28	0.0807	0.2510	-1.785	4.283	1.169E-28
2.818	245.8	1.963E-27	8	0.686	0.310	65.68	9.113E-28	6.974E-28	0.1471	0.4296	-1.503	4.196	9.342E-29
3.959	174.0	1.990E-27	4	0.685	0.600	48.80	5.457E-28	5.030E-28	0.2848	0.7773	-0.935	4.004	6.737E-29
3.416	206.4	1.760E-27	6	0.700	0.460	56.70	6.072E-28	5.157E-28	0.2183	0.6064	-1.225	4.094	6.907E-29
3.352	307.6	3.254E-28	7	1.015	0.220	77.76	9.273E-29	9.716E-29	0.1044	0.2132	-3.342	3.842	1.301E-29
3.669	280.3	3.103E-28	7	1.015	0.320	72.50	7.885E-29	8.463E-29	0.1518	0.3074	-3.005	3.809	1.134E-29
4.009	256.0	2.941E-28	7	1.015	0.420	67.53	6.629E-29	7.340E-29	0.1993	0.3991	-2.709	3.764	9.832E-30
4.369	234.4	2.809E-28	6	1.015	0.520	62.86	5.600E-29	6.433E-29	0.2468	0.4881	-2.450	3.708	8.616E-30
4.748	215.4	2.499E-28	5	1.015	0.620	58.58	4.397E-29	5.266E-29	0.2942	0.5733	-2.229	3.641	7.053E-30
5.143	198.6	2.104E-28	5	1.015	0.720	54.64	3.269E-29	4.093E-29	0.3417	0.6551	-2.037	3.565	5.482E-30
5.553	183.8	1.814E-28	5	1.015	0.820	51.06	2.490E-29	3.268E-29	0.3891	0.7330	-1.873	3.479	4.377E-30
6.018	169.5	1.410E-28	5	1.015	0.930	47.51	1.694E-29	2.344E-29	0.4413	0.8144	-1.720	3.374	3.139E-30
6.452	158.0	1.031E-28	5	1.015	1.030	44.59	1.100E-29	1.598E-29	0.4888	0.8849	-1.601	3.270	2.141E-30
6.895	147.7	6.973E-29	5	1.015	1.130	41.92	6.632E-30	1.012E-29	0.5363	0.9526	-1.497	3.156	1.355E-30
7.346	138.6	4.327E-29	5	1.015	1.230	39.53	3.680E-30	5.891E-30	0.5837	1.0164	-1.411	3.034	7.891E-31
7.804	130.4	2.523E-29	5	1.015	1.330	37.34	1.926E-30	3.233E-30	0.6312	1.0775	-1.337	2.901	4.331E-31
8.268	123.1	1.263E-29	6	1.015	1.430	35.37	8.684E-31	1.528E-30	0.6786	1.1353	-1.275	2.759	2.046E-31
8.737	116.4	5.487E-30	7	1.015	1.530	33.55	3.411E-31	6.281E-31	0.7262	1.1914	-1.221	2.605	8.413E-32

$p_1 = 12.880$ GeV; $p_1^* = 2.370$ GeV/c; $E_1^* + E_2^* = 5.098$ GeV; $s = 25.993$ GeV²; $\gamma_{c.m.} = 2.7171$; $\beta_{c.m.} = 0.92981$; $y_1^* = 1.6570$; $p_{3,\,max}^* = 2.148$ GeV/c; [69 S 1]

p_3	θ_3	$\dfrac{d^2\sigma}{d\Omega_3\,dp_3}$	Error	p_T	p_L^*	θ_3^*	$\dfrac{1}{p_3^{*\,2}}\cdot\dfrac{d^2\sigma}{d\Omega_3^*\,dp_3^*}$	$\dfrac{E_3}{p_3^2}\cdot\dfrac{d^2\sigma}{d\Omega_3\,dp_3}$	x^*	y_3^*	t	m_4	$\dfrac{d^2\sigma}{dt\,dm_4^2}$
10.036	5.0	5.282E-27	57	0.050	1.911	1.50	2.746E-28	5.264E-28	0.8898	3.2510	0.186	2.543	6.842E-29
7.432	6.7	6.944E-27	37	0.050	1.414	2.03	6.575E-28	9.345E-28	0.6581	2.9507	0.354	3.394	1.215E-28
4.961	10.1	1.531E-26	17	0.050	0.940	3.04	3.243E-27	3.087E-27	0.4379	2.5466	0.504	4.038	4.013E-28
2.837	17.6	2.209E-26	9	0.050	0.531	5.38	1.414E-26	7.796E-27	0.2473	1.9880	0.606	4.515	1.013E-27
1.461	34.2	1.641E-26	8	0.050	0.259	10.91	3.775E-26	1.128E-26	0.1208	1.3259	0.606	4.792	1.467E-27
0.832	60.1	1.678E-26	7	0.050	0.125	21.76	1.054E-25	2.045E-26	0.0583	0.7670	0.504	4.902	2.658E-27
0.551	90.8	1.338E-26	7	0.050	0.055	42.28	1.584E-25	2.505E-26	0.0256	0.3625	0.354	4.940	3.256E-27
0.404	124.1	1.124E-26	8	0.050	0.009	79.36	1.981E-25	2.944E-26	0.0044	0.0633	0.187	4.949	3.826E-27
10.387	14.4	1.922E-27	58	0.150	1.976	4.34	9.317E-29	1.851E-28	0.9198	2.9620	0.139	2.400	2.406E-29
7.880	19.0	4.256E-27	30	0.150	1.496	5.73	3.578E-28	5.402E-28	0.6964	2.6856	0.293	3.258	7.022E-29
5.553	27.0	1.197E-26	13	0.150	1.049	8.14	2.017E-27	2.156E-27	0.4884	2.3356	0.423	3.887	2.803E-28
3.587	41.8	1.652E-26	8	0.150	0.669	12.64	6.590E-27	4.609E-27	0.3113	1.8986	0.504	4.345	5.991E-28
2.212	67.8	2.290E-26	6	0.150	0.397	20.70	2.322E-26	1.037E-26	0.1848	1.4152	0.504	4.632	1.348E-27
1.429	105.1	2.170E-26	5	0.150	0.234	32.65	4.907E-26	1.526E-26	0.1089	0.9783	0.423	4.779	1.983E-27
1.008	149.4	2.035E-26	5	0.150	0.137	47.51	8.260E-26	2.038E-26	0.0640	0.6284	0.293	4.847	2.649E-27

p_3	θ_3	$\dfrac{d^2\sigma}{d\Omega_3\,dp_3}$	Error	p_T	p_L^*	θ_3^*	$\dfrac{1}{p_3^{*2}}\cdot\dfrac{d^2\sigma}{d\Omega_3^*\,dp_3^*}$	$\dfrac{E_3}{p_3^2}\cdot\dfrac{d^2\sigma}{d\Omega_3\,dp_3}$	x^*	y_3^*	t	m_4	$\dfrac{d^2\sigma}{dt\,dm_4^2}$
GeV/c	mrad	cm² sr⁻¹ GeV⁻¹ c	%	GeV/c	GeV/c	deg	cm² sr⁻¹ GeV⁻³ c³	cm² sr⁻¹ GeV⁻² c³			GeV²	GeV	cm²/GeV⁴

$p_1 = 12.880$ GeV; $p_1^* = 2.370$ GeV/c; $E_1^* + E_2^* = 5.098$ GeV; $s = 25.993$ GeV²; $\gamma_{c.m.} = 2.7171$; $\beta_{c.m.} = 0.92981$; $y_1^* = 1.6570$; $p_{3,\,max}^* = 2.148$ GeV/c; [69 S 1] (cont.)

p_3	θ_3	$\dfrac{d^2\sigma}{d\Omega_3\,dp_3}$	Error	p_T	p_L^*	θ_3^*	$\dfrac{1}{p_3^{*2}}\cdot\dfrac{d^2\sigma}{d\Omega_3^*\,dp_3^*}$	$\dfrac{E_3}{p_3^2}\cdot\dfrac{d^2\sigma}{d\Omega_3\,dp_3}$	x^*	y_3^*	t	m_4	$\dfrac{d^2\sigma}{dt\,dm_4^2}$
0.765	197.4	1.619E-26	6	0.150	0.074	63.86	9.879E-26	2.151E-26	0.0343	0.3519	0.139	4.878	2.796E-27
0.613	247.4	1.438E-26	6	0.150	0.027	79.97	1.164E-25	2.406E-26	0.0124	0.1291	-0.028	4.889	3.127E-27
8.663	28.9	1.216E-27	50	0.250	1.640	8.67	8.435E-29	1.404E-28	0.7633	2.4457	0.186	3.007	1.825E-29
6.507	38.4	4.217E-27	22	0.250	1.224	11.54	5.156E-28	6.482E-28	0.5699	2.1594	0.293	3.632	8.426E-29
4.685	53.4	9.923E-27	11	0.250	0.870	16.03	2.313E-27	2.119E-27	0.4051	1.8307	0.354	4.083	2.754E-28
3.312	75.6	1.671E-26	7	0.250	0.599	22.67	7.611E-27	5.050E-27	0.2787	1.4833	0.354	4.387	6.564E-28
2.386	105.0	1.841E-26	6	0.250	0.409	31.44	1.548E-26	7.729E-27	0.1904	1.1543	0.293	4.574	1.005E-27
1.795	139.8	1.609E-26	6	0.250	0.281	41.67	2.240E-26	8.991E-27	0.1309	0.8680	0.186	4.682	1.169E-27
1.413	177.9	1.584E-26	6	0.250	0.192	52.55	3.270E-26	1.126E-26	0.0892	0.6269	0.051	4.743	1.464E-27
1.156	218.1	1.425E-26	6	0.250	0.125	63.48	3.976E-26	1.243E-26	0.0581	0.4231	-0.102	4.778	1.615E-27
0.975	259.4	1.356E-26	6	0.250	0.072	73.90	4.757E-26	1.405E-26	0.0336	0.2495	-0.264	4.796	1.826E-27
0.842	301.5	1.008E-26	8	0.250	0.028	83.53	4.217E-26	1.213E-26	0.0132	0.0988	-0.433	4.804	1.577E-27
9.656	36.3	8.636E-28	57	0.350	1.822	10.88	4.808E-29	8.945E-29	0.8481	2.2794	0.051	2.654	1.163E-29
7.641	45.8	3.452E-27	24	0.350	1.432	13.73	3.051E-28	4.518E-28	0.6668	2.0453	0.139	3.303	5.874E-29
5.917	59.2	4.397E-27	18	0.350	1.096	17.71	6.413E-28	7.433E-28	0.5103	1.7892	0.186	3.767	9.662E-29
4.544	77.1	6.555E-27	13	0.350	0.825	23.00	1.592E-27	1.443E-27	0.3839	1.5246	0.186	4.095	1.876E-28
3.521	99.6	9.537E-27	10	0.350	0.617	29.56	3.749E-27	2.711E-27	0.2873	1.2687	0.139	4.317	3.524E-28
2.789	125.8	1.028E-26	9	0.350	0.463	37.07	6.181E-27	3.691E-27	0.2157	1.0347	0.051	4.464	4.797E-28
2.270	154.8	1.224E-26	8	0.350	0.348	45.14	1.053E-26	5.402E-27	0.1622	0.8269	-0.065	4.558	7.022E-28
1.896	185.7	9.867E-27	9	0.350	0.260	53.39	1.140E-26	5.218E-27	0.1211	0.6445	-0.202	4.620	6.783E-28
1.620	217.7	1.055E-26	8	0.350	0.190	61.51	1.549E-26	6.536E-27	0.0884	0.4849	-0.351	4.659	8.497E-28
1.411	250.6	1.010E-26	8	0.350	0.132	69.34	1.802E-26	7.193E-27	0.0614	0.3434	-0.510	4.684	9.350E-28
1.249	284.0	8.766E-27	9	0.350	0.083	76.72	1.831E-26	7.062E-27	0.0385	0.2175	-0.675	4.699	9.180E-28
1.120	317.7	8.579E-27	9	0.350	0.039	83.57	2.038E-26	7.719E-27	0.0184	0.1045	-0.844	4.706	1.003E-27
1.016	351.7	7.859E-27	9	0.350	0.001	89.89	2.072E-26	7.808E-27	0.0003	0.0018	-1.018	4.708	1.015E-27
8.864	50.8	6.742E-28	57	0.450	1.657	15.20	4.417E-29	7.607E-29	0.7713	1.9701	-0.028	2.907	9.888E-30
7.207	62.5	2.596E-27	25	0.450	1.333	18.66	2.548E-28	3.603E-28	0.6205	1.7629	0.011	3.405	4.683E-29
5.834	77.2	2.981E-27	21	0.450	1.061	22.98	4.402E-28	5.111E-28	0.4941	1.5511	0.011	3.765	6.644E-29
4.745	95.0	2.737E-27	21	0.450	0.842	28.13	5.983E-28	5.771E-28	0.3918	1.3438	-0.028	4.022	7.501E-29
3.907	115.4	6.114E-27	13	0.450	0.668	33.95	1.915E-27	1.566E-27	0.3111	1.1488	-0.101	4.204	2.035E-28
3.271	138.0	7.327E-27	11	0.450	0.532	40.23	3.156E-27	2.242E-27	0.2476	0.9696	-0.202	4.332	2.914E-28
2.787	162.2	7.664E-27	11	0.450	0.423	46.76	4.347E-27	2.753E-27	0.1971	0.8076	-0.323	4.422	3.579E-28
2.414	187.5	6.821E-27	11	0.450	0.335	53.31	4.895E-27	2.830E-27	0.1561	0.6621	-0.459	4.485	3.679E-28
2.122	213.7	7.757E-27	10	0.450	0.262	59.80	6.795E-27	3.663E-27	0.1220	0.5307	-0.607	4.529	4.762E-28
1.890	240.4	7.204E-27	11	0.450	0.200	66.06	7.469E-27	3.822E-27	0.0930	0.4122	-0.762	4.560	4.968E-28
1.702	267.5	6.386E-27	11	0.450	0.146	72.06	7.635E-27	3.765E-27	0.0678	0.3046	-0.923	4.581	4.894E-28
1.548	294.9	6.274E-27	11	0.450	0.098	77.74	8.458E-27	4.069E-27	0.0455	0.2061	-1.090	4.594	5.290E-28
1.420	322.5	6.240E-27	12	0.450	0.055	83.08	9.308E-27	4.416E-27	0.0254	0.1156	-1.260	4.602	5.740E-28
1.311	350.3	4.843E-27	13	0.450	0.015	88.09	7.882E-27	3.715E-27	0.0070	0.0318	-1.432	4.605	4.829E-28

p_3	θ_3	$\dfrac{\mathrm{d}^2\sigma}{\mathrm{d}\Omega_3\,\mathrm{d}p_3}$	Error	p_T	p_L^*	θ_3^*	$\dfrac{1}{p_3^{*2}}\dfrac{\mathrm{d}^2\sigma}{\mathrm{d}\Omega_3^*\,\mathrm{d}p_3^*}$	$\dfrac{E_3}{p_3^2}\dfrac{\mathrm{d}^2\sigma}{\mathrm{d}\Omega_3\,\mathrm{d}p_3}$	x^*	y_3^*	t	m_4	$\dfrac{\mathrm{d}^2\sigma}{\mathrm{d}t\,\mathrm{d}m_4^2}$
GeV/c	mrad	cm² sr⁻¹ GeV⁻¹ c	%	GeV/c	GeV/c	deg	cm² sr⁻¹ GeV⁻³ c³	cm² sr⁻¹ GeV⁻² c³			GeV²	GeV	cm²/GeV⁴
$p_1 = 12.880$ GeV; $p_1^* = 2.370$ GeV/c; $E_1^* + E_2^* = 5.098$ GeV; $s = 25.993$ GeV²; $\gamma_{\mathrm{c.m.}} = 2.7171$; $\beta_{\mathrm{c.m.}} = 0.92981$; $y_1^* = 1.6570$; $p_{3,\mathrm{max}}^* = 2.148$ GeV/c; [69 S 1] (cont.)													
7.153	77.0	1.368E-27	35	0.550	1.303	22.88	1.346E-28	1.913E-28	0.6067	1.5690	-0.169	3.394	2.487E-29
6.018	91.5	3.117E-27	21	0.550	1.075	27.09	4.261E-28	5.181E-28	0.5006	1.3956	-0.202	3.690	6.735E-29
5.101	108.0	2.745E-27	21	0.550	0.887	31.79	5.112E-28	5.383E-28	0.4131	1.2298	-0.263	3.908	6.998E-29
4.370	126.2	3.286E-27	19	0.550	0.733	36.87	8.113E-28	7.523E-28	0.3414	1.0738	-0.351	4.069	9.779E-29
3.788	145.7	3.729E-27	18	0.550	0.607	42.18	1.186E-27	9.851E-28	0.2825	0.9297	-0.459	4.188	1.281E-28
3.324	166.2	3.569E-27	18	0.550	0.502	47.60	1.418E-27	1.075E-27	0.2338	0.7977	-0.583	4.276	1.397E-28
2.950	187.5	5.849E-27	13	0.550	0.414	53.04	2.827E-27	1.985E-27	0.1926	0.6766	-0.719	4.342	2.580E-28
2.645	209.4	4.889E-27	14	0.550	0.338	58.41	2.803E-27	1.851E-27	0.1574	0.5655	-0.864	4.391	2.406E-28
2.394	231.8	3.923E-27	16	0.550	0.272	63.66	2.608E-27	1.641E-27	0.1268	0.4632	-1.018	4.427	2.134E-28
2.185	254.4	3.561E-27	16	0.550	0.214	68.70	2.693E-27	1.633E-27	0.0998	0.3694	-1.176	4.453	2.123E-28
2.009	277.3	3.807E-27	16	0.550	0.162	73.55	3.218E-27	1.900E-27	0.0756	0.2824	-1.340	4.471	2.469E-28
1.859	300.4	3.611E-27	17	0.550	0.115	78.18	3.364E-27	1.948E-27	0.0536	0.2015	-1.507	4.484	2.532E-28
1.730	323.6	2.603E-27	19	0.550	0.072	82.57	2.639E-27	1.510E-27	0.0334	0.1261	-1.677	4.492	1.962E-28
1.618	346.9	2.425E-27	19	0.550	0.032	86.72	2.647E-27	1.504E-27	0.0147	0.0555	-1.849	4.496	1.955E-28
7.321	88.9	8.855E-28	44	0.650	1.314	26.32	8.213E-29	1.210E-28	0.6119	1.4332	-0.379	3.316	1.573E-29
6.343	102.7	1.274E-27	35	0.650	1.115	30.25	1.547E-28	2.009E-28	0.5191	1.2888	-0.434	3.574	2.611E-29
5.535	117.7	1.160E-27	35	0.650	0.947	34.46	1.812E-28	2.096E-28	0.4409	1.1522	-0.510	3.770	2.725E-29
4.871	133.9	1.507E-27	30	0.650	0.805	38.92	2.963E-28	3.095E-28	0.3750	1.0230	-0.608	3.919	4.023E-29
4.324	150.9	1.315E-27	31	0.650	0.685	43.48	3.186E-28	3.043E-28	0.3191	0.9031	-0.719	4.034	3.955E-29
3.872	168.7	2.026E-27	26	0.650	0.583	48.13	5.922E-28	5.236E-28	0.2713	0.7912	-0.845	4.123	6.806E-29
3.497	187.0	3.515E-27	18	0.650	0.494	52.76	1.214E-27	1.006E-27	0.2301	0.6878	-0.981	4.191	1.308E-28
3.182	205.8	2.520E-27	21	0.650	0.417	57.35	1.010E-27	7.927E-28	0.1940	0.5914	-1.126	4.244	1.030E-28
2.915	224.9	3.514E-27	18	0.650	0.348	61.84	1.608E-27	1.207E-27	0.1620	0.5021	-1.277	4.285	1.569E-28
2.688	244.2	3.390E-27	18	0.650	0.287	66.19	1.744E-27	1.263E-27	0.1335	0.4191	-1.432	4.316	1.642E-28
2.493	263.8	3.117E-27	19	0.650	0.231	70.42	1.779E-27	1.252E-27	0.1077	0.3412	-1.593	4.340	1.628E-28
2.323	283.5	2.412E-27	21	0.650	0.180	74.48	1.510E-27	1.040E-27	0.0840	0.2684	-1.756	4.358	1.352E-28
2.176	303.4	2.280E-27	22	0.650	0.134	78.38	1.548E-27	1.050E-27	0.0622	0.1997	-1.925	4.370	1.365E-28
2.046	323.3	2.352E-27	22	0.650	0.090	82.10	1.717E-27	1.152E-27	0.0420	0.1352	-2.093	4.378	1.498E-28
1.931	343.3	2.466E-27	21	0.650	0.049	85.65	1.921E-27	1.280E-27	0.0230	0.0742	-2.265	4.383	1.664E-28
1.829	363.3	1.867E-27	25	0.650	0.011	89.03	1.540E-27	1.024E-27	0.0051	0.0166	-2.438	4.385	1.331E-28
$p_1 = 18.000$ GeV/c; $p_1^* = 2.831$ GeV/c; $E_1^* + E_2^* = 5.965$ GeV; $s = 35.581$ GeV²; $\gamma_{\mathrm{c.m.}} = 3.1790$; $\beta_{\mathrm{c.m.}} = 0.94924$; $y_1^* = 1.8240$; $p_{3,\mathrm{max}}^* = 2.640$ GeV/c; [69 S 1]													
10.380	4.8	2.977E-27	58	0.050	1.672	1.71	1.709E-28	2.868E-28	0.6334	3.1177	0.354	3.947	2.668E-29
6.929	7.2	1.628E-26	17	0.050	1.113	2.57	2.092E-27	2.350E-27	0.4218	2.7136	0.504	4.712	2.186E-28
3.963	12.6	3.028E-26	8	0.050	0.631	4.53	1.179E-26	7.645E-27	0.2391	2.1550	0.606	5.279	7.111E-28
2.042	24.5	2.659E-26	6	0.050	0.313	9.07	3.766E-26	1.305E-26	0.1186	1.4928	0.606	5.610	1.214E-27
1.166	42.9	2.515E-26	6	0.050	0.160	17.39	9.971E-26	2.172E-26	0.0605	0.9345	0.504	5.745	2.021E-27
0.775	64.5	2.009E-26	6	0.050	0.082	31.27	1.553E-25	2.634E-26	0.0312	0.5299	0.354	5.795	2.450E-27
0.571	87.7	1.805E-26	6	0.050	0.034	55.46	2.138E-25	3.254E-26	0.0130	0.2301	0.187	5.812	3.027E-27

$p_1 = 18.000$ GeV/c; $p_1^* = 2.831$ GeV/c; $E_1^* + E_2^* = 5.965$ GeV; $s = 35.581$ GeV^2; $\gamma_{\text{c.m.}} = 3.1790$; $\beta_{\text{c.m.}} = 0.94924$; $y_1^* = 1.8240$; $p_{3,\,\text{max}}^* = 2.640$ GeV/c; [69 S 1] (cont.)

p_3	θ_3	$\dfrac{\mathrm{d}^2\sigma}{\mathrm{d}\Omega_3\,\mathrm{d}p_3}$	Error	p_T	p_L^*	θ_3^*	$\dfrac{1}{p_3^{*2}}\cdot\dfrac{\mathrm{d}^2\sigma}{\mathrm{d}\Omega_3^*\,\mathrm{d}p_3^*}$	$\dfrac{E_3}{p_3^2}\cdot\dfrac{\mathrm{d}^2\sigma}{\mathrm{d}\Omega_3\,\mathrm{d}p_3}$	x^*	y_3^*	t	m_4	$\dfrac{\mathrm{d}^2\sigma}{\mathrm{d}t\,\mathrm{d}m_4^2}$
GeV/c	mrad	cm² sr⁻¹ GeV⁻¹ c	%	GeV/c	GeV/c	deg	cm² sr⁻¹ GeV⁻³ c³	cm² sr⁻¹ GeV⁻² c³			GeV²	GeV	cm²/GeV⁴
11.006	13.6	6.209E-27	25	0.150	1.770	4.84	3.166E-28	5.642E-28	0.6706	2.8527	0.293	3.787	5.248E-29
7.755	19.3	1.588E-26	11	0.150	1.243	6.88	1.626E-27	2.048E-27	0.4709	2.5028	0.423	4.535	1.905E-28
5.010	30.0	2.622E-26	6	0.150	0.796	10.68	6.373E-27	5.236E-27	0.3014	2.0656	0.504	5.079	4.870E-28
3.090	48.6	3.195E-26	5	0.150	0.478	17.44	1.992E-26	1.035E-26	0.1809	1.5823	0.504	5.422	9.627E-28
1.996	75.2	2.940E-26	4	0.150	0.289	27.39	4.163E-26	1.477E-26	0.1097	1.1453	0.423	5.601	1.373E-27
1.407	106.8	2.562E-26	5	0.150	0.181	39.69	6.697E-26	1.830E-26	0.0685	0.7955	0.293	5.687	1.702E-27
1.068	141.0	2.351E-26	5	0.150	0.111	53.45	9.519E-26	2.220E-26	0.0421	0.5191	0.138	5.729	2.065E-27
0.854	176.5	1.821E-26	5	0.150	0.061	67.72	1.010E-25	2.161E-26	0.0233	0.2956	-0.028	5.749	2.010E-27
0.710	212.8	1.788E-26	5	0.150	0.023	81.41	1.245E-25	2.567E-26	0.0086	0.1103	-0.202	5.757	2.387E-27
12.098	20.7	3.601E-27	28	0.250	1.942	7.34	1.517E-28	2.977E-28	0.7356	2.6126	0.186	3.491	2.769E-29
9.087	27.5	7.479E-27	16	0.250	1.452	9.77	5.561E-28	8.231E-28	0.5502	2.3263	0.293	4.236	7.656E-29
6.542	38.2	1.247E-26	10	0.250	1.036	13.56	1.774E-27	1.907E-27	0.3925	1.9978	0.354	4.773	1.773E-28
4.623	54.1	2.005E-26	6	0.250	0.718	19.19	5.612E-27	4.339E-27	0.2721	1.6502	0.354	5.136	4.036E-28
3.329	75.2	2.440E-26	5	0.250	0.499	26.63	1.276E-26	7.336E-27	0.1889	1.3215	0.293	5.361	6.824E-28
2.502	100.1	2.437E-26	5	0.250	0.352	35.37	2.149E-26	9.755E-27	0.1334	1.0350	0.186	5.494	9.074E-28
1.968	127.4	2.226E-26	5	0.250	0.252	44.78	2.973E-26	1.134E-26	0.0955	0.7938	0.051	5.572	1.055E-27
1.607	156.2	2.281E-26	5	0.250	0.179	54.41	4.220E-26	1.425E-26	0.0678	0.5899	-0.101	5.619	1.325E-27
1.353	185.8	1.863E-26	5	0.250	0.123	63.86	4.445E-26	1.384E-26	0.0465	0.4163	-0.264	5.647	1.288E-27
1.167	216.0	1.666E-26	6	0.250	0.077	72.89	4.847E-26	1.438E-26	0.0292	0.2657	-0.434	5.662	1.337E-27
1.024	246.5	1.486E-26	6	0.250	0.038	81.29	5.072E-26	1.465E-26	0.0145	0.1333	-0.606	5.671	1.362E-27
0.913	277.3	1.482E-26	6	0.250	0.004	88.98	5.735E-26	1.642E-26	0.0017	0.0155	-0.783	5.673	1.527E-27
13.483	26.0	1.400E-27	44	0.350	2.159	9.21	4.737E-29	1.038E-28	0.8180	2.4465	0.051	3.074	9.659E-30
10.669	32.8	3.640E-27	23	0.350	1.701	11.63	1.959E-28	3.412E-28	0.6443	2.2123	0.139	3.850	3.174E-29
8.261	42.4	6.282E-27	15	0.350	1.306	15.00	5.595E-28	7.605E-28	0.4947	1.9563	0.186	4.403	7.074E-29
6.342	55.2	1.014E-26	10	0.350	0.988	19.51	1.512E-27	1.599E-27	0.3743	1.6916	0.186	4.794	1.488E-28
4.912	71.3	1.313E-26	8	0.350	0.747	25.11	3.196E-27	2.674E-27	0.2830	1.4357	0.139	5.062	2.487E-28
3.889	90.1	1.386E-26	8	0.350	0.570	31.56	5.220E-27	3.566E-27	0.2159	1.2015	0.051	5.239	3.317E-28
3.162	110.9	1.788E-26	6	0.350	0.439	38.54	9.781E-27	5.660E-27	0.1664	0.9939	-0.065	5.357	5.265E-28
2.639	133.0	1.695E-26	6	0.350	0.341	45.77	1.266E-26	6.432E-27	0.1290	0.8120	-0.201	5.435	5.983E-28
2.252	156.1	1.744E-26	6	0.350	0.263	53.05	1.688E-26	7.759E-27	0.0998	0.6516	-0.352	5.488	7.217E-28
1.959	179.7	1.689E-26	6	0.350	0.201	60.16	2.024E-26	8.644E-27	0.0761	0.5104	-0.511	5.523	8.040E-28
1.730	203.7	1.425E-26	7	0.350	0.149	67.01	2.040E-26	8.264E-27	0.0563	0.3846	-0.675	5.547	7.687E-28
1.549	228.0	1.389E-26	7	0.350	0.104	73.52	2.303E-26	9.003E-27	0.0392	0.2715	-0.845	5.562	8.375E-28
1.401	252.5	1.319E-26	7	0.350	0.064	79.65	2.475E-26	9.461E-27	0.0242	0.1689	-1.018	5.571	8.800E-28
1.279	277.1	1.016E-26	8	0.350	0.028	85.36	2.115E-26	7.991E-27	0.0108	0.0753	-1.193	5.576	7.433E-28
12.376	36.4	6.601E-28	57	0.450	1.969	12.87	2.635E-29	5.334E-29	0.7458	2.1371	-0.028	3.384	4.961E-30
10.061	44.7	2.428E-27	26	0.450	1.589	15.81	1.456E-28	2.414E-28	0.6018	1.9300	0.011	3.979	2.245E-29
8.142	55.3	4.979E-27	16	0.450	1.271	19.50	4.513E-28	6.116E-28	0.4814	1.7180	0.011	4.408	5.689E-29
6.620	68.0	7.003E-27	13	0.450	1.015	23.91	9.454E-28	1.058E-27	0.3846	1.5107	-0.028	4.717	9.842E-29
5.449	82.7	7.912E-27	11	0.450	0.815	28.91	1.543E-27	1.452E-27	0.3087	1.3156	-0.101	4.937	1.351E-28

p_3	θ_3	$\dfrac{d^2\sigma}{d\Omega_3\,dp_3}$	Error	p_T	p_L^*	θ_3^*	$\dfrac{1}{p_3^{*2}}\cdot\dfrac{d^2\sigma}{d\Omega_3^*\,dp_3^*}$	$\dfrac{E_3}{p_3^2}\cdot\dfrac{d^2\sigma}{d\Omega_3\,dp_3}$	x^*	y_3^*	t	m_4	$\dfrac{d^2\sigma}{dt\,dm_4^2}$
GeV/c	mrad	cm² sr⁻¹ GeV⁻¹ c	%	GeV/c	GeV/c	deg	cm² sr⁻¹ GeV⁻³ c³	cm² sr⁻¹ GeV⁻² c³			GeV²	GeV	cm²/GeV⁴

$p_1 = 18.000$ GeV/c; $p_1^* = 2.831$ GeV/c; $E_1^* + E_2^* = 5.965$ GeV; $s = 35.581$ GeV²; $\gamma_{c.m.} = 3.1790$; $\beta_{c.m.} = 0.94924$; $y_1^* = 1.8240$; $p_{3,max}^* = 2.640$ GeV/c; [69 S 1] (cont.)

p_3	θ_3	$\dfrac{d^2\sigma}{d\Omega_3\,dp_3}$	Error	p_T	p_L^*	θ_3^*	$\dfrac{1}{p_3^{*2}}\cdot\dfrac{d^2\sigma}{d\Omega_3^*\,dp_3^*}$	$\dfrac{E_3}{p_3^2}\cdot\dfrac{d^2\sigma}{d\Omega_3\,dp_3}$	x^*	y_3^*	t	m_4	$\dfrac{d^2\sigma}{dt\,dm_4^2}$
4.559	98.9	8.135E-27	11	0.450	0.659	34.35	2.205E-27	1.785E-27	0.2495	1.1366	-0.202	5.093	1.661E-28
3.881	116.2	1.041E-26	9	0.450	0.536	40.04	3.763E-27	2.684E-27	0.2029	0.9748	-0.323	5.205	2.497E-28
3.358	134.4	1.169E-26	8	0.450	0.437	45.84	5.423E-27	3.484E-27	0.1655	0.8291	-0.459	5.285	3.241E-28
2.949	153.2	1.087E-26	9	0.450	0.356	51.64	6.248E-27	3.690E-27	0.1349	0.6978	-0.607	5.344	3.432E-28
2.623	172.4	1.034E-26	9	0.450	0.288	57.33	7.146E-27	3.948E-27	0.1093	0.5794	-0.762	5.386	3.672E-28
2.359	191.9	8.620E-27	9	0.450	0.231	62.86	6.979E-27	3.660E-27	0.0873	0.4718	-0.923	5.417	3.405E-28
2.142	211.7	1.102E-26	9	0.450	0.180	68.21	1.022E-26	5.156E-27	0.0682	0.3731	-1.090	5.439	4.796E-28
1.960	231.6	9.311E-27	9	0.450	0.135	73.30	9.719E-27	4.763E-27	0.0511	0.2827	-1.259	5.455	4.430E-28
1.807	251.7	8.433E-27	9	0.450	0.094	78.16	9.741E-27	4.681E-27	0.0357	0.1989	-1.432	5.465	4.354E-28
1.676	271.8	8.128E-27	10	0.450	0.057	82.73	1.025E-26	4.866E-27	0.0217	0.1215	-1.606	5.472	4.527E-28
1.563	292.0	7.992E-27	10	0.450	0.023	87.06	1.088E-26	5.134E-27	0.0088	0.0491	-1.783	5.475	4.775E-28
9.983	55.1	1.522E-27	33	0.550	1.560	19.42	9.186E-29	1.525E-28	0.5909	1.7360	-0.169	3.975	1.418E-29
8.397	65.6	2.331E-27	25	0.550	1.294	23.02	1.965E-28	2.776E-28	0.4903	1.5626	-0.202	4.329	2.582E-29
7.115	77.4	4.404E-27	17	0.550	1.076	27.07	5.088E-28	6.191E-28	0.4078	1.3966	-0.264	4.592	5.758E-29
6.092	90.4	4.582E-27	16	0.550	0.899	31.45	7.076E-28	7.523E-28	0.3406	1.2408	-0.351	4.787	6.998E-29
5.278	104.4	5.750E-27	14	0.550	0.755	36.08	1.154E-27	1.090E-27	0.2859	1.0967	-0.459	4.933	1.014E-28
4.628	119.1	6.770E-27	12	0.550	0.636	40.83	1.717E-27	1.464E-27	0.2410	0.9648	-0.583	5.043	1.361E-28
4.104	134.4	8.550E-27	11	0.550	0.537	45.66	2.667E-27	2.085E-27	0.2036	0.8436	-0.719	5.126	1.939E-28
3.676	150.2	5.898E-27	13	0.550	0.454	50.49	2.210E-27	1.606E-27	0.1719	0.7322	-0.865	5.190	1.493E-28
3.324	166.2	7.071E-27	11	0.550	0.382	55.22	3.113E-27	2.129E-27	0.1447	0.6306	-1.017	5.238	1.980E-28
3.029	182.6	7.994E-27	11	0.550	0.319	59.88	4.058E-27	2.642E-27	0.1209	0.5361	-1.177	5.276	2.457E-28
2.781	199.1	6.331E-27	12	0.550	0.264	64.40	3.643E-27	2.279E-27	0.0999	0.4492	-1.340	5.304	2.120E-28
2.569	215.7	4.998E-27	14	0.550	0.214	68.74	3.214E-27	1.948E-27	0.0810	0.3686	-1.506	5.326	1.812E-28
2.387	232.5	7.857E-27	11	0.550	0.169	72.94	5.570E-27	3.297E-27	0.0639	0.2931	-1.676	5.342	3.067E-28
2.229	249.4	5.467E-27	13	0.550	0.127	76.97	4.225E-27	2.457E-27	0.0482	0.2224	-1.849	5.354	2.286E-28
2.090	266.3	5.074E-27	13	0.550	0.089	80.80	4.236E-27	2.433E-27	0.0337	0.1563	-2.022	5.362	2.263E-28
1.968	283.2	4.642E-27	14	0.550	0.053	84.45	4.149E-27	2.365E-27	0.0203	0.0941	-2.197	5.367	2.199E-28
1.860	300.2	4.167E-27	15	0.550	0.020	87.92	3.957E-27	2.247E-27	0.0076	0.0351	-2.374	5.369	2.090E-28
11.848	54.9	1.402E-27	37	0.650	1.853	19.33	6.012E-29	1.183E-28	0.7019	1.7488	-0.351	3.481	1.101E-29
10.215	63.7	5.113E-28	58	0.650	1.580	22.37	2.921E-29	5.006E-29	0.5985	1.6002	-0.379	3.893	4.656E-30
8.848	73.5	2.200E-27	26	0.650	1.349	25.73	1.654E-28	2.487E-28	0.5109	1.4563	-0.433	4.203	2.313E-29
7.718	84.3	2.165E-27	25	0.650	1.155	29.38	2.106E-28	2.806E-28	0.4374	1.3192	-0.510	4.439	2.610E-29
6.788	95.9	3.002E-27	21	0.650	0.992	33.23	3.704E-28	4.423E-28	0.3758	1.1904	-0.607	4.621	4.114E-29
6.023	108.1	4.389E-27	17	0.650	0.855	37.22	6.729E-28	7.289E-28	0.3240	1.0705	-0.719	4.762	6.780E-29
5.390	120.9	3.005E-27	20	0.650	0.739	41.32	5.609E-28	5.577E-28	0.2801	0.9584	-0.845	4.872	5.187E-29
4.864	134.0	3.281E-27	20	0.650	0.640	45.42	7.312E-28	6.748E-28	0.2426	0.8552	-0.980	4.959	6.277E-29
4.422	147.5	4.145E-27	16	0.650	0.554	49.54	1.084E-27	9.378E-28	0.2100	0.7590	-1.124	5.027	8.723E-29
4.048	161.3	3.790E-27	18	0.650	0.479	53.62	1.143E-27	9.368E-28	0.1814	0.6692	-1.277	5.082	8.714E-29
3.728	175.2	3.685E-27	18	0.650	0.412	57.60	1.265E-27	9.892E-28	0.1562	0.5862	-1.431	5.125	9.201E-29
3.453	189.4	5.050E-27	15	0.650	0.352	61.54	1.945E-27	1.464E-27	0.1335	0.5079	-1.594	5.160	1.361E-28
3.215	203.6	3.173E-27	19	0.650	0.299	65.33	1.355E-27	9.879E-28	0.1131	0.4352	-1.757	5.187	9.189E-29

p_3	θ_3	$\dfrac{d^2\sigma}{d\Omega_3\,dp_3}$	Error	p_T	p_L^*	θ_3^*	$\dfrac{1}{p_3^{*2}}\cdot\dfrac{d^2\sigma}{d\Omega_3^*\,dp_3^*}$	$\dfrac{E_3}{p_3^2}\cdot\dfrac{d^2\sigma}{d\Omega_3\,dp_3}$	x^*	y_3^*	t	m_4	$\dfrac{d^2\sigma}{dt\,dm_4^2}$
GeV/c	mrad	cm² sr⁻¹ GeV⁻¹ c	%	GeV/c	GeV/c	deg	cm² sr⁻¹ GeV⁻³ c³	cm² sr⁻¹ GeV⁻² c³			GeV²	GeV	cm²/GeV⁴

$p_1 = 18.000$ GeV/c; $p_1^* = 2.831$ GeV/c; $E_1^* + E_2^* = 5.965$ GeV; $s = 35.581$ GeV²; $\gamma_{\text{c.m.}} = 3.1790$; $\beta_{\text{c.m.}} = 0.94924$; $y_1^* = 1.8240$; $p_{3,\text{max}}^* = 2.640$ GeV/c; [69 S 1] (cont.)

p_3	θ_3	$\dfrac{d^2\sigma}{d\Omega_3\,dp_3}$	Error	p_T	p_L^*	θ_3^*	$\dfrac{1}{p_3^{*2}}\cdot\dfrac{d^2\sigma}{d\Omega_3^*\,dp_3^*}$	$\dfrac{E_3}{p_3^2}\cdot\dfrac{d^2\sigma}{d\Omega_3\,dp_3}$	x^*	y_3^*	t	m_4	$\dfrac{d^2\sigma}{dt\,dm_4^2}$
3.006	217.9	4.039E-27	16	0.650	0.249	69.01	1.895E-27	1.345E-27	0.0945	0.3669	-1.923	5.209	1.251E-28
2.823	232.4	4.735E-27	15	0.650	0.204	72.59	2.414E-27	1.679E-27	0.0772	0.3020	-2.095	5.225	1.562E-28
2.660	246.8	3.401E-27	17	0.650	0.162	76.00	1.871E-27	1.280E-27	0.0614	0.2414	-2.264	5.238	1.191E-28
2.516	261.4	3.446E-27	17	0.650	0.123	79.32	2.028E-27	1.372E-27	0.0465	0.1834	-2.440	5.247	1.276E-28
2.386	275.9	3.221E-27	18	0.650	0.086	82.47	2.017E-27	1.352E-27	0.0325	0.1288	-2.613	5.254	1.258E-28
2.269	290.5	3.652E-27	17	0.650	0.051	85.51	2.419E-27	1.613E-27	0.0193	0.0766	-2.788	5.258	1.500E-28
2.164	305.1	1.977E-27	23	0.650	0.018	88.42	1.376E-27	9.155E-28	0.0068	0.0270	-2.966	5.260	8.515E-29

$p_1 = 21.080$ GeV/c; $p_1^* = 3.075$ GeV/c; $E_1^* + E_2^* = 6.431$ GeV; $s = 41.354$ GeV²; $\gamma_{\text{c.m.}} = 3.4272$; $\beta_{\text{c.m.}} = 0.95648$; $y_1^* = 1.9029$; $p_{3,\text{max}}^* = 2.898$ GeV/c; [69 S 1]

p_3	θ_3	$\dfrac{d^2\sigma}{d\Omega_3\,dp_3}$	Error	p_T	p_L^*	θ_3^*	$\dfrac{1}{p_3^{*2}}\cdot\dfrac{d^2\sigma}{d\Omega_3^*\,dp_3^*}$	$\dfrac{E_3}{p_3^2}\cdot\dfrac{d^2\sigma}{d\Omega_3\,dp_3}$	x^*	y_3^*	t	m_4	$\dfrac{d^2\sigma}{dt\,dm_4^2}$
12.154	4.1	9.389E-27	35	0.050	1.810	1.58	4.255E-28	7.726E-28	0.6246	3.1966	0.354	4.245	6.136E-29
8.113	6.2	1.824E-26	17	0.050	1.205	2.38	1.851E-27	2.249E-27	0.4160	2.7924	0.504	5.075	1.786E-28
4.641	10.8	3.119E-26	9	0.050	0.684	4.18	9.602E-27	6.724E-27	0.2362	2.2341	0.606	5.689	5.340E-28
2.392	20.9	3.212E-26	6	0.050	0.342	8.33	3.612E-26	1.345E-26	0.1179	1.5718	0.606	6.048	1.068E-27
1.366	36.6	3.030E-26	6	0.050	0.177	15.75	9.648E-26	2.230E-26	0.0612	1.0131	0.504	6.197	1.771E-27
0.910	55.0	2.545E-26	6	0.050	0.096	27.50	1.601E-25	2.829E-26	0.0332	0.6096	0.354	6.253	2.247E-27
0.671	74.6	2.086E-26	6	0.050	0.047	47.04	2.043E-25	3.175E-26	0.0161	0.3091	0.187	6.275	2.522E-27
0.527	95.0	1.483E-26	7	0.050	0.011	77.73	1.958E-25	2.911E-26	0.0038	0.0733	0.010	6.282	2.312E-27
16.985	8.8	4.605E-27	41	0.150	2.529	3.39	1.069E-28	2.711E-28	0.8728	3.2078	0.139	2.956	2.153E-29
12.886	11.6	6.165E-27	28	0.150	1.916	4.48	2.483E-28	4.785E-28	0.6614	2.9316	0.293	4.073	3.800E-29
9.081	16.5	1.297E-26	14	0.150	1.347	6.36	1.049E-27	1.428E-27	0.4647	2.5816	0.423	4.884	1.135E-28
5.866	25.6	2.769E-26	7	0.150	0.863	9.86	5.324E-27	4.722E-27	0.2978	2.1445	0.504	5.474	3.750E-28
3.618	41.5	3.730E-26	5	0.150	0.520	16.09	1.846E-26	1.032E-26	0.1795	1.6613	0.504	5.847	8.194E-28
2.337	64.2	3.903E-26	4	0.150	0.318	25.23	4.419E-26	1.673E-26	0.1099	1.2242	0.423	6.042	1.329E-27
1.647	91.2	3.445E-26	4	0.150	0.203	36.49	7.281E-26	2.099E-26	0.0700	0.8741	0.293	6.137	1.667E-27
1.250	120.3	2.852E-26	5	0.150	0.130	49.09	9.461E-26	2.296E-26	0.0449	0.5980	0.139	6.185	1.823E-27
1.000	150.5	2.783E-26	5	0.150	0.079	62.33	1.281E-25	2.810E-26	0.0271	0.3749	-0.028	6.209	2.232E-27
0.831	181.5	2.290E-26	5	0.150	0.039	75.43	1.340E-25	2.794E-26	0.0135	0.1891	-0.202	6.220	2.219E-27
0.710	212.8	1.910E-26	6	0.150	0.006	87.54	1.338E-25	2.742E-26	0.0022	0.0314	-0.379	6.224	2.178E-27
14.166	17.6	4.336E-27	28	0.250	2.103	6.78	1.442E-28	3.061E-28	0.7258	2.6916	0.186	3.752	2.431E-29
10.639	23.5	9.742E-27	15	0.250	1.574	9.03	5.725E-28	9.158E-28	0.5431	2.4053	0.293	4.561	7.273E-29
7.660	32.6	1.750E-26	9	0.250	1.124	12.54	1.970E-27	2.285E-27	0.3880	2.0768	0.354	5.143	1.815E-28
5.412	46.2	2.548E-26	6	0.250	0.781	17.74	5.659E-27	4.710E-27	0.2697	1.7291	0.354	5.538	3.741E-28
3.897	64.2	2.697E-26	5	0.250	0.545	24.62	1.124E-26	6.925E-27	0.1883	1.4004	0.293	5.784	5.500E-28
2.929	85.5	2.980E-26	5	0.250	0.389	32.71	2.108E-26	1.019E-26	0.1344	1.1141	0.186	5.929	8.090E-28
2.302	108.8	2.845E-26	5	0.250	0.283	41.47	3.077E-26	1.238E-26	0.0976	0.8727	0.051	6.016	9.834E-28
1.880	133.4	2.686E-26	5	0.250	0.206	50.49	4.060E-26	1.433E-26	0.0712	0.6689	-0.101	6.069	1.138E-27
1.582	158.7	2.516E-26	5	0.250	0.148	59.43	4.956E-26	1.597E-26	0.0510	0.4952	-0.264	6.102	1.268E-27
1.363	184.5	2.115E-26	6	0.250	0.101	68.08	5.139E-26	1.560E-26	0.0347	0.3445	-0.433	6.121	1.239E-27
1.196	210.6	1.791E-26	6	0.250	0.061	76.25	5.149E-26	1.508E-26	0.0211	0.2121	-0.607	6.132	1.197E-27
1.065	236.9	1.755E-26	6	0.250	0.027	83.83	5.780E-26	1.662E-26	0.0093	0.0943	-0.783	6.138	1.320E-27

$p_1 = 21.080$ GeV/c; $p_1^* = 3.075$ GeV/c; $E_1^* + E_2^* = 6.431$ GeV; $s = 41.354$ GeV²; $\gamma_{c.m.} = 3.4272$; $\beta_{c.m.} = 0.95648$; $y_1^* = 1.9029$; $p_{3,\,max}^* = 2.898$ GeV/c; [69 S 1] (cont.)

p_3	θ_3	$\dfrac{d^2\sigma}{d\Omega_3\,dp_3}$	Error	p_T	p_L^*	θ_3^*	$\dfrac{1}{p_3^{*2}}\cdot\dfrac{d^2\sigma}{d\Omega_3^*\,dp_3^*}$	$\dfrac{E_3}{p_3^2}\cdot\dfrac{d^2\sigma}{d\Omega_3\,dp_3}$	x^*	y_3^*	t	m_4	$\dfrac{d^2\sigma}{dt\,dm_4^2}$
GeV/c	mrad	cm² sr⁻¹ GeV⁻¹ c	%	GeV/c	GeV/c	deg	cm² sr⁻¹ GeV⁻³ c³	cm² sr⁻¹ GeV⁻² c³			GeV²	GeV	cm²/GeV⁴
12.491	28.0	5.079E-27	22	0.350	1.844	10.75	2.161E-28	4.066E-28	0.6362	2.2912	0.139	4.144	3.230E-29
9.672	36.2	8.525E-27	14	0.350	1.417	13.87	6.010E-28	8.815E-28	0.4892	2.0351	0.186	4.744	7.001E-29
7.424	47.2	1.110E-26	11	0.350	1.075	18.04	1.313E-27	1.495E-27	0.3709	1.7706	0.186	5.170	1.188E-28
5.750	60.9	1.352E-26	9	0.350	0.815	23.23	2.618E-27	2.352E-27	0.2814	1.5147	0.139	5.461	1.868E-28
4.551	77.0	1.946E-26	7	0.350	0.626	29.23	5.858E-27	4.278E-27	0.2159	1.2804	0.051	5.655	3.398E-28
3.700	94.7	1.889E-26	7	0.350	0.486	35.74	8.304E-27	5.109E-27	0.1678	1.0727	-0.065	5.785	4.058E-28
3.087	113.6	2.045E-26	6	0.350	0.382	42.50	1.236E-26	6.631E-27	0.1318	0.8909	-0.201	5.871	5.267E-28
2.633	133.3	1.929E-26	7	0.350	0.301	49.35	1.522E-26	7.337E-27	0.1037	0.7308	-0.351	5.931	5.827E-28
2.289	153.5	1.955E-26	7	0.350	0.235	56.10	1.926E-26	8.557E-27	0.0812	0.5895	-0.510	5.972	6.796E-28
2.021	174.0	1.635E-26	7	0.350	0.181	62.64	1.940E-26	8.109E-27	0.0625	0.4638	-0.674	6.000	6.441E-28
1.808	194.8	1.810E-26	7	0.350	0.135	68.93	2.509E-26	1.004E-26	0.0465	0.3506	-0.845	6.019	7.975E-28
1.635	215.8	1.559E-26	7	0.350	0.094	74.91	2.463E-26	9.570E-27	0.0326	0.2478	-1.018	6.031	7.601E-28
1.491	236.9	1.386E-26	7	0.350	0.058	80.55	2.449E-26	9.336E-27	0.0201	0.1541	-1.193	6.039	7.415E-28
1.372	258.0	1.348E-26	8	0.350	0.026	85.79	2.614E-26	9.876E-27	0.0089	0.0683	-1.371	6.043	7.844E-28
14.490	31.1	1.596E-27	41	0.450	2.135	11.90	5.038E-29	1.102E-28	0.7368	2.2161	-0.028	3.641	8.749E-30
11.779	38.2	3.296E-27	25	0.450	1.725	14.62	1.565E-28	2.798E-28	0.5952	2.0089	0.011	4.287	2.223E-29
9.532	47.2	5.031E-27	19	0.450	1.382	18.04	3.616E-28	5.279E-28	0.4769	1.7969	0.011	4.754	4.192E-29
7.749	58.1	8.197E-27	13	0.450	1.107	22.12	8.796E-28	1.058E-27	0.3820	1.5897	-0.028	5.090	8.403E-29
6.377	70.6	9.802E-27	11	0.450	0.892	26.78	1.525E-27	1.537E-27	0.3077	1.3945	-0.101	5.330	1.221E-28
5.335	84.4	1.218E-26	10	0.450	0.725	31.85	2.643E-27	2.284E-27	0.2500	1.2155	-0.202	5.501	1.814E-28
4.540	99.3	1.213E-26	9	0.450	0.593	37.17	3.528E-27	2.673E-27	0.2048	1.0536	-0.323	5.624	2.123E-28
3.928	114.8	1.388E-26	9	0.450	0.489	42.61	5.207E-27	3.536E-27	0.1688	0.9081	-0.459	5.713	2.808E-28
3.448	130.9	1.159E-26	9	0.450	0.404	48.10	5.421E-27	3.364E-27	0.1394	0.7766	-0.607	5.779	2.672E-28
3.066	147.3	1.194E-26	9	0.450	0.333	53.49	6.756E-27	3.898E-27	0.1149	0.6583	-0.762	5.827	3.096E-28
2.756	164.0	1.201E-26	9	0.450	0.273	58.78	8.016E-27	4.363E-27	0.0941	0.5506	-0.923	5.863	3.466E-28
2.501	180.9	1.114E-26	9	0.450	0.220	63.91	8.577E-27	4.461E-27	0.0761	0.4522	-1.090	5.889	3.543E-28
2.288	198.0	1.022E-26	10	0.450	0.174	68.85	8.908E-27	4.475E-27	0.0601	0.3615	-1.260	5.909	3.554E-28
2.108	215.2	1.225E-26	9	0.450	0.133	73.59	1.190E-26	5.824E-27	0.0458	0.2778	-1.433	5.923	4.626E-28
1.954	232.4	1.054E-26	10	0.450	0.095	78.07	1.125E-26	5.408E-27	0.0328	0.2004	-1.607	5.932	4.295E-28
1.821	249.7	8.171E-27	11	0.450	0.061	82.34	9.474E-27	4.500E-27	0.0209	0.1281	-1.783	5.938	3.574E-28
1.705	267.1	8.613E-27	10	0.450	0.028	86.39	1.074E-26	5.069E-27	0.0098	0.0602	-1.961	5.942	4.026E-28
11.687	47.1	1.609E-27	35	0.550	1.696	17.97	7.699E-29	1.377E-28	0.5853	1.8148	-0.169	4.286	1.094E-29
9.829	56.0	3.155E-27	23	0.550	1.410	21.31	2.112E-28	3.210E-28	0.4866	1.6416	-0.201	4.672	2.550E-29
8.328	66.1	4.482E-27	18	0.550	1.176	25.07	4.123E-28	5.383E-28	0.4058	1.4755	-0.264	4.958	4.275E-29
7.129	77.2	5.707E-27	15	0.550	0.986	29.15	7.039E-28	8.007E-28	0.3403	1.3197	-0.351	5.171	6.359E-29
6.176	89.2	8.346E-27	13	0.550	0.832	33.48	1.342E-27	1.352E-27	0.2871	1.1756	-0.459	5.331	1.074E-28
5.414	101.8	7.735E-27	12	0.550	0.705	37.95	1.578E-27	1.429E-27	0.2435	1.0431	-0.584	5.452	1.135E-28
4.799	114.9	7.187E-27	13	0.550	0.601	42.49	1.813E-27	1.498E-27	0.2073	0.9219	-0.720	5.545	1.190E-28
4.298	128.3	8.243E-27	12	0.550	0.513	47.02	2.510E-27	1.919E-27	0.1769	0.8113	-0.865	5.616	1.524E-28
3.885	142.1	7.635E-27	12	0.550	0.437	51.54	2.745E-27	1.967E-27	0.1508	0.7089	-1.019	5.671	1.562E-28
3.540	156.0	7.645E-27	12	0.550	0.372	55.96	3.186E-27	2.161E-27	0.1283	0.6153	-1.177	5.714	1.717E-28

p_3	θ_3	$\dfrac{d^2\sigma}{d\Omega_3\,dp_3}$	Error	p_T	p_L^*	θ_3^*	$\dfrac{1}{p_3^{*2}}\cdot\dfrac{d^2\sigma}{d\Omega_3^*\,dp_3^*}$	$\dfrac{E_3}{p_3^2}\cdot\dfrac{d^2\sigma}{d\Omega_3\,dp_3}$	x^*	y_3^*	t	m_4	$\dfrac{d^2\sigma}{dt\,dm_4^2}$
GeV/c	mrad	cm² sr⁻¹ GeV⁻¹ c	%	GeV/c	GeV/c	deg	cm² sr⁻¹ GeV⁻³ c³	cm² sr⁻¹ GeV⁻² c³			GeV²	GeV	cm²/GeV⁴
$p_1 = 21.080$ GeV/c; $p_1^* = 3.075$ GeV/c; $E_1^* + E_2^* = 6.431$ GeV; $s = 41.354$ GeV²; $\gamma_\mathrm{c.m.} = 3.4272$; $\beta_\mathrm{c.m.} = 0.95648$; $y_1^* = 1.9029$; $p_{3,\,\mathrm{max}}^* = 2.898$ GeV/c; [69 S 1] (cont.)													
3.248	170.1	7.005E−27	12	0.550	0.314	60.28	3.330E−27	2.159E−27	0.1083	0.5285	−1.339	5.748	1.715E−28
3.000	184.4	7.394E−27	12	0.550	0.262	64.49	3.946E−27	2.467E−27	0.0906	0.4474	−1.507	5.773	1.960E−28
2.785	198.8	8.509E−27	11	0.550	0.216	68.57	5.039E−27	3.059E−27	0.0745	0.3718	−1.677	5.794	2.430E−28
2.599	213.2	6.270E−27	14	0.550	0.174	72.47	4.072E−27	2.416E−27	0.0599	0.3015	−1.848	5.809	1.919E−28
2.436	227.7	5.754E−27	14	0.550	0.135	76.24	4.057E−27	2.366E−27	0.0465	0.2353	−2.022	5.820	1.879E−28
2.293	242.2	6.286E−27	13	0.550	0.099	79.83	4.769E−27	2.746E−27	0.0341	0.1731	−2.197	5.828	2.181E−28
2.165	256.8	6.164E−27	14	0.550	0.065	83.28	4.996E−27	2.853E−27	0.0224	0.1140	−2.374	5.833	2.266E−28
2.052	271.4	7.954E−27	12	0.550	0.033	86.56	6.834E−27	3.885E−27	0.0114	0.0582	−2.553	5.836	3.086E−28
1.949	286.0	5.553E−27	14	0.550	0.003	89.69	5.035E−27	2.856E−27	0.0010	0.0053	−2.730	5.838	2.269E−28
11.957	54.4	1.257E−27	41	0.650	1.720	20.70	5.701E−29	1.051E−28	0.5936	1.6791	−0.379	4.202	8.350E−30
10.357	62.8	2.810E−27	26	0.650	1.472	23.83	1.680E−28	2.713E−28	0.5079	1.5352	−0.433	4.539	2.155E−29
9.033	72.0	3.988E−27	20	0.650	1.263	27.23	3.093E−28	4.415E−28	0.4360	1.3982	−0.510	4.797	3.507E−29
7.944	81.9	3.091E−27	23	0.650	1.089	30.82	3.049E−28	3.892E−28	0.3760	1.2693	−0.607	4.996	3.091E−29
7.047	92.4	3.260E−27	21	0.650	0.943	34.56	4.009E−28	4.627E−28	0.3256	1.1491	−0.719	5.151	3.675E−29
6.305	103.3	4.951E−27	17	0.650	0.820	38.41	7.439E−28	7.854E−28	0.2830	1.0371	−0.846	5.272	6.238E−29
5.688	114.5	3.489E−27	20	0.650	0.715	42.27	6.285E−28	6.136E−28	0.2468	0.9339	−0.980	5.368	4.873E−29
5.170	126.1	5.053E−27	16	0.650	0.624	46.17	1.072E−27	9.777E−28	0.2154	0.8373	−1.126	5.445	7.766E−29
4.731	137.8	5.159E−27	16	0.650	0.545	50.01	1.272E−27	1.093E−27	0.1881	0.7483	−1.275	5.506	8.682E−29
4.357	149.8	4.837E−27	16	0.650	0.475	53.84	1.359E−27	1.111E−27	0.1640	0.6646	−1.434	5.555	8.822E−29
4.034	161.8	3.868E−27	18	0.650	0.413	57.55	1.226E−27	9.594E−28	0.1426	0.5872	−1.592	5.595	7.620E−29
3.754	174.0	5.397E−27	15	0.650	0.357	61.21	1.907E−27	1.439E−27	0.1232	0.5142	−1.756	5.627	1.143E−28
3.509	186.3	4.199E−27	18	0.650	0.306	64.78	1.636E−27	1.198E−27	0.1057	0.4456	−1.924	5.653	9.512E−29
3.294	198.6	5.013E−27	16	0.650	0.260	68.22	2.134E−27	1.523E−27	0.0896	0.3813	−2.092	5.674	1.210E−28
3.103	211.1	5.056E−27	16	0.650	0.216	71.59	2.332E−27	1.631E−27	0.0747	0.3199	−2.266	5.690	1.295E−28
2.932	223.5	4.338E−27	17	0.650	0.176	74.81	2.154E−27	1.481E−27	0.0609	0.2625	−2.437	5.703	1.176E−28
2.780	236.0	3.818E−27	18	0.650	0.139	77.93	2.025E−27	1.375E−27	0.0480	0.2076	−2.613	5.713	1.092E−28
2.643	248.5	3.493E−27	20	0.650	0.104	80.92	1.967E−27	1.323E−27	0.0358	0.1556	−2.789	5.720	1.051E−28
2.518	261.1	3.044E−27	20	0.650	0.070	83.82	1.811E−27	1.211E−27	0.0243	0.1057	−2.967	5.725	9.616E−29
2.406	273.6	2.604E−27	22	0.650	0.039	86.58	1.628E−27	1.084E−27	0.0134	0.0584	−3.144	5.728	8.611E−29
2.303	286.2	3.166E−27	20	0.650	0.009	89.25	2.071E−27	1.377E−27	0.0030	0.0129	−3.324	5.729	1.094E−28
$p_1 = 24.120$ GeV/c; $p_1^* = 3.299$ GeV/c; $E_1^* + E_2^* = 6.860$ GeV; $s = 47.053$ GeV²; $\gamma_\mathrm{c.m.} = 3.6557$; $\beta_\mathrm{c.m.} = 0.96186$; $y_1 = 1.9702$; $p_{3,\,\mathrm{max}}^* = 3.132$ GeV/c; [69 S 1]													
9.282	5.4	2.119E−26	20	0.050	1.290	2.22	1.758E−27	2.283E−27	0.4119	2.8597	0.504	5.409	1.585E−28
5.310	9.4	3.966E−26	9	0.050	0.733	3.90	9.990F−27	7.472E−27	0.2341	2.3013	0.606	6.067	5.186E−28
2.737	18.3	4.064E−26	7	0.050	0.367	7.75	3.752E−26	1.487E−26	0.1173	1.6390	0.606	6.453	1.032E−27
1.564	32.0	3.854E−26	6	0.050	0.193	14.50	1.016E−25	2.474E−26	0.0617	1.0806	0.504	6.613	1.717E−27
1.042	48.0	3.091E−26	7	0.050	0.108	24.81	1.631E−25	2.993E−26	0.0345	0.6766	0.354	6.675	2.078E−27
0.769	65.1	2.207E−26	7	0.050	0.057	41.21	1.836E−25	2.917E−26	0.0182	0.3760	0.186	6.700	2.025E−27
0.605	82.7	2.215E−26	8	0.050	0.021	67.31	2.509E−25	3.757E−26	0.0067	0.1405	0.010	6.710	2.608E−27

p_3	θ_3	$\dfrac{d^2\sigma}{d\Omega_3\,dp_3}$	Error	p_T	p_L^*	θ_3^*	$\dfrac{1}{p_3^{*2}}\cdot\dfrac{d^2\sigma}{d\Omega_3^*\,dp_3^*}$	$\dfrac{E_3}{p_3^2}\cdot\dfrac{d^2\sigma}{d\Omega_3\,dp_3}$	x^*	y_3^*	t	m_4	$\dfrac{d^2\sigma}{dt\,dm_4^2}$
GeV/c	mrad	cm² sr⁻¹ GeV⁻¹ c	%	GeV/c	GeV/c	deg	cm² sr⁻¹ GeV⁻³ c³	cm² sr⁻¹ GeV⁻² c³			GeV²	GeV	cm²/GeV⁴
$p_1 = 24.120$ GeV/c; $p_1^* = 3.299$ GeV/c; $E_1^* + E_2^* = 6.860$ GeV; $s = 47.053$ GeV²; $\gamma_{c.m.} = 3.6557$; $\beta_{c.m.} = 0.96186$; $y_1 = 1.9702$; $p_{3,\max}^* = 3.132$ GeV/c; [69 S 1] (cont.)													
19.432	7.7	4.348E-27	50	0.150	2.706	3.17	8.247E-29	2.238E-28	0.8639	3.2751	0.139	3.138	1.553E-29
14.743	10.2	1.163E-26	26	0.150	2.051	4.18	3.828E-28	7.889E-28	0.6547	2.9992	0.293	4.336	5.476E-29
10.389	14.4	1.777E-26	15	0.150	1.441	5.94	1.175E-27	1.711E-27	0.4602	2.6488	0.423	5.206	1.187E-28
6.711	22.4	3.206E-26	8	0.150	0.924	9.21	5.046E-27	4.778E-27	0.2952	2.2120	0.504	5.838	3.317E-28
4.139	36.2	3.939E-26	6	0.150	0.559	15.02	1.600E-26	9.522E-27	0.1785	1.7285	0.504	6.237	6.610E-28
2.674	56.1	3.511E-26	5	0.150	0.345	23.52	3.279E-26	1.315E-26	0.1100	1.2916	0.423	6.448	9.127E-28
1.884	79.7	3.667E-26	5	0.150	0.223	33.96	6.450E-26	1.952E-26	0.0711	0.9414	0.293	6.551	1.355E-27
1.430	105.1	3.215E-26	5	0.150	0.147	45.65	8.965E-26	2.259E-26	0.0468	0.6654	0.139	6.604	1.568E-27
1.144	131.5	3.065E-26	5	0.150	0.094	58.05	1.198E-25	2.699E-26	0.0299	0.4421	-0.028	6.632	1.874E-27
0.951	158.4	2.548E-26	6	0.150	0.053	70.46	1.279E-25	2.708E-26	0.0170	0.2570	-0.201	6.646	1.880E-27
0.812	185.8	2.240E-26	6	0.150	0.020	82.32	1.359E-25	2.799E-26	0.0065	0.0986	-0.379	6.652	1.943E-27
16.206	15.4	3.196E-27	41	0.250	2.250	6.34	8.693E-29	1.972E-28	0.7186	2.7588	0.186	3.994	1.369E-29
12.172	20.5	1.353E-26	15	0.250	1.685	8.44	6.504E-28	1.112E-27	0.5380	2.4726	0.293	4.861	7.716E-29
8.763	28.5	1.978E-26	10	0.250	1.205	11.72	1.823E-27	2.258E-27	0.3847	2.1440	0.354	5.485	1.567E-28
6.191	40.4	2.508E-26	8	0.250	0.839	16.59	4.570E-27	4.052E-27	0.2680	1.7964	0.354	5.908	2.813E-28
4.458	56.1	2.980E-26	6	0.250	0.588	23.03	1.022E-26	6.688E-27	0.1878	1.4677	0.293	6.172	4.642E-28
3.350	74.7	3.397E-26	6	0.250	0.423	30.61	1.988E-26	1.015E-26	0.1350	1.1814	0.186	6.330	7.045E-28
2.633	95.1	3.020E-26	6	0.250	0.311	38.83	2.719E-26	1.149E-26	0.0992	0.9400	0.051	6.425	7.973E-28
2.149	116.6	2.917E-26	6	0.250	0.230	47.34	3.701E-26	1.360E-26	0.0736	0.7362	-0.101	6.483	9.442E-28
1.808	138.7	2.654E-26	6	0.250	0.170	55.83	4.424E-26	1.472E-26	0.0542	0.5626	-0.264	6.520	1.022E-27
1.557	151.3	2.426E-26	6	0.250	0.121	64.13	5.030E-26	1.564E-26	0.0387	0.4117	-0.434	6.543	1.086E-27
1.366	184.1	2.313E-26	6	0.250	0.081	72.04	5.719E-26	1.702E-26	0.0259	0.2794	-0.607	6.557	1.181E-27
1.216	207.1	2.301E-26	6	0.250	0.046	79.47	6.565E-26	1.905E-26	0.0148	0.1616	-0.784	6.565	1.322E-27
1.096	230.2	1.813E-26	7	0.250	0.016	86.34	5.813E-26	1.668E-26	0.0051	0.0558	-0.962	6.568	1.158E-27
14.290	24.5	3.584E-27	31	0.350	1.974	10.05	1.248E-28	2.508E-28	0.6304	2.3586	0.139	4.416	1.741E-29
11.064	31.6	9.562E-27	17	0.350	1.519	12.97	5.521E-28	8.643E-28	0.4851	2.1024	0.186	5.059	6.000E-29
8.493	41.2	1.270E-26	12	0.350	1.154	16.87	1.232E-27	1.496E-27	0.3684	1.8379	0.186	5.516	1.038E-28
6.577	53.2	1.946E-26	9	0.350	0.878	21.74	3.098E-27	2.959E-27	0.2803	1.5820	0.139	5.828	2.054E-28
5.205	67.3	2.295E-26	8	0.350	0.676	27.37	5.699E-27	4.411E-27	0.2159	1.3477	0.051	6.038	3.062E-28
4.231	82.8	2.369E-26	7	0.350	0.529	33.50	8.628E-27	5.602E-27	0.1689	1.1400	-0.065	6.178	3.889E-28
3.529	99.3	2.503E-26	7	0.350	0.419	39.89	1.260E-26	7.098E-27	0.1337	0.9579	-0.202	6.273	4.927E-28
3.010	116.5	2.370E-26	7	0.350	0.334	46.35	1.566E-26	7.882E-27	0.1066	0.7984	-0.351	6.338	5.471E-28
2.616	134.2	2.258E-26	7	0.350	0.266	52.80	1.875E-26	8.644E-27	0.0848	0.6568	-0.510	6.383	6.000E-28
2.309	152.2	1.896E-26	8	0.350	0.210	59.10	1.908E-26	8.226E-27	0.0669	0.5307	-0.676	6.415	5.710E-28
2.064	170.4	2.051E-26	7	0.350	0.162	65.17	2.428E-26	9.960E-27	0.0517	0.4175	-0.845	6.438	6.913E-28
1.866	188.7	1.941E-26	8	0.350	0.121	70.97	2.636E-26	1.043E-26	0.0386	0.3152	-1.018	6.453	7.241E-28
1.702	207.2	1.682E-26	8	0.350	0.084	76.49	2.567E-26	9.916E-27	0.0269	0.2214	-1.194	6.463	6.883E-28
1.564	225.7	1.582E-26	8	0.350	0.051	81.68	2.670E-26	1.016E-26	0.0163	0.1355	-1.371	6.470	7.049E-28
1.447	244.3	1.441E-26	9	0.350	0.021	86.56	2.651E-26	1.000E-26	0.0067	0.0559	-1.550	6.473	6.945E-28

p_3	θ_3	$\dfrac{d^2\sigma}{d\Omega_3\,dp_3}$	Error	p_T	p_L^*	θ_3^*	$\dfrac{1}{p_3^{*2}}\cdot\dfrac{d^2\sigma}{d\Omega_3^*\,dp_3^*}$	$\dfrac{E_3}{p_3^3}\cdot\dfrac{d^2\sigma}{d\Omega_3\,dp_3}$	x^*	y_3^*	t	m_4	$\dfrac{d^2\sigma}{dt\,dm_4^2}$
GeV/c	mrad	cm²sr⁻¹GeV⁻¹c	%	GeV/c	GeV/c	deg	cm²sr⁻¹GeV⁻³c³	cm²sr⁻¹GeV⁻²c³			GeV²	GeV	cm²/GeV⁴

$p_1 = 24.120$ GeV/c; $p_1^* = 3.299$ GeV/c; $E_1^* + E_2^* = 6.860$ GeV; $s = 47.053$ GeV²; $\gamma_{c.m.} = 3.6557$; $\beta_{c.m.} = 0.96186$; $y_1 = 1.9702$; $p_{3,\,max}^* = 3.132$ GeV/c; [69 S 1] (cont.)

p_3	θ_3	$\dfrac{d^2\sigma}{d\Omega_3\,dp_3}$	Error	p_T	p_L^*	θ_3^*	$\dfrac{1}{p_3^{*2}}\cdot\dfrac{d^2\sigma}{d\Omega_3^*\,dp_3^*}$	$\dfrac{E_3}{p_3^3}\cdot\dfrac{d^2\sigma}{d\Omega_3\,dp_3}$	x^*	y_3^*	t	m_4	$\dfrac{d^2\sigma}{dt\,dm_4^2}$
13.475	33.4	4.854E-27	25	0.450	1.849	13.68	1.888E-28	3.602E-28	0.5903	2.0762	0.011	4.571	2.501E-29
10.904	41.3	5.207E-27	22	0.450	1.483	16.88	3.069E-28	4.776E-28	0.4736	1.8643	0.011	5.072	3.315E-29
8.864	50.8	8.990E-27	15	0.450	1.190	20.71	7.924E-28	1.014E-27	0.3801	1.6569	-0.028	5.432	7.041E-29
7.294	61.7	1.163E-26	13	0.450	0.962	25.08	1.489E-27	1.595E-27	0.3070	1.4618	-0.101	5.691	1.107E-28
6.101	73.8	1.361E-26	11	0.450	0.784	29.85	2.439E-27	2.231E-27	0.2504	1.2827	-0.202	5.875	1.549E-28
5.192	86.8	1.531E-26	10	0.450	0.646	34.86	3.690E-27	2.950E-27	0.2062	1.1210	-0.323	6.009	2.048E-28
4.491	100.4	1.207E-26	11	0.450	0.536	40.03	3.768E-27	2.689E-27	0.1711	0.9750	-0.460	6.106	1.866E-28
3.941	114.4	1.823E-26	9	0.450	0.447	45.21	7.131E-27	4.629E-27	0.1426	0.8443	-0.606	6.178	3.213E-28
3.503	128.8	1.201E-26	11	0.450	0.373	50.37	5.713E-27	3.431E-27	0.1190	0.7255	-0.762	6.232	2.382E-28
3.148	143.4	1.654E-26	9	0.450	0.310	55.44	9.327E-27	5.259E-27	0.0990	0.6179	-0.923	6.272	3.651E-28
2.856	158.2	1.353E-26	10	0.450	0.256	60.38	8.848E-27	4.743E-27	0.0817	0.5194	-1.090	6.302	3.292E-28
2.612	173.2	1.067E-26	11	0.450	0.208	65.18	7.940E-27	4.091E-27	0.0665	0.4286	-1.261	6.325	2.840E-28
2.405	188.2	1.467E-26	10	0.450	0.166	69.77	1.223E-26	6.110E-27	0.0530	0.3452	-1.432	6.342	4.241E-28
2.229	203.3	1.035E-26	11	0.450	0.128	74.17	9.530E-27	4.652E-27	0.0408	0.2676	-1.607	6.354	3.229E-28
2.076	218.5	9.428E-27	12	0.450	0.093	78.38	9.479E-27	4.552E-27	0.0295	0.1951	-1.783	6.363	3.160E-28
1.943	233.7	1.343E-26	10	0.450	0.060	82.38	1.459E-26	6.930E-27	0.0192	0.1275	-1.961	6.368	4.810E-28
1.826	248.9	9.339E-27	12	0.450	0.030	86.16	1.087E-26	5.129E-27	0.0096	0.0640	-2.139	6.372	3.561E-28
1.723	264.2	1.204E-26	10	0.450	0.002	89.77	1.488E-26	7.011E-27	0.0006	0.0039	-2.320	6.373	4.866E-28
15.941	34.5	1.774E-27	45	0.550	2.186	14.12	4.928E-29	1.113E-28	0.6979	2.0582	-0.169	4.011	7.725E-30
13.370	41.2	2.380E-27	35	0.550	1.820	16.81	9.337E-29	1.780E-28	0.5812	1.8822	-0.169	4.573	1.236E-29
11.244	48.9	2.282E-27	33	0.550	1.516	19.95	1.254E-28	2.030E-28	0.4839	1.7088	-0.202	4.987	1.409E-29
9.525	57.8	4.785E-27	21	0.550	1.266	23.47	3.620E-28	5.024E-28	0.4044	1.5428	-0.264	5.295	3.487E-29
8.154	67.5	6.999E-27	18	0.550	1.065	27.32	7.115E-28	8.585E-28	0.3400	1.3869	-0.351	5.524	5.959E-29
7.062	78.0	8.953E-27	15	0.550	0.901	31.39	1.190E-27	1.268E-27	0.2878	1.2429	-0.459	5.697	8.802E-29
6.190	89.0	7.996E-27	15	0.550	0.768	35.61	1.353E-27	1.292E-27	0.2452	1.1107	-0.583	5.829	8.969E-29
5.487	100.4	1.130E-26	12	0.550	0.658	39.90	2.371E-27	2.060E-27	0.2100	0.9897	-0.719	5.929	1.430E-28
4.913	112.2	9.089E-27	14	0.550	0.565	44.23	2.311E-27	1.851E-27	0.1804	0.8784	-0.865	6.007	1.285E-28
4.440	124.2	8.858E-27	14	0.550	0.486	48.52	2.671E-27	1.996E-27	0.1553	0.7766	-1.018	6.068	1.386E-28
4.044	136.4	8.024E-27	15	0.550	0.418	52.75	2.817E-27	1.985E-27	0.1335	0.6826	-1.176	6.116	1.378E-28
3.710	148.8	8.259E-27	15	0.550	0.358	56.93	3.320E-27	2.228E-27	0.1144	0.5954	-1.340	6.154	1.546E-28
3.425	161.3	1.048E-26	12	0.550	0.305	60.99	4.753E-27	3.062E-27	0.0974	0.5145	-1.507	6.183	2.126E-28
3.180	173.8	6.941E-27	15	0.550	0.257	64.91	3.507E-27	2.185E-27	0.0822	0.4396	-1.675	6.207	1.517E-28
2.966	186.5	9.104E-27	13	0.550	0.214	68.74	5.067E-27	3.073E-27	0.0683	0.3687	-1.848	6.225	2.133E-28
2.780	199.2	7.255E-27	15	0.550	0.174	72.42	4.401E-27	2.613E-27	0.0557	0.3025	-2.023	6.239	1.814E-28
2.615	211.9	7.233E-27	15	0.550	0.138	75.94	4.744E-27	2.770E-27	0.0440	0.2404	-2.197	6.250	1.923E-28
2.468	224.7	8.772E-27	14	0.550	0.103	79.35	6.173E-27	3.560E-27	0.0330	0.1813	-2.374	6.258	2.471E-28
2.338	237.5	7.205E-27	15	0.550	0.071	82.60	5.397E-27	3.087E-27	0.0228	0.1255	-2.553	6.263	2.143E-28
2.220	250.3	8.189E-27	14	0.550	0.041	85.71	6.498E-27	3.696E-27	0.0132	0.0726	-2.730	6.267	2.566E-28
2.114	263.2	5.889E-27	16	0.550	0.012	88.71	4.919E-27	2.792E-27	0.0040	0.0219	-2.912	6.268	1.938E-28
15.867	41.0	1.062E-27	57	0.650	2.162	16.74	2.960E-29	6.693E-29	0.6902	1.8950	-0.351	4.006	4.646E-30
13.678	47.5	1.232E-27	50	0.650	1.848	19.38	4.586E-29	9.008E-29	0.5901	1.7464	-0.379	4.486	6.253E-30

p_3	θ_3	$\dfrac{\mathrm{d}^2\sigma}{\mathrm{d}\Omega_3\,\mathrm{d}p_3}$	Error	p_T	p_L^*	θ_3^*	$\dfrac{1}{p_3^{*2}}\cdot\dfrac{\mathrm{d}^2\sigma}{\mathrm{d}\Omega_3^*\,\mathrm{d}p_3^*}$	$\dfrac{E_3}{p_3^2}\cdot\dfrac{\mathrm{d}^2\sigma}{\mathrm{d}\Omega_3\,\mathrm{d}p_3}$	x^*	y_3^*	t	m_4	$\dfrac{\mathrm{d}^2\sigma}{\mathrm{d}t\,\mathrm{d}m_4^2}$
GeV/c	mrad	cm² sr⁻¹ GeV⁻¹ c	%	GeV/c	GeV/c	deg	cm² sr⁻¹ GeV⁻³ c³	cm² sr⁻¹ GeV⁻² c³			GeV²	GeV	cm²/GeV⁴

$p_1 = 24.120$ GeV/c; $p_1^* = 3.299$ GeV/c; $E_1^* + E_2^* = 6.860$ GeV; $s = 47.053$ GeV²; $\gamma_{\mathrm{c.m.}} = 3.6557$; $\beta_{\mathrm{c.m.}} = 0.96186$; $y_1 = 1.9702$; $p_{3,\,\mathrm{max}}^* = 3.132$ GeV/c; [69 S 1] (cont.)

p_3	θ_3	$\dfrac{\mathrm{d}^2\sigma}{\mathrm{d}\Omega_3\,\mathrm{d}p_3}$	Error	p_T	p_L^*	θ_3^*	$\dfrac{1}{p_3^{*2}}\dfrac{\mathrm{d}^2\sigma}{\mathrm{d}\Omega_3^*\,\mathrm{d}p_3^*}$	$\dfrac{E_3}{p_3^2}\dfrac{\mathrm{d}^2\sigma}{\mathrm{d}\Omega_3\,\mathrm{d}p_3}$	x^*	y_3^*	t	m_4	$\dfrac{\mathrm{d}^2\sigma}{\mathrm{d}t\,\mathrm{d}m_4^2}$
11.846	54.9	2.225E-27	35	0.650	1.584	22.32	1.094E-28	1.878E-28	0.5056	1.6024	-0.433	4.849	1.304E-29
10.331	63.0	2.062E-27	35	0.650	1.362	25.51	1.317E-28	1.996E-28	0.4350	1.4654	-0.510	5.126	1.386E-29
9.085	71.6	5.000E-27	21	0.650	1.178	28.89	4.070E-28	5.504E-28	0.3761	1.3366	-0.607	5.340	3.821E-29
8.058	80.7	4.341E-27	23	0.650	1.023	32.42	4.415E-28	5.388E-28	0.3267	1.2163	-0.719	5.507	3.740E-29
7.209	90.3	8.205E-27	17	0.650	0.893	36.05	1.022E-27	1.138E-27	0.2852	1.1047	-0.845	5.639	7.902E-29
6.503	100.1	6.302E-27	19	0.650	0.782	39.71	9.441E-28	9.693E-28	0.2498	1.0013	-0.980	5.744	6.728E-29
5.910	110.2	5.776E-27	20	0.650	0.687	43.41	1.022E-27	9.776E-28	0.2194	0.9050	-1.125	5.827	6.786E-29
5.407	120.5	6.369E-27	18	0.650	0.604	47.09	1.312E-27	1.178E-27	0.1929	0.8155	-1.276	5.895	8.179E-29
4.978	130.9	4.089E-27	22	0.650	0.532	50.72	9.656E-28	8.217E-28	0.1697	0.7325	-1.431	5.950	5.704E-29
4.609	141.5	4.495E-27	21	0.650	0.467	54.32	1.201E-27	9.757E-28	0.1491	0.6544	-1.593	5.994	6.773E-29
4.288	152.2	5.274E-27	19	0.650	0.409	57.85	1.577E-27	1.231E-27	0.1305	0.5813	-1.758	6.030	8.542E-29
4.007	162.9	5.755E-27	19	0.650	0.356	61.27	1.906E-27	1.437E-27	0.1137	0.5131	-1.923	6.060	9.976E-29
3.760	173.7	5.944E-27	18	0.650	0.308	64.62	2.159E-27	1.582E-27	0.0984	0.4487	-2.092	6.084	1.098E-28
3.541	184.6	4.895E-27	20	0.650	0.264	67.89	1.934E-27	1.383E-27	0.0843	0.3875	-2.264	6.104	9.603E-29
3.346	195.5	4.513E-27	21	0.650	0.223	71.04	1.925E-27	1.350E-27	0.0713	0.3299	-2.438	6.120	9.371E-29
3.171	206.5	5.299E-27	19	0.650	0.185	74.11	2.423E-27	1.673E-27	0.0591	0.2748	-2.614	6.132	1.161E-28
3.013	217.5	5.070E-27	20	0.650	0.149	77.07	2.472E-27	1.685E-27	0.0477	0.2226	-2.791	6.142	1.169E-28
2.870	228.4	5.062E-27	20	0.650	0.116	79.90	2.617E-27	1.766E-27	0.0370	0.1733	-2.964	6.150	1.226E-28
2.741	239.5	4.622E-27	20	0.650	0.084	82.67	2.519E-27	1.688E-27	0.0267	0.1255	-3.146	6.154	1.172E-28
2.622	250.5	4.401E-27	21	0.650	0.053	85.31	2.520E-27	1.681E-27	0.0170	0.0802	-3.323	6.158	1.167E-28
2.514	261.5	2.884E-27	25	0.650	0.024	87.84	1.727E-27	1.149E-27	0.0078	0.0368	-3.502	6.160	7.975E-29

$p_1 = 28.440$ GeV/c; $p_1^* = 3.593$ GeV/c; $E_1^* + E_2^* = 7.427$ GeV; $s = 55.154$ GeV²; $\gamma_{\mathrm{c.m.}} = 3.9579$; $\beta_{\mathrm{c.m.}} = 0.96756$; $y_1^* = 2.0525$; $p_{3,\,\mathrm{max}}^* = 3.438$ GeV/c; [69 S 1]

p_3	θ_3	$\dfrac{\mathrm{d}^2\sigma}{\mathrm{d}\Omega_3\,\mathrm{d}p_3}$	Error	p_T	p_L^*	θ_3^*	$\dfrac{1}{p_3^{*2}}\dfrac{\mathrm{d}^2\sigma}{\mathrm{d}\Omega_3^*\,\mathrm{d}p_3^*}$	$\dfrac{E_3}{p_3^2}\dfrac{\mathrm{d}^2\sigma}{\mathrm{d}\Omega_3\,\mathrm{d}p_3}$	x^*	y_3^*	t	m_4	$\dfrac{\mathrm{d}^2\sigma}{\mathrm{d}t\,\mathrm{d}m_4^2}$
16.394	3.0	1.100E-26	44	0.050	2.103	1.36	3.183E-28	6.710E-28	0.6116	3.3462	0.354	4.885	3.950E-29
10.943	4.6	3.159E-26	18	0.050	1.401	2.04	2.049E-27	2.887E-27	0.4076	2.9420	0.504	5.852	1.700E-28
6.260	8.0	4.552E-26	10	0.050	0.797	3.59	8.971E-27	7.273E-27	0.2318	2.3836	0.606	6.567	4.282E-28
3.228	15.5	4.666E-26	7	0.050	0.401	7.10	3.381E-26	1.447E-26	0.1168	1.7215	0.606	6.987	8.517E-28
1.845	27.1	4.033E-26	7	0.050	0.214	13.15	8.419E-26	2.192E-26	0.0623	1.1630	0.504	7.163	1.291E-27
1.230	40.7	3.578E-26	7	0.050	0.124	22.02	1.516E-25	2.928E-26	0.0360	0.7591	0.354	7.232	1.723E-27
0.909	55.1	2.912E-26	7	0.050	0.070	35.37	1.974E-25	3.241E-26	0.0205	0.4589	0.187	7.262	1.908E-27
0.716	69.9	2.580E-26	8	0.050	0.033	56.27	2.415E-25	3.671E-26	0.0097	0.2233	0.011	7.274	2.161E-27
0.588	85.1	1.998E-26	9	0.050	0.004	84.86	2.354E-25	3.492E-26	0.0013	0.0303	-0.170	7.278	2.056E-27
17.382	8.6	2.623E-27	58	0.150	2.227	3.85	6.747E-29	1.509E-28	0.6478	3.0812	0.293	4.685	8.884E-30
12.248	12.2	1.419E-26	18	0.150	1.566	5.47	7.336E-28	1.159E-27	0.4555	2.7311	0.423	5.631	6.821E-29
7.912	19.0	3.381E-26	8	0.150	1.006	8.48	4.164E-27	4.274E-27	0.2925	2.2942	0.504	6.319	2.516E-28
4.880	30.7	4.448E-26	6	0.150	0.610	13.82	1.417E-26	9.118E-27	0.1774	1.8108	0.504	6.754	5.368E-28
3.152	47.6	4.645E-26	5	0.150	0.379	21.60	3.425E-26	1.475E-26	0.1102	1.3738	0.423	6.984	8.684E-28
2.221	67.6	4.524E-26	5	0.150	0.248	31.13	6.339E-26	2.041E-26	0.0722	1.0237	0.293	7.099	1.202E-27
1.685	89.1	4.088E-26	5	0.150	0.168	41.78	9.193E-26	2.434E-26	0.0488	0.7475	0.139	7.158	1.433E-27
1.349	111.5	3.667E-26	6	0.150	0.112	53.15	1.169E-25	2.733E-26	0.0327	0.5244	-0.028	7.190	1.609E-27
1.121	134.3	3.029E-26	6	0.150	0.071	64.73	1.256E-25	2.723E-26	0.0206	0.3391	-0.202	7.208	1.603E-27

$p_1 = 28.440$ GeV/c; $p_1^* = 3.593$ GeV/c; $E_1^* + E_2^* = 7.427$ GeV; $\gamma_{c.m.} = 3.9579$; $\beta_{c.m.} = 0.96756$; $y_1^* = 2.0525$; $p_{3,\,max}^* = 3.438$ GeV/c; $s = 55.154$ GeV²; [69 S 1] (cont.)

p_3	θ_3	$\dfrac{d^2\sigma}{d\Omega_3\,dp_3}$	Error	p_T	p_L^*	θ_3^*	$\dfrac{1}{p_3^{*2}}\dfrac{d^2\sigma}{d\Omega_3^*\,dp_3^*}$	$\dfrac{E_3}{p_3^{*2}}\dfrac{d^2\sigma}{d\Omega_3^*\,dp_3^*}$	x^*	y_3^*	t	m_4	$\dfrac{d^2\sigma}{dt\,dm_4^2}$
GeV/c	mrad	cm² sr⁻¹ GeV⁻¹ c	%	GeV/c	GeV/c	deg	cm² sr⁻¹ GeV⁻³ c³	cm² sr⁻¹ GeV⁻² c³			GeV²	GeV	cm²/GeV⁴
0.957	157.4	2.660E-26	6	0.150	0.037	76.04	1.349E-25	2.809E-26	0.0108	0.1809	-0.379	7.217	1.654E-27
0.834	180.7	2.451E-26	6	0.150	0.009	86.59	1.453E-25	2.980E-26	0.0026	0.0435	-0.559	7.220	1.754E-27
19.107	13.1	3.306E-27	44	0.250	2.445	5.84	7.028E-29	1.730E-28	0.7112	2.8415	0.187	4.314	1.019E-29
14.350	17.4	1.536E-26	16	0.250	1.832	7.77	5.774E-28	1.070E-27	0.5327	2.5550	0.293	5.257	6.302E-29
10.331	24.2	2.007E-26	11	0.250	1.311	10.80	1.448E-27	1.943E-27	0.3813	2.2263	0.354	5.936	1.144E-28
7.299	34.3	2.998E-26	8	0.250	0.915	15.28	4.284E-27	4.108E-27	0.2662	1.8787	0.354	6.398	2.418E-28
5.255	47.6	3.584E-26	6	0.250	0.644	21.21	9.678E-27	6.823E-27	0.1874	1.5501	0.293	6.686	4.016E-28
3.948	63.4	4.034E-26	6	0.250	0.466	28.20	1.869E-26	1.022E-26	0.1356	1.2638	0.187	6.859	6.019E-28
3.102	80.7	3.819E-26	6	0.250	0.346	35.82	2.742E-26	1.232E-26	0.1007	1.0223	0.051	6.964	7.259E-28
2.532	98.9	3.563E-26	6	0.250	0.261	43.71	3.635E-26	1.409E-26	0.0760	0.8186	-0.101	7.030	8.297E-28
2.130	117.7	3.515E-26	6	0.250	0.198	51.68	4.752E-26	1.654E-26	0.0575	0.6446	-0.264	7.071	9.736E-28
1.833	136.8	3.042E-26	6	0.250	0.147	59.49	5.169E-26	1.664E-26	0.0428	0.4940	-0.433	7.099	9.798E-28
1.608	156.1	2.546E-26	7	0.250	0.106	67.03	5.206E-26	1.589E-26	0.0308	0.3620	-0.606	7.116	9.356E-28
1.431	175.6	2.468E-26	7	0.250	0.071	74.22	5.876E-26	1.733E-26	0.0205	0.2443	-0.783	7.127	1.020E-27
1.288	195.3	2.525E-26	7	0.250	0.043	81.00	6.823E-26	1.972E-26	0.0115	0.1379	-0.962	7.133	1.161E-27
1.172	215.0	2.076E-26	8	0.250	0.012	87.26	6.224E-26	1.784E-26	0.0035	0.0418	-1.143	7.136	1.050E-27
16.847	20.8	4.909E-27	30	0.350	2.147	9.26	1.337E-28	2.914E-28	0.6244	2.4406	0.139	4.775	1.715E-29
13.044	26.8	1.364E-26	15	0.350	1.654	11.95	6.166E-28	1.046E-27	0.4809	2.1847	0.186	5.476	6.156E-29
10.012	35.0	1.185E-26	14	0.350	1.258	15.55	9.016E-28	1.184E-27	0.3658	1.9203	0.187	5.973	6.968E-29
7.753	45.2	2.109E-26	10	0.350	0.959	20.04	2.639E-27	2.721E-27	0.2791	1.6643	0.139	6.314	1.602E-28
6.135	57.1	2.541E-26	8	0.350	0.742	25.25	4.977E-27	4.143E-27	0.2159	1.4300	0.051	6.543	2.439E-28
4.987	70.2	2.879E-26	7	0.350	0.584	30.93	8.307E-27	5.775E-27	0.1699	1.2223	-0.065	6.697	3.400E-28
4.158	84.3	2.651E-26	8	0.350	0.467	36.87	1.064E-26	6.379E-27	0.1357	1.0402	-0.202	6.802	3.755E-28
3.546	98.9	2.949E-26	7	0.350	0.376	42.93	1.563E-26	8.323E-27	0.1094	0.8804	-0.351	6.875	4.900E-28
3.081	113.9	2.697E-26	7	0.350	0.305	48.99	1.808E-26	8.763E-27	0.0886	0.7388	-0.511	6.926	5.159E-28
2.718	129.1	2.223E-26	8	0.350	0.246	54.92	1.821E-26	8.190E-27	0.0715	0.6132	-0.675	6.964	4.828E-28
2.430	144.6	2.031E-26	8	0.350	0.196	60.72	1.970E-26	8.372E-27	0.0571	0.4998	-0.846	6.990	4.928E-28
2.195	160.1	2.128E-26	8	0.350	0.154	66.28	2.387E-26	9.714E-27	0.0447	0.3976	-1.017	7.009	5.719E-28
2.001	175.8	1.939E-26	9	0.350	0.116	71.62	2.463E-26	9.714E-27	0.0338	0.3039	-1.193	7.023	5.718E-28
1.838	191.6	1.790E-26	9	0.350	0.083	76.72	2.532E-26	9.767E-27	0.0240	0.2176	-1.372	7.032	5.750E-28
1.700	207.4	1.951E-26	9	0.350	0.052	81.52	3.027E-26	1.152E-26	0.0152	0.1381	-1.551	7.037	6.779E-28
1.581	223.2	1.822E-26	9	0.350	0.024	86.04	3.064E-26	1.157E-26	0.0071	0.0643	-1.730	7.040	6.811E-28
19.543	23.0	1.504E-27	58	0.450	2.487	10.26	3.040E-29	7.696E-29	0.7234	2.3656	-0.028	4.192	4.531E-30
15.885	28.3	3.846E-27	31	0.450	2.012	12.60	1.172E-28	2.421E-28	0.5853	2.1585	0.011	4.947	1.425E-29
12.854	35.0	5.600E-27	23	0.450	1.617	15.56	2.588E-28	4.357E-28	0.4702	1.9465	0.011	5.492	2.565E-29
10.449	43.1	9.684E-27	17	0.450	1.300	19.10	6.704E-28	9.269E-28	0.3781	1.7393	-0.028	5.885	5.456E-29
8.598	52.4	1.484E-26	12	0.450	1.053	23.14	1.496E-27	1.726E-27	0.3063	1.5441	-0.101	6.167	1.016E-28
7.191	62.6	1.694E-26	11	0.450	0.862	27.55	2.398E-27	2.356E-27	0.2508	1.3651	-0.202	6.370	1.387E-28
6.118	73.6	1.741E-26	10	0.450	0.714	32.22	3.328E-27	2.846E-27	0.2077	1.2032	-0.323	6.517	1.676E-28
5.291	85.2	1.721E-26	11	0.450	0.597	37.03	4.281E-27	3.254E-27	0.1735	1.0576	-0.459	6.624	1.916E-28

Diddens/Schlüpmann

p_3	θ_3	$\dfrac{d^2\sigma}{d\Omega_3\,dp_3}$	Error	p_T	p_L^*	θ_3^*	$\dfrac{1}{p_3^{*2}}\cdot\dfrac{d^2\sigma}{d\Omega_3^*\,dp_3^*}$	$\dfrac{E_3}{p_3^2}\cdot\dfrac{d^2\sigma}{d\Omega_3\,dp_3}$	x^*	y_3^*	t	m_4	$\dfrac{d^2\sigma}{dt\,dm_4^2}$
GeV/c	mrad	cm² sr⁻¹ GeV⁻¹ c	%	GeV/c	GeV/c	deg	cm² sr⁻¹ GeV⁻³ c³	cm² sr⁻¹ GeV⁻² c³			GeV²	GeV	cm²/GeV⁴

$p_1 = 28.440$ GeV/c; $p_1^* = 3.593$ GeV/c; $E_1^* + E_2^* = 7.427$ GeV; $s = 55.154$ GeV²; $\gamma_{c.m.} = 3.9579$; $\beta_{c.m.} = 0.96756$; $y_1^* = 2.0525$; $p_{3,max}^* = 3.438$ GeV/c; [69 S 1] (cont.)

p_3	θ_3	$\dfrac{d^2\sigma}{d\Omega_3\,dp_3}$	Error	p_T	p_L^*	θ_3^*	$\dfrac{1}{p_3^{*2}}\cdot\dfrac{d^2\sigma}{d\Omega_3^*\,dp_3^*}$	$\dfrac{E_3}{p_3^2}\cdot\dfrac{d^2\sigma}{d\Omega_3\,dp_3}$	x^*	y_3^*	t	m_4	$\dfrac{d^2\sigma}{dt\,dm_4^2}$
4.643	97.1	2.023E-26	10	0.450	0.502	41.89	6.334E-27	4.359E-27	0.1459	0.9265	-0.607	6.705	2.566E-28
4.126	109.3	1.796E-26	10	0.450	0.423	46.75	6.876E-27	4.355E-27	0.1231	0.8077	-0.763	6.765	2.564E-28
3.707	121.7	1.802E-26	10	0.450	0.357	51.54	8.225E-27	4.865E-27	0.1040	0.7000	-0.924	6.811	2.864E-28
3.362	134.3	1.896E-26	10	0.450	0.301	56.25	1.010E-26	5.644E-27	0.0875	0.6013	-1.091	6.846	3.323E-28
3.073	146.9	1.431E-26	11	0.450	0.251	60.79	8.731E-27	4.661E-27	0.0731	0.5114	-1.258	6.873	2.744E-28
2.829	159.7	1.428E-26	11	0.450	0.208	65.23	9.818E-27	5.054E-27	0.0604	0.4276	-1.431	6.894	2.975E-28
2.621	172.6	1.545E-26	11	0.450	0.168	69.51	1.180E-26	5.903E-27	0.0489	0.3497	-1.608	6.909	3.475E-28
2.440	185.5	1.275E-26	12	0.450	0.132	73.61	1.069E-26	5.234E-27	0.0385	0.2774	-1.784	6.921	3.081E-28
2.283	198.4	1.290E-26	11	0.450	0.100	77.52	1.176E-26	5.661E-27	0.0290	0.2098	-1.961	6.930	3.333E-28
2.145	211.4	8.523E-27	14	0.450	0.069	81.28	8.360E-27	3.982E-27	0.0201	0.1461	-2.141	6.935	2.344E-28
2.023	224.4	1.053E-26	13	0.450	0.041	84.85	1.103E-26	5.218E-27	0.0118	0.0861	-2.322	6.939	3.072E-28
1.914	237.4	8.625E-27	14	0.450	0.014	88.24	9.583E-27	4.518E-27	0.0040	0.0294	-2.502	6.941	2.660E-28
18.792	29.3	1.332E-27	58	0.550	2.379	13.02	2.898E-29	7.088E-29	0.6920	2.1406	-0.169	4.341	4.173E-30
15.761	34.9	2.624E-27	38	0.550	1.984	15.50	8.070E-29	1.665E-28	0.5769	1.9646	-0.169	4.953	9.801E-30
13.254	41.5	3.275E-27	31	0.550	1.654	18.39	1.413E-28	2.471E-28	0.4811	1.7911	-0.202	5.404	1.455E-29
11.227	49.0	5.237E-27	23	0.550	1.385	21.66	3.117E-28	4.665E-28	0.4028	1.6250	-0.264	5.740	2.746E-29
9.610	57.3	9.825E-27	17	0.550	1.168	25.22	7.875E-28	1.022E-27	0.3397	1.4694	-0.351	5.991	6.019E-29
8.323	56.1	8.982E-27	16	0.550	0.992	29.00	9.442E-28	1.079E-27	0.2886	1.3253	-0.459	6.180	6.354E-29
7.294	75.5	1.101E-26	15	0.550	0.849	32.92	1.478E-27	1.510E-27	0.2470	1.1931	-0.583	6.325	8.888E-29
6.465	85.2	1.347E-26	13	0.550	0.732	36.93	2.251E-27	2.084E-27	0.2128	1.0719	-0.719	6.436	1.227E-28
5.788	95.2	1.173E-26	14	0.550	0.633	40.98	2.384E-27	2.027E-27	0.1842	0.9609	-0.865	6.523	1.193E-28
5.229	105.4	9.352E-27	15	0.550	0.549	45.03	2.265E-27	1.789E-27	0.1598	0.8587	-1.019	6.591	1.053E-28
4.762	115.8	1.014E-26	15	0.550	0.477	49.05	2.872E-27	2.130E-27	0.1389	0.7644	-1.178	6.645	1.254E-28
4.368	126.3	1.078E-26	14	0.550	0.415	53.00	3.513E-27	2.469E-27	0.1206	0.6775	-1.341	6.688	1.454E-28
4.031	136.9	9.494E-27	15	0.550	0.359	56.86	3.509E-27	2.357E-27	0.1044	0.5967	-1.508	6.723	1.387E-28
3.742	147.5	9.728E-27	15	0.550	0.310	60.61	4.024E-27	2.601E-27	0.0901	0.5219	-1.675	6.751	1.531E-28
3.489	158.3	1.263E-26	13	0.550	0.265	64.30	5.786E-27	3.623E-27	0.0770	0.4510	-1.848	6.773	2.133E-28
3.269	169.1	8.534E-27	15	0.550	0.224	67.86	4.283E-27	2.613E-27	0.0651	0.3848	-2.023	6.791	1.538E-28
3.074	179.9	1.137E-26	13	0.550	0.186	71.29	6.199E-27	3.703E-27	0.0542	0.3226	-2.197	6.805	2.180E-28
2.901	190.8	6.948E-27	17	0.550	0.151	74.62	4.082E-27	2.398E-27	0.0440	0.2635	-2.376	6.815	1.412E-28
2.746	201.7	1.044E-26	14	0.550	0.119	77.82	6.565E-27	3.807E-27	0.0345	0.2077	-2.553	6.824	2.241E-28
2.607	212.6	8.381E-27	15	0.550	0.088	80.89	5.606E-27	3.219E-27	0.0256	0.1547	-2.732	6.830	1.895E-28
2.481	223.5	9.755E-27	14	0.550	0.059	83.84	6.904E-27	3.938E-27	0.0173	0.1044	-2.910	6.834	2.318E-28
2.368	234.4	6.692E-27	18	0.550	0.032	86.67	4.981E-27	2.831E-27	0.0093	0.0564	-3.091	6.836	1.667E-28
2.264	245.4	8.215E-27	15	0.550	0.006	89.40	6.406E-27	3.635E-27	0.0017	0.0102	-3.273	6.837	2.140E-28
13.963	46.6	1.736E-27	44	0.650	1.730	20.59	6.707E-29	1.243E-28	0.5033	1.6847	-0.433	5.257	7.320E-30
12.177	53.4	4.144E-27	27	0.650	1.492	23.54	2.084E-28	3.403E-28	0.4339	1.5478	-0.510	5.560	2.004E-29
10.707	60.7	4.209E-27	27	0.650	1.293	26.69	2.704E-28	3.931E-28	0.3761	1.4188	-0.607	5.794	2.314E-29
9.496	68.5	6.419E-27	21	0.650	1.127	29.97	5.166E-28	6.760E-28	0.3279	1.2987	-0.719	5.978	3.980E-29
8.495	76.6	6.091E-27	22	0.650	0.988	33.34	6.022E-28	7.171E-28	0.2873	1.1870	-0.845	6.123	4.222E-29
7.662	84.9	6.955E-27	20	0.650	0.870	36.78	8.293E-28	9.079E-28	0.2530	1.0834	-0.981	6.238	5.345E-29

p_3	θ_3	$\dfrac{d^2\sigma}{d\Omega_3\,dp_3}$	Error	p_T	p_L^*	θ_3^*	$\dfrac{1}{p_3^{*2}}\cdot\dfrac{d^2\sigma}{d\Omega_3^*\,dp_3^*}$	$\dfrac{E_3}{p_3^2}\cdot\dfrac{d^2\sigma}{d\Omega_3\,dp_3}$	x^*	y_3^*	t	m_4	$\dfrac{d^2\sigma}{dt\,dm_4^2}$
GeV/c	mrad	cm²sr⁻¹GeV⁻¹c	%	GeV/c	GeV/c	deg	cm²sr⁻¹GeV⁻³c³	cm²sr⁻¹GeV⁻²c³			GeV²	GeV	cm²/GeV⁴
$p_1 = 28.440$ GeV/c; $p_1^* = 3.593$ GeV/c; $E_1^* + E_2^* = 7.427$ GeV; $s = 55.154$ GeV²; $\gamma_{c.m.} = 3.9579$; $\beta_{c.m.} = 0.96756$; $y_1^* = 2.0525$; $p_{3,max}^* = 3.438$ GeV/c; [69 S 1] (cont.)													
6.962	93.5	7.816E-27	18	0.650	0.768	40.24	1.105E-27	1.123E-27	0.2235	0.9872	-1.125	6.331	6.610E-29
6.369	102.2	8.858E-27	17	0.650	0.680	43.68	1.463E-27	1.391E-27	0.1979	0.8982	-1.275	6.407	8.190E-29
5.862	111.1	8.204E-27	17	0.650	0.603	47.13	1.559E-27	1.400E-27	0.1755	0.8145	-1.432	6.468	8.241E-29
5.426	120.1	6.087E-27	20	0.650	0.535	50.54	1.315E-27	1.122E-27	0.1557	0.7365	-1.594	6.519	6.606E-29
5.047	129.1	5.772E-27	21	0.650	0.474	53.86	1.401E-27	1.144E-27	0.1380	0.6640	-1.755	6.561	6.735E-29
4.716	138.3	7.121E-27	18	0.650	0.419	57.17	1.921E-27	1.511E-27	0.1220	0.5951	-1.925	6.595	8.893E-29
4.424	147.5	4.922E-27	22	0.650	0.370	60.39	1.463E-27	1.113E-27	0.1075	0.5305	-2.095	6.624	6.553E-29
4.165	156.7	4.555E-27	24	0.650	0.324	63.51	1.480E-27	1.094E-27	0.0942	0.4698	-2.265	6.648	6.443E-29
3.934	166.0	6.337E-27	20	0.650	0.282	66.57	2.232E-27	1.612E-27	0.0819	0.4119	-2.439	6.667	9.489E-29
3.727	175.3	5.463E-27	21	0.650	0.243	69.54	2.073E-27	1.467E-27	0.0705	0.3571	-2.613	6.683	8.635E-29
3.541	184.6	6.046E-27	19	0.650	0.206	72.41	2.455E-27	1.709E-27	0.0599	0.3052	-2.788	6.696	1.006E-28
3.372	194.0	5.014E-27	22	0.650	0.172	75.21	2.167E-27	1.488E-27	0.0499	0.2553	-2.967	6.706	8.761E-29
3.219	203.3	4.157E-27	24	0.650	0.139	77.89	1.903E-27	1.293E-27	0.0405	0.2082	-3.143	6.715	7.610E-29
3.079	212.7	4.406E-27	23	0.650	0.109	80.51	2.127E-27	1.432E-27	0.0316	0.1627	-3.322	6.721	8.433E-29
2.950	222.1	4.148E-27	24	0.650	0.079	83.04	2.103E-27	1.408E-27	0.0231	0.1192	-3.501	6.725	8.287E-29
2.833	231.5	3.130E-27	27	0.650	0.052	85.47	1.659E-27	1.106E-27	0.0150	0.0774	-3.682	6.728	6.512E-29
2.724	240.9	4.397E-27	23	0.650	0.025	87.82	2.430E-27	1.616E-27	0.0072	0.0373	-3.862	6.730	9.515E-29
$p_1 = 19.200$ GeV/c; $p_1^* = 2.929$ GeV/c; $E_1^* + E_2^* = 6.151$ GeV; $s = 37.830$ GeV²; $\gamma_{c.m.} = 3.2779$; $\beta_{c.m.} = 0.95233$; $y_1^* = 1.8562$; $p_{3,max}^* = 2.743$ GeV/c; [71 B 1, 71 B 2]													
3.581	14.0	3.323E-26	29	0.050	0.550	5.19	1.630E-26	9.286E-27	0.2005	2.0216	0.618	5.554	8.097E-28
2.954	16.9	4.754E-26	20	0.050	0.450	6.34	3.400E-26	1.611E-26	0.1641	1.8294	0.621	5.659	1.405E-27
2.334	21.4	3.459E-26	20	0.050	0.350	8.13	3.905E-26	1.484E-26	0.1276	1.5941	0.612	5.760	1.294E-27
1.728	29.0	4.955E-26	15	0.050	0.250	11.31	9.899E-26	2.877E-26	0.0911	1.2937	0.576	5.854	2.509E-27
1.151	43.5	4.098E-26	15	0.050	0.150	18.44	1.700E-25	3.586E-26	0.0547	0.8895	0.481	5.938	3.127E-27
0.654	76.5	2.058E-26	10	0.050	0.050	45.00	2.055E-25	3.216E-26	0.0182	0.3311	0.230	5.994	2.805E-27
5.811	17.2	1.996E-26	21	0.100	0.900	6.34	3.750E-27	3.436E-27	0.3281	2.3586	0.536	5.155	2.996E-28
4.546	22.0	2.783E-26	14	0.100	0.700	8.13	8.498E-27	6.125E-27	0.2552	2.1131	0.567	5.384	5.341E-28
3.638	41.2	3.467E-26	17	0.150	0.550	15.25	1.625E-26	9.537E-27	0.2005	1.7135	0.511	5.534	8.316E-28
3.022	49.7	3.739E-26	13	0.150	0.450	18.43	2.505E-26	1.238E-26	0.1641	1.5281	0.494	5.636	1.080E-27
2.418	62.1	3.882E-26	11	0.150	0.350	23.20	3.965E-26	1.608E-26	0.1276	1.3049	0.455	5.732	1.402E-27
1.835	81.8	3.193E-26	10	0.150	0.250	30.96	5.399E-26	1.745E-26	0.0911	1.0288	0.376	5.820	1.522E-27
1.293	116.3	2.579E-26	10	0.150	0.150	45.01	7.899E-26	2.006E-26	0.0547	0.6785	0.216	5.893	1.749E-27
0.836	180.4	2.035E-26	7	0.150	0.050	71.56	1.170E-25	2.468E-26	0.0182	0.2417	-0.105	5.938	2.152E-27
3.747	66.8	3.541E-26	13	0.250	0.550	24.44	1.525E-26	9.457E-27	0.2005	1.4077	0.307	5.497	8.247E-28
3.150	79.4	3.231E-26	12	0.250	0.450	29.05	1.925E-26	1.027E-26	0.1641	1.2338	0.255	5.594	8.953E-28
2.571	97.4	2.525E-26	13	0.250	0.350	35.54	2.175E-26	9.835E-27	0.1276	1.0302	0.168	5.682	8.576E-28
2.022	124.0	3.005E-26	10	0.250	0.250	45.01	3.920E-26	1.490E-26	0.0911	0.7886	0.026	5.760	1.299E-27
1.521	165.1	2.326E-26	9	0.250	0.150	59.05	4.749E-26	1.535E-26	0.0547	0.5023	-0.210	5.820	1.339E-27
1.100	229.3	1.856E-26	7	0.250	0.050	78.70	5.851E-26	1.701E-26	0.0182	0.1736	-0.595	5.854	1.483E-27

p_3	θ_3	$\dfrac{d^2\sigma}{d\Omega_3\,dp_3}$	Error	p_T	p_L^*	θ_3^*	$\dfrac{1}{p_3^{*2}}\cdot\dfrac{d^2\sigma}{d\Omega_3^*\,dp_3^*}$	$\dfrac{E_3}{p_3^2}\cdot\dfrac{d^2\sigma}{d\Omega_3\,dp_3}$	x^*	y_3^*	t	m_4	$\dfrac{d^2\sigma}{dt\,dm_4^2}$
GeV/c	mrad	cm² sr⁻¹ GeV⁻¹ c	%	GeV/c	GeV/c	deg	cm² sr⁻¹ GeV⁻³ c³	cm² sr⁻¹ GeV⁻² c³			GeV²	GeV	cm²/GeV⁴
$p_1 = 19.200$ GeV/c; $p_1^* = 2.929$ GeV/c; $E_1^* + E_2^* = 6.151$ GeV; $s = 37.830$ GeV²; $\gamma_{c.m.} = 3.2779$; $\beta_{c.m.} = 0.95233$; $y_1^* = 1.8562$; $p_{3,max}^* = 2.743$ GeV/c; [71 B 1, 71 B 2] (cont.)													
5.951	50.4	1.227E-26	16	0.300	0.900	18.43	2.151E-27	2.062E-27	0.3281	1.7260	0.274	5.104	1.798E-28
4.721	63.6	2.192E-26	9	0.300	0.700	23.20	5.999E-27	4.645E-27	0.2552	1.4941	0.238	5.322	4.051E-28
3.900	89.9	1.260E-26	20	0.350	0.550	32.47	4.849E-27	3.233E-27	0.2005	1.1721	0.021	5.445	2.819E-28
3.326	105.4	1.667E-26	15	0.350	0.450	37.86	8.548E-27	5.017E-27	0.1641	1.0126	-0.074	5.534	4.375E-28
2.775	126.5	1.803E-26	14	0.350	0.350	45.01	1.265E-26	6.506E-27	0.1276	0.8300	-0.214	5.615	5.673E-28
2.258	155.6	1.876E-26	10	0.350	0.250	54.46	1.840E-26	8.323E-27	0.0911	0.6225	-0.417	5.682	7.257E-28
1.792	196.6	1.297E-26	10	0.350	0.150	66.81	1.790E-26	7.259E-27	0.0547	0.3880	-0.717	5.732	6.330E-28
1.395	253.6	1.348E-26	10	0.350	0.050	81.87	2.555E-26	9.711E-27	0.0182	0.1323	-1.146	5.760	8.468E-28
4.088	110.3	1.036E-26	21	0.450	0.550	39.29	3.501E-27	2.535E-27	0.2005	0.9948	-0.333	5.380	2.211E-28
3.538	127.5	1.221E-26	17	0.450	0.450	44.98	5.302E-27	3.454E-27	0.1641	0.8496	-0.471	5.462	3.012E-28
3.013	149.9	1.281E-26	14	0.450	0.350	52.12	7.251E-27	4.256E-27	0.1276	0.6874	-0.660	5.534	3.711E-28
2.525	179.2	1.217E-26	13	0.450	0.250	60.95	9.050E-27	4.827E-27	0.0911	0.5083	-0.917	5.594	4.210E-28
2.084	217.6	1.008E-26	12	0.450	0.150	71.56	9.802E-27	4.847E-27	0.0547	0.3133	-1.263	5.636	4.226E-28
1.703	267.3	9.251E-27	11	0.450	0.050	83.65	1.150E-26	5.449E-27	0.0183	0.1061	-1.721	5.659	4.752E-28
6.214	80.6	6.616E-27	18	0.500	0.900	29.06	1.025E-27	1.065E-27	0.3281	1.3177	-0.219	5.007	9.287E-29
5.040	99.4	7.454E-27	14	0.500	0.700	35.54	1.700E-27	1.482E-27	0.2552	1.1077	-0.360	5.209	1.292E-28
4.305	128.1	9.181E-27	21	0.550	0.550	45.00	2.700E-27	2.134E-27	0.2005	0.8595	-0.739	5.304	1.861E-28
3.776	146.2	4.782E-27	27	0.550	0.450	50.72	1.750E-27	1.267E-27	0.1640	0.7271	-0.919	5.380	1.105E-28
3.275	168.7	1.025E-26	16	0.550	0.350	57.52	4.699E-27	3.133E-27	0.1276	0.5834	-1.150	5.445	2.732E-28
2.809	197.0	8.178E-27	16	0.550	0.250	65.55	4.701E-27	2.914E-27	0.0912	0.4276	-1.449	5.498	2.541E-28
2.388	232.4	6.751E-27	16	0.550	0.150	74.75	4.825E-27	2.832E-27	0.0547	0.2614	-1.832	5.534	2.469E-28
2.019	276.0	5.965E-27	15	0.550	0.050	84.81	5.200E-27	2.962E-27	0.0182	0.0879	-2.312	5.554	2.583E-28
6.573	106.7	2.944E-27	25	0.700	0.900	37.88	3.900E-28	4.480E-28	0.3281	1.0543	-0.894	4.870	3.906E-29
5.460	128.5	3.602E-27	19	0.700	0.700	44.98	6.601E-28	6.599E-28	0.2552	0.8680	-1.148	5.055	5.754E-29
4.415	159.2	3.173E-27	17	0.700	0.500	54.46	8.251E-28	7.190E-28	0.1823	0.6532	-1.531	5.209	6.270E-29
3.472	203.0	3.760E-27	13	0.700	0.300	66.79	1.400E-27	1.084E-27	0.1094	0.4089	-2.105	5.322	9.452E-29
2.671	265.2	2.980E-27	12	0.700	0.100	81.88	1.550E-27	1.117E-27	0.0364	0.1395	-2.949	5.384	9.742E-29
7.005	128.8	6.725E-28	50	0.900	0.900	44.99	7.501E-29	9.603E-29	0.3281	0.8732	-1.703	4.701	8.374E-30
5.949	151.9	9.222E-28	38	0.900	0.700	52.13	1.350E-28	1.551E-28	0.2552	0.7078	-2.066	4.870	1.352E-29
4.964	182.3	1.160E-27	29	0.900	0.500	60.95	2.250E-28	2.338E-28	0.1823	0.5246	-2.562	5.007	2.038E-29
4.077	222.6	7.424E-28	32	0.900	0.300	71.57	1.900E-28	1.822E-28	0.1093	0.3236	-3.242	5.104	1.589E-29
3.313	275.2	1.031E-27	24	0.900	0.100	83.67	3.400E-28	3.115E-28	0.0364	0.1094	-4.152	5.155	2.717E-29
$p_1 = 19.200$ GeV/c; $p_1^* = 2.929$ GeV/c; $E_1^* + E_2^* = 6.151$ GeV; $s = 37.830$ GeV²; $\gamma_{c.m.} = 3.2779$; $\beta_{c.m.} = 0.95233$; $y_1^* = 1.8562$; $p_{3,max}^* = 2.743$ GeV/c; [70 A 1]													
4.500	12.5	5.449E-26	3	0.056	0.695	4.63	1.703E-26	1.211E-26	0.2535	2.2350	0.597	5.394	1.056E-27
4.500	20.0	5.105E-26	3	0.090	0.693	7.39	1.592E-26	1.135E-26	0.2528	2.1363	0.576	5.392	9.897E-28
4.500	30.0	5.284E-26	3	0.135	0.690	11.07	1.639E-26	1.175E-26	0.2515	1.9800	0.533	5.388	1.024E-27

$p_1 = 19.200$ GeV/c; $p_1^* = 2.929$ GeV/c; $E_1^* + E_2^* = 6.151$ GeV; $s = 37.830$ GeV²; $\gamma_{\text{c.m.}} = 3.2779$; $\beta_{\text{c.m.}} = 0.95233$; $y_1^* = 1.8562$; $p_{3,\,\text{max}}^* = 2.743$ GeV/c; [70 A 1] (cont.)

p_3	θ_3	$\dfrac{d^2\sigma}{d\Omega_3\,dp_3}$	Error	p_T	p_L^*	θ_3^*	$\dfrac{1}{p_3^{*2}}\cdot\dfrac{d^2\sigma}{d\Omega_3^*\,dp_3^*}$	$\dfrac{E_3}{p_3^2}\cdot\dfrac{d^2\sigma}{d\Omega_3\,dp_3}$	x^*	y_3^*	t	m_4	$\dfrac{d^2\sigma}{dt\,dm_4^2}$
GeV/c	mrad	cm² sr⁻¹ GeV⁻¹ c	%	GeV/c	GeV/c	deg	cm² sr⁻¹ GeV⁻³ c³	cm² sr⁻¹ GeV⁻² c³			GeV²	GeV	cm²/GeV⁴
4.500	40.0	4.512E-26	3	0.180	0.685	14.73	1.390E-26	1.003E-26	0.2496	1.8203	0.472	5.383	8.748E-28
4.500	50.0	4.098E-26	3	0.225	0.678	18.35	1.252E-26	9.111E-27	0.2472	1.6697	0.394	5.376	7.945E-28
4.500	60.0	3.436E-26	3	0.270	0.670	21.94	1.039E-26	7.639E-27	0.2442	1.5317	0.299	5.367	6.662E-28
4.500	70.0	2.804E-26	3	0.315	0.660	25.49	8.372E-27	6.234E-27	0.2407	1.4062	0.187	5.356	5.436E-28
6.000	12.5	4.072E-26	3	0.075	0.931	4.61	7.188E-27	6.789E-27	0.3394	2.4710	0.544	5.122	5.920E-28
6.000	20.0	3.799E-26	3	0.120	0.929	7.36	6.690E-27	6.333E-27	0.3385	2.3211	0.516	5.119	5.523E-28
6.000	30.0	3.020E-26	3	0.180	0.924	11.03	5.292E-27	5.035E-27	0.3367	2.1080	0.459	5.113	4.390E-28
6.000	40.0	2.430E-26	5	0.240	0.917	14.67	4.229E-27	4.051E-27	0.3342	1.9100	0.378	5.106	3.533E-28
6.000	50.0	2.062E-26	3	0.300	0.908	18.28	3.557E-27	3.438E-27	0.3310	1.7345	0.274	5.095	2.998E-28
6.000	60.0	1.643E-26	3	0.360	0.897	21.85	2.805E-27	2.739E-27	0.3271	1.5800	0.148	5.083	2.389E-28
6.000	70.0	1.314E-26	3	0.420	0.884	25.39	2.215E-27	2.191E-27	0.3224	1.4434	-0.002	5.068	1.910E-28
8.000	12.5	2.179E-26	3	0.100	1.244	4.59	2.169E-27	2.724E-27	0.4536	2.6783	0.462	4.733	2.376E-28
8.000	20.0	1.929E-26	3	0.160	1.241	7.35	1.915E-27	2.412E-27	0.4524	2.4659	0.425	4.729	2.103E-28
8.000	30.0	1.499E-26	3	0.240	1.234	11.00	1.481E-27	1.874E-27	0.4501	2.1977	0.348	4.721	1.634E-28
8.000	40.0	1.167E-26	3	0.320	1.225	14.63	1.145E-27	1.459E-27	0.4467	1.9686	0.241	4.709	1.272E-28
8.000	50.0	8.735E-27	3	0.400	1.214	18.24	8.496E-28	1.092E-27	0.4424	1.7750	0.102	4.695	9.523E-29
8.000	60.0	6.384E-27	3	0.480	1.199	21.80	6.144E-28	7.981E-28	0.4372	1.6095	-0.066	4.677	6.960E-29
8.000	70.0	4.596E-27	3	0.560	1.182	25.33	4.369E-28	5.746E-28	0.4309	1.4657	-0.266	4.655	5.010E-29
10.000	12.5	1.210E-26	3	0.125	1.557	4.59	7.716E-28	1.210E-27	0.5676	2.8141	0.374	4.308	1.055E-28
10.000	20.0	1.024E-26	3	0.200	1.553	7.34	6.514E-28	1.024E-27	0.5662	2.5505	0.327	4.303	8.930E-29
10.000	30.0	7.816E-27	3	0.300	1.545	10.99	4.948E-28	7.817E-28	0.5632	2.2454	0.231	4.291	6.816E-29
10.000	35.0	6.300E-27	3	0.350	1.540	12.81	3.975E-28	6.301E-28	0.5613	2.1154	0.169	4.284	5.494E-29
10.000	40.0	5.753E-27	3	0.400	1.533	14.62	3.617E-28	5.754E-28	0.5590	1.9982	0.097	4.276	5.017E-29
10.000	50.0	3.454E-27	3	0.500	1.519	18.22	2.152E-28	3.454E-28	0.5536	1.7949	-0.076	4.255	3.012E-29
10.000	60.0	2.032E-27	3	0.600	1.501	21.78	1.253E-28	2.032E-28	0.5471	1.6237	-0.287	4.231	1.772E-29
10.000	70.0	1.350E-27	3	0.699	1.479	25.31	8.221E-29	1.350E-28	0.5393	1.4763	-0.536	4.201	1.177E-29
11.000	12.5	7.838E-27	3	0.137	1.713	4.59	4.132E-28	7.126E-28	0.6246	2.8648	0.329	4.079	6.214E-29
11.000	20.0	6.987E-27	3	0.220	1.709	7.34	3.675E-28	6.352E-28	0.6230	2.5798	0.277	4.072	5.539E-29
11.000	30.0	5.634E-27	3	0.330	1.700	10.98	2.948E-28	5.122E-28	0.6197	2.2611	0.172	4.060	4.467E-29
11.000	35.0	4.490E-27	2	0.385	1.694	12.80	2.342E-28	4.082E-28	0.6176	2.1275	0.103	4.051	3.560E-29
11.000	40.0	3.577E-27	3	0.440	1.687	14.61	1.859E-28	3.252E-28	0.6151	2.0077	0.024	4.041	2.836E-29
11.000	50.0	1.737E-27	3	0.550	1.671	18.21	8.949E-29	1.579E-28	0.6092	1.8012	-0.166	4.018	1.377E-29
11.000	60.0	1.049E-27	3	0.660	1.651	21.77	5.347E-29	9.537E-29	0.6020	1.6282	-0.398	3.989	8.317E-30
11.000	70.0	6.090E-28	3	0.769	1.628	25.30	3.066E-29	5.537E-29	0.5935	1.4796	-0.673	3.954	4.828E-30
12.000	12.5	4.614E-27	3	0.150	1.870	4.59	2.045E-28	3.845E-28	0.6816	2.9070	0.283	3.836	3.353E-29
12.000	20.0	4.630E-27	3	0.240	1.865	7.33	2.047E-28	3.859E-28	0.6798	2.6032	0.227	3.829	3.365E-29
12.000	30.0	3.288E-27	3	0.360	1.855	10.98	1.446E-28	2.740E-28	0.6762	2.2734	0.111	3.814	2.389E-29
12.000	35.0	2.120E-27	3	0.420	1.849	12.80	9.295E-29	1.767E-28	0.6739	2.1368	0.036	3.804	1.541E-29

Diddens/Schlüpmann

p_3	θ_3	$\dfrac{\mathrm{d}^2\sigma}{\mathrm{d}\Omega_3\,\mathrm{d}p_3}$	Error	p_T	p_L^*	θ_3^*	$\dfrac{1}{p_3^{*2}}\cdot\dfrac{\mathrm{d}^2\sigma}{\mathrm{d}\Omega_3^*\,\mathrm{d}p_3^*}$	$\dfrac{E_3}{p_3^2}\cdot\dfrac{\mathrm{d}^2\sigma}{\mathrm{d}\Omega_3\,\mathrm{d}p_3}$	x^*	y_3^*	t	m_4	$\dfrac{\mathrm{d}^2\sigma}{\mathrm{d}t\,\mathrm{d}m_4^2}$
GeV/c	mrad	cm² sr⁻¹ GeV⁻¹ c	%	GeV/c	GeV/c	deg	cm² sr⁻¹ GeV⁻³ c³	cm² sr⁻¹ GeV⁻² c³			GeV²	GeV	cm²/GeV⁴

$p_1 = 19.200$ GeV/c; $p_1^* = 2.929$ GeV/c; $E_1^* + E_2^* = 6.151$ GeV; $s = 37.830$ GeV²; $\gamma_\mathrm{c.m.} = 3.2779$; $\beta_\mathrm{c.m.} = 0.95233$; $y_1^* = 1.8562$; $p_{3,\,\mathrm{max}}^* = 2.743$ GeV/c; [70 A 1] (cont.)

p_3	θ_3	$\dfrac{\mathrm{d}^2\sigma}{\mathrm{d}\Omega_3\,\mathrm{d}p_3}$	Error	p_T	p_L^*	θ_3^*	$\dfrac{1}{p_3^{*2}}\cdot\dfrac{\mathrm{d}^2\sigma}{\mathrm{d}\Omega_3^*\,\mathrm{d}p_3^*}$	$\dfrac{E_3}{p_3^2}\cdot\dfrac{\mathrm{d}^2\sigma}{\mathrm{d}\Omega_3\,\mathrm{d}p_3}$	x^*	y_3^*	t	m_4	$\dfrac{\mathrm{d}^2\sigma}{\mathrm{d}t\,\mathrm{d}m_4^2}$
12.000	40.0	1.561E-27	3	0.480	1.841	14.61	6.819E-29	1.301E-28	0.6712	2.0151	-0.050	3.792	1.134E-29
12.000	50.0	8.066E-28	3	0.600	1.823	18.21	3.493E-29	6.722E-29	0.6648	1.8061	-0.257	3.765	5.862E-30
12.000	60.0	4.617E-28	3	0.720	1.802	21.77	1.978E-29	3.848E-29	0.6569	1.6316	-0.511	3.731	3.355E-30
12.000	70.0	2.411E-28	3	0.839	1.776	25.29	1.020E-29	2.009E-29	0.6476	1.4822	-0.810	3.691	1.752E-30
13.000	12.5	2.194E-27	3	0.162	2.026	4.59	8.286E-29	1.688E-28	0.7385	2.9426	0.236	3.577	1.472E-29
13.000	20.0	2.070E-27	3	0.260	2.021	7.33	7.798E-29	1.592E-28	0.7366	2.6223	0.175	3.568	1.389E-29
13.000	30.0	1.159E-27	3	0.390	2.010	10.98	4.345E-29	8.916E-29	0.7327	2.2831	0.051	3.551	7.775E-30
13.000	35.0	7.990E-28	3	0.455	2.003	12.80	2.986E-29	6.147E-29	0.7302	2.1443	-0.030	3.539	5.360E-30
13.000	40.0	5.937E-28	3	0.520	1.995	14.61	2.210E-29	4.567E-29	0.7273	2.0209	-0.124	3.526	3.983E-30
13.000	50.0	3.050E-28	3	0.650	1.976	18.20	1.126E-29	2.346E-29	0.7203	1.8099	-0.349	3.494	2.046E-30
13.000	60.0	1.720E-28	3	0.780	1.952	21.77	6.280E-30	1.323E-29	0.7118	1.6343	-0.623	3.454	1.154E-30
13.000	70.0	7.879E-29	3	0.909	1.925	25.29	2.841E-30	6.061E-30	0.7017	1.4842	-0.947	3.407	5.285E-31
14.000	12.5	1.140E-27	3	0.175	2.182	4.59	3.713E-29	8.143E-29	0.7955	2.9727	0.190	3.297	7.101E-30
14.000	20.0	6.989E-28	3	0.280	2.176	7.33	2.271E-29	4.992E-29	0.7934	2.6379	0.124	3.287	4.353E-30
14.000	30.0	3.850E-28	3	0.420	2.165	10.98	1.245E-29	2.750E-29	0.7892	2.2910	-0.010	3.266	2.398E-30
14.000	35.0	2.664E-28	3	0.490	2.157	12.79	8.585E-30	1.903E-29	0.7865	2.1502	-0.098	3.253	1.659E-30
14.000	40.0	1.970E-28	3	0.560	2.149	14.60	6.325E-30	1.407E-29	0.7834	2.0255	-0.199	3.237	1.227E-30
14.000	50.0	9.882E-29	3	0.700	2.128	18.20	3.145E-30	7.059E-30	0.7759	1.8130	-0.440	3.200	6.155E-31
14.000	60.0	4.768E-29	3	0.839	2.103	21.76	1.501E-30	3.406E-30	0.7667	1.6364	-0.736	3.153	2.970E-31
14.000	70.0	2.090E-29	3	0.979	2.073	25.28	6.500E-31	1.493E-30	0.7558	1.4857	-1.085	3.097	1.302E-31
15.000	35.0	9.857E-29	3	0.525	2.312	12.79	2.767E-30	6.572E-30	0.8428	2.1551	-0.165	2.939	5.731E-31
16.000	35.0	2.680E-29	3	0.560	2.466	12.79	6.614E-31	1.675E-30	0.8991	2.1591	-0.233	2.587	1.461E-31
14.500	12.5	7.213E-28	3	0.181	2.260	4.59	2.190E-29	4.975E-29	0.8239	2.9861	0.166	3.148	4.338E-30
15.000	12.5	3.940E-28	3	0.187	2.338	4.58	1.118E-29	2.627E-29	0.8524	2.9984	0.142	2.991	2.291E-30
15.100	12.5	3.308E-28	3	0.189	2.354	4.58	9.262E-30	2.191E-29	0.8581	3.0008	0.138	2.959	1.910E-30
15.400	12.5	2.432E-28	3	0.192	2.401	4.58	6.547E-30	1.579E-29	0.8752	3.0077	0.124	2.859	1.377E-30
15.700	12.5	1.825E-28	3	0.196	2.447	4.58	4.727E-30	1.162E-29	0.8922	3.0142	0.109	2.757	1.014E-30
16.000	12.5	1.294E-28	3	0.200	2.494	4.58	3.227E-30	8.088E-30	0.9093	3.0205	0.095	2.650	7.053E-31
16.300	12.5	8.796E-29	3	0.204	2.541	4.58	2.114E-30	5.397E-30	0.9264	3.0265	0.081	2.539	4.706E-31
16.600	12.5	5.895E-29	3	0.207	2.588	4.58	1.366E-30	3.551E-30	0.9435	3.0323	0.067	2.422	3.097E-31
16.750	12.5	4.000E-29	7	0.209	2.611	4.58	9.103E-31	2.388E-30	0.9520	3.0350	0.060	2.362	2.082E-31
16.830	12.5	3.510E-29	7	0.210	2.624	4.58	7.912E-31	2.086E-30	0.9565	3.0365	0.056	2.329	1.819E-31
16.900	12.5	3.110E-29	7	0.211	2.635	4.58	6.953E-31	1.840E-30	0.9605	3.0378	0.053	2.300	1.605E-31
16.970	12.5	2.550E-29	7	0.212	2.646	4.58	5.654E-31	1.503E-30	0.9645	3.0390	0.049	2.271	1.310E-31
17.030	12.5	2.310E-29	8	0.213	2.655	4.58	5.086E-31	1.356E-30	0.9679	3.0401	0.046	2.245	1.183E-31
17.050	12.5	2.340E-29	8	0.213	2.658	4.58	5.140E-31	1.372E-30	0.9691	3.0404	0.045	2.236	1.197E-31
17.130	12.5	1.890E-29	8	0.214	2.671	4.58	4.113E-31	1.103E-30	0.9736	3.0418	0.042	2.202	9.621E-32
17.200	12.5	1.450E-29	8	0.215	2.682	4.58	3.130E-31	8.431E-31	0.9776	3.0431	0.038	2.171	7.352E-32
17.270	12.5	9.000E-30	10	0.216	2.692	4.58	1.927E-31	5.212E-31	0.9816	3.0443	0.035	2.140	4.545E-32

p_3	θ_3	$\dfrac{d^2\sigma}{d\Omega_3\,dp_3}$	Error	p_T	p_L^*	θ_3^*	$\dfrac{1}{p_3^{*2}}\cdot\dfrac{d^2\sigma}{d\Omega_3^*\,dp_3^*}$	$\dfrac{E_3}{p_3^2}\cdot\dfrac{d^2\sigma}{d\Omega_3\,dp_3}$	x^*	y_3^*	t	m_4	$\dfrac{d^2\sigma}{dt\,dm_4^2}$
GeV/c	mrad	cm² sr⁻¹ GeV⁻¹ c	%	GeV/c	GeV/c	deg	cm² sr⁻¹ GeV⁻³ c³	cm² sr⁻¹ GeV⁻² c³			GeV²	GeV	cm²/GeV⁴

$p_1 = 19.200$ GeV/c; $p_1^* = 2.929$ GeV/c; $E_1^* + E_2^* = 6.151$ GeV; $s = 37.830$ GeV²; $\gamma_{c.m.} = 3.2779$; $\beta_{c.m.} = 0.95233$; $y_1^* = 1.8562$; $p_{3,max}^* = 2.743$ GeV/c; [70 A 1] (cont.)

p_3	θ_3	$\dfrac{d^2\sigma}{d\Omega_3\,dp_3}$	Error	p_T	p_L^*	θ_3^*	$\dfrac{1}{p_3^{*2}}\cdot\dfrac{d^2\sigma}{d\Omega_3^*\,dp_3^*}$	$\dfrac{E_3}{p_3^2}\cdot\dfrac{d^2\sigma}{d\Omega_3\,dp_3}$	x^*	y_3^*	t	m_4	$\dfrac{d^2\sigma}{dt\,dm_4^2}$
17.330	12.5	5.600E-30	15	0.217	2.702	4.58	1.191E-31	3.231E-31	0.9850	3.0453	0.032	2.113	2.818E-32
17.350	12.5	5.190E-30	15	0.217	2.705	4.58	1.101E-31	2.991E-31	0.9861	3.0456	0.031	2.103	2.609E-32
17.420	12.5	1.510E-30	50	0.218	2.716	4.58	3.177E-32	8.668E-32	0.9901	3.0468	0.028	2.071	7.559E-33

$p_1 = 24.00$ GeV/c; $p_1^* = 3.290$ GeV/c; $E_1^* + E_2^* = 6.843$ GeV; $s = 46.828$ GeV²; $\gamma_{c.m.} = 3.6469$; $\beta_{c.m.} = 0.96167$; $y_1^* = 1.9677$; $p_{3,max}^* = 3.123$ GeV/c; [72 A 1]

p_3	θ_3	$\dfrac{d^2\sigma}{d\Omega_3\,dp_3}$	Error	p_T	p_L^*	θ_3^*	$\dfrac{1}{p_3^{*2}}\cdot\dfrac{d^2\sigma}{d\Omega_3^*\,dp_3^*}$	$\dfrac{E_3}{p_3^2}\cdot\dfrac{d^2\sigma}{d\Omega_3\,dp_3}$	x^*	y_3^*	t	m_4	$\dfrac{d^2\sigma}{dt\,dm_4^2}$
4.000	17.0	6.098E-26	3	0.068	0.548	7.07	2.676E-26	1.525E-26	0.1756	1.9745	0.608	6.249	1.064E-27
4.000	37.0	5.309E-26	3	0.148	0.541	15.31	2.299E-26	1.328E-26	0.1731	1.7042	0.505	6.240	9.265E-28
4.000	57.0	4.111E-26	3	0.228	0.527	23.39	1.741E-26	1.028E-26	0.1687	1.4309	0.324	6.226	7.174E-28
4.000	67.0	3.554E-26	3	0.268	0.518	27.35	1.483E-26	8.890E-27	0.1658	1.3083	0.205	6.216	6.202E-28
4.000	87.0	2.502E-26	3	0.348	0.495	35.05	1.008E-26	6.259E-27	0.1586	1.0922	-0.090	6.193	4.366E-28
4.000	107.0	1.763E-26	3	0.427	0.467	42.44	6.803E-27	4.410E-27	0.1496	0.9090	-0.462	6.162	3.077E-28
4.000	127.0	1.107E-26	3	0.507	0.433	49.47	4.066E-27	2.769E-27	0.1387	0.7514	-0.910	6.126	1.932E-28
4.000	147.0	5.828E-27	3	0.586	0.393	56.13	2.027E-27	1.458E-27	0.1259	0.6137	-1.435	6.083	1.017E-28
6.000	17.0	5.039E-26	3	0.102	0.830	7.01	9.911E-27	8.401E-27	0.2657	2.2724	0.560	5.937	5.860E-28
6.000	27.0	4.522E-26	3	0.162	0.825	11.11	8.845E-27	7.539E-27	0.2642	2.0597	0.497	5.932	5.259E-28
6.000	37.0	3.763E-26	3	0.222	0.818	15.18	7.303E-27	6.273E-27	0.2619	1.8557	0.405	5.924	4.376E-28
6.000	57.0	2.413E-26	3	0.342	0.797	23.20	4.578E-27	4.023E-27	0.2554	1.5129	0.134	5.901	2.806E-28
6.000	67.0	1.966E-26	3	0.402	0.784	27.13	3.675E-27	3.278E-27	0.2510	1.3713	-0.044	5.886	2.286E-28
6.000	87.0	1.173E-26	3	0.521	0.750	34.80	2.116E-27	1.956E-27	0.2402	1.1322	-0.488	5.848	1.364E-28
6.000	107.0	6.451E-27	3	0.641	0.708	42.15	1.115E-27	1.075E-27	0.2267	0.9364	-1.045	5.800	7.503E-29
6.000	127.0	3.351E-27	3	0.760	0.657	49.17	5.509E-28	5.587E-28	0.2103	0.7712	-1.718	5.742	3.897E-29
6.000	147.0	1.352E-27	3	0.879	0.597	55.81	2.103E-28	2.254E-28	0.1912	0.6287	-2.504	5.673	1.572E-29
8.000	17.0	3.563E-26	3	0.136	1.110	6.99	3.953E-27	4.454E-27	0.3554	2.4402	0.492	5.606	3.107E-28
8.000	27.0	2.715E-26	3	0.216	1.103	11.08	2.996E-27	3.394E-27	0.3533	2.1628	0.408	5.598	2.368E-28
8.000	37.0	2.062E-26	3	0.296	1.094	15.14	2.258E-27	2.578E-27	0.3503	1.9218	0.285	5.587	1.798E-28
8.000	47.0	1.564E-26	3	0.376	1.082	19.16	1.695E-27	1.955E-27	0.3464	1.7183	0.124	5.573	1.364E-28
8.000	57.0	1.200E-26	3	0.456	1.067	23.14	1.284E-27	1.500E-27	0.3415	1.5451	-0.076	5.555	1.047E-28
8.000	67.0	8.972E-27	3	0.536	1.049	27.06	9.461E-28	1.122E-27	0.3357	1.3954	-0.314	5.534	7.825E-29
8.000	87.0	4.197E-27	3	0.695	1.004	34.71	4.270E-28	5.247E-28	0.3214	1.1470	-0.904	5.480	3.660E-29
8.000	107.0	1.789E-27	3	0.854	0.947	42.05	1.743E-28	2.237E-28	0.3033	0.9463	-1.648	5.412	1.560E-29
8.000	127.0	6.969E-28	3	1.013	0.879	49.06	6.460E-29	8.713E-29	0.2815	0.7784	-2.545	5.328	6.078E-30
8.000	147.0	2.207E-28	3	1.172	0.799	55.70	1.936E-29	2.759E-29	0.2559	0.6340	-3.594	5.229	1.925E-30
10.000	17.0	2.187E-26	3	0.170	1.389	6.98	1.555E-27	2.187E-27	0.4448	2.5423	0.417	5.254	1.526E-28
10.000	27.0	1.617E-26	3	0.270	1.381	11.06	1.144E-27	1.617E-27	0.4422	2.2189	0.311	5.243	1.128E-28
10.000	37.0	1.119E-26	3	0.370	1.369	15.12	7.851E-28	1.119E-27	0.4385	1.9557	0.158	5.229	7.807E-29
10.000	47.0	7.830E-27	3	0.470	1.354	19.13	5.438E-28	7.831E-28	0.4336	1.7406	-0.044	5.209	5.463E-29
10.000	57.0	5.342E-27	3	0.570	1.335	23.11	3.664E-28	5.343E-28	0.4275	1.5608	-0.293	5.185	3.727E-29
10.000	67.0	3.604E-27	3	0.669	1.313	27.02	2.435E-28	3.604E-28	0.4203	1.4069	-0.591	5.157	2.514E-29
10.000	87.0	1.222E-27	3	0.859	1.256	34.67	7.967E-29	1.222E-28	0.4023	1.1540	-1.329	5.085	8.526E-30

p_3	θ_3	$\dfrac{d^2\sigma}{d\Omega_3\,dp_3}$	Error	p_T	p_L^*	θ_3^*	$\dfrac{1}{p_3^{*2}}\cdot\dfrac{d^2\sigma}{d\Omega_3^*\,dp_3^*}$	$\dfrac{E_3}{p_3^2}\cdot\dfrac{d^2\sigma}{d\Omega_3\,dp_3}$	x^*	y_3^*	t	m_4	$\dfrac{d^2\sigma}{dt\,dm_4^2}$
GeV/c	mrad	cm² sr⁻¹ GeV⁻¹ c	%	GeV/c	GeV/c	deg	cm² sr⁻¹ GeV⁻³ c³	cm² sr⁻¹ GeV⁻² c³			GeV²	GeV	cm²/GeV⁴

$p_1 = 24.00$ GeV/c; $p_1^* = 3.290$ GeV/c; $E_1^* + E_2^* = 6.843$ GeV; $s = 46.828$ GeV²; $\gamma_{c.m.} = 3.6469$; $\beta_{c.m.} = 0.96167$; $y_1^* = 1.9677$; $p_{3,\,max}^* = 3.123$ GeV/c; [72 A 1] (cont.)

p_3	θ_3	$\dfrac{d^2\sigma}{d\Omega_3\,dp_3}$	Error	p_T	p_L^*	θ_3^*	$\dfrac{1}{p_3^{*2}}\cdot\dfrac{d^2\sigma}{d\Omega_3^*\,dp_3^*}$	$\dfrac{E_3}{p_3^2}\cdot\dfrac{d^2\sigma}{d\Omega_3\,dp_3}$	x^*	y_3^*	t	m_4	$\dfrac{d^2\sigma}{dt\,dm_4^2}$
10.000	107.0	3.873E-28	3	1.068	1.186	42.01	2.418E-29	3.873E-29	0.3797	0.9510	-2.259	4.992	2.702E-30
10.000	127.0	1.104E-28	3	1.267	1.101	49.01	6.557E-30	1.104E-29	0.3524	0.7817	-3.379	4.879	7.702E-31
10.000	147.0	2.712E-29	3	1.465	1.001	55.65	1.524E-30	2.712E-30	0.3206	0.6365	-4.691	4.743	1.892E-31
12.000	17.0	1.317E-26	3	0.204	1.668	6.97	6.508E-28	1.098E-27	0.5342	2.6079	0.338	4.875	7.657E-29
12.000	27.0	9.171E-27	3	0.324	1.659	11.05	4.507E-28	7.643E-28	0.5311	2.2522	0.211	4.862	5.332E-29
12.000	37.0	5.829E-27	3	0.444	1.645	15.11	2.842E-28	4.858E-28	0.5266	1.9751	0.027	4.843	3.389E-29
12.000	47.0	3.208E-27	3	0.564	1.626	19.12	1.548E-28	2.674E-28	0.5207	1.7532	-0.215	4.818	1.865E-29
12.000	57.0	1.972E-27	3	0.684	1.603	23.09	9.398E-29	1.643E-28	0.5134	1.5695	-0.515	4.787	1.146E-29
12.000	67.0	1.100E-27	3	0.803	1.576	27.01	5.165E-29	9.167E-29	0.5048	1.4133	-0.872	4.750	6.395E-30
12.000	77.0	5.602E-28	3	0.923	1.545	30.86	2.586E-29	4.669E-29	0.4947	1.2776	-1.286	4.706	3.257E-30
12.000	87.0	2.828E-28	3	1.043	1.509	34.64	1.281E-29	2.357E-29	0.4832	1.1578	-1.758	4.655	1.644E-30
12.000	97.0	1.376E-28	3	1.162	1.469	38.35	6.106E-30	1.147E-29	0.4703	1.0506	-2.287	4.598	8.000E-31
12.000	107.0	6.408E-29	3	1.282	1.424	41.98	2.780E-30	5.340E-30	0.4561	0.9536	-2.873	4.534	3.725E-31
12.000	117.0	2.930E-29	3	1.401	1.375	45.53	1.241E-30	2.442E-30	0.4404	0.8650	-3.517	4.462	1.703E-31
12.000	127.0	1.387E-29	3	1.520	1.322	48.98	5.724E-31	1.156E-30	0.4233	0.7835	-4.218	4.383	8.064E-32
12.000	137.0	6.605E-30	4	1.639	1.264	52.35	2.653E-31	5.505E-31	0.4049	0.7081	-4.976	4.296	3.840E-32
14.000	17.0	7.128E-27	3	0.238	1.947	6.97	2.589E-28	5.092E-28	0.6235	2.6521	0.256	4.465	3.552E-29
14.000	27.0	5.186E-27	3	0.378	1.936	11.05	1.873E-28	3.704E-28	0.6199	2.2734	0.108	4.448	2.584E-29
14.000	37.0	2.244E-27	3	0.518	1.920	15.10	8.043E-29	1.603E-28	0.6147	1.9871	-0.107	4.424	1.118E-29
14.000	47.0	1.143E-27	3	0.658	1.898	19.11	4.055E-29	8.165E-29	0.6078	1.7609	-0.389	4.392	5.696E-30
14.000	57.0	5.817E-28	3	0.798	1.872	23.08	2.038E-29	4.155E-29	0.5993	1.5748	-0.738	4.352	2.899E-30
14.000	67.0	2.558E-28	3	0.937	1.840	27.00	8.829E-30	1.827E-29	0.5892	1.4172	-1.155	4.304	1.275E-30
14.000	77.0	1.153E-28	3	1.077	1.803	30.85	3.913E-30	8.236E-30	0.5774	1.2806	-1.638	4.247	5.746E-31
14.000	87.0	4.736E-29	3	1.216	1.761	34.63	1.577E-30	3.383E-30	0.5640	1.1602	-2.189	4.182	2.360E-31
14.000	97.0	1.978E-29	3	1.356	1.714	38.34	6.451E-31	1.413E-30	0.5490	1.0525	-2.806	4.107	9.857E-32
14.000	107.0	7.560E-30	4	1.495	1.662	41.97	2.411E-31	5.400E-31	0.5323	0.9551	-3.490	4.023	3.767E-32
14.000	117.0	3.278E-30	4	1.634	1.605	45.51	1.020E-31	2.342E-31	0.5141	0.8663	-4.241	3.929	1.633E-32
14.000	127.0	1.397E-30	5	1.773	1.543	48.97	4.238E-32	9.979E-32	0.4942	0.7846	-5.059	3.823	6.962E-33
16.000	17.0	2.779E-27	3	0.272	2.226	6.97	7.731E-29	1.737E-28	0.7128	2.6831	0.173	4.012	1.212E-29
16.000	27.0	1.349E-27	3	0.432	2.213	11.04	3.732E-29	8.432E-29	0.7087	2.2877	0.004	3.991	5.882E-30
16.000	37.0	5.781E-28	3	0.592	2.194	15.09	1.587E-29	3.613E-29	0.7027	1.9951	-0.242	3.960	2.521E-30
16.000	47.0	2.779E-28	3	0.752	2.170	19.11	7.550E-30	1.737E-29	0.6948	1.7659	-0.564	3.919	1.212E-30
16.000	57.0	1.184E-28	3	0.912	2.140	23.08	3.176E-30	7.400E-30	0.6851	1.5783	-0.963	3.868	5.163E-31
16.000	67.0	4.622E-29	3	1.071	2.103	26.99	1.222E-30	2.889E-30	0.6735	1.4197	-1.439	3.806	2.015E-31
16.000	77.0	1.699E-29	3	1.231	2.061	30.84	4.416E-31	1.062E-30	0.6601	1.2825	-1.992	3.733	7.408E-32
16.000	87.0	5.840E-30	4	1.390	2.014	34.62	1.489E-31	3.650E-31	0.6448	1.1617	-2.621	3.647	2.546E-32
16.000	97.0	1.973E-30	5	1.550	1.960	38.33	4.928E-32	1.233E-31	0.6276	1.0537	-3.326	3.549	8.603E-33
16.000	107.0	6.883E-31	7	1.709	1.901	41.96	1.681E-32	4.302E-32	0.6086	0.9561	-4.108	3.437	3.001E-33
16.000	117.0	2.330E-31	11	1.868	1.835	45.50	5.553E-33	1.456E-32	0.5877	0.8671	-4.967	3.310	1.016E-33
16.000	127.0	6.642E-32	21	2.027	1.764	48.96	1.543E-33	4.151E-33	0.5650	0.7853	-5.901	3.166	2.896E-34

p_3	θ_3	$\dfrac{d^2\sigma}{d\Omega_3\,dp_3}$	Error	p_T	p_L^*	θ_3^*	$\dfrac{1}{p_3^{*2}}\cdot\dfrac{d^2\sigma}{d\Omega_3^*\,dp_3^*}$	$\dfrac{E_3}{p_3^2}\cdot\dfrac{d^2\sigma}{d\Omega_3\,dp_3}$	x^*	y_3^*	t	m_4	$\dfrac{d^2\sigma}{dt\,dm_4^2}$
GeV/c	mrad	cm² sr⁻¹ GeV⁻¹ c	%	GeV/c	GeV/c	deg	cm² sr⁻¹ GeV⁻³ c³	cm² sr⁻¹ GeV⁻² c³			GeV²	GeV	cm²/GeV⁴

$p_1 = 24.00$ GeV/c; $p_1^* = 3.290$ GeV/c; $E_1^* + E_2^* = 6.843$ GeV; $s = 46.828$ GeV²; $\gamma_{\mathrm{c.m.}} = 3.6469$; $\beta_{\mathrm{c.m.}} = 0.96167$; $y_1^* = 1.9677$; $p_{3,\max}^* = 3.123$ GeV/c; [72 A 1] (cont.)

p_3	θ_3	$\dfrac{d^2\sigma}{d\Omega_3\,dp_3}$	Error	p_T	p_L^*	θ_3^*	$\dfrac{1}{p_3^{*2}}\cdot\dfrac{d^2\sigma}{d\Omega_3^*\,dp_3^*}$	$\dfrac{E_3}{p_3^2}\cdot\dfrac{d^2\sigma}{d\Omega_3\,dp_3}$	x^*	y_3^*	t	m_4	$\dfrac{d^2\sigma}{dt\,dm_4^2}$
18.000	17.0	5.930E-28	3	0.306	2.505	6.97	1.304E-29	3.295E-29	0.8020	2.7054	0.089	3.502	2.298E-30
18.000	27.0	2.780E-28	3	0.486	2.490	11.04	6.078E-30	1.544E-29	0.7974	2.2977	-0.101	3.474	1.077E-30
18.000	37.0	1.133E-28	3	0.666	2.469	15.09	2.458E-30	6.295E-30	0.7907	2.0007	-0.378	3.434	4.391E-31
18.000	47.0	4.466E-29	3	0.846	2.442	19.10	9.588E-31	2.481E-30	0.7818	1.7695	-0.740	3.381	1.731E-31
18.000	57.0	1.664E-29	3	1.025	2.408	23.07	3.528E-31	9.245E-31	0.7709	1.5807	-1.189	3.314	6.449E-32
18.000	67.0	5.958E-30	5	1.205	2.367	26.98	1.245E-31	3.310E-31	0.7579	1.4215	-1.725	3.232	2.309E-32
18.000	77.0	1.526E-30	5	1.385	2.320	30.83	3.134E-32	8.478E-32	0.7428	1.2839	-2.346	3.135	5.914E-33
18.000	87.0	4.163E-31	8	1.564	2.266	34.62	8.390E-33	2.313E-32	0.7256	1.1627	-3.054	3.020	1.613E-33
18.000	97.0	1.290E-31	14	1.743	2.206	38.32	2.546E-33	7.167E-33	0.7062	1.0545	-3.848	2.885	5.000E-34
20.000	17.0	1.151E-28	3	0.340	2.783	6.96	2.068E-30	5.805E-30	0.8913	2.7221	0.004	2.902	4.050E-31
20.000	27.0	4.827E-29	3	0.540	2.767	11.04	8.550E-31	2.414E-30	0.8861	2.3050	-0.207	2.866	1.684E-31
20.000	37.0	1.542E-29	4	0.740	2.744	15.09	2.710E-31	7.710E-31	0.8787	2.0047	-0.514	2.812	5.379E-32
20.000	47.0	4.615E-30	4	0.940	2.713	19.10	8.027E-32	2.308E-31	0.8688	1.7720	-0.917	2.739	1.610E-32
20.000	57.0	1.242E-30	6	1.139	2.675	23.07	2.133E-32	6.210E-32	0.8567	1.5825	-1.416	2.646	4.332E-33
20.000	67.0	2.944E-31	9	1.339	2.630	26.98	4.982E-33	1.472E-32	0.8422	1.4228	-2.011	2.531	1.027E-33
20.000	77.0	2.967E-32	26	1.538	2.578	30.83	4.937E-34	1.484E-33	0.8254	1.2848	-2.701	2.391	1.035E-34
22.000	27.0	7.070E-31	15	0.594	3.044	11.04	1.035E-32	3.214E-32	0.9748	2.3104	-0.313	2.087	2.242E-33
16.000	22.0	1.816E-27	3	0.352	2.220	9.01	5.040E-29	1.135E-28	0.7109	2.4691	0.098	4.003	7.918E-30
20.000	22.0	7.050E-29	3	0.440	2.776	9.01	1.253E-30	3.525E-30	0.8890	2.4942	-0.089	2.886	2.459E-31
20.500	22.0	4.640E-29	3	0.451	2.846	9.00	7.847E-31	2.263E-30	0.9112	2.4964	-0.113	2.714	1.579E-31
21.000	22.0	2.470E-29	3	0.462	2.915	9.00	3.981E-31	1.176E-30	0.9335	2.4985	-0.136	2.531	8.205E-32
21.500	22.0	6.270E-30	5	0.473	2.985	9.00	9.640E-32	2.916E-31	0.9557	2.5004	-0.160	2.333	2.034E-32
22.000	22.0	1.730E-30	20	0.484	3.054	9.00	2.540E-32	7.864E-32	0.9780	2.5022	-0.184	2.117	5.486E-33

$p_1 = 29.700$ GeV/c; $p_1^* = 3.674$ GeV/c; $E_1^* = E_2^* = 7.584$ GeV; $s = 57.517$ GeV²; $\gamma_{\mathrm{c.m.}} = 4.0418$; $\beta_{\mathrm{c.m.}} = 0.96891$; $y_1 = 2.0742$; $p_{3,\max}^* = 3.523$ GeV/c; [67 A 1]

p_3	θ_3	$\dfrac{d^2\sigma}{d\Omega_3\,dp_3}$	Error	p_T	p_L^*	θ_3^*	$\dfrac{1}{p_3^{*2}}\cdot\dfrac{d^2\sigma}{d\Omega_3^*\,dp_3^*}$	$\dfrac{E_3}{p_3^2}\cdot\dfrac{d^2\sigma}{d\Omega_3\,dp_3}$	x^*	y_3^*	t	m_4	$\dfrac{d^2\sigma}{dt\,dm_4^2}$
4.508	15.0	7.495E-26	5	0.068	0.556	6.93	2.882E-26	1.663E-26	0.1578	1.9886	0.608	6.984	9.377E-28
5.113	15.0	7.616E-26	5	0.077	0.633	6.91	2.284E-26	1.490E-26	0.1796	2.0880	0.601	6.902	8.400E-28
5.772	15.0	8.147E-26	5	0.087	0.716	6.89	1.922E-26	1.412E-26	0.2033	2.1783	0.590	6.811	7.959E-28
6.510	15.0	6.754E-26	5	0.098	0.809	6.88	1.255E-26	1.038E-26	0.2297	2.2622	0.574	6.708	5.850E-28
7.101	15.0	5.764E-26	5	0.107	0.884	6.87	9.011E-27	8.119E-27	0.2509	2.3188	0.560	6.623	4.577E-28
7.695	15.0	5.547E-26	5	0.115	0.959	6.87	7.391E-27	7.210E-27	0.2721	2.3681	0.545	6.538	4.064E-28
8.292	15.0	5.382E-26	5	0.124	1.034	6.86	6.180E-27	6.492E-27	0.2934	2.4111	0.529	6.450	3.659E-28
8.898	15.0	4.723E-26	5	0.133	1.110	6.86	4.713E-27	5.309E-27	0.3151	2.4492	0.512	6.360	2.993E-28
9.489	15.0	4.218E-26	5	0.142	1.184	6.85	3.702E-27	4.446E-27	0.3361	2.4817	0.494	6.271	2.506E-28
10.100	15.0	3.471E-26	5	0.151	1.261	6.85	2.690E-27	3.437E-27	0.3579	2.5114	0.476	6.177	1.938E-28
10.700	15.0	3.132E-26	5	0.160	1.336	6.85	2.164E-27	2.927E-27	0.3793	2.5370	0.457	6.084	1.650E-28
11.300	15.0	2.785E-26	5	0.169	1.411	6.85	1.726E-27	2.465E-27	0.4007	2.5598	0.438	5.989	1.389E-28
11.900	15.0	2.641E-26	5	0.178	1.487	6.85	1.476E-27	2.219E-27	0.4221	2.5801	0.419	5.893	1.251E-28
12.490	15.0	2.422E-26	5	0.187	1.561	6.84	1.229E-27	1.939E-27	0.4431	2.5979	0.400	5.796	1.093E-28

p_3	θ_3	$\dfrac{d^2\sigma}{d\Omega_3\,dp_3}$	Error	p_T	p_L^*	θ_3^*	$\dfrac{1}{p_3^{*2}}\cdot\dfrac{d^2\sigma}{d\Omega_3^*\,dp_3^*}$	$\dfrac{E_3}{p_3^2}\cdot\dfrac{d^2\sigma}{d\Omega_3\,dp_3}$	x^*	y_3^*	t	m_4	$\dfrac{d^2\sigma}{dt\,dm_4^2}$
GeV/c	mrad	cm² sr⁻¹ GeV⁻¹ c	%	GeV/c	GeV/c	deg	cm² sr⁻¹ GeV⁻³ c³	cm² sr⁻¹ GeV⁻² c³			GeV²	GeV	cm²/GeV⁴

$p_1 = 29.700$ GeV/c; $p_1^* = 3.674$ GeV/c; $E_1^* + E_2^* = 7.584$ GeV; $s = 57.517$ GeV²; $\gamma_{c.m.} = 4.0418$; $\beta_{c.m.} = 0.96891$; $y_1^* = 2.0742$; $p_{3,max}^* = 3.523$ GeV/c; [67 A 1] (cont.)

p_3	θ_3	$\dfrac{d^2\sigma}{d\Omega_3\,dp_3}$	Error	p_T	p_L^*	θ_3^*	$\dfrac{1}{p_3^{*2}}\cdot\dfrac{d^2\sigma}{d\Omega_3^*\,dp_3^*}$	$\dfrac{E_3}{p_3^2}\cdot\dfrac{d^2\sigma}{d\Omega_3\,dp_3}$	x^*	y_3^*	t	m_4	$\dfrac{d^2\sigma}{dt\,dm_4^2}$
13.110	15.0	1.919E−26	5	0.197	1.639	6.84	8.839E−28	1.464E−27	0.4652	2.6146	0.379	5.693	8.252E−29
13.720	15.0	1.658E−26	5	0.206	1.715	6.84	6.974E−28	1.209E−27	0.4869	2.6294	0.359	5.590	6.813E−29
14.320	15.0	1.390E−26	5	0.215	1.790	6.84	5.367E−28	9.707E−28	0.5082	2.6425	0.339	5.487	5.472E−29
14.920	15.0	1.183E−26	5	0.224	1.866	6.84	4.209E−28	7.929E−28	0.5296	2.6544	0.319	5.381	4.470E−29
15.510	15.0	1.025E−26	5	0.233	1.939	6.84	3.375E−28	6.609E−28	0.5506	2.6649	0.299	5.276	3.726E−29
16.110	15.0	8.753E−27	5	0.242	2.015	6.84	2.671E−28	5.433E−28	0.5719	2.6747	0.279	5.166	3.063E−29
16.720	15.0	8.214E−27	5	0.251	2.091	6.84	2.328E−28	4.913E−28	0.5936	2.6837	0.258	5.052	2.770E−29
17.320	15.0	7.564E−27	6	0.260	2.166	6.84	1.998E−28	4.367E−28	0.6150	2.6918	0.237	4.937	2.462E−29
17.930	15.0	5.889E−27	6	0.269	2.243	6.84	1.451E−28	3.285E−28	0.6367	2.6994	0.216	4.818	1.852E−29
18.520	15.0	5.468E−27	6	0.278	2.317	6.84	1.263E−28	2.953E−28	0.6577	2.7061	0.196	4.699	1.664E−29
19.080	15.0	3.655E−27	7	0.286	2.387	6.84	7.955E−29	1.916E−28	0.6776	2.7120	0.177	4.584	1.080E−29
19.490	15.0	2.890E−27	8	0.292	2.438	6.84	6.028E−29	1.483E−28	0.6922	2.7160	0.162	4.497	8.359E−30
20.420	15.0	1.785E−27	9	0.306	2.555	6.84	3.392E−29	8.742E−29	0.7253	2.7243	0.130	4.295	4.928E−30
21.600	15.0	9.309E−28	11	0.324	2.703	6.84	1.581E−29	4.310E−29	0.7672	2.7335	0.089	4.024	2.430E−30
22.810	15.0	3.875E−28	18	0.342	2.854	6.84	5.903E−30	1.699E−29	0.8103	2.7417	0.046	3.726	9.577E−31
24.010	15.0	1.787E−28	25	0.360	3.005	6.83	2.457E−30	7.443E−30	0.8530	2.7487	0.004	3.404	4.196E−31
24.800	15.0	1.390E−28	50	0.372	3.104	6.83	1.791E−30	5.605E−30	0.8811	2.7528	−0.024	3.175	3.160E−31
6.696	96.0	8.460E−27	7	0.642	0.711	42.07	1.305E−27	1.264E−27	0.2019	0.9386	−1.217	6.546	7.124E−29
7.103	96.0	6.568E−27	4	0.681	0.755	42.04	9.013E−28	9.249E−28	0.2143	0.9411	−1.335	6.479	5.214E−29
7.697	96.0	4.117E−27	5	0.738	0.819	42.01	4.815E−28	5.350E−28	0.2325	0.9441	−1.509	6.378	3.016E−29
8.291	96.0	3.151E−27	6	0.795	0.883	41.99	3.178E−28	3.801E−28	0.2507	0.9465	−1.683	6.276	2.143E−29
8.893	96.0	2.206E−27	7	0.852	0.948	41.97	1.935E−28	2.481E−28	0.2690	0.9484	−1.861	6.171	1.399E−29
9.477	96.0	1.644E−27	8	0.908	1.010	41.95	1.270E−28	1.735E−28	0.2869	0.9500	−2.034	6.068	9.780E−30
10.100	96.0	1.207E−27	8	0.968	1.077	41.94	8.213E−29	1.195E−28	0.3059	0.9514	−2.219	5.955	6.737E−30
10.690	96.0	8.259E−28	10	1.025	1.141	41.93	5.024E−29	7.736E−29	0.3239	0.9525	−2.395	5.847	4.361E−30
11.200	96.0	5.450E−28	15	1.074	1.196	41.92	3.017E−29	4.866E−29	0.3394	0.9533	−2.547	5.751	2.743E−30
13.610	96.0	8.892E−29	22	1.305	1.454	41.90	3.336E−30	6.534E−30	0.4128	0.9560	−3.269	5.275	3.683E−31
14.420	96.0	4.549E−29	20	1.382	1.541	41.89	1.521E−30	3.155E−30	0.4375	0.9566	−3.512	5.106	1.778E−31
2.899	160.0	1.576E−26	5	0.462	0.201	66.43	1.041E−26	5.443E−27	0.0572	0.4063	−1.585	7.043	3.068E−28
3.485	160.0	1.131E−26	4	0.555	0.247	66.01	5.209E−27	3.248E−27	0.0701	0.4192	−2.014	6.934	1.831E−28
4.105	160.0	7.083E−27	4	0.654	0.295	65.75	2.363E−27	1.726E−27	0.0836	0.4274	−2.477	6.816	9.732E−29
4.696	160.0	4.824E−27	5	0.748	0.340	65.59	1.233E−27	1.028E−27	0.0964	0.4325	−2.926	6.700	5.793E−29
5.795	160.0	1.723E−27	6	0.923	0.422	65.41	2.902E−28	2.974E−28	0.1199	0.4383	−3.769	6.480	1.677E−29
6.984	160.0	6.094E−28	7	1.113	0.512	65.31	7.081E−29	8.727E−29	0.1452	0.4417	−4.689	6.232	4.920E−30
8.182	160.0	2.368E−28	11	1.304	0.601	65.24	2.007E−29	2.895E−29	0.1706	0.4438	−5.621	5.972	1.632E−30
9.390	160.0	6.326E−29	21	1.496	0.691	65.20	4.074E−30	6.738E−30	0.1962	0.4451	−6.564	5.697	3.798E−31

3.6.2 pp→π⁺X

$p_1 = 3.670$ GeV/c; $p_1^* = 1.156$ GeV/c; $E_1^* + E_2^* = 2.978$ GeV; $s = 8.868$ GeV²; $\gamma_{c.m.} = 1.5871$; $\beta_{c.m.} = 0.77652$; $y_1^* = 1.0365$; $p_{3,max}^* = 0.890$ GeV/c; [62 M 1]

p_3	θ_3	$\dfrac{d^2\sigma}{d\Omega_3\,dp_3}$	Error	p_T	p_L^*	θ_3^*	$\dfrac{1}{p_3^{*2}}\cdot\dfrac{d^2\sigma}{d\Omega_3^*\,dp_3^*}$	$\dfrac{E_3}{p_3^2}\cdot\dfrac{d^2\sigma}{d\Omega_3\,dp_3}$	x^*	y_3^*	t	m_4	$\dfrac{d^2\sigma}{dt\,dm_4^2}$
GeV/c	mrad	cm²sr⁻¹GeV⁻¹c	%	GeV/c	GeV/c	deg	cm²sr⁻¹GeV⁻³c³	cm²sr⁻¹GeV⁻²c³			GeV²	GeV	cm²/GeV⁴
0.437	0.0	5.195E-27	5	0.000	0.128	0.00	6.580E-26	1.247E-26	0.1440	0.8228	0.632	2.785	5.690E-27
0.604	0.0	1.028E-26	5	0.000	0.195	0.00	7.287E-26	1.746E-26	0.2187	1.1350	0.636	2.731	7.965E-27
0.759	0.0	1.590E-26	5	0.000	0.254	0.00	7.351E-26	2.129E-26	0.2849	1.3588	0.624	2.676	9.711E-27
0.910	0.0	2.585E-26	5	0.000	0.310	0.00	8.465E-26	2.875E-26	0.3476	1.5369	0.604	2.620	1.311E-26
1.057	0.0	4.035E-26	5	0.000	0.364	0.00	9.881E-26	3.849E-26	0.4084	1.6856	0.581	2.563	1.756E-26
1.203	0.0	4.731E-26	5	0.000	0.417	0.00	9.005E-26	3.958E-26	0.4681	1.8139	0.555	2.504	1.806E-26
1.348	0.0	3.715E-26	5	0.000	0.469	0.00	5.659E-26	2.770E-26	0.5270	1.9269	0.527	2.444	1.264E-26
1.492	0.0	1.938E-26	5	0.000	0.521	0.00	2.418E-26	1.305E-26	0.5853	2.0279	0.498	2.382	5.951E-27
1.636	0.0	1.138E-26	5	0.000	0.573	0.00	1.184E-26	6.983E-27	0.6432	2.1194	0.469	2.319	3.186E-27
1.779	0.0	7.955E-27	5	0.000	0.624	0.00	7.015E-27	4.486E-27	0.7009	2.2030	0.438	2.254	2.047E-27
1.922	0.0	4.257E-27	5	0.000	0.675	0.00	3.221E-27	2.221E-27	0.7583	2.2800	0.408	2.187	1.013E-27
2.064	0.0	1.701E-27	5	0.000	0.726	0.00	1.117E-27	8.260E-28	0.8156	2.3514	0.377	2.117	3.768E-28
0.384	246.8	5.628E-27	5	0.094	0.087	47.01	8.223E-26	1.559E-26	0.0982	0.4990	0.537	2.785	7.112E-27
0.526	274.3	1.096E-26	5	0.142	0.133	47.00	9.003E-26	2.157E-26	0.1491	0.6246	0.493	2.731	9.841E-27
0.658	285.8	1.675E-26	5	0.186	0.173	47.00	8.983E-26	2.601E-26	0.1943	0.6893	0.437	2.676	1.187E-26
0.787	291.9	1.702E-26	5	0.226	0.211	47.01	6.470E-26	2.197E-26	0.2371	0.7277	0.377	2.620	1.002E-26
0.913	295.5	1.316E-26	5	0.265	0.248	47.00	3.741E-26	1.458E-26	0.2786	0.7527	0.313	2.563	6.649E-27
1.039	297.9	1.037E-26	5	0.305	0.284	47.00	2.292E-26	1.007E-26	0.3192	0.7696	0.248	2.504	4.596E-27
1.163	299.5	8.120E-27	5	0.343	0.320	46.99	1.436E-26	7.032E-27	0.3594	0.7818	0.182	2.444	3.208E-27
1.287	300.7	6.382E-27	5	0.381	0.355	47.00	9.246E-27	4.989E-27	0.3992	0.7907	0.115	2.382	2.276E-27
1.410	301.6	4.900E-27	5	0.419	0.391	47.00	5.923E-27	3.492E-27	0.4387	0.7975	0.047	2.319	1.593E-27
1.533	302.3	3.670E-27	5	0.456	0.426	47.00	3.758E-27	2.403E-27	0.4780	0.8027	-0.021	2.254	1.096E-27
1.656	302.8	2.688E-27	5	0.494	0.461	47.00	2.362E-27	1.629E-27	0.5172	0.8070	-0.089	2.187	7.431E-28
1.779	303.3	1.953E-27	5	0.531	0.495	47.01	1.489E-27	1.101E-27	0.5562	0.8101	-0.157	2.117	5.024E-28
1.901	303.6	9.759E-28	5	0.568	0.530	47.00	6.519E-28	5.147E-28	0.5952	0.8130	-0.226	2.046	2.348E-28
0.321	390.2	8.254E-27	5	0.122	0.040	71.96	1.478E-25	2.804E-26	0.0447	0.2128	0.427	2.785	1.279E-26
0.432	442.5	1.226E-26	5	0.185	0.060	72.01	1.246E-25	2.982E-26	0.0675	0.2564	0.326	2.732	1.361E-26
0.538	464.7	1.376E-26	5	0.241	0.078	72.01	9.130E-26	2.642E-26	0.0879	0.2775	0.219	2.677	1.205E-26
0.642	476.5	1.155E-26	5	0.294	0.096	71.99	5.421E-26	1.841E-26	0.1075	0.2896	0.110	2.620	8.399E-27
0.744	483.7	9.044E-27	5	0.346	0.112	72.00	3.174E-26	1.237E-26	0.1262	0.2969	-0.001	2.563	5.642E-27
0.845	488.4	7.200E-27	5	0.396	0.129	72.00	1.964E-26	8.636E-27	0.1446	0.3018	-0.112	2.504	3.940E-27
0.946	491.6	5.382E-27	5	0.447	0.145	72.00	1.174E-26	5.751E-27	0.1630	0.3054	-0.223	2.443	2.624E-27
1.046	493.9	3.977E-27	5	0.496	0.161	71.99	7.106E-27	3.836E-27	0.1810	0.3080	-0.335	2.382	1.750E-27
1.145	495.6	2.978E-27	5	0.545	0.177	71.99	4.446E-27	2.620E-27	0.1988	0.3099	-0.446	2.319	1.195E-27
1.245	497.0	2.378E-27	5	0.594	0.193	72.00	3.005E-27	1.922E-27	0.2166	0.3113	-0.559	2.254	8.768E-28
1.344	498.1	1.783E-27	5	0.642	0.209	72.00	1.935E-27	1.334E-27	0.2343	0.3123	-0.671	2.187	6.085E-28
1.444	498.9	1.187E-27	5	0.691	0.225	72.00	1.116E-27	8.259E-28	0.2521	0.3134	-0.784	2.117	3.768E-28
1.543	499.6	7.909E-28	5	0.739	0.240	72.00	6.517E-28	5.147E-28	0.2697	0.3141	-0.897	2.046	2.348E-28
1.642	500.2	3.952E-28	5	0.788	0.256	72.00	2.877E-28	2.416E-28	0.2873	0.3146	-1.009	1.971	1.102E-28

p_3	θ_3	$\dfrac{d^2\sigma}{d\Omega_3\,dp_3}$	Error	p_T	p_L^*	θ_3^*	$\dfrac{1}{p_3^{*2}}\cdot\dfrac{d^2\sigma}{d\Omega_3^*\,dp_3^*}$	$\dfrac{E_3}{p_3^2}\cdot\dfrac{d^2\sigma}{d\Omega_3\,dp_3}$	x^*	y_3^*	t	m_4	$\dfrac{d^2\sigma}{dt\,dm_4^2}$
GeV/c	mrad	cm² sr⁻¹ GeV⁻¹ c	%	GeV/c	GeV/c	deg	cm² sr⁻¹ GeV⁻³ c³	cm² sr⁻¹ GeV⁻² c³			GeV²	GeV	cm²/GeV⁴

$p_1 = 3.670$ GeV/c; $p_1^* = 1.156$ GeV/c; $E_1^* + E_2^* = 2.978$ GeV; $s = 8.868$ GeV²; $\gamma_{c.m.} = 1.5871$; $\beta_{c.m.} = 0.77652$; $y_1^* = 1.0365$; $p_{3,\,max}^* = 0.890$ GeV/c; [62 M 1] (cont.)

p_3	θ_3	$\dfrac{d^2\sigma}{d\Omega_3\,dp_3}$	Error	p_T	p_L^*	θ_3^*	$\dfrac{1}{p_3^{*2}}\cdot\dfrac{d^2\sigma}{d\Omega_3^*\,dp_3^*}$	$\dfrac{E_3}{p_3^2}\cdot\dfrac{d^2\sigma}{d\Omega_3\,dp_3}$	x^*	y_3^*	t	m_4	$\dfrac{d^2\sigma}{dt\,dm_4^2}$
0.288	457.8	8.084E-27	5	0.127	0.016	83.01	1.645E-25	3.118E-26	0.0175	0.0825	0.371	2.785	1.422E-26
0.385	526.0	1.134E-26	5	0.193	0.024	83.01	1.308E-25	3.134E-26	0.0266	0.0993	0.241	2.731	1.430E-26
0.478	555.3	9.759E-27	5	0.252	0.031	83.01	7.349E-26	2.128E-26	0.0347	0.1071	0.109	2.676	9.710E-27
0.569	571.0	7.313E-27	5	0.307	0.038	83.01	3.901E-26	1.325E-26	0.0423	0.1114	-0.024	2.620	6.043E-27
0.658	580.5	5.311E-27	5	0.361	0.044	83.00	2.117E-26	8.247E-27	0.0497	0.1142	-0.158	2.563	3.762E-27
0.747	586.7	4.053E-27	5	0.414	0.051	83.00	1.255E-26	5.517E-27	0.0570	0.1161	-0.292	2.504	2.517E-27
0.836	591.0	3.162E-27	5	0.466	0.057	83.00	7.834E-27	3.835E-27	0.0642	0.1174	-0.426	2.444	1.750E-27
0.924	594.1	2.279E-27	5	0.517	0.064	83.00	4.623E-27	2.494E-27	0.0714	0.1183	-0.560	2.382	1.138E-27
1.012	596.5	1.750E-27	5	0.569	0.070	83.00	2.961E-27	1.745E-27	0.0784	0.1189	-0.695	2.319	7.963E-28
1.100	598.3	1.224E-27	5	0.620	0.076	83.00	1.754E-27	1.122E-27	0.0854	0.1194	-0.829	2.254	5.117E-28
1.188	599.7	8.733E-28	5	0.670	0.082	83.00	1.074E-27	7.404E-28	0.0924	0.1198	-0.964	2.186	3.378E-28
1.275	600.8	5.236E-28	5	0.721	0.088	83.00	5.585E-28	4.131E-28	0.0994	0.1202	-1.098	2.117	1.884E-28
1.363	601.7	1.744E-28	5	0.771	0.095	83.00	1.629E-28	1.287E-28	0.1064	0.1205	-1.233	2.046	5.869E-29

$p_1 = 12.500$ GeV/c; $p_1^* = 2.332$ GeV/c; $E_1^* + E_2^* = 5.028$ GeV; $s = 25.281$ GeV²; $\gamma_{c.m.} = 2.6796$; $\beta_{c.m.} = 0.92776$; $y_1^* = 1.6420$; $p_{3,\,max}^* = 2.161$ GeV/c; [71 A 1]

p_3	θ_3	$\dfrac{d^2\sigma}{d\Omega_3\,dp_3}$	Error	p_T	p_L^*	θ_3^*	$\dfrac{1}{p_3^{*2}}\cdot\dfrac{d^2\sigma}{d\Omega_3^*\,dp_3^*}$	$\dfrac{E_3}{p_3^2}\cdot\dfrac{d^2\sigma}{d\Omega_3\,dp_3}$	x^*	y_3^*	t	m_4	$\dfrac{d^2\sigma}{dt\,dm_4^2}$
3.139	0.0	1.020E-25	7	0.000	0.600	0.00	5.280E-26	3.253E-26	0.2776	2.1645	0.601	4.371	4.357E-27
3.182	44.5	6.769E-26	7	0.141	0.600	13.26	3.369E-26	2.129E-26	0.2776	1.8245	0.521	4.353	2.852E-27
3.224	62.1	4.981E-26	7	0.200	0.600	18.43	2.388E-26	1.546E-26	0.2776	1.6323	0.442	4.335	2.071E-27
3.285	80.6	3.864E-26	7	0.265	0.600	23.79	1.756E-26	1.177E-26	0.2776	1.4462	0.328	4.308	1.577E-27
3.325	90.3	3.075E-26	7	0.300	0.600	26.56	1.351E-26	9.256E-27	0.2776	1.3569	0.253	4.291	1.240E-27
3.383	102.6	2.433E-26	7	0.346	0.600	30.00	1.018E-26	7.198E-27	0.2776	1.2524	0.145	4.265	9.641E-28
3.564	132.0	1.274E-26	6	0.469	0.600	38.02	4.620E-27	3.577E-27	0.2776	1.0326	-0.195	4.185	4.792E-28
3.746	154.0	8.546E-27	6	0.575	0.600	43.77	2.710E-27	2.283E-27	0.2775	0.8916	-0.537	4.102	3.058E-28
3.900	168.9	5.726E-27	6	0.656	0.600	47.53	1.633E-27	1.469E-27	0.2776	0.8053	-0.825	4.032	1.968E-28
4.045	181.0	3.773E-27	6	0.728	0.600	50.51	9.785E-28	9.333E-28	0.2776	0.7399	-1.097	3.964	1.250E-28
4.195	191.9	2.515E-27	6	0.800	0.600	53.13	5.941E-28	5.999E-28	0.2776	0.6841	-1.379	3.892	8.035E-29
4.325	200.3	1.646E-27	6	0.861	0.600	55.12	3.598E-28	3.808E-28	0.2776	0.6430	-1.623	3.829	5.100E-29
4.461	208.2	1.132E-27	6	0.922	0.600	56.95	2.289E-28	2.539E-28	0.2776	0.6057	-1.877	3.762	3.400E-29
4.579	214.5	7.896E-28	6	0.975	0.600	58.39	1.496E-28	1.725E-28	0.2776	0.5768	-2.099	3.702	2.311E-29
4.637	217.4	6.644E-28	4	1.000	0.600	59.04	1.220E-28	1.433E-28	0.2776	0.5638	-2.208	3.673	1.920E-29
4.693	220.1	5.370E-28	6	1.025	0.600	59.65	9.575E-29	1.145E-28	0.2776	0.5518	-2.313	3.644	1.533E-29
4.814	225.6	3.901E-28	6	1.077	0.600	60.87	6.535E-29	8.107E-29	0.2776	0.5277	-2.539	3.581	1.086E-29
4.857	227.5	3.583E-28	4	1.095	0.600	61.29	5.872E-29	7.380E-29	0.2775	0.5195	-2.621	3.558	9.885E-30
4.921	230.1	2.819E-28	6	1.122	0.600	61.87	4.476E-29	5.731E-29	0.2777	0.5085	-2.738	3.525	7.676E-30
5.034	234.7	2.120E-28	6	1.171	0.600	62.87	3.185E-29	4.213E-29	0.2775	0.4891	-2.953	3.464	5.643E-30
5.064	235.8	2.011E-28	4	1.183	0.600	63.11	2.978E-29	3.973E-29	0.2776	0.4845	-3.008	3.448	5.321E-30
5.134	238.4	1.636E-28	6	1.212	0.600	63.67	2.344E-29	3.188E-29	0.2776	0.4738	-3.139	3.410	4.270E-30
5.260	242.9	1.076E-28	4	1.265	0.600	64.63	1.454E-29	2.046E-29	0.2775	0.4554	-3.377	3.339	2.741E-30
5.446	248.9	6.102E-29	4	1.342	0.600	65.90	7.592E-30	1.121E-29	0.2776	0.4314	-3.724	3.233	1.501E-30
5.623	254.2	3.605E-29	4	1.414	0.600	67.01	4.158E-30	6.413E-30	0.2776	0.4106	-4.057	3.129	8.590E-31
5.794	258.9	2.122E-29	5	1.483	0.600	67.98	2.281E-30	3.663E-30	0.2776	0.3926	-4.377	3.025	4.907E-31
5.957	263.1	1.101E-29	5	1.549	0.600	68.84	1.109E-30	1.849E-30	0.2775	0.3766	-4.685	2.922	2.476E-31
6.115	266.8	6.239E-30	5	1.612	0.600	69.58	5.913E-31	1.021E-30	0.2776	0.3628	-4.979	2.819	1.367E-31

p_3	θ_3	$\dfrac{d^2\sigma}{d\Omega_3\,dp_3}$	Error	p_T	p_L^*	θ_3^*	$\dfrac{1}{p_3^{*2}}\cdot\dfrac{d^2\sigma}{d\Omega_3^*\,dp_3^*}$	$\dfrac{E_3}{p_3^2}\cdot\dfrac{d^2\sigma}{d\Omega_3\,dp_3}$	x^*	y_3^*	t	m_4	$\dfrac{d^2\sigma}{dt\,dm_4^2}$
GeV/c	mrad	cm² sr⁻¹ GeV⁻¹ c	%	GeV/c	GeV/c	deg	cm² sr⁻¹ GeV⁻³ c³	cm² sr⁻¹ GeV⁻² c³			GeV²	GeV	cm²/GeV⁴
$p_1 = 12.500$ GeV/c; $p_1^* = 2.332$ GeV/c; $E_1^* + E_2^* = 5.028$ GeV; $s = 25.281$ GeV²; $\gamma_{c.m.} = 2.6796$; $\beta_{c.m.} = 0.92776$; $y_1^* = 1.6420$; $p_{3,\max}^* = 2.161$ GeV/c; [71 A 1] (cont.)													
6.268	270.2	3.890E-30	5	1.673	0.600	70.27	3.482E-31	6.208E-31	0.2776	0.3503	-5.265	2.715	8.315E-32
6.416	273.3	2.401E-30	5	1.732	0.600	70.89	2.036E-31	3.743E-31	0.2777	0.3389	-5.543	2.611	5.014E-32
6.560	276.2	1.377E-30	6	1.789	0.600	71.46	1.110E-31	2.100E-31	0.2776	0.3284	-5.815	2.505	2.812E-32
6.700	278.8	8.035E-31	5	1.844	0.600	71.97	6.170E-32	1.200E-31	0.2777	0.3191	-6.076	2.398	1.607E-32
6.836	281.3	4.751E-31	6	1.898	0.600	72.46	3.484E-32	6.951E-32	0.2775	0.3102	-6.334	2.289	9.311E-33
6.968	283.5	2.712E-31	8	1.949	0.600	72.89	1.904E-32	3.893E-32	0.2776	0.3025	-6.579	2.178	5.214E-33
7.098	285.6	1.628E-31	8	2.000	0.600	73.29	1.096E-32	2.294E-32	0.2777	0.2951	-6.822	2.064	3.073E-33
1.980	239.2	1.342E-26	11	0.469	0.220	64.87	1.266E-26	6.795E-27	0.1018	0.4356	-0.772	4.461	9.101E-28
2.358	200.2	1.426E-26	10	0.469	0.320	55.69	1.036E-26	6.058E-27	0.1480	0.6146	-0.547	4.407	8.114E-28
2.769	170.2	1.444E-26	9	0.469	0.420	48.15	8.096E-27	5.222E-27	0.1943	0.7777	-0.383	4.338	6.994E-28
3.248	144.9	1.374E-26	8	0.469	0.530	41.50	5.869E-27	4.234E-27	0.2452	0.9391	-0.255	4.248	5.671E-28
3.701	127.1	1.248E-26	7	0.469	0.630	36.68	4.230E-27	3.374E-27	0.2914	1.0705	-0.173	4.157	4.520E-28
4.167	112.8	1.157E-26	6	0.469	0.730	32.73	3.161E-27	2.778E-27	0.3377	1.1900	-0.114	4.058	3.721E-28
4.643	101.2	1.018E-26	6	0.469	0.830	29.47	2.277E-27	2.194E-27	0.3840	1.2987	-0.073	3.951	2.938E-28
5.175	90.8	9.190E-27	5	0.469	0.940	26.52	1.676E-27	1.776E-27	0.4349	1.4077	-0.044	3.827	2.379E-28
5.664	82.9	8.000E-27	5	0.469	1.040	24.28	1.229E-27	1.413E-27	0.4812	1.4983	-0.028	3.707	1.892E-28
6.157	76.3	6.935E-27	5	0.469	1.140	22.37	9.081E-28	1.127E-27	0.5274	1.5819	-0.020	3.581	1.509E-28
6.703	70.0	6.056E-27	5	0.469	1.250	20.57	6.732E-28	9.037E-28	0.5783	1.6672	-0.019	3.435	1.210E-28
7.203	65.2	4.946E-27	5	0.469	1.350	19.16	4.783E-28	6.868E-28	0.6246	1.7392	-0.023	3.295	9.199E-29
7.704	60.9	3.893E-27	5	0.469	1.450	17.93	3.303E-28	5.054E-28	0.6708	1.8066	-0.031	3.148	6.769E-29
8.208	57.2	2.823E-27	5	0.469	1.550	16.84	2.116E-28	3.440E-28	0.7171	1.8700	-0.043	2.992	4.607E-29
2.149	310.1	4.456E-27	11	0.656	0.130	78.78	3.042E-27	2.078E-27	0.0602	0.1928	-1.927	4.293	2.783E-28
2.467	269.0	4.953E-27	10	0.656	0.230	70.67	2.837E-27	2.011E-27	0.1064	0.3367	-1.591	4.263	2.693E-28
2.819	234.7	5.310E-27	9	0.656	0.330	63.28	2.524E-27	1.886E-27	0.1527	0.4744	-1.317	4.218	2.526E-28
3.200	206.4	5.549E-27	8	0.656	0.430	56.75	2.179E-27	1.736E-27	0.1989	0.6038	-1.100	4.158	2.325E-28
3.605	182.9	5.760E-27	7	0.656	0.530	51.05	1.871E-27	1.599E-27	0.2452	0.7253	-0.925	4.087	2.142E-28
4.029	163.5	5.600E-27	6	0.656	0.630	46.15	1.512E-27	1.391E-27	0.2914	0.8380	-0.788	4.006	1.863E-28
4.459	147.3	5.333E-27	6	0.655	0.730	41.94	1.204E-27	1.194E-27	0.3377	0.9427	-0.679	3.916	1.599E-28
4.966	132.4	4.792E-27	5	0.656	0.840	37.97	8.983E-28	9.653E-28	0.3886	1.0496	-0.585	3.807	1.293E-28
5.429	121.1	4.384E-27	5	0.656	0.940	34.90	6.996E-28	8.078E-28	0.4349	1.1390	-0.521	3.700	1.082E-28
5.900	111.4	3.800E-27	5	0.656	1.040	32.24	5.206E-28	6.442E-28	0.4812	1.2227	-0.471	3.586	8.629E-29
2.272	292.8	3.377E-27	13	0.656	0.170	75.46	2.153E-27	1.489E-27	0.0787	0.2510	-1.785	4.283	1.995E-28
2.818	245.8	3.599E-27	8	0.686	0.310	65.68	1.671E-27	1.279E-27	0.1434	0.4296	-1.503	4.196	1.713E-28
3.959	174.0	4.560E-27	4	0.685	0.600	48.80	1.251E-27	1.153E-27	0.2776	0.7773	-0.935	4.004	1.544E-28
3.416	206.4	3.569E-27	6	0.700	0.460	56.70	1.231E-27	1.046E-27	0.2128	0.6064	-1.225	4.094	1.401E-28
3.451	302.9	6.054E-28	7	1.029	0.240	76.87	1.647E-28	1.756E-28	0.1111	0.2291	-3.341	3.818	2.352E-29
3.772	276.5	5.889E-28	6	1.030	0.340	71.73	1.429E-28	1.562E-28	0.1572	0.3215	-3.012	3.782	2.093E-29
4.115	252.9	5.885E-28	6	1.030	0.440	66.86	1.268E-28	1.431E-28	0.2035	0.4117	-2.721	3.736	1.917E-29
4.478	232.0	5.661E-28	6	1.030	0.540	62.32	1.080E-28	1.265E-28	0.2498	0.4988	-2.469	3.678	1.694E-29

p_3	θ_3	$\dfrac{\mathrm{d}^2\sigma}{\mathrm{d}\Omega_3\,\mathrm{d}p_3}$	Error	p_T	p_L^*	θ_3^*	$\dfrac{1}{p_3^{*2}}\cdot\dfrac{\mathrm{d}^2\sigma}{\mathrm{d}\Omega_3^*\,\mathrm{d}p_3^*}$	$\dfrac{E_3}{p_3^2}\cdot\dfrac{\mathrm{d}^2\sigma}{\mathrm{d}\Omega_3^*\,\mathrm{d}p_3}$	x^*	y_3^*	t	m_4	$\dfrac{\mathrm{d}^2\sigma}{\mathrm{d}t\,\mathrm{d}m_4^2}$
GeV/c	mrad	cm² sr⁻¹ GeV⁻¹ c	%	GeV/c	GeV/c	deg	cm² sr⁻¹ GeV⁻³ c³	cm² sr⁻¹ GeV⁻² c³			GeV²	GeV	cm²/GeV⁴
$p_1 = 12.500$ GeV/c; $p_1^* = 2.332$ GeV/c; $E_1^* + E_2^* = 5.028$ GeV; $s = 25.281$ GeV²; $\gamma_\mathrm{c.m.} = 2.6796$; $\beta_\mathrm{c.m.} = 0.92776$; $y_1^* = 1.6420$; $p_{3,\max}^* = 2.161$ GeV/c; [71 A 1] (cont.)													
4.859	213.5	5.324E−28	5	1.030	0.640	58.13	8.983E−29	1.096E−28	0.2961	0.5825	−2.250	3.610	1.468E−29
5.256	197.2	5.044E−28	5	1.030	0.740	54.30	7.525E−29	9.600E−29	0.3423	0.6625	−2.063	3.532	1.286E−29
5.667	182.7	4.458E−28	5	1.030	0.840	50.79	5.889E−29	7.869E−29	0.3886	0.7393	−1.900	3.445	1.054E−29
6.090	169.9	3.902E−28	5	1.030	0.940	47.61	4.574E−29	6.409E−29	0.4349	0.8123	−1.761	3.348	8.584E−30
6.523	158.5	3.317E−28	5	1.030	1.040	44.71	3.460E−29	5.086E−29	0.4811	0.8820	−1.641	3.243	6.813E−30
7.011	147.4	2.736E−28	5	1.030	1.150	41.84	2.518E−29	3.903E−29	0.5321	0.9549	−1.529	3.117	5.228E−30
7.462	138.4	2.067E−28	5	1.029	1.250	39.47	1.704E−29	2.771E−29	0.5783	1.0181	−1.442	2.993	3.711E−30
7.920	130.4	1.516E−28	5	1.030	1.350	37.34	1.124E−29	1.914E−29	0.6245	1.0778	−1.369	2.858	2.564E−30
8.384	123.1	1.031E−28	5	1.029	1.450	35.37	6.895E−30	1.230E−29	0.6709	1.1356	−1.305	2.713	1.647E−30
8.900	115.9	7.304E−29	5	1.029	1.560	33.41	4.379E−30	8.208E−30	0.7218	1.1960	−1.246	2.540	1.099E−30
$p_1 = 14.250$ GeV/c; $p_1^* = 2.502$ GeV/c; $E_1^* + E_2^* = 5.344$ GeV; $s = 28.557$ GeV²; $\gamma_\mathrm{c.m.} = 2.8479$; $\beta_\mathrm{c.m.} = 0.93633$; $y_1^* = 1.7074$; $p_{3,\max}^* = 2.340$ GeV/c; [72 A 1]													
4.500	12.0	8.195E−26	3	0.054	0.809	3.82	2.214E−26	1.822E−26	0.3458	2.3893	0.551	4.447	2.141E−27
4.750	12.0	8.092E−26	3	0.057	0.855	3.81	1.963E−26	1.704E−26	0.3653	2.4360	0.538	4.393	2.002E−27
5.000	12.0	7.625E−26	3	0.060	0.900	3.81	1.671E−26	1.526E−26	0.3848	2.4796	0.525	4.338	1.793E−27
5.120	12.0	7.233E−26	3	0.061	0.922	3.81	1.512E−26	1.413E−26	0.3941	2.4995	0.519	4.311	1.660E−27
5.250	12.0	6.881E−26	3	0.063	0.946	3.81	1.368E−26	1.311E−26	0.4042	2.5204	0.512	4.282	1.540E−27
5.500	12.0	5.973E−26	3	0.066	0.992	3.81	1.083E−26	1.086E−26	0.4237	2.5587	0.498	4.225	1.276E−27
5.750	12.0	5.228E−26	3	0.069	1.037	3.81	8.673E−27	9.095E−27	0.4431	2.5948	0.485	4.168	1.069E−27
6.000	12.0	4.593E−26	3	0.072	1.082	3.81	7.000E−27	7.657E−27	0.4625	2.6287	0.471	4.109	8.996E−28
6.250	12.0	3.827E−26	3	0.075	1.128	3.80	5.377E−27	6.125E−27	0.4820	2.6606	0.457	4.050	7.196E−28
6.500	12.0	3.465E−26	3	0.078	1.173	3.80	4.503E−27	5.332E−27	0.5014	2.6908	0.442	3.990	6.265E−28
6.750	12.0	3.283E−26	3	0.081	1.219	3.80	3.957E−27	4.865E−27	0.5208	2.7193	0.428	3.929	5.716E−28
7.000	12.0	3.030E−26	3	0.084	1.264	3.80	3.396E−27	4.329E−27	0.5402	2.7462	0.414	3.867	5.087E−28
7.340	12.0	2.772E−26	3	0.088	1.326	3.80	2.827E−27	3.777E−27	0.5666	2.7806	0.394	3.781	4.438E−28
7.520	12.0	2.573E−26	3	0.090	1.359	3.80	2.500E−27	3.422E−27	0.5806	2.7978	0.383	3.735	4.021E−28
7.710	12.0	2.331E−26	3	0.093	1.393	3.80	2.155E−27	3.024E−27	0.5953	2.8152	0.372	3.685	3.553E−28
7.900	12.0	2.260E−26	3	0.095	1.428	3.80	1.990E−27	2.861E−27	0.6100	2.8320	0.361	3.635	3.362E−28
8.090	12.0	2.101E−26	3	0.097	1.462	3.80	1.765E−27	2.597E−27	0.6248	2.8482	0.350	3.584	3.052E−28
8.290	12.0	1.871E−26	3	0.099	1.498	3.80	1.497E−27	2.257E−27	0.6403	2.8645	0.338	3.529	2.652E−28
8.490	12.0	1.615E−26	3	0.102	1.535	3.80	1.232E−27	1.902E−27	0.6558	2.8802	0.326	3.474	2.235E−28
8.700	12.0	1.366E−26	3	0.104	1.573	3.80	9.923E−28	1.570E−27	0.6721	2.8960	0.313	3.415	1.845E−28
8.920	12.0	1.239E−26	3	0.107	1.613	3.80	8.563E−28	1.389E−27	0.6891	2.9118	0.300	3.352	1.632E−28
9.140	12.0	1.219E−26	3	0.110	1.653	3.80	8.025E−28	1.334E−27	0.7062	2.9271	0.287	3.288	1.567E−28
9.360	12.0	1.134E−26	3	0.112	1.693	3.80	7.119E−28	1.212E−27	0.7233	2.9416	0.273	3.223	1.424E−28
9.590	12.0	1.049E−26	3	0.115	1.734	3.80	6.274E−28	1.094E−27	0.7411	2.9562	0.259	3.153	1.285E−28
9.830	12.0	9.108E−27	3	0.118	1.778	3.80	5.185E−28	9.266E−28	0.7597	2.9708	0.245	3.078	1.089E−28
10.320	12.0	7.733E−27	3	0.124	1.867	3.80	3.994E−28	7.494E−28	0.7977	2.9986	0.215	2.920	8.805E−29
10.570	12.0	6.405E−27	3	0.127	1.912	3.80	3.154E−28	6.060E−28	0.8170	3.0118	0.199	2.836	7.120E−29
10.830	12.0	4.375E−27	3	0.130	1.959	3.79	2.052E−28	4.040E−28	0.8372	3.0250	0.184	2.745	4.747E−29
11.100	12.0	2.529E−27	3	0.133	2.008	3.79	1.129E−28	2.279E−28	0.8581	3.0380	0.167	2.648	2.677E−29
11.370	12.0	1.610E−27	3	0.136	2.057	3.79	6.853E−29	1.416E−28	0.8790	3.0505	0.150	2.548	1.664E−29
11.510	12.0	1.422E−27	3	0.138	2.083	3.79	5.906E−29	1.236E−28	0.8899	3.0567	0.142	2.494	1.452E−29

p_3	θ_3	$\dfrac{d^2\sigma}{d\Omega_3\,dp_3}$	Error	p_T	p_L^*	θ_3^*	$\dfrac{1}{p_3^{*2}}\cdot\dfrac{d^2\sigma}{d\Omega_3^*\,dp_3^*}$	$\dfrac{E_3}{p_3^2}\cdot\dfrac{d^2\sigma}{d\Omega_3\,dp_3}$	x^*	y_3^*	t	m_4	$\dfrac{d^2\sigma}{dt\,dm_4^2}$
GeV/c	mrad	cm² sr⁻¹ GeV⁻¹ c	%	GeV/c	GeV/c	deg	cm² sr⁻¹ GeV⁻³ c³	cm² sr⁻¹ GeV⁻² c³			GeV²	GeV	cm²/GeV⁴

$p_1 = 14.250$ GeV/c; $p_1^* = 2.502$ GeV/c; $E_1^* + E_2^* = 5.344$ GeV; $s = 28.557$ GeV²; $\gamma_{c.m.} = 2.8479$; $\beta_{c.m.} = 0.93633$; $y_1^* = 1.7074$; $p_{3,\mathrm{max}}^* = 2.340$ GeV/c; [72 A 1] (cont.)

p_3	θ_3	$\dfrac{d^2\sigma}{d\Omega_3\,dp_3}$	Error	p_T	p_L^*	θ_3^*	$\dfrac{1}{p_3^{*2}}\cdot\dfrac{d^2\sigma}{d\Omega_3^*\,dp_3^*}$	$\dfrac{E_3}{p_3^2}\cdot\dfrac{d^2\sigma}{d\Omega_3\,dp_3}$	x^*	y_3^*	t	m_4	$\dfrac{d^2\sigma}{dt\,dm_4^2}$
11.650	12.0	1.405E-27	3	0.140	2.108	3.79	5.697E-29	1.206E-28	0.9007	3.0628	0.133	2.439	1.417E-29
11.800	12.0	1.431E-27	3	0.142	2.135	3.79	5.656E-29	1.213E-28	0.9123	3.0691	0.124	2.379	1.425E-29
11.940	12.0	1.447E-27	3	0.143	2.161	3.79	5.586E-29	1.212E-28	0.9232	3.0749	0.115	2.321	1.424E-29
12.090	12.0	1.150E-27	3	0.145	2.188	3.79	4.330E-29	9.513E-29	0.9348	3.0809	0.106	2.257	1.118E-29
12.240	12.0	8.285E-28	3	0.147	2.215	3.79	3.043E-29	6.769E-29	0.9464	3.0868	0.097	2.192	7.953E-30
12.390	12.0	6.080E-28	3	0.149	2.242	3.79	2.180E-29	4.907E-29	0.9581	3.0926	0.087	2.125	5.766E-30
12.690	12.0	3.019E-28	3	0.152	2.297	3.79	1.032E-29	2.379E-29	0.9813	3.1036	0.069	1.983	2.795E-30
12.840	12.0	1.910E-28	3	0.154	2.324	3.79	6.376E-30	1.488E-29	0.9929	3.1090	0.059	1.908	1.748E-30

$p_1 = 19.200$ GeV/c; $p_1^* = 2.929$ GeV/c; $E_1^* + E_2^* = 6.151$ GeV; $s = 37.830$ GeV²; $\gamma_{c.m.} = 3.2779$; $\beta_{c.m.} = 0.95233$; $y_1^* = 1.8562$; $p_{3,\mathrm{max}}^* = 2.787$ GeV/c; [70 A 1]

p_3	θ_3	$\dfrac{d^2\sigma}{d\Omega_3\,dp_3}$	Error	p_T	p_L^*	θ_3^*	$\dfrac{1}{p_3^{*2}}\cdot\dfrac{d^2\sigma}{d\Omega_3^*\,dp_3^*}$	$\dfrac{E_3}{p_3^2}\cdot\dfrac{d^2\sigma}{d\Omega_3\,dp_3}$	x^*	y_3^*	t	m_4	$\dfrac{d^2\sigma}{dt\,dm_4^2}$
4.500	12.5	1.217E-25	3	0.056	0.695	4.63	3.804E-26	2.706E-26	0.2494	2.2350	0.597	5.394	2.359E-27
4.500	20.0	1.155E-25	3	0.090	0.693	7.39	3.601E-26	2.568E-26	0.2488	2.1363	0.576	5.392	2.239E-27
4.500	30.0	1.096E-25	3	0.135	0.690	11.07	3.400E-26	2.437E-26	0.2475	1.9800	0.533	5.388	2.125E-27
4.500	40.0	9.714E-26	3	0.180	0.685	14.73	2.993E-26	2.160E-26	0.2456	1.8203	0.472	5.383	1.883E-27
4.500	50.0	7.821E-26	3	0.225	0.678	18.35	2.389E-26	1.739E-26	0.2432	1.6697	0.394	5.376	1.516E-27
4.500	60.0	6.509E-26	3	0.270	0.670	21.94	1.967E-26	1.447E-26	0.2403	1.5317	0.299	5.367	1.262E-27
4.500	70.0	5.087E-26	3	0.315	0.660	25.49	1.519E-26	1.131E-26	0.2369	1.4062	0.187	5.356	9.862E-28
6.000	12.5	1.237E-25	3	0.075	0.931	4.61	2.184E-26	2.062E-26	0.3340	2.4710	0.544	5.122	1.798E-27
6.000	20.0	1.123E-25	3	0.120	0.929	7.36	1.978E-26	1.872E-26	0.3331	2.3211	0.516	5.119	1.633E-27
6.000	30.0	8.639E-26	3	0.180	0.924	11.03	1.514E-26	1.440E-26	0.3314	2.1080	0.459	5.113	1.256E-27
6.000	40.0	6.366E-26	3	0.240	0.917	14.67	1.108E-26	1.061E-26	0.3289	1.9100	0.378	5.106	9.255E-28
6.000	50.0	4.671E-26	3	0.300	0.908	18.28	8.059E-27	7.787E-27	0.3257	1.7345	0.274	5.095	6.790E-28
6.000	60.0	3.596E-26	3	0.360	0.897	21.85	6.139E-27	5.995E-27	0.3219	1.5800	0.148	5.083	5.228E-28
6.000	70.0	2.804E-26	3	0.420	0.884	25.39	4.728E-27	4.675E-27	0.3173	1.4434	-0.002	5.068	4.076E-28
8.000	12.5	5.734E-26	3	0.100	1.244	4.59	5.707E-27	7.169E-27	0.4464	2.6783	0.462	4.733	6.251E-28
8.000	20.0	5.024E-26	3	0.160	1.241	7.35	4.989E-27	6.281E-27	0.4452	2.4659	0.425	4.729	5.477E-28
8.000	30.0	3.789E-26	3	0.240	1.234	11.00	3.744E-27	4.737E-27	0.4429	2.1977	0.348	4.721	4.131E-28
8.000	40.0	2.800E-26	3	0.320	1.225	14.63	2.748E-27	3.501E-27	0.4396	1.9686	0.241	4.709	3.053E-28
8.000	50.0	2.018E-26	5	0.400	1.214	18.24	1.963E-27	2.523E-27	0.4354	1.7750	0.102	4.695	2.200E-28
8.000	60.0	1.444E-26	3	0.480	1.199	21.80	1.390E-27	1.805E-27	0.4302	1.6095	-0.066	4.677	1.574E-28
8.000	70.0	1.056E-26	3	0.560	1.182	25.33	1.004E-27	1.320E-27	0.4241	1.4657	-0.266	4.655	1.151E-28
10.000	12.5	3.695E-26	3	0.125	1.557	4.59	2.356E-27	3.695E-27	0.5586	2.8141	0.374	4.308	3.222E-28
10.000	20.0	3.059E-26	3	0.200	1.553	7.34	1.946E-27	3.059E-27	0.5572	2.5505	0.327	4.303	2.668E-28
10.000	30.0	2.131E-26	3	0.300	1.545	10.99	1.349E-27	2.131E-27	0.5542	2.2454	0.231	4.291	1.858E-28
10.000	40.0	1.368E-26	3	0.400	1.533	14.62	8.600E-28	1.368E-27	0.5501	1.9982	0.097	4.276	1.193E-28
10.000	50.0	8.758E-27	3	0.500	1.519	18.22	5.458E-28	8.759E-28	0.5448	1.7949	-0.076	4.255	7.638E-29
10.000	60.0	5.915E-27	3	0.600	1.501	21.78	3.647E-28	5.916E-28	0.5384	1.6237	-0.287	4.231	5.158E-29
10.000	70.0	4.037E-27	3	0.699	1.479	25.31	2.458E-28	4.037E-28	0.5307	1.4763	-0.536	4.201	3.521E-29

p_3	θ_3	$\dfrac{\mathrm{d}^2\sigma}{\mathrm{d}\Omega_3\,\mathrm{d}p_3}$	Error	p_{T}	p_{L}^*	θ_3^*	$\dfrac{1}{p_3^{*2}}\cdot\dfrac{\mathrm{d}^2\sigma}{\mathrm{d}\Omega_3^*\,\mathrm{d}p_3^*}$	$\dfrac{E_3}{p_3^{*2}}\cdot\dfrac{\mathrm{d}^2\sigma}{\mathrm{d}\Omega_3\,\mathrm{d}p_3}$	x^*	y_3^*	t	m_4	$\dfrac{\mathrm{d}^2\sigma}{\mathrm{d}t\,\mathrm{d}m_4^2}$
GeV/c	mrad	cm² sr⁻¹ GeV⁻¹ c	%	GeV/c	GeV/c	deg	cm² sr⁻¹ GeV⁻³ c³	cm² sr⁻¹ GeV⁻² c³			GeV²	GeV	cm²/GeV⁴
11.000	12.5	2.663E−26	3	0.137	1.713	4.59	1.404E−27	2.421E−27	0.6147	2.8648	0.329	4.079	2.111E−28
11.000	20.0	2.058E−26	3	0.220	1.709	7.34	1.082E−27	1.871E−27	0.6131	2.5798	0.277	4.072	1.632E−28
11.000	30.0	1.386E−26	3	0.330	1.700	10.98	7.253E−28	1.260E−27	0.6099	2.2611	0.172	4.060	1.099E−28
11.000	40.0	8.568E−27	3	0.440	1.687	14.61	4.453E−28	7.790E−28	0.6053	2.0077	0.024	4.041	6.793E−29
11.000	50.0	5.095E−27	4	0.550	1.671	18.21	2.625E−28	4.632E−28	0.5995	1.8012	−0.166	4.018	4.039E−29
11.000	60.0	3.361E−27	4	0.660	1.651	21.77	1.713E−28	3.056E−28	0.5924	1.6282	−0.398	3.989	2.665E−29
11.000	70.0	2.105E−27	3	0.769	1.628	25.30	1.060E−28	1.914E−28	0.5840	1.4796	−0.673	3.954	1.669E−29
12.000	12.5	1.716E−26	3	0.150	1.870	4.59	7.604E−28	1.430E−27	0.6707	2.9070	0.283	3.836	1.247E−28
12.000	20.0	1.353E−26	3	0.240	1.865	7.33	5.981E−28	1.128E−27	0.6690	2.6032	0.227	3.829	9.833E−29
12.000	30.0	8.170E−27	3	0.360	1.855	10.98	3.594E−28	6.809E−28	0.6655	2.2734	0.111	3.814	5.937E−29
12.000	40.0	4.801E−27	3	0.480	1.841	14.61	2.097E−28	4.001E−28	0.6605	2.0151	−0.050	3.792	3.489E−29
12.000	50.0	2.829E−27	3	0.600	1.823	18.21	1.225E−28	2.358E−28	0.6542	1.8061	−0.257	3.765	2.056E−29
12.000	60.0	1.643E−27	3	0.720	1.802	21.77	7.039E−29	1.369E−28	0.6464	1.6316	−0.511	3.731	1.194E−29
12.000	70.0	8.685E−28	3	0.839	1.776	25.29	3.675E−29	7.238E−29	0.6373	1.4822	−0.810	3.691	6.312E−30
13.000	12.5	1.285E−26	3	0.162	2.026	4.59	4.853E−28	9.885E−28	0.7268	2.9426	0.236	3.577	8.620E−29
13.000	20.0	8.170E−27	3	0.260	2.021	7.33	3.078E−28	6.285E−28	0.7249	2.6223	0.175	3.568	5.481E−29
13.000	30.0	4.434E−27	3	0.390	2.010	10.98	1.662E−28	3.411E−28	0.7211	2.2831	0.051	3.551	2.974E−29
13.000	40.0	2.477E−27	3	0.520	1.995	14.61	9.222E−29	1.905E−28	0.7157	2.0209	−0.124	3.526	1.662E−29
13.000	50.0	1.192E−27	3	0.650	1.976	18.20	4.399E−29	9.170E−29	0.7089	1.8099	−0.349	3.494	7.996E−30
13.000	60.0	7.201E−28	3	0.780	1.952	21.77	2.629E−29	5.540E−29	0.7005	1.6343	−0.623	3.454	4.831E−30
13.000	70.0	3.534E−28	3	0.909	1.925	25.29	1.274E−29	2.719E−29	0.6905	1.4842	−0.947	3.407	2.371E−30
14.000	12.5	7.122E−27	5	0.175	2.182	4.59	2.319E−28	5.087E−28	0.7828	2.9727	0.190	3.297	4.436E−29
14.000	20.0	4.194E−27	3	0.280	2.176	7.33	1.363E−28	2.996E−28	0.7808	2.6379	0.124	3.287	2.612E−29
14.000	30.0	1.962E−27	3	0.420	2.165	10.98	6.343E−29	1.401E−28	0.7767	2.2910	−0.010	3.266	1.222E−29
14.000	40.0	1.061E−27	3	0.560	2.149	14.60	3.406E−29	7.579E−29	0.7709	2.0255	−0.199	3.237	6.609E−30
14.000	50.0	5.581E−28	4	0.700	2.128	18.20	1.776E−29	3.987E−29	0.7635	1.8130	−0.440	3.200	3.476E−30
14.000	60.0	2.947E−28	3	0.839	2.103	21.76	9.279E−30	2.105E−29	0.7545	1.6364	−0.736	3.153	1.836E−30
14.000	70.0	1.262E−28	3	0.979	2.073	25.28	3.925E−30	9.015E−30	0.7438	1.4857	−1.085	3.097	7.861E−31
15.000	12.5	2.545E−27	5	0.187	2.338	4.58	7.221E−29	1.697E−28	0.8388	2.9984	0.142	2.991	1.480E−29
15.000	20.0	1.625E−27	3	0.300	2.332	7.33	4.600E−29	1.083E−28	0.8367	2.6509	0.072	2.979	9.447E−30
15.000	30.0	8.475E−28	3	0.450	2.320	10.98	2.387E−29	5.650E−29	0.8323	2.2975	−0.072	2.955	4.927E−30
15.000	40.0	4.462E−28	3	0.600	2.303	14.60	1.248E−29	2.975E−29	0.8261	2.0293	−0.273	2.921	2.594E−30
15.000	50.0	2.318E−28	3	0.750	2.280	18.20	6.427E−30	1.545E−29	0.8182	1.8154	−0.532	2.876	1.348E−30
15.000	60.0	1.034E−28	3	0.899	2.253	21.76	2.837E−30	6.894E−30	0.8085	1.6382	−0.849	2.820	6.011E−31
15.000	70.0	3.764E−29	3	1.049	2.221	25.28	1.020E−30	2.509E−30	0.7970	1.4870	−1.223	2.753	2.188E−31
16.000	12.5	1.588E−27	5	0.200	2.494	4.58	3.961E−29	9.925E−29	0.8948	3.0205	0.095	2.650	8.655E−30
16.000	20.0	8.088E−28	3	0.320	2.488	7.33	2.012E−29	5.055E−29	0.8925	2.6618	0.020	2.636	4.408E−30
16.000	30.0	3.106E−28	3	0.480	2.475	10.98	7.689E−30	1.941E−29	0.8878	2.3028	−0.133	2.606	1.693E−30

$p_1 = 19.200$ GeV/c; $p_1^* = 2.929$ GeV/c; $E_1^* + E_2^* = 6.151$ GeV; $s = 37.830$ GeV²; $\gamma_{\mathrm{c.m.}} = 3.2779$; $\beta_{\mathrm{c.m.}} = 0.95233$; $y_1^* = 1.8562$; $p_{3,\mathrm{max}}^* = 2.787$ GeV/c; [70 A 1] (cont.)

p_3	θ_3	$\dfrac{\mathrm{d}^2\sigma}{\mathrm{d}\Omega_3\,\mathrm{d}p_3}$	Error	p_T	p_L^*	θ_3^*	$\dfrac{1}{p_3^{*2}}\cdot\dfrac{\mathrm{d}^2\sigma}{\mathrm{d}\Omega_3^*\,\mathrm{d}p_3^*}$	$\dfrac{E_3}{p_3^2}\cdot\dfrac{\mathrm{d}^2\sigma}{\mathrm{d}\Omega_3\,\mathrm{d}p_3}$	x^*	y_3^*	t	m_4	$\dfrac{\mathrm{d}^2\sigma}{\mathrm{d}t\,\mathrm{d}m_4^2}$
GeV/c	mrad	cm² sr⁻¹ GeV⁻¹ c	%	GeV/c	GeV/c	deg	cm² sr⁻¹ GeV⁻³ c²	cm² sr⁻¹ GeV⁻² c³			GeV²	GeV	cm²/GeV⁴
$p_1 = 19.200$ GeV/c; $p_1^* = 2.929$ GeV/c; $E_1^* + E_2^* = 6.151$ GeV; $s = 37.830$ GeV²; $\gamma_\mathrm{c.m.} = 3.2779$; $\beta_\mathrm{c.m.} = 0.95233$; $y_1^* = 1.8562$; $p_{3,\max}^* = 2.787$ GeV/c; [70 A 1] (cont.)													
16.000	40.0	1.518E−28	3	0.640	2.456	14.60	3.732E−30	9.488E−30	0.8813	2.0324	−0.348	2.565	8.274E−31
16.000	50.0	6.880E−29	5	0.800	2.433	18.20	1.677E−30	4.300E−30	0.8728	1.8175	−0.625	2.510	3.750E−31
16.000	60.0	2.331E−29	5	0.959	2.404	21.76	5.621E−31	1.457E−30	0.8624	1.6396	−0.962	2.442	1.270E−31
16.000	70.0	6.813E−30	3	1.119	2.370	25.28	1.623E−31	4.258E−31	0.8502	1.4881	−1.361	2.359	3.713E−32
13.500	12.5	9.160E−27	3	0.169	2.104	4.59	3.208E−28	6.786E−28	0.7548	2.9583	0.213	3.440	5.917E−29
13.800	12.5	7.944E−27	3	0.172	2.151	4.59	2.663E−28	5.757E−28	0.7716	2.9671	0.199	3.355	5.020E−29
14.100	12.5	6.372E−27	3	0.176	2.198	4.59	2.046E−28	4.519E−28	0.7884	2.9755	0.185	3.267	3.941E−29
14.400	12.5	4.686E−27	3	0.180	2.244	4.59	1.443E−28	3.254E−28	0.8052	2.9835	0.171	3.178	2.838E−29
14.700	12.5	3.266E−27	3	0.184	2.291	4.59	9.649E−29	2.222E−28	0.8220	2.9911	0.157	3.086	1.937E−29
15.000	12.5	2.361E−27	3	0.187	2.338	4.58	6.699E−29	1.574E−28	0.8388	2.9984	0.142	2.991	1.373E−29
15.300	12.5	1.856E−27	3	0.191	2.385	4.58	5.062E−29	1.213E−28	0.8556	3.0054	0.128	2.893	1.058E−29
15.600	12.5	1.631E−27	3	0.195	2.432	4.58	4.279E−29	1.046E−28	0.8724	3.0121	0.114	2.791	9.117E−30
15.900	12.5	1.516E−27	3	0.199	2.479	4.58	3.829E−29	9.535E−29	0.8892	3.0184	0.100	2.686	8.315E−30
16.200	12.5	1.193E−27	3	0.202	2.525	4.58	2.902E−29	7.364E−29	0.9060	3.0245	0.086	2.576	6.422E−30
16.500	12.5	8.180E−28	3	0.206	2.572	4.58	1.918E−29	4.958E−29	0.9228	3.0304	0.072	2.462	4.323E−30
16.800	12.5	5.570E−28	3	0.210	2.619	4.58	1.260E−29	3.316E−29	0.9396	3.0360	0.057	2.342	2.891E−30
17.100	12.5	3.670E−28	3	0.214	2.666	4.58	8.014E−30	2.146E−29	0.9564	3.0413	0.043	2.215	1.872E−30
17.400	12.5	1.864E−28	3	0.217	2.713	4.58	3.931E−30	1.071E−29	0.9733	3.0465	0.029	2.080	9.342E−31
17.700	12.5	8.283E−29	5	0.221	2.760	4.58	1.688E−30	4.680E−30	0.9901	3.0514	0.014	1.937	4.081E−31
18.000	12.5	7.100E−30	20	0.225	2.806	4.58	1.399E−31	3.945E−31	1.0069	3.0561	0.000	1.782	3.440E−32
13.500	20.0	6.151E−27	3	0.270	2.098	7.33	2.149E−28	4.557E−28	0.7529	2.6305	0.150	3.430	3.973E−29
13.800	20.0	5.062E−27	3	0.276	2.145	7.33	1.693E−28	3.668E−28	0.7696	2.6350	0.134	3.345	3.199E−29
14.100	20.0	3.921E−27	3	0.282	2.192	7.33	1.256E−28	2.781E−28	0.7864	2.6393	0.119	3.257	2.425E−29
14.400	20.0	2.862E−27	3	0.288	2.239	7.33	8.789E−29	1.988E−28	0.8031	2.6434	0.103	3.167	1.733E−29
14.700	20.0	2.101E−27	3	0.294	2.285	7.33	6.192E−29	1.429E−28	0.8199	2.6473	0.088	3.075	1.246E−29
15.000	20.0	1.600E−27	3	0.300	2.332	7.33	4.529E−29	1.067E−28	0.8367	2.6509	0.072	2.979	9.302E−30
15.300	20.0	1.258E−27	3	0.306	2.379	7.33	3.423E−29	8.223E−29	0.8534	2.6544	0.057	2.880	7.170E−30
15.600	20.0	1.043E−27	3	0.312	2.426	7.33	2.730E−29	6.686E−29	0.8702	2.6577	0.041	2.778	5.830E−30
15.900	20.0	8.529E−28	3	0.318	2.472	7.33	2.149E−29	5.364E−29	0.8870	2.6608	0.026	2.672	4.678E−30
16.200	20.0	6.218E−28	3	0.324	2.519	7.33	1.509E−29	3.838E−29	0.9037	2.6638	0.010	2.561	3.347E−30
16.500	20.0	4.150E−28	3	0.330	2.566	7.33	9.709E−30	2.515E−29	0.9205	2.6666	−0.006	2.446	2.193E−30
16.800	20.0	2.740E−28	3	0.336	2.612	7.33	6.184E−30	1.631E−29	0.9372	2.6693	−0.021	2.325	1.422E−30
17.100	20.0	1.610E−28	3	0.342	2.659	7.33	3.507E−30	9.416E−30	0.9540	2.6719	−0.037	2.197	8.210E−31
17.400	20.0	8.420E−29	3	0.348	2.706	7.33	1.772E−30	4.839E−30	0.9708	2.6743	−0.053	2.061	4.220E−31
17.700	20.0	3.473E−29	3	0.354	2.753	7.33	7.062E−31	1.962E−30	0.9875	2.6766	−0.068	1.915	1.711E−31
18.000	20.0	2.347E−30	10	0.360	2.799	7.33	4.614E−32	1.304E−31	1.0043	2.6789	−0.084	1.758	1.137E−32
13.500	30.0	3.131E−27	3	0.405	2.087	10.98	1.088E−28	2.319E−28	0.7489	2.2873	0.020	3.411	2.023E−29
13.800	30.0	2.427E−27	3	0.414	2.134	10.98	8.075E−29	1.759E−28	0.7656	2.2896	0.002	3.325	1.534E−29
14.100	30.0	1.827E−27	3	0.423	2.180	10.98	5.823E−29	1.296E−28	0.7822	2.2917	−0.017	3.237	1.130E−29
14.400	30.0	1.371E−27	3	0.432	2.227	10.98	4.190E−29	9.521E−29	0.7989	2.2938	−0.035	3.145	8.303E−30

p_3	θ_3	$\dfrac{d^2\sigma}{d\Omega_3\,dp_3}$	Error	p_T	p_L^*	θ_3^*	$\dfrac{1}{p_3^{*2}}\cdot\dfrac{d^2\sigma}{d\Omega_3^*\,dp_3^*}$	$\dfrac{E_3}{p_3^2}\cdot\dfrac{d^2\sigma}{d\Omega_3\,dp_3}$	x^*	y_3^*	t	m_4	$\dfrac{d^2\sigma}{dt\,dm_4^2}$
GeV/c	mrad	cm² sr⁻¹ GeV⁻¹ c	%	GeV/c	GeV/c	deg	cm² sr⁻¹ GeV⁻³ c³	cm² sr⁻¹ CeV⁻² c³			GeV²	GeV	cm²/GeV⁴
$p_1 = 19.200$ GeV/c; $p_1^* = 2.929$ GeV/c; $E_1^* + E_2^* = 6.151$ GeV; $s = 37.830$ GeV²; $\gamma_{c.m.} = 3.2779$; $\beta_{c.m.} = 0.9523$; $y_1^* = 1.8562$; $p_{3,\,max}^* = 2.787$ GeV/c; [70 A 1] (cont.)													
14.700	30.0	1.055E-27	3	0.441	2.273	10.98	3.094E-29	7.177E-29	0.8156	2.2957	-0.053	3.052	6.259E-30
15.000	30.0	8.258E-28	3	0.450	2.320	10.98	2.326E-29	5.506E-29	0.8323	2.2975	-0.072	2.955	4.801E-30
15.300	30.0	6.320E-28	3	0.459	2.366	10.98	1.711E-29	4.131E-29	0.8489	2.2992	-0.090	2.855	3.602E-30
15.600	30.0	4.750E-28	3	0.468	2.413	10.98	1.237E-29	3.045E-29	0.8656	2.3008	-0.109	2.751	2.655E-30
15.900	30.0	3.481E-28	3	0.477	2.459	10.98	8.726E-30	2.189E-29	0.8823	2.3023	-0.127	2.643	1.909E-30
16.200	30.0	2.394E-28	3	0.486	2.506	10.98	5.781E-30	1.478E-29	0.8990	2.3038	-0.146	2.531	1.289E-30
16.500	30.0	1.581E-28	3	0.495	2.552	10.97	3.681E-30	9.582E-30	0.9156	2.3051	-0.164	2.413	8.356E-31
16.800	30.0	9.750E-29	3	0.504	2.599	10.97	2.190E-30	5.804E-30	0.9323	2.3065	-0.183	2.290	5.061E-31
17.100	30.0	5.500E-29	3	0.513	2.645	10.97	1.192E-30	3.216E-30	0.9490	2.3077	-0.201	2.159	2.805E-31
17.400	30.0	2.503E-29	3	0.522	2.692	10.97	5.240E-31	1.439E-30	0.9656	2.3089	-0.220	2.020	1.254E-31
17.700	30.0	7.845E-30	3	0.531	2.738	10.97	1.587E-31	4.432E-31	0.9823	2.3100	-0.238	1.870	3.865E-32
$p_1 = 24.000$ GeV/c; $p_1^* = 3.290$ GeV/c; $E_1^* + E_2^* = 6.843$ GeV; $s = 46.828$ GeV²; $\gamma_{c.m.} = 3.6469$; $\beta_{c.m.} = 0.96167$; $y_1^* = 1.9677$; $p_{3,\,max}^* = 3.163$ GeV/c; [72 A 1]													
4.000	17.0	9.513E-26	3	0.068	0.548	7.07	4.175E-26	2.380E-26	0.1734	1.9745	0.608	6.249	1.660E-27
4.000	27.0	9.523E-26	3	0.108	0.545	11.20	4.157E-26	2.382E-26	0.1724	1.8464	0.566	6.245	1.662E-27
4.000	37.0	9.358E-26	3	0.148	0.541	15.31	4.053E-26	2.341E-26	0.1709	1.7042	0.505	6.240	1.633E-27
4.000	47.0	8.829E-26	3	0.188	0.534	19.37	3.785E-26	2.209E-26	0.1690	1.5635	0.424	6.234	1.541E-27
4.000	57.0	7.809E-26	3	0.228	0.527	23.39	3.307E-26	1.953E-26	0.1666	1.4309	0.324	6.226	1.363E-27
4.000	67.0	6.556E-26	3	0.268	0.518	27.35	2.736E-26	1.640E-26	0.1637	1.3083	0.205	6.216	1.144E-27
4.000	77.0	5.253E-26	3	0.308	0.507	31.24	2.156E-26	1.314E-26	0.1604	1.1956	0.067	6.205	9.167E-28
4.000	87.0	4.261E-26	3	0.348	0.495	35.05	1.716E-26	1.066E-26	0.1566	1.0922	-0.090	6.193	7.436E-28
4.000	97.0	3.565E-26	3	0.387	0.482	38.79	1.407E-26	8.918E-27	0.1524	0.9970	-0.266	6.178	6.221E-28
4.000	107.0	2.977E-26	3	0.427	0.467	42.44	1.149E-26	7.447E-27	0.1477	0.9090	-0.462	6.162	5.195E-28
4.000	117.0	2.454E-26	3	0.467	0.451	46.00	9.246E-27	6.139E-27	0.1425	0.8274	-0.677	6.145	4.282E-28
4.000	127.0	2.006E-26	3	0.507	0.433	49.47	7.369E-27	5.018E-27	0.1369	0.7514	-0.910	6.126	3.501E-28
4.000	137.0	1.604E-26	3	0.546	0.414	52.85	5.737E-27	4.012E-27	0.1309	0.6803	-1.163	6.105	2.799E-28
4.000	147.0	1.272E-26	3	0.586	0.393	56.13	4.424E-27	3.182E-27	0.1243	0.6137	-1.435	6.083	2.220E-28
5.000	17.0	1.177E-25	3	0.085	0.689	7.03	3.324E-26	2.355E-26	0.2180	2.1463	0.588	6.095	1.643E-27
5.000	27.0	1.125E-25	3	0.135	0.685	11.14	3.162E-26	2.253E-26	0.2167	1.9738	0.535	6.091	1.572E-27
5.000	37.0	1.008E-25	3	0.185	0.680	15.23	2.809E-26	2.017E-26	0.2149	1.7969	0.459	6.084	1.407E-27
5.000	47.0	8.286E-26	3	0.235	0.672	19.27	2.286E-26	1.658E-26	0.2124	1.6319	0.358	6.076	1.157E-27
5.000	57.0	6.541E-26	3	0.285	0.662	23.27	1.782E-26	1.309E-26	0.2095	1.4825	0.233	6.066	9.130E-28
5.000	67.0	5.031E-26	3	0.335	0.651	27.21	1.350E-26	1.007E-26	0.2059	1.3482	0.084	6.054	7.022E-28
5.000	77.0	3.941E-26	3	0.385	0.638	31.08	1.040E-26	7.885E-27	0.2017	1.2272	-0.088	6.039	5.501E-28
5.000	87.0	3.188E-26	3	0.434	0.623	34.89	8.259E-27	6.378E-27	0.1970	1.1177	-0.285	6.023	4.450E-28
5.000	97.0	2.590E-26	3	0.484	0.606	38.61	6.573E-27	5.182E-27	0.1917	1.0180	-0.505	6.005	3.615E-28
5.000	107.0	2.070E-26	3	0.534	0.588	42.25	5.137E-27	4.142E-27	0.1858	0.9266	-0.750	5.984	2.889E-28
5.000	117.0	1.603E-26	3	0.584	0.567	45.81	3.883E-27	3.207E-27	0.1794	0.8423	-1.018	5.962	2.237E-28
5.000	127.0	1.223E-26	3	0.633	0.545	49.27	2.888E-27	2.447E-27	0.1724	0.7642	-1.310	5.937	1.707E-28
5.000	137.0	9.054E-27	3	0.683	0.521	52.65	2.081E-27	1.812E-27	0.1648	0.6914	-1.626	5.911	1.264E-28
5.000	147.0	6.722E-27	3	0.732	0.495	55.92	1.502E-27	1.345E-27	0.1566	0.6233	-1.966	5.882	9.382E-29

Diddens/Schlüpmann

p_3	θ_3	$\dfrac{d^2\sigma}{d\Omega_3\,dp_3}$	Error	p_T	p_L^*	θ_3^*	$\dfrac{1}{p_3^{*2}}\cdot\dfrac{d^2\sigma}{d\Omega_3^*\,dp_3^*}$	$\dfrac{E_3}{p_3^2}\cdot\dfrac{d^2\sigma}{d\Omega_3\,dp_3}$	x^*	y_3^*	t	m_4	$\dfrac{d^2\sigma}{dt\,dm_4^2}$
GeV/c	mrad	cm² sr⁻¹ GeV⁻¹ c	%	GeV/c	GeV/c	deg	cm² sr⁻¹ GeV⁻³ c³	cm² sr⁻¹ GeV⁻² c³			GeV²	GeV	cm²/GeV⁴
					$p_1 = 24.000$ GeV/c; $p_1^* = 3.290$ GeV/c; $E_1^* + E_2^* = 6.843$ GeV; $s = 46.828$ GeV²; $\gamma_{c.m.} = 3.6469$; $\beta_{c.m.} = 0.96167$; $y_1^* = 1.9677$; $p_{3,max}^* = 3.163$ GeV/c; [72 A 1] (cont.)							
6.000	17.0	1.320E-25	3	0.102	0.830	7.01	2.596E-26	2.201E-26	0.2624	2.2724	0.560	5.937	1.535E-27
6.000	27.0	1.118E-25	3	0.152	0.825	11.11	2.187E-26	1.864E-26	0.2609	2.0597	0.497	5.932	1.300E-27
6.000	37.0	8.694E-26	3	0.222	0.818	15.18	1.687E-26	1.449E-26	0.2586	1.8557	0.405	5.924	1.011E-27
6.000	47.0	6.402E-26	3	0.282	0.809	19.21	1.230E-26	1.067E-26	0.2557	1.6734	0.284	5.914	7.446E-28
6.000	57.0	4.871E-26	3	0.342	0.797	23.20	9.241E-27	8.121E-27	0.2521	1.5129	0.134	5.901	5.665E-28
6.000	67.0	3.740E-26	3	0.402	0.784	27.13	6.991E-27	6.235E-27	0.2478	1.3713	-0.044	5.886	4.350E-28
6.000	77.0	2.934E-26	3	0.462	0.768	31.00	5.393E-27	4.891E-27	0.2429	1.2453	-0.252	5.868	3.412E-28
6.000	87.0	2.300E-26	3	0.521	0.750	34.80	4.149E-27	3.834E-27	0.2372	1.1322	-0.488	5.848	2.675E-28
6.000	97.0	1.802E-25	3	0.581	0.730	38.52	3.184E-27	3.004E-27	0.2308	1.0298	-0.752	5.825	2.096E-28
6.000	107.0	1.314E-26	3	0.641	0.708	42.15	2.270E-27	2.191E-27	0.2238	0.9364	-1.045	5.800	1.528E-28
6.000	117.0	9.437E-27	3	0.700	0.683	45.70	1.592E-27	1.573E-27	0.2161	0.8506	-1.367	5.772	1.098E-28
6.000	127.0	6.576E-27	3	0.760	0.657	49.17	1.081E-27	1.096E-27	0.2077	0.7712	-1.718	5.742	7.648E-29
6.000	137.0	4.615E-27	3	0.819	0.628	52.54	7.385E-28	7.694E-28	0.1985	0.6975	-2.097	5.709	5.367E-29
6.000	147.0	3.293E-27	3	0.879	0.597	55.81	5.123E-28	5.490E-28	0.1888	0.6287	-2.504	5.673	3.830E-29
7.000	17.0	1.286E-25	3	0.119	0.970	6.99	1.862E-26	1.838E-26	0.3067	2.3673	0.528	5.774	1.282E-27
7.000	27.0	9.359E-26	3	0.189	0.964	11.09	1.347E-26	1.337E-26	0.3049	2.1197	0.454	5.768	9.329E-28
7.000	37.0	6.385E-26	3	0.259	0.956	15.15	9.122E-27	9.125E-27	0.3023	1.8947	0.346	5.758	6.365E-28
7.000	47.0	4.955E-26	3	0.329	0.945	19.18	7.005E-27	7.080E-27	0.2989	1.7002	0.205	5.746	4.939E-28
7.000	57.0	3.650E-26	3	0.399	0.932	23.16	5.096E-27	5.215E-27	0.2947	1.5322	0.031	5.731	3.638E-28
7.000	67.0	2.663E-26	3	0.469	0.916	27.09	3.664E-27	3.805E-27	0.2897	1.3857	-0.178	5.713	2.654E-28
7.000	77.0	2.065E-26	3	0.538	0.898	30.95	2.793E-27	2.951E-27	0.2839	1.2565	-0.419	5.691	2.058E-28
7.000	87.0	1.569E-26	3	0.608	0.877	34.74	2.083E-27	2.242E-27	0.2773	1.1411	-0.695	5.667	1.564E-28
7.000	97.0	1.117E-26	3	0.678	0.854	38.46	1.452E-27	1.596E-27	0.2699	1.0371	-1.003	5.640	1.113E-28
7.000	107.0	7.617E-27	3	0.748	0.828	42.09	9.683E-28	1.088E-27	0.2617	0.9424	-1.345	5.609	7.593E-29
7.000	117.0	5.032E-27	3	0.817	0.799	45.64	6.245E-28	7.190E-28	0.2526	0.8556	-1.721	5.576	5.016E-29
7.000	127.0	3.298E-27	3	0.887	0.768	49.10	3.989E-28	4.712E-28	0.2428	0.7755	-2.130	5.539	3.287E-29
7.000	137.0	2.170E-27	3	0.956	0.734	52.47	2.555E-28	3.101E-28	0.2322	0.7012	-2.572	5.499	2.163E-29
7.000	147.0	1.501E-27	3	1.025	0.698	55.74	1.718E-28	2.145E-28	0.2208	0.6319	-3.048	5.456	1.496E-29
8.000	17.0	1.039E-25	3	0.136	1.110	6.99	1.153E-26	1.299E-26	0.3509	2.4402	0.492	5.606	9.062E-28
8.000	27.0	6.743E-26	3	0.216	1.103	11.08	7.441E-27	8.430E-27	0.3489	2.1628	0.408	5.598	5.881E-28
8.000	37.0	4.738E-26	3	0.296	1.094	15.14	5.187E-27	5.923E-27	0.3459	1.9218	0.285	5.587	4.132E-28
8.000	47.0	3.523E-26	3	0.376	1.082	19.16	3.818E-27	4.404E-27	0.3420	1.7183	0.124	5.573	3.073E-28
8.000	57.0	2.557E-26	3	0.456	1.067	23.14	2.736E-27	3.197E-27	0.3372	1.5451	-0.076	5.555	2.230E-28
8.000	67.0	1.879E-26	3	0.536	1.049	27.06	1.981E-27	2.349E-27	0.3315	1.3954	-0.314	5.534	1.639E-28
8.000	77.0	1.405E-26	3	0.615	1.028	30.92	1.457E-27	1.757E-27	0.3249	1.2639	-0.590	5.509	1.225E-28
8.000	87.0	9.730E-27	3	0.695	1.004	34.71	9.899E-28	1.216E-27	0.3173	1.1470	-0.904	5.480	8.486E-29
8.000	97.0	6.355E-27	3	0.775	0.977	38.42	6.333E-28	7.945E-28	0.3088	1.0418	-1.257	5.448	5.542E-29
8.000	107.0	3.947E-27	3	0.854	0.947	42.05	3.846E-28	4.935E-28	0.2995	0.9463	-1.648	5.412	3.442E-29
8.000	117.0	2.443E-27	3	0.934	0.915	45.60	2.323E-28	3.054E-28	0.2891	0.8589	-2.077	5.372	2.131E-29
8.000	127.0	1.515E-27	3	1.013	0.879	49.06	1.404E-28	1.894E-28	0.2779	0.7784	-2.545	5.328	1.321E-29
8.000	137.0	9.531E-28	3	1.093	0.841	52.43	8.600E-29	1.192E-28	0.2658	0.7036	-3.050	5.281	8.312E-30
8.000	147.0	5.923E-28	3	1.172	0.799	55.70	5.195E-29	7.405E-29	0.2527	0.6340	-3.594	5.229	5.166E-30

$p_1 = 24.000$ GeV/c; $p_1^* = 3.290$ GeV/c; $E_1^* + E_2^* = 6.843$ GeV; $s = 46.828$ GeV²; $\gamma_{c.m.} = 3.6469$; $\beta_{c.m.} = 0.9616^7$; $y_1^* = 1.9677$; $p_{3,\,max}^* = 3.163$ GeV/c; [72 A 1] (cont.)

p_3	θ_3	$\dfrac{d^2\sigma}{d\Omega_3\,dp_3}$	Error	p_T	p_L^*	θ_3^*	$\dfrac{1}{p_3^{*2}}\cdot\dfrac{d^2\sigma}{d\Omega_3^*\,dp_3^*}$	$\dfrac{E_3}{p_3^2}\cdot\dfrac{d^2\sigma}{d\Omega_3\,dp_3}$	x^*	y_3^*	t	m_4	$\dfrac{d^2\sigma}{dt\,dm_4^2}$
GeV/c	mrad	cm² sr⁻¹ GeV⁻¹ c	%	GeV/c	GeV/c	deg	cm² sr⁻¹ GeV⁻³ c³	cm² sr⁻¹ GeV⁻² c³			GeV²	GeV	cm²/GeV⁴
9.000	17.0	7.318E-26	3	0.153	1.249	6.98	6.421E-27	8.132E-27	0.3951	2.4972	0.455	5.433	5.673E-28
9.000	27.0	4.950E-26	3	0.243	1.242	11.07	4.319E-27	5.501E-27	0.3928	2.1947	0.360	5.424	3.837E-28
9.000	37.0	3.521E-26	3	0.333	1.232	15.12	3.048E-27	3.913E-27	0.3895	1.9413	0.222	5.411	2.730E-28
9.000	47.0	2.444E-26	3	0.423	1.218	19.15	2.094E-27	2.716E-27	0.3851	1.7312	0.041	5.394	1.895E-28
9.000	57.0	1.800E-26	3	0.513	1.201	23.12	1.523E-27	2.000E-27	0.3797	1.5542	-0.184	5.374	1.395E-28
9.000	67.0	1.303E-26	3	0.603	1.181	27.04	1.086E-27	1.448E-27	0.3733	1.4021	-0.452	5.349	1.010E-28
9.000	77.0	8.971E-27	3	0.692	1.157	30.90	7.355E-28	9.969E-28	0.3658	1.2690	-0.762	5.319	6.954E-29
9.000	87.0	5.620E-27	3	0.782	1.130	34.68	4.521E-28	6.245E-28	0.3573	1.1511	-1.116	5.286	4.357E-29
9.000	97.0	3.338E-27	3	0.872	1.100	38.39	2.630E-28	3.709E-28	0.3478	1.0451	-1.513	5.248	2.588E-29
9.000	107.0	1.929E-27	3	0.961	1.067	42.03	1.486E-28	2.144E-28	0.3372	0.9491	-1.953	5.206	1.495E-29
9.000	117.0	1.086E-27	3	1.051	1.030	45.57	8.166E-29	1.207E-28	0.3256	0.8612	-2.436	5.160	8.419E-30
9.000	127.0	6.408E-28	3	1.140	0.990	49.03	4.697E-29	7.121E-29	0.3130	0.7803	-2.961	5.109	4.968E-30
9.000	137.0	3.759E-28	3	1.229	0.947	52.40	2.682E-29	4.177E-29	0.2993	0.7053	-3.530	5.053	2.914E-30
9.000	147.0	2.213E-28	3	1.318	0.900	55.67	1.535E-29	2.459E-29	0.2846	0.6355	-4.141	4.992	1.716E-30
10.000	17.0	5.514E-26	3	0.170	1.389	6.98	3.921E-27	5.515E-27	0.4392	2.5423	0.417	5.254	3.847E-28
10.000	27.0	3.902E-26	3	0.270	1.381	11.06	2.760E-27	3.902E-27	0.4367	2.2189	0.311	5.243	2.722E-28
10.000	37.0	2.628E-26	3	0.370	1.369	15.12	1.844E-27	2.628E-27	0.4330	1.9557	0.158	5.229	1.833E-28
10.000	47.0	1.738E-26	3	0.470	1.354	19.13	1.207E-27	1.738E-27	0.4281	1.7406	-0.044	5.209	1.213E-28
10.000	57.0	1.241E-26	3	0.570	1.335	23.11	8.511E-28	1.241E-27	0.4221	1.5608	-0.293	5.185	8.658E-29
10.000	67.0	8.697E-27	3	0.669	1.313	27.02	5.877E-28	8.698E-28	0.4150	1.4069	-0.591	5.157	6.068E-29
10.000	77.0	5.341E-27	3	0.769	1.286	30.88	3.549E-28	5.342E-28	0.4067	1.2728	-0.936	5.123	3.726E-29
10.000	87.0	3.046E-27	3	0.859	1.256	34.67	1.986E-28	3.046E-28	0.3973	1.1540	-1.329	5.085	2.125E-29
10.000	97.0	1.627E-27	3	0.968	1.223	38.38	1.039E-28	1.627E-28	0.3867	1.0475	-1.770	5.041	1.135E-29
10.000	107.0	8.698E-28	3	1.068	1.186	42.01	5.430E-29	8.699E-29	0.3749	0.9510	-2.259	4.992	6.068E-30
10.000	117.0	4.694E-28	3	1.167	1.145	45.55	2.861E-29	4.694E-29	0.3620	0.8628	-2.795	4.938	3.275E-30
10.000	127.0	2.578E-28	3	1.267	1.101	49.01	1.531E-29	2.578E-29	0.3480	0.7817	-3.379	4.879	1.799E-30
10.000	137.0	1.427E-28	3	1.366	1.053	52.38	8.250E-30	1.427E-29	0.3328	0.7065	-4.011	4.814	9.956E-31
10.000	147.0	7.907E-29	3	1.465	1.001	55.65	4.444E-30	7.908E-30	0.3165	0.6365	-4.691	4.743	5.517E-31
11.000	17.0	4.454E-26	3	0.187	1.529	6.97	2.619E-27	4.049E-27	0.4833	2.5785	0.378	5.068	2.825E-28
11.000	27.0	3.059E-26	3	0.297	1.520	11.06	1.789E-27	2.781E-27	0.4805	2.2375	0.261	5.056	1.940E-28
11.000	37.0	1.954E-26	3	0.407	1.507	15.11	1.134E-27	1.777E-27	0.4765	1.9666	0.093	5.040	1.239E-28
11.000	47.0	1.225E-26	3	0.517	1.490	19.13	7.034E-28	1.114E-27	0.4712	1.7477	-0.129	5.018	7.769E-29
11.000	57.0	8.354E-27	3	0.627	1.469	23.10	4.742E-28	7.604E-28	0.4646	1.5657	-0.404	4.990	5.305E-29
11.000	67.0	5.474E-27	3	0.736	1.444	27.01	3.058E-28	4.977E-28	0.4567	1.4105	-0.731	4.957	3.472E-29
11.000	77.0	3.057E-27	3	0.846	1.416	30.87	1.679E-28	2.779E-28	0.4476	1.2755	-1.111	4.919	1.939E-29
11.000	87.0	1.556E-27	3	0.956	1.383	34.65	8.387E-29	1.415E-28	0.4372	1.1562	-1.543	4.875	9.869E-30
11.000	97.0	7.755E-28	3	1.055	1.346	38.36	4.094E-29	7.051E-29	0.4255	1.0492	-2.028	4.825	4.919E-30
11.000	107.0	3.894E-28	3	1.175	1.305	41.99	2.010E-29	3.540E-29	0.4126	0.9525	-2.566	4.769	2.470E-30
11.000	117.0	1.986E-28	3	1.284	1.260	45.54	1.001E-29	1.806E-29	0.3985	0.8641	-3.156	4.706	1.260E-30
11.000	127.0	1.031E-28	3	1.393	1.211	48.99	5.063E-30	9.373E-30	0.3830	0.7827	-3.798	4.638	6.539E-31
11.000	137.0	5.401E-29	3	1.502	1.159	52.36	2.581E-30	4.910E-30	0.3663	0.7074	-4.493	4.562	3.426E-31
11.000	147.0	2.937E-29	3	1.611	1.102	55.63	1.365E-30	2.670E-30	0.3484	0.6373	-5.241	4.479	1.863E-31

$p_1 = 24.000$ GeV/c; $p_1^* = 3.290$ GeV/c; $E_1^* + E_2^* = 6.843$ GeV; $s = 46.828$ GeV²; $\gamma_{c.m.} = 3.6469$; $\beta_{c.m.} = 0.96167$; $y_1^* = 1.9677$; $p_{3,max}^* = 3.163$ GeV/c; [72 A 1] (cont.)

p_3	θ_3	$\dfrac{d^2\sigma}{d\Omega_3\,dp_3}$	Error	p_T	p_L^*	θ_3^*	$\dfrac{1}{p_3^{*2}}\cdot\dfrac{d^2\sigma}{d\Omega_3^*\,dp_3^*}$	$\dfrac{E_3}{p_3^2}\cdot\dfrac{d^2\sigma}{d\Omega_3\,dp_3}$	x^*	y_3^*	t	m_4	$\dfrac{d^2\sigma}{dt\,dm_4^2}$
GeV/c	mrad	cm² sr⁻¹ GeV⁻¹ c	%	GeV/c	GeV/c	deg	cm² sr⁻¹ GeV⁻³ c³	cm² sr⁻¹ GeV⁻² c³			GeV²	GeV	cm²/GeV⁴
12.000	17.0	3.605E-26	3	0.204	1.668	6.97	1.782E-27	3.004E-27	0.5274	2.6079	0.338	4.875	2.096E-28
12.000	27.0	2.334E-26	3	0.324	1.659	11.05	1.147E-27	1.945E-27	0.5244	2.2522	0.211	4.862	1.357E-28
12.000	37.0	1.398E-26	3	0.444	1.645	15.11	6.817E-28	1.165E-27	0.5200	1.9751	0.027	4.843	8.128E-29
12.000	47.0	8.477E-27	3	0.564	1.626	19.12	4.091E-28	7.065E-28	0.5142	1.7532	-0.215	4.818	4.928E-29
12.000	57.0	5.510E-27	3	0.684	1.603	23.09	2.626E-28	4.592E-28	0.5070	1.5695	-0.515	4.787	3.203E-29
12.000	67.0	3.343E-27	3	0.803	1.576	27.01	1.570E-28	2.786E-28	0.4984	1.4133	-0.872	4.750	1.944E-29
12.000	77.0	1.674E-27	3	0.923	1.545	30.86	7.729E-29	1.395E-28	0.4884	1.2776	-1.286	4.706	9.732E-30
12.000	87.0	7.681E-28	3	1.043	1.509	34.64	3.480E-29	6.401E-29	0.4771	1.1578	-1.758	4.655	4.466E-30
12.000	97.0	3.399E-28	3	1.162	1.469	38.35	1.508E-29	2.833E-29	0.4644	1.0506	-2.287	4.598	1.976E-30
12.000	107.0	1.594E-28	3	1.282	1.424	41.98	6.915E-30	1.328E-29	0.4503	0.9536	-2.873	4.534	9.267E-31
12.000	117.0	7.809E-29	3	1.401	1.375	45.53	3.307E-30	6.508E-30	0.4348	0.8650	-3.517	4.462	4.540E-31
12.000	127.0	3.683E-29	3	1.520	1.322	48.98	1.520E-30	3.069E-30	0.4180	0.7835	-4.218	4.383	2.141E-31
12.000	137.0	1.782E-29	3	1.639	1.264	52.35	7.158E-31	1.485E-30	0.3998	0.7081	-4.976	4.296	1.036E-31
12.000	142.0	1.249E-29	4	1.698	1.234	54.00	4.947E-31	1.041E-30	0.3902	0.6724	-5.377	4.249	7.261E-32
13.000	17.0	3.035E-26	3	0.221	1.808	6.97	1.278E-27	2.335E-27	0.5715	2.6321	0.297	4.674	1.629E-28
13.000	27.0	1.819E-26	3	0.351	1.797	11.05	7.619E-28	1.399E-27	0.5682	2.2639	0.160	4.660	9.762E-29
13.000	37.0	9.770E-27	3	0.481	1.782	15.10	4.060E-28	7.516E-28	0.5635	1.9818	-0.040	4.638	5.243E-29
13.000	47.0	6.034E-27	3	0.611	1.762	19.12	2.482E-28	4.642E-28	0.5572	1.7575	-0.302	4.610	3.238E-29
13.000	57.0	3.498E-27	3	0.741	1.738	23.09	1.421E-28	2.691E-28	0.5494	1.5725	-0.626	4.575	1.877E-29
13.000	67.0	1.773E-27	3	0.870	1.708	27.00	7.096E-29	1.364E-28	0.5401	1.4155	-1.013	4.532	9.515E-30
13.000	77.0	7.786E-28	3	1.000	1.674	30.85	3.064E-29	5.990E-29	0.5293	1.2793	-1.462	4.482	4.178E-30
13.000	87.0	3.165E-28	3	1.130	1.635	34.64	1.222E-29	2.435E-29	0.5170	1.1591	-1.973	4.425	1.699E-30
13.000	97.0	1.335E-28	3	1.259	1.592	38.34	5.049E-30	1.027E-29	0.5032	1.0516	-2.546	4.360	7.164E-31
13.000	107.0	5.927E-29	3	1.388	1.543	41.97	2.191E-30	4.559E-30	0.4880	0.9544	-3.182	4.286	3.181E-31
13.000	117.0	2.739E-29	3	1.518	1.490	45.52	9.885E-31	2.107E-30	0.4712	0.8657	-3.879	4.204	1.470E-31
13.000	127.0	1.175E-29	3	1.647	1.433	48.97	4.133E-31	9.039E-31	0.4530	0.7841	-4.638	4.113	6.306E-32
13.000	137.0	5.330E-30	4	1.775	1.370	52.34	1.825E-31	4.100E-31	0.4333	0.7086	-5.460	4.012	2.860E-32
14.000	17.0	2.157E-26	3	0.238	1.947	6.97	7.835E-28	1.541E-27	0.6156	2.6521	0.256	4.465	1.075E-28
14.000	27.0	1.239E-26	3	0.378	1.936	11.05	4.476E-28	8.850E-28	0.6121	2.2734	0.108	4.448	6.174E-29
14.000	37.0	6.440E-27	3	0.518	1.920	15.10	2.308E-28	4.600E-28	0.6069	1.9871	-0.107	4.424	3.209E-29
14.000	57.0	1.916E-27	3	0.798	1.872	23.08	6.712E-29	1.369E-28	0.5917	1.5748	-0.738	4.352	9.548E-30
14.000	67.0	8.422E-28	3	0.937	1.840	27.00	2.907E-29	6.016E-29	0.5817	1.4172	-1.155	4.304	4.197E-30
14.000	77.0	3.305E-28	3	1.077	1.803	30.85	1.122E-29	2.361E-29	0.5701	1.2806	-1.638	4.247	1.647E-30
14.000	87.0	1.222E-28	3	1.216	1.761	34.63	4.069E-30	8.729E-30	0.5569	1.1602	-2.189	4.182	6.089E-31
14.000	97.0	4.854E-29	3	1.355	1.714	38.34	1.583E-30	3.467E-30	0.5421	1.0525	-2.806	4.107	2.419E-31
14.000	107.0	1.985E-29	3	1.495	1.662	41.97	6.329E-31	1.418E-30	0.5256	0.9551	-3.490	4.023	9.892E-32
14.000	117.0	8.542E-30	4	1.634	1.605	45.51	2.659E-31	6.102E-31	0.5076	0.8663	-4.241	3.929	4.257E-32
14.000	127.0	3.826E-30	4	1.773	1.543	48.97	1.161E-31	2.733E-31	0.4880	0.7846	-5.059	3.823	1.907E-32
14.000	137.0	1.678E-30	6	1.912	1.476	52.33	4.954E-32	1.199E-31	0.4667	0.7090	-5.944	3.706	8.362E-33
15.000	17.0	1.584E-26	3	0.255	2.087	6.97	5.013E-28	1.056E-27	0.6597	2.6689	0.215	4.244	7.367E-29
15.000	27.0	8.007E-27	3	0.405	2.074	11.05	2.520E-28	5.338E-28	0.6559	2.2812	0.056	4.226	3.724E-29

Diddens/Schlüpmann

p_3	θ_3	$\dfrac{d^2\sigma}{d\Omega_3\,dp_3}$	Error	p_T	p_L^*	θ_3^*	$\dfrac{1}{p_3^{*2}}\cdot\dfrac{d^2\sigma}{d\Omega_3^*\,dp_3^*}$	$\dfrac{E_3}{p_3^2}\cdot\dfrac{d^2\sigma}{d\Omega_3\,dp_3}$	x^*	y_3^*	t	m_4	$\dfrac{d^2\sigma}{dt\,dm_4^2}$
GeV/c	mrad	cm² sr⁻¹ GeV⁻¹ c	%	GeV/c	GeV/c	deg	cm² sr⁻¹ GeV⁻³ c³	cm² sr⁻¹ GeV⁻² c³			GeV²	GeV	cm²/GeV⁴

$p_1 = 24.000$ GeV/c; $p_1^* = 3.290$ GeV/c; $E_1^* + E_2^* = 6.843$ GeV; $s = 46.828$ GeV²; $\gamma_{c.m.} = 3.6469$; $\beta_{c.m.} = 0.96167$; $y_1^* = 1.9677$; $p_{3,max}^* = 3.163$ GeV/c; [72 A 1] (cont.)

p_3	θ_3	$\dfrac{d^2\sigma}{d\Omega_3\,dp_3}$	Error	p_T	p_L^*	θ_3^*	$\dfrac{1}{p_3^{*2}}\dfrac{d^2\sigma}{d\Omega_3^*dp_3^*}$	$\dfrac{E_3}{p_3^2}\dfrac{d^2\sigma}{d\Omega_3dp_3}$	x^*	y_3^*	t	m_4	$\dfrac{d^2\sigma}{dt\,dm_4^2}$
15.000	37.0	4.182E-27	3	0.555	2.057	15.10	1.306E-28	2.788E-28	0.6504	1.9915	-0.174	4.198	1.945E-29
15.000	47.0	2.181E-27	3	0.705	2.034	19.11	6.741E-29	1.454E-28	0.6431	1.7637	-0.477	4.162	1.014E-29
15.000	57.0	1.020E-27	3	0.855	2.006	23.08	3.113E-29	6.800E-29	0.6341	1.5767	-0.851	4.117	4.744E-30
15.000	67.0	4.091E-28	3	1.004	1.972	26.99	1.230E-29	2.727E-29	0.6234	1.4186	-1.297	4.063	1.903E-30
15.000	77.0	1.398E-28	3	1.154	1.932	30.84	4.133E-30	9.320E-30	0.6110	1.2817	-1.815	3.998	6.502E-31
15.000	87.0	4.857E-29	3	1.303	1.888	34.63	1.409E-30	3.238E-30	0.5968	1.1610	-2.405	3.924	2.259E-31
15.000	97.0	1.784E-29	4	1.453	1.837	38.33	5.069E-31	1.189E-30	0.5809	1.0531	-3.066	3.839	8.297E-32
15.000	107.0	6.959E-30	4	1.602	1.782	41.96	1.933E-31	4.640E-31	0.5633	0.9557	-3.799	3.742	3.237E-32
15.000	117.0	2.791E-30	5	1.751	1.720	45.50	7.568E-32	1.861E-31	0.5440	0.8668	-4.604	3.633	1.298E-32
15.000	127.0	1.028E-30	7	1.900	1.654	48.96	2.717E-32	6.854E-32	0.5229	0.7850	-5.480	3.510	4.781E-33
16.000	17.0	1.125E-26	3	0.272	2.226	6.97	3.130E-28	7.032E-28	0.7038	2.6831	0.173	4.012	4.905E-29
16.000	27.0	5.217E-27	3	0.432	2.213	11.04	1.443E-28	3.261E-28	0.6997	2.2877	0.004	3.991	2.275E-29
16.000	37.0	2.491E-27	3	0.592	2.194	15.09	6.837E-29	1.557E-28	0.6938	1.9951	-0.242	3.960	1.086E-29
16.000	57.0	5.228E-28	3	0.912	2.140	23.08	1.403E-29	3.268E-29	0.6765	1.5783	-0.963	3.868	2.280E-30
16.000	67.0	1.937E-28	3	1.071	2.103	26.99	5.120E-30	1.211E-29	0.6651	1.4197	-1.439	3.806	8.446E-31
16.000	77.0	5.863E-29	3	1.231	2.061	30.84	1.524E-30	3.665E-30	0.6518	1.2825	-1.992	3.733	2.556E-31
16.000	87.0	1.814E-29	3	1.390	2.014	34.62	4.626E-31	1.134E-30	0.6367	1.1617	-2.621	3.647	7.909E-32
16.000	97.0	6.555E-30	3	1.550	1.960	38.33	1.637E-31	4.097E-31	0.6197	1.0537	-3.326	3.549	2.858E-32
16.000	107.0	2.402E-30	5	1.709	1.901	41.96	5.865E-32	1.501E-31	0.6009	0.9561	-4.108	3.437	1.047E-32
16.000	117.0	9.283E-31	8	1.868	1.835	45.50	2.213E-32	5.802E-32	0.5803	0.8671	-4.967	3.310	4.048E-33
16.000	127.0	3.256E-31	14	2.027	1.764	48.96	7.564E-33	2.035E-32	0.5579	0.7853	-5.901	3.166	1.420E-33
17.000	17.0	6.454E-27	3	0.289	2.365	6.97	1.591E-28	3.797E-28	0.7479	2.6951	0.131	3.766	2.649E-29
17.000	27.0	2.786E-27	3	0.459	2.352	11.04	6.828E-29	1.639E-28	0.7435	2.2931	-0.048	3.742	1.143E-29
17.000	37.0	1.255E-27	3	0.629	2.332	15.09	3.052E-29	7.383E-29	0.7373	1.9981	-0.310	3.707	5.150E-30
17.000	47.0	5.842E-28	3	0.799	2.306	19.11	1.406E-29	3.437E-29	0.7290	1.7679	-0.652	3.660	2.397E-30
17.000	57.0	2.263E-28	3	0.968	2.274	23.07	5.378E-30	1.331E-29	0.7189	1.5796	-1.076	3.602	9.287E-31
17.000	67.0	7.121E-29	3	1.138	2.235	26.99	1.668E-30	4.189E-30	0.7067	1.4207	-1.582	3.531	2.922E-31
17.000	77.0	1.937E-29	3	1.308	2.191	30.84	4.460E-31	1.139E-30	0.6926	1.2833	-2.169	3.447	7.949E-32
17.000	87.0	5.525E-30	5	1.477	2.140	34.62	1.248E-31	3.250E-31	0.6765	1.1622	-2.837	3.348	2.267E-32
17.000	97.0	1.879E-30	6	1.646	2.083	38.33	4.158E-32	1.105E-31	0.6585	1.0542	-3.587	3.234	7.711E-33
17.000	107.0	6.935E-31	9	1.816	2.020	41.95	1.500E-32	4.080E-32	0.6386	0.9565	-4.418	3.103	2.846E-33
17.000	117.0	1.993E-31	13	1.984	1.950	45.50	4.208E-33	1.172E-32	0.6167	0.8674	-5.330	2.953	8.179E-34
17.000	127.0	8.019E-32	21	2.153	1.875	48.95	1.650E-33	4.717E-33	0.5928	0.7856	-6.323	2.779	3.291E-34
18.000	17.0	2.832E-27	3	0.306	2.505	6.97	6.226E-29	1.573E-28	0.7919	2.7054	0.089	3.502	1.098E-29
18.000	27.0	1.266E-27	3	0.486	2.490	11.04	2.768E-29	7.034E-29	0.7874	2.2977	-0.101	3.474	4.907E-30
18.000	37.0	6.178E-28	3	0.666	2.469	15.09	1.340E-29	3.432E-29	0.7807	2.0007	-0.378	3.434	2.394E-30
18.000	47.0	2.727E-28	3	0.846	2.442	19.10	5.855E-30	1.515E-29	0.7720	1.7695	-0.740	3.381	1.057E-30
18.000	57.0	9.759E-29	3	1.025	2.408	23.07	2.069E-30	5.422E-30	0.7612	1.5807	-1.189	3.314	3.782E-31
18.000	67.0	2.448E-29	3	1.205	2.367	26.98	5.114E-31	1.360E-30	0.7483	1.4215	-1.725	3.232	9.488E-32
18.000	77.0	5.469E-30	4	1.385	2.320	30.83	1.123E-31	3.038E-31	0.7334	1.2839	-2.346	3.135	2.120E-32

p_3	θ_3	$\dfrac{d^2\sigma}{d\Omega_3\,dp_3}$	Error	p_T	p_L^*	θ_3^*	$\dfrac{1}{p_3^{*2}}\cdot\dfrac{d^2\sigma}{d\Omega_3^*\,dp_3^*}$	$\dfrac{E_3}{p_3^2}\cdot\dfrac{d^2\sigma}{d\Omega_3\,dp_3}$	x^*	y_3^*	t	m_4	$\dfrac{d^2\sigma}{dt\,dm_4^2}$
GeV/c	mrad	cm²sr⁻¹GeV⁻¹c	%	GeV/c	GeV/c	deg	cm²sr⁻¹GeV⁻³c²	cm²sr⁻¹GeV⁻²c³			GeV²	GeV	cm²/GeV⁴
$p_1 = 24.000$ GeV/c; $p_1^* = 3.290$ GeV/c; $E_1^* + E_2^* = 6.843$ GeV; $s = 46.828$ GeV²; $\gamma_{c.m.} = 3.6469$; $\beta_{c.m.} = 0.96167$; $y_1^* = 1.9677$; $p_{3,\,max}^* = 3.163$ GeV/c; [72 A 1] (cont.)													
18.000	87.0	1.366E−30	6	1.564	2.266	34.62	2.753E−32	7.589E−32	0.7164	1.1627	−3.054	3.020	5.294E−33
18.000	97.0	3.662E−31	10	1.743	2.206	38.32	7.228E−33	2.035E−32	0.6973	1.0545	−3.848	2.885	1.419E−33
18.000	107.0	1.154E−31	17	1.922	2.139	41.95	2.227E−33	6.411E−33	0.6762	0.9568	−4.727	2.728	4.473E−34
18.000	117.0	5.310E−32	25	2.101	2.065	45.49	1.000E−33	2.950E−33	0.6530	0.8677	−5.693	2.545	2.058E−34
18.000	127.0	1.628E−32	44	2.280	1.985	48.95	2.989E−34	9.045E−34	0.6278	0.7858	−6.745	2.330	6.310E−35
19.000	17.0	1.658E−27	3	0.323	2.644	6.96	3.272E−29	8.727E−29	0.8360	2.7144	0.047	3.216	6.088E−30
19.000	27.0	6.648E−28	3	0.513	2.629	11.04	1.305E−29	3.499E−29	0.8312	2.3016	−0.154	3.185	2.441E−30
19.000	37.0	3.067E−28	3	0.703	2.607	15.09	5.971E−30	1.614E−29	0.8241	2.0029	−0.446	3.138	1.126E−30
19.000	47.0	1.199E−28	3	0.893	2.577	19.10	2.311E−30	6.311E−30	0.8150	1.7708	−0.829	3.077	4.402E−31
19.000	57.0	3.269E−29	3	1.082	2.541	23.07	6.221E−31	1.721E−30	0.8036	1.5817	−1.303	2.999	1.200E−31
19.000	67.0	7.762E−30	4	1.272	2.499	26.98	1.455E−31	4.085E−31	0.7900	1.4222	−1.868	2.903	2.850E−32
19.000	77.0	1.625E−30	4	1.462	2.449	30.83	2.996E−32	8.553E−32	0.7742	1.2844	−2.524	2.788	5.967E−33
19.000	87.0	4.133E−31	6	1.651	2.392	34.61	7.476E−33	2.175E−32	0.7563	1.1631	−3.271	2.650	1.518E−33
19.000	97.0	7.239E−32	17	1.840	2.328	38.32	1.282E−33	3.810E−33	0.7362	1.0549	−4.109	2.487	2.658E−34
19.000	107.0	2.818E−32	33	2.029	2.258	41.95	4.881E−34	1.483E−33	0.7138	0.9571	−5.037	2.293	1.035E−34
19.000	117.0	9.206E−33	57	2.218	2.180	45.49	1.556E−34	4.845E−34	0.6894	0.8679	−6.057	2.059	3.380E−35
20.000	17.0	1.077E−27	3	0.340	2.783	6.96	1.918E−29	5.385E−29	0.8800	2.7221	0.004	2.902	3.757E−30
20.000	27.0	3.603E−28	3	0.540	2.767	11.04	6.382E−30	1.802E−29	0.8750	2.3050	−0.207	2.866	1.257E−30
20.000	37.0	1.512E−28	3	0.740	2.744	15.09	2.657E−30	7.560E−30	0.8676	2.0047	−0.514	2.812	5.274E−31
20.000	47.0	4.783E−29	3	0.940	2.713	19.10	8.319E−31	2.392E−30	0.8579	1.7720	−0.917	2.739	1.668E−31
20.000	57.0	1.078E−29	3	1.139	2.675	23.07	1.851E−31	5.390E−31	0.8459	1.5825	−1.416	2.646	3.760E−32
20.000	67.0	1.894E−30	5	1.339	2.630	26.98	3.205E−32	9.470E−32	0.8316	1.4228	−2.011	2.531	6.607E−33
20.000	77.0	3.685E−31	7	1.538	2.578	30.83	6.131E−33	1.843E−32	0.8150	1.2848	−2.701	2.391	1.285E−33
20.000	87.0	6.423E−32	22	1.738	2.518	34.61	1.049E−33	3.212E−33	0.7961	1.1635	−3.488	2.221	2.240E−34
20.000	97.0	1.203E−32	70	1.937	2.451	38.32	1.924E−34	6.015E−34	0.7750	1.0551	−4.370	2.012	4.196E−35
4.000	12.0	8.994E−26	3	0.048	0.550	4.99	3.954E−26	2.250E−26	0.1737	2.0251	0.622	6.250	1.570E−27
4.500	12.0	1.048E−25	3	0.054	0.620	4.98	3.652E−26	2.330E−26	0.1961	2.1290	0.615	6.174	1.625E−27
5.000	12.0	1.162E−25	3	0.060	0.691	4.96	3.287E−26	2.325E−26	0.2184	2.2193	0.606	6.097	1.622E−27
5.500	12.0	1.270E−25	3	0.066	0.761	4.96	2.974E−26	2.310E−26	0.2407	2.2984	0.594	6.018	1.611E−27
6.000	12.0	1.357E−25	3	0.072	0.831	4.95	2.674E−26	2.262E−26	0.2629	2.3684	0.581	5.939	1.578E−27
6.500	12.0	1.420E−25	3	0.078	0.902	4.94	2.386E−26	2.185E−26	0.2851	2.4305	0.567	5.858	1.524E−27
7.000	12.0	1.427E−25	3	0.084	0.972	4.94	2.069E−26	2.039E−26	0.3072	2.4859	0.552	5.776	1.422E−27
7.500	12.0	1.379E−25	3	0.090	1.042	4.94	1.743E−26	1.839E−26	0.3294	2.5356	0.536	5.693	1.283E−27
8.000	12.0	1.266E−25	3	0.096	1.112	4.93	1.407E−26	1.583E−26	0.3515	2.5803	0.520	5.608	1.104E−27
8.500	12.0	1.110E−25	3	0.102	1.182	4.93	1.093E−26	1.306E−26	0.3737	2.6206	0.504	5.523	9.111E−28
9.000	12.0	9.212E−26	3	0.108	1.252	4.93	8.097E−27	1.024E−26	0.3958	2.6572	0.487	5.436	7.141E−28
9.500	12.0	7.680E−26	3	0.114	1.322	4.93	6.061E−27	8.085E−27	0.4179	2.6903	0.469	5.347	5.640E−28
10.000	12.0	6.516E−26	3	0.120	1.392	4.93	4.642E−27	6.517E−27	0.4400	2.7204	0.452	5.257	4.546E−28
10.500	12.0	5.755E−26	3	0.126	1.462	4.93	3.719E−27	5.481E−27	0.4622	2.7479	0.434	5.165	3.824E−28
11.000	12.0	5.201E−26	3	0.132	1.532	4.93	3.063E−27	4.729E−27	0.4843	2.7730	0.416	5.072	3.299E−28

p_3	θ_3	$\dfrac{\mathrm{d}^2\sigma}{\mathrm{d}\Omega_3\,\mathrm{d}p_3}$	Error	p_T	p_L^*	θ_3^*	$\dfrac{1}{p_3^{*2}}\cdot\dfrac{\mathrm{d}^2\sigma}{\mathrm{d}\Omega_3^*\,\mathrm{d}p_3^*}$	$\dfrac{E_3}{p_3^2}\cdot\dfrac{\mathrm{d}^2\sigma}{\mathrm{d}\Omega_3\,\mathrm{d}p_3}$	x^*	y_3^*	t	m_4	$\dfrac{\mathrm{d}^2\sigma}{\mathrm{d}t\,\mathrm{d}m_4^2}$
GeV/c	mrad	cm² sr⁻¹ GeV⁻¹ c	%	GeV/c	GeV/c	deg	cm² sr⁻¹ GeV⁻³ c³	cm² sr⁻¹ GeV⁻² c³			GeV²	GeV	cm²/GeV⁴

$p_1 = 24.000$ GeV/c; $p_1^* = 3.290$ GeV/c; $E_1^* + E_2^* = 6.843$ GeV; $s = 46.828$ GeV²; $\gamma_\mathrm{c.m.} = 3.6469$; $\beta_\mathrm{c.m.} = 0.96167$; $y_1^* = 1.9677$; $p_{3,\mathrm{max}}^* = 3.163$ GeV/c; [72 A 1] (cont.)

p_3	θ_3	$\mathrm{d}^2\sigma/\mathrm{d}\Omega_3\,\mathrm{d}p_3$	Error	p_T	p_L^*	θ_3^*	$\frac{1}{p_3^{*2}}\frac{\mathrm{d}^2\sigma}{\mathrm{d}\Omega_3^*\mathrm{d}p_3^*}$	$\frac{E_3}{p_3^2}\frac{\mathrm{d}^2\sigma}{\mathrm{d}\Omega_3\mathrm{d}p_3}$	x^*	y_3^*	t	m_4	$\mathrm{d}^2\sigma/\mathrm{d}t\,\mathrm{d}m_4^2$
11.500	12.0	4.794E−26	3	0.138	1.601	4.92	2.584E−27	4.169E−27	0.5064	2.7960	0.398	4.977	2.908E−28
12.000	12.0	4.396E−26	3	0.144	1.671	4.92	2.176E−27	3.664E−27	0.5284	2.8170	0.379	4.879	2.556E−28
12.500	12.0	3.924E−26	3	0.150	1.741	4.92	1.791E−27	3.139E−27	0.5505	2.8364	0.361	4.780	2.190E−28
13.000	12.0	3.455E−26	3	0.156	1.811	4.92	1.458E−27	2.658E−27	0.5726	2.8542	0.342	4.679	1.854E−28
13.500	12.0	2.978E−26	3	0.162	1.881	4.92	1.165E−27	2.206E−27	0.5947	2.8706	0.323	4.576	1.539E−28
14.000	12.0	2.521E−26	3	0.168	1.951	4.92	9.174E−28	1.801E−27	0.6168	2.8858	0.305	4.470	1.256E−28
14.500	12.0	2.130E−26	3	0.174	2.021	4.92	7.226E−28	1.469E−27	0.6389	2.8998	0.286	4.362	1.025E−28
15.000	12.0	1.821E−26	3	0.180	2.090	4.92	5.773E−28	1.214E−27	0.6610	2.9128	0.267	4.251	8.469E−29
15.500	12.0	1.611E−26	3	0.186	2.160	4.92	4.784E−28	1.039E−27	0.6830	2.9249	0.248	4.136	7.251E−29
16.000	12.0	1.415E−26	3	0.192	2.230	4.92	3.943E−28	8.844E−28	0.7051	2.9361	0.229	4.019	6.170E−29
16.500	12.0	1.082E−26	3	0.198	2.300	4.92	2.836E−28	6.558E−28	0.7272	2.9465	0.209	3.898	4.575E−29
17.000	12.0	8.748E−27	3	0.204	2.370	4.92	2.160E−28	5.146E−28	0.7493	2.9563	0.190	3.773	3.590E−29
17.500	12.0	6.600E−27	3	0.210	2.440	4.92	1.538E−28	3.772E−28	0.7713	2.9653	0.171	3.644	2.631E−29
18.000	12.0	4.509E−27	3	0.216	2.509	4.92	9.931E−29	2.505E−28	0.7934	2.9738	0.152	3.510	1.748E−29
18.500	12.0	3.501E−27	3	0.222	2.579	4.92	7.300E−29	1.892E−28	0.8155	2.9817	0.132	3.371	1.320E−29
19.230	12.0	2.248E−27	3	0.231	2.681	4.92	4.338E−29	1.169E−28	0.8477	2.9924	0.104	3.157	8.155E−30
19.620	12.0	1.888E−27	3	0.235	2.736	4.92	3.500E−29	9.623E−29	0.8649	2.9977	0.089	3.037	6.713E−30
20.020	12.0	1.585E−27	3	0.240	2.791	4.92	2.822E−29	7.917E−29	0.8826	3.0028	0.073	2.908	5.523E−30
20.430	12.0	1.161E−27	3	0.245	2.849	4.92	1.985E−29	5.683E−29	0.9007	3.0079	0.057	2.769	3.964E−30
20.840	12.0	7.583E−28	3	0.250	2.906	4.92	1.246E−29	3.639E−29	0.9188	3.0127	0.041	2.624	2.538E−30
21.270	12.0	5.113E−28	3	0.255	2.966	4.92	8.066E−30	2.404E−29	0.9378	3.0175	0.024	2.462	1.677E−30
21.700	12.0	3.009E−28	3	0.260	3.026	4.92	4.561E−30	1.387E−29	0.9567	3.0220	0.008	2.288	9.674E−31
22.150	12.0	1.282E−28	3	0.266	3.089	4.92	1.865E−30	5.788E−30	0.9766	3.0265	−0.010	2.092	4.038E−31
22.600	12.0	3.345E−29	3	0.271	3.152	4.92	4.675E−31	1.480E−30	0.9965	3.0308	−0.028	1.874	1.033E−31

$p_1 = 29.700$ GeV/c; $p_1^* = 3.674$ GeV/c; $E_1^* + E_2^* = 7.584$ GeV; $s = 57.517$ GeV²; $\gamma_\mathrm{c.m.} = 4.0418$; $\beta_\mathrm{c.m.} = 0.96891$; $y_1^* = 2.0742$; $p_{3,\mathrm{max}}^* = 3.559$ GeV/c; [67 A 1]

p_3	θ_3	$\mathrm{d}^2\sigma/\mathrm{d}\Omega_3\,\mathrm{d}p_3$	Error	p_T	p_L^*	θ_3^*	$\frac{1}{p_3^{*2}}\frac{\mathrm{d}^2\sigma}{\mathrm{d}\Omega_3^*\mathrm{d}p_3^*}$	$\frac{E_3}{p_3^2}\frac{\mathrm{d}^2\sigma}{\mathrm{d}\Omega_3\mathrm{d}p_3}$	x^*	y_3^*	t	m_4	$\mathrm{d}^2\sigma/\mathrm{d}t\,\mathrm{d}m_4^2$
3.906	15.0	8.784E−26	9	0.059	0.479	6.97	4.477E−26	2.250E−26	0.1347	1.8696	0.610	7.065	1.269E−27
4.524	15.0	1.079E−25	6	0.068	0.558	6.93	4.120E−26	2.386E−26	0.1568	1.9915	0.607	6.982	1.345E−27
5.127	15.0	1.220E−25	5	0.077	0.634	6.91	3.639E−26	2.380E−26	0.1783	2.0901	0.601	6.900	1.342E−27
5.861	15.0	1.317E−25	4	0.088	0.727	6.89	3.014E−26	2.248E−26	0.2044	2.1893	0.588	6.799	1.267E−27
6.520	15.0	1.329E−25	4	0.098	0.810	6.88	2.462E−26	2.039E−26	0.2278	2.2632	0.574	6.706	1.149E−27
7.110	15.0	1.274E−25	4	0.107	0.885	6.87	1.987E−26	1.792E−26	0.2487	2.3196	0.560	6.622	1.010E−27
7.704	15.0	1.396E−25	4	0.116	0.960	6.87	1.856E−26	1.812E−26	0.2697	2.3688	0.545	6.536	1.022E−27
8.311	15.0	1.468E−25	4	0.125	1.036	6.86	1.678E−26	1.767E−26	0.2911	2.4124	0.528	6.447	9.959E−28
8.913	15.0	1.470E−25	4	0.134	1.112	6.86	1.462E−26	1.649E−26	0.3124	2.4501	0.511	6.358	9.299E−28
9.506	15.0	1.236E−25	4	0.143	1.186	6.85	1.081E−26	1.300E−26	0.3333	2.4826	0.494	6.268	7.331E−28
10.120	15.0	9.367E−26	4	0.152	1.263	6.85	7.232E−27	9.257E−27	0.3550	2.5123	0.475	6.174	5.218E−28
10.720	15.0	8.428E−26	4	0.161	1.339	6.85	5.801E−27	7.863E−27	0.3762	2.5378	0.456	6.081	4.432E−28
11.320	15.0	7.758E−26	5	0.170	1.414	6.85	4.790E−27	6.854E−27	0.3973	2.5605	0.438	5.986	3.864E−28
11.920	15.0	7.411E−26	5	0.179	1.489	6.85	4.127E−27	6.218E−27	0.4185	2.5807	0.418	5.890	3.505E−28
12.510	15.0	6.152E−26	5	0.188	1.563	6.84	3.111E−27	4.918E−27	0.4393	2.5984	0.399	5.793	2.772E−28
13.130	15.0	6.139E−26	4	0.197	1.641	6.84	2.819E−27	4.676E−27	0.4612	2.6151	0.379	5.690	2.636E−28

p_3	θ_3	$\dfrac{d^2\sigma}{d\Omega_3\,dp_3}$	Error	p_T	p_L^*	θ_3^*	$\dfrac{1}{p_3^{*2}}\cdot\dfrac{d^2\sigma}{d\Omega_3^*\,dp_3^*}$	$\dfrac{E_3}{p_3^2}\cdot\dfrac{d^2\sigma}{d\Omega_3\,dp_3}$	x^*	y_3^*	t	m_4	$\dfrac{d^2\sigma}{dt\,dm_4^2}$
GeV/c	mrad	cm² sr⁻¹ GeV⁻¹ c	%	GeV/c	GeV/c	deg	cm² sr⁻¹ GeV⁻³ c³	cm² sr⁻¹ GeV⁻² c³			GeV²	GeV	cm²/GeV⁴

$p_1 = 29.700$ GeV/c; $p_1^* = 3.674$ GeV/c; $E_1^* + E_2^* = 7.584$ GeV; $s = 57.517$ GeV²; $\gamma_{c.m.} = 4.0418$; $\beta_{c.m.} = 0.96891$; $y_1^* = 2.0742$; $p_{3,max}^* = 3.559$ GeV/c; [67 A 1] (cont.)

p_3	θ_3	$\dfrac{d^2\sigma}{d\Omega_3\,dp_3}$	Error	p_T	p_L^*	θ_3^*	$\dfrac{1}{p_3^{*2}}\cdot\dfrac{d^2\sigma}{d\Omega_3^*\,dp_3^*}$	$\dfrac{E_3}{p_3^2}\cdot\dfrac{d^2\sigma}{d\Omega_3\,dp_3}$	x^*	y_3^*	t	m_4	$\dfrac{d^2\sigma}{dt\,dm_4^2}$
13.740	15.0	5.323E-26	4	0.206	1.718	6.84	2.232E-27	3.874E-27	0.4827	2.6299	0.359	5.587	2.184E-28
14.340	15.0	4.704E-26	4	0.215	1.793	6.84	1.811E-27	3.280E-27	0.5038	2.6429	0.339	5.483	1.849E-28
14.940	15.0	4.469E-26	4	0.224	1.858	6.84	1.586E-27	2.991E-27	0.5249	2.6547	0.318	5.378	1.686E-28
15.530	15.0	3.711E-26	4	0.233	1.942	6.84	1.219E-27	2.390E-27	0.5457	2.6653	0.298	5.272	1.347E-28
16.120	15.0	3.670E-26	4	0.242	2.016	6.84	1.119E-27	2.277E-27	0.5665	2.6748	0.278	5.164	1.283E-28
16.740	15.0	3.195E-26	5	0.251	2.094	6.84	9.032E-28	1.909E-27	0.5884	2.6840	0.257	5.048	1.076E-28
17.350	15.0	3.095E-26	5	0.260	2.170	6.84	8.145E-28	1.784E-27	0.6098	2.6922	0.236	4.931	1.006E-28
17.950	15.0	2.935E-26	5	0.269	2.245	6.84	7.217E-28	1.635E-27	0.6310	2.6996	0.216	4.814	9.218E-29
18.540	15.0	2.706E-26	5	0.278	2.319	6.84	6.237E-28	1.460E-27	0.6517	2.7063	0.195	4.695	8.228E-29
19.130	15.0	2.282E-26	5	0.287	2.393	6.84	4.941E-28	1.193E-27	0.6725	2.7125	0.175	4.573	6.725E-29
6.679	96.0	1.297E-26	4	0.640	0.709	42.07	2.011E-27	1.942E-27	0.1993	0.9385	-1.212	6.549	1.095E-28
7.290	96.0	9.121E-27	4	0.699	0.775	42.03	1.188E-27	1.251E-27	0.2178	0.9421	-1.390	6.447	7.054E-29
7.887	96.0	6.751E-27	4	0.756	0.839	42.00	7.521E-28	8.561E-28	0.2359	0.9449	-1.565	6.346	4.826E-29
8.486	96.0	4.802E-27	5	0.813	0.904	41.98	4.624E-28	5.659E-28	0.2540	0.9472	-1.741	6.243	3.190E-29
9.081	96.0	3.367E-27	5	0.870	0.968	41.96	2.832E-28	3.708E-28	0.2720	0.9490	-1.917	6.138	2.090E-29
9.690	96.0	2.381E-27	7	0.929	1.033	41.95	1.760E-28	2.457E-28	0.2904	0.9505	-2.097	6.030	1.385E-29
10.300	96.0	1.666E-27	7	0.987	1.099	41.94	1.090E-28	1.618E-28	0.3088	0.9518	-2.279	5.919	9.119E-30
10.880	96.0	1.232E-27	8	1.043	1.161	41.93	7.227E-29	1.132E-28	0.3263	0.9528	-2.452	5.811	6.384E-30
14.810	96.0	1.084E-28	16	1.420	1.583	41.89	3.435E-30	7.320E-30	0.4448	0.9568	-3.629	5.022	4.126E-31
16.010	96.0	3.532E-29	20	1.535	1.711	41.88	9.580E-31	2.206E-30	0.4810	0.9575	-3.990	4.754	1.244E-31
3.494	160.0	1.403E-26	4	0.557	0.248	66.01	6.429E-27	4.019E-27	0.0696	0.4193	-2.020	6.932	2.265E-28
4.112	160.0	9.338E-27	4	0.655	0.295	65.75	3.104E-27	2.272E-27	0.0829	0.4275	-2.483	6.814	1.281E-28
4.699	160.0	5.909E-27	4	0.749	0.340	65.59	1.509E-27	1.258E-27	0.0955	0.4326	-2.928	6.700	7.092E-29
5.598	160.0	2.955E-27	5	0.892	0.408	65.44	5.331E-28	5.280E-28	0.1146	0.4375	-3.617	6.520	2.977E-29
6.791	160.0	1.184E-27	6	1.082	0.497	65.32	1.455E-28	1.744E-28	0.1397	0.4413	-4.539	6.273	9.831E-30
7.986	160.0	4.258E-28	9	1.272	0.586	65.25	3.788E-29	5.333E-29	0.1648	0.4435	-5.468	6.015	3.006E-30
9.192	160.0	1.594E-28	16	1.454	0.676	65.21	1.071E-29	1.734E-29	0.1901	0.4450	-6.410	5.743	9.777E-31

3.6.3 pp→K⁻X

$p_1 = 12.500$ GeV/c; $p_1^* = 2.332$ GeV/c; $E_1^* = E_2^* = 5.028$ GeV; $s = 25.281$ GeV²; $\gamma_{c.m.} = 2.6796$; $\beta_{c.m.} = 0.92776$; $y_1^* = 1.6420$; $p_{3,max}^* = 1.917$ GeV/c; [71 A 1]

p_3	θ_3	$\dfrac{d^2\sigma}{d\Omega_3\,dp_3}$	Error	p_T	p_L^*	θ_3^*	$\dfrac{1}{p_3^{*2}}\cdot\dfrac{d^2\sigma}{d\Omega_3^*\,dp_3^*}$	$\dfrac{E_3}{p_3^2}\cdot\dfrac{d^2\sigma}{d\Omega_3\,dp_3}$	x^*	y_3^*	t	m_4	$\dfrac{d^2\sigma}{dt\,dm_4^2}$
4.009	137.1	1.060E-28	11	0.548	0.600	42.40	2.802E-29	2.664E-29	0.3129	0.7431	-0.858	3.995	3.568E-30
4.147	153.1	6.623E-29	11	0.632	0.600	46.51	1.605E-29	1.608E-29	0.3130	0.6914	-1.115	3.931	2.154E-30
4.279	166.0	4.777E-29	11	0.707	0.600	49.68	1.070E-29	1.124E-29	0.3130	0.6493	-1.359	3.868	1.505E-30
4.404	176.8	2.851E-29	14	0.775	0.600	52.24	5.937E-30	6.514E-30	0.3130	0.6139	-1.594	3.807	8.725E-31
4.525	186.0	1.938E-29	14	0.837	0.600	54.36	3.773E-30	4.308E-30	0.3130	0.5838	-1.819	3.747	5.771E-31
4.640	194.0	1.131E-29	15	0.895	0.600	56.15	2.069E-30	2.451E-30	0.3129	0.5577	-2.035	3.689	3.283E-31
4.752	201.0	6.348E-30	16	0.949	0.600	57.69	1.095E-30	1.343E-30	0.3129	0.5350	-2.243	3.632	1.799E-31
4.860	207.2	5.583E-30	17	1.000	0.600	59.03	9.119E-31	1.155E-30	0.3130	0.5151	-2.444	3.576	1.547E-31
4.965	212.8	5.168E-30	17	1.049	0.600	60.22	8.014E-31	1.046E-30	0.3130	0.4971	-2.639	3.521	1.401E-31

p_3	θ_3	$\dfrac{\mathrm{d}^2\sigma}{\mathrm{d}\Omega_3\,\mathrm{d}p_3}$	Error	p_T	p_L^*	θ_3^*	$\dfrac{1}{p_3^{*2}}\cdot\dfrac{\mathrm{d}^2\sigma}{\mathrm{d}\Omega_3^*\,\mathrm{d}p_3^*}$	$\dfrac{E_3}{p_3^2}\cdot\dfrac{\mathrm{d}^2\sigma}{\mathrm{d}\Omega_3\,\mathrm{d}p_3}$	x^*	y_3^*	t	m_4	$\dfrac{\mathrm{d}^2\sigma}{\mathrm{d}t\,\mathrm{d}m_4^2}$
GeV/c	mrad	cm² sr⁻¹ GeV⁻¹ c	%	GeV/c	GeV/c	deg	cm² sr⁻¹ GeV⁻³ c³	cm² sr⁻¹ GeV⁻² c³			GeV²	GeV	cm²/GeV⁴
$p_1 = 12.500$ GeV/c; $p_1^* = 2.332$ GeV/c; $E_1^* = E_2^* = 5.028$ GeV; $s = 25.281$ GeV²; $\gamma_\mathrm{c.m.} = 2.6796$; $\beta_\mathrm{c.m.} = 0.92776$; $y_1^* = 1.6420$; $p_{3,\max}^* = 1.917$ GeV/c; [71 A 1] (cont.)													
5.067	217.9	4.255E-30	20	1.095	0.600	61.28	6.282E-31	8.437E-31	0.3130	0.4808	-2.829	3.467	1.130E-31
5.165	222.6	2.821E-30	28	1.140	0.600	62.25	3.976E-31	5.487E-31	0.3129	0.4657	-3.016	3.413	7.349E-32
3.001	212.3	7.817E-29	13	0.632	0.300	64.61	3.082E-29	2.640E-29	0.1565	0.3658	-1.783	4.112	3.536E-30
3.361	189.3	8.305E-29	11	0.632	0.400	57.68	2.786E-29	2.498E-29	0.2087	0.4799	-1.518	4.063	3.345E-30
3.744	169.7	6.263E-29	11	0.632	0.500	51.66	1.785E-29	1.687E-29	0.2608	0.5887	-1.297	4.002	2.260E-30
4.147	153.1	6.623E-29	14	0.632	0.600	46.51	1.605E-29	1.608E-29	0.3130	0.6914	-1.115	3.931	2.154E-30
4.567	138.9	4.346E-29	12	0.632	0.700	42.09	8.989E-30	9.572E-30	0.3652	0.7884	-0.964	3.849	1.282E-30
5.001	126.8	4.717E-29	13	0.632	0.800	38.33	8.365E-30	9.478E-30	0.4173	0.8793	-0.841	3.759	1.269E-30
$p_1 = 19.200$ GeV/c; $p_1^* = 2.929$ GeV/c; $E_1^* + E_2^* = 6.151$ GeV; $s = 37.830$ GeV²; $\gamma_\mathrm{c.m.} = 3.2779$; $\beta_\mathrm{c.m.} = 0.95233$; $y_1^* = 1.8562$; $p_{3,\max}^* = 2.592$ GeV/c; [70 A 1]													
4.500	12.5	1.808E-27	3	0.055	0.618	5.20	5.098E-28	4.042E-28	0.2383	1.0431	-0.134	5.322	3.525E-29
4.500	20.0	1.596E-27	3	0.090	0.616	8.31	4.491E-28	3.568E-28	0.2376	1.0332	-0.155	5.320	3.111E-29
4.500	30.0	1.710E-27	3	0.135	0.612	12.43	4.790E-28	3.823E-28	0.2362	1.0134	-0.198	5.316	3.334E-29
4.500	40.0	1.477E-27	3	0.180	0.607	16.51	4.112E-28	3.302E-28	0.2342	0.9869	-0.259	5.310	2.879E-29
4.500	50.0	1.438E-27	3	0.225	0.600	20.53	3.972E-28	3.215E-28	0.2317	0.9547	-0.337	5.303	2.803E-29
4.500	60.0	1.371E-27	3	0.270	0.592	24.49	3.751E-28	3.065E-28	0.2285	0.9181	-0.432	5.294	2.673E-29
4.500	70.0	1.179E-27	4	0.315	0.583	28.37	3.190E-28	2.636E-28	0.2248	0.8779	-0.544	5.283	2.298E-29
6.000	12.5	1.245E-27	3	0.075	0.873	4.91	2.071E-28	2.082E-28	0.3367	1.3246	0.051	5.070	1.816E-29
6.000	20.0	1.196E-27	3	0.120	0.870	7.85	1.985E-28	2.000E-28	0.3358	1.3072	0.023	5.067	1.744E-29
6.000	30.0	1.023E-27	3	0.180	0.865	11.75	1.690E-28	1.711E-28	0.3339	1.2734	-0.034	5.062	1.492E-29
6.000	40.0	9.018E-28	3	0.240	0.859	15.61	1.480E-28	1.508E-28	0.3312	1.2296	-0.115	5.054	1.315E-29
6.000	50.0	8.001E-28	3	0.300	0.850	19.44	1.302E-28	1.338E-28	0.3278	1.1784	-0.219	5.043	1.167E-29
6.000	60.0	6.360E-28	3	0.360	0.839	23.21	1.025E-28	1.064E-28	0.3236	1.1222	-0.345	5.031	9.275E-30
6.000	70.0	5.022E-28	3	0.420	0.826	26.93	7.999E-29	8.398E-29	0.3187	1.0630	-0.495	5.016	7.323E-30
8.000	12.5	6.679E-28	3	0.100	1.201	4.76	6.425E-29	8.365E-29	0.4632	1.6028	0.148	4.697	7.294E-30
8.000	20.0	5.929E-28	3	0.160	1.197	7.61	5.690E-29	7.425E-29	0.4619	1.5729	0.111	4.693	6.475E-30
8.000	30.0	4.970E-28	3	0.240	1.191	11.39	4.747E-29	6.224E-29	0.4594	1.5167	0.034	4.684	5.428E-30
8.000	40.0	3.576E-28	3	0.320	1.182	15.15	3.393E-29	4.479E-29	0.4559	1.4473	-0.074	4.673	3.905E-30
8.000	50.0	2.745E-28	3	0.400	1.170	18.87	2.582E-29	3.438E-29	0.4513	1.3702	-0.212	4.658	2.998E-30
8.000	60.0	1.922E-28	3	0.480	1.155	22.55	1.790E-29	2.407E-29	0.4458	1.2897	-0.381	4.640	2.099E-30
8.000	70.0	1.419E-28	3	0.560	1.138	26.18	1.306E-29	1.777E-29	0.4392	1.2088	-0.580	4.618	1.550E-30
10.000	12.5	2.241E-28	3	0.125	1.522	4.69	1.398E-29	2.244E-29	0.5872	1.8147	0.167	4.281	1.957E-30
10.000	20.0	1.972E-28	3	0.200	1.518	7.50	1.227E-29	1.974E-29	0.5857	1.7697	0.121	4.276	1.722E-30
10.000	30.0	1.588E-28	3	0.300	1.510	11.24	9.835E-30	1.590E-29	0.5825	1.6885	0.025	4.265	1.386E-30
10.000	35.0	1.278E-28	3	0.350	1.505	13.09	7.890E-30	1.280E-29	0.5805	1.6420	-0.038	4.257	1.116E-30
10.000	40.0	1.150E-28	3	0.400	1.498	14.94	7.075E-30	1.151E-29	0.5781	1.5932	-0.110	4.249	1.004E-30
10.000	50.0	7.253E-29	3	0.500	1.484	18.62	4.424E-30	7.262E-30	0.5724	1.4925	-0.282	4.229	6.332E-31

p_3	θ_3	$\dfrac{d^2\sigma}{d\Omega_3\,dp_3}$	Error	p_T	p_L^*	θ_3^*	$\dfrac{1}{p_3^{*2}}\cdot\dfrac{d^2\sigma}{d\Omega_3^*\,dp_3^*}$	$\dfrac{E_3}{p_3^2}\cdot\dfrac{d^2\sigma}{d\Omega_3\,dp_3}$	x^*	y_3^*	t	m_4	$\dfrac{d^2\sigma}{dt\,dm_4^2}$
GeV/c	mrad	cm² sr⁻¹ GeV⁻¹ c	%	GeV/c	GeV/c	deg	cm² sr⁻¹ GeV⁻³ c³	cm² sr⁻¹ GeV⁻² c³			GeV²	GeV	cm²/GeV⁴

$p_1 = 19.200$ GeV/c; $p_1^* = 2.929$ GeV/c; $E_1^* + E_2^* = 6.151$ GeV; $s = 37.830$ GeV²; $\gamma_{c.m.} = 3.2779$; $\beta_{c.m.} = 0.95233$; $y_1^* = 1.8562$; $p_{3,max}^* = 2.592$ GeV/c; [70 A 1] (cont.)

p_3	θ_3	$\dfrac{d^2\sigma}{d\Omega_3\,dp_3}$	Error	p_T	p_L^*	θ_3^*	$\dfrac{1}{p_3^{*2}}\cdot\dfrac{d^2\sigma}{d\Omega_3^*\,dp_3^*}$	$\dfrac{E_3}{p_3^2}\cdot\dfrac{d^2\sigma}{d\Omega_3\,dp_3}$	x^*	y_3^*	t	m_4	$\dfrac{d^2\sigma}{dt\,dm_4^2}$
10.000	60.0	4.463E-29	3	0.600	1.466	22.25	2.694E-30	4.468E-30	0.5655	1.3918	-0.494	4.204	3.897E-31
10.000	70.0	2.850E-29	3	0.699	1.444	25.84	1.700E-30	2.853E-30	0.5572	1.2942	-0.743	4.174	2.488E-31
11.000	12.5	1.370E-28	3	0.137	1.681	4.67	7.092E-30	1.247E-29	0.6487	1.9036	0.161	4.056	1.087E-30
11.000	20.0	1.212E-28	3	0.220	1.677	7.47	6.259E-30	1.103E-29	0.6470	1.8503	0.110	4.050	9.618E-31
11.000	30.0	8.614E-29	3	0.330	1.668	11.19	4.427E-30	7.839E-30	0.6436	1.7562	0.004	4.036	6.836E-31
11.000	35.0	6.350E-29	4	0.385	1.662	13.04	3.253E-30	5.779E-30	0.6413	1.7033	-0.065	4.028	5.039E-31
11.000	40.0	5.361E-29	3	0.440	1.655	14.88	2.737E-30	4.879E-30	0.6387	1.6485	-0.144	4.018	4.254E-31
11.000	50.0	2.903E-29	3	0.550	1.639	18.54	1.469E-30	2.642E-30	0.6324	1.5372	-0.334	3.994	2.304E-31
11.000	60.0	1.630E-29	3	0.660	1.619	22.16	8.164E-31	1.483E-30	0.6248	1.4281	-0.566	3.965	1.293E-31
11.000	70.0	9.374E-30	3	0.769	1.596	25.74	4.638E-31	8.530E-31	0.6158	1.3238	-0.840	3.930	7.439E-32
12.000	12.5	6.260E-29	3	0.150	1.840	4.66	2.732E-30	5.221E-30	0.7100	1.9837	0.148	3.816	4.553E-31
12.000	20.0	5.211E-29	3	0.240	1.836	7.45	2.268E-30	4.346E-30	0.7082	1.9218	0.092	3.809	3.790E-31
12.000	30.0	3.277E-29	3	0.360	1.826	11.15	1.420E-30	2.733E-30	0.7044	1.8146	-0.024	3.794	2.383E-31
12.000	35.0	2.150E-29	3	0.420	1.819	13.00	9.285E-31	1.793E-30	0.7019	1.7555	-0.098	3.784	1.564E-31
12.000	40.0	1.696E-29	3	0.480	1.812	14.83	7.297E-31	1.415E-30	0.6991	1.6950	-0.185	3.772	1.233E-31
12.000	50.0	8.702E-30	3	0.600	1.794	18.48	3.712E-31	7.258E-31	0.6923	1.5741	-0.392	3.745	6.329E-32
12.000	60.0	4.702E-30	3	0.720	1.773	22.09	1.985E-31	3.922E-31	0.6839	1.4575	-0.645	3.711	3.420E-32
12.000	70.0	2.504E-30	4	0.839	1.747	25.66	1.044E-31	2.088E-31	0.6741	1.3476	-0.945	3.670	1.821E-32
13.000	12.5	1.598E-29	3	0.162	1.999	4.65	5.956E-31	1.230E-30	0.7712	2.0564	0.129	3.559	1.073E-31
13.000	20.0	1.388E-29	3	0.260	1.994	7.43	5.161E-31	1.068E-30	0.7692	1.9854	0.068	3.551	9.317E-32
13.000	30.0	8.294E-30	3	0.390	1.983	11.13	3.069E-31	6.385E-31	0.7651	1.8653	-0.057	3.533	5.567E-32
13.000	35.0	5.520E-30	3	0.455	1.976	12.96	2.036E-31	4.249E-31	0.7624	1.8003	-0.138	3.522	3.705E-32
13.000	40.0	4.023E-30	3	0.520	1.968	14.80	1.478E-31	3.097E-31	0.7593	1.7345	-0.231	3.508	2.700E-32
13.000	50.0	1.950E-30	5	0.650	1.949	18.44	7.105E-32	1.501E-31	0.7519	1.6049	-0.456	3.476	1.309E-32
13.000	60.0	1.170E-30	5	0.780	1.925	22.04	4.218E-32	9.006E-32	0.7429	1.4817	-0.730	3.436	7.854E-33
13.000	70.0	5.583E-31	7	0.909	1.898	25.60	1.988E-32	4.298E-32	0.7322	1.3669	-1.055	3.389	3.748E-33
14.000	12.5	2.975E-30	3	0.175	2.157	4.64	9.580E-32	2.126E-31	0.8322	2.1227	0.106	3.282	1.854E-32
14.000	20.0	2.180E-30	4	0.280	2.151	7.42	7.003E-32	1.558E-31	0.8300	2.0424	0.040	3.272	1.359E-32
14.000	30.0	1.183E-30	5	0.420	2.140	11.10	3.782E-32	8.455E-32	0.8256	1.9096	-0.094	3.251	7.373E-33
14.000	35.0	8.320E-31	7	0.490	2.132	12.94	2.651E-32	5.947E-32	0.8227	1.8390	-0.181	3.238	5.185E-33
14.000	40.0	5.933E-31	6	0.560	2.124	14.77	1.884E-32	4.240E-32	0.8194	1.7682	-0.282	3.222	3.698E-33
14.000	50.0	2.523E-31	9	0.700	2.103	18.40	7.941E-33	1.803E-32	0.8114	1.6307	-0.524	3.184	1.572E-33
14.000	60.0	1.279E-31	14	0.839	2.078	22.00	3.983E-33	9.141E-33	0.8017	1.5018	-0.819	3.138	7.971E-34
15.000	35.0	8.403E-32	33	0.525	2.288	12.92	2.336E-33	5.605E-33	0.8829	1.8725	-0.228	2.926	4.888E-34
13.600	12.5	6.422E-30	3	0.170	2.094	4.64	2.190E-31	4.725E-31	0.8078	2.0969	0.116	3.396	4.120E-32
13.900	12.5	3.900E-30	4	0.174	2.141	4.64	1.274E-31	2.808E-31	0.8261	2.1164	0.108	3.311	2.448E-32
14.200	12.5	2.027E-30	4	0.177	2.188	4.64	6.347E-32	1.428E-31	0.8444	2.1353	0.101	3.223	1.246E-32

Diddens/Schlüpmann

p_3	θ_3	$\dfrac{d^2\sigma}{d\Omega_3\,dp_3}$	Error	p_T	p_L^*	θ_3^*	$\dfrac{1}{p_3^{*2}}\cdot\dfrac{d^2\sigma}{d\Omega_3^*\,dp_3^*}$	$\dfrac{E_3}{p_3^2}\cdot\dfrac{d^2\sigma}{d\Omega_3\,dp_3}$	x^*	y_3^*	t	m_4	$\dfrac{d^2\sigma}{dt\,dm_4^2}$
GeV/c	mrad	cm² sr⁻¹ GeV⁻¹ c	%	GeV/c	GeV/c	deg	cm² sr⁻¹ GeV⁻³ c³	cm² sr⁻¹ GeV⁻² c³			GeV²	GeV	cm²/GeV⁴

$p_1 = 19.200$ GeV/c; $p_1^* = 2.929$ GeV/c; $E_1^* + E_2^* = 6.151$ GeV; $s = 37.830$ GeV²; $\gamma_{c.m.} = 3.2779$; $\beta_{c.m.} = 0.95233$; $y_1^* = 1.8562$; $p_{3,\,max}^* = 2.592$ GeV/c; [70 A 1] (cont.)

p_3	θ_3	$\dfrac{d^2\sigma}{d\Omega_3\,dp_3}$	Error	p_T	p_L^*	θ_3^*	$\dfrac{1}{p_3^{*2}}\cdot\dfrac{d^2\sigma}{d\Omega_3^*\,dp_3^*}$	$\dfrac{E_3}{p_3^2}\cdot\dfrac{d^2\sigma}{d\Omega_3\,dp_3}$	x^*	y_3^*	t	m_4	$\dfrac{d^2\sigma}{dt\,dm_4^2}$
14.500	12.5	1.160E-30	3	0.181	2.236	4.63	3.485E-32	8.005E-32	0.8626	2.1538	0.093	3.134	6.980E-33
14.800	12.5	6.251E-31	6	0.185	2.283	4.63	1.803E-32	4.226E-32	0.8809	2.1717	0.085	3.041	3.685E-33
15.100	12.5	2.436E-31	8	0.189	2.330	4.63	6.754E-33	1.614E-32	0.8991	2.1893	0.077	2.946	1.408E-33
15.400	12.5	1.009E-31	13	0.192	2.378	4.63	2.691E-33	6.555E-33	0.9174	2.2064	0.068	2.847	5.716E-34
15.700	12.5	3.449E-32	21	0.196	2.425	4.63	8.853E-34	2.198E-33	0.9356	2.2231	0.059	2.745	1.917E-34
16.000	12.5	2.267E-32	40	0.200	2.472	4.62	5.605E-34	1.418E-33	0.9539	2.2394	0.050	2.639	1.236E-34

$p_1 = 24.000$ GeV/c; $p_1^* = 3.290$ GeV/c; $E_1^* + E_2^* = 6.843$ GeV; $s = 46.828$ GeV²; $\gamma_{c.m.} = 3.6469$; $\beta_{c.m.} = 0.96167$; $y_1^* = 1.9677$; $p_{3,\,max}^* = 2.988$ GeV/c; [72 A 1]

p_3	θ_3	$\dfrac{d^2\sigma}{d\Omega_3\,dp_3}$	Error	p_T	p_L^*	θ_3^*	$\dfrac{1}{p_3^{*2}}\cdot\dfrac{d^2\sigma}{d\Omega_3^*\,dp_3^*}$	$\dfrac{E_3}{p_3^2}\cdot\dfrac{d^2\sigma}{d\Omega_3\,dp_3}$	x^*	y_3^*	t	m_4	$\dfrac{d^2\sigma}{dt\,dm_4^2}$
4.000	17.0	2.130E-27	3	0.058	0.451	8.58	7.986E-28	5.365E-28	0.1508	0.8117	-0.509	6.154	3.743E-29
4.000	37.0	1.903E-27	4	0.148	0.443	18.48	7.055E-28	4.794E-28	0.1481	0.7778	-0.613	6.146	3.344E-29
4.000	57.0	1.549E-27	4	0.228	0.429	27.98	5.633E-28	3.902E-28	0.1435	0.7238	-0.793	6.131	2.722E-29
4.000	67.0	1.458E-27	4	0.268	0.420	32.53	5.237E-28	3.673E-28	0.1405	0.6911	-0.912	6.122	2.562E-29
4.000	87.0	1.033E-27	3	0.348	0.397	41.17	3.599E-28	2.602E-28	0.1330	0.6181	-1.207	6.097	1.815E-29
4.000	107.0	8.257E-28	4	0.427	0.369	49.16	2.773E-28	2.080E-28	0.1235	0.5390	-1.579	6.067	1.451E-29
4.000	127.0	5.267E-28	4	0.507	0.335	56.52	1.695E-28	1.327E-28	0.1121	0.4576	-2.028	6.030	9.256E-30
4.000	147.0	2.780E-28	5	0.586	0.295	63.25	8.528E-29	7.003E-29	0.0988	0.3764	-2.552	5.986	4.885E-30
6.000	17.0	1.810E-27	3	0.102	0.764	7.60	3.306E-28	3.027E-28	0.2558	1.2036	-0.112	5.877	2.112E-29
6.000	27.0	1.712E-27	3	0.162	0.760	12.04	3.111E-28	2.863E-28	0.2542	1.1733	-0.175	5.872	1.997E-29
6.000	37.0	1.513E-27	3	0.222	0.753	16.43	2.729E-28	2.530E-28	0.2518	1.1322	-0.267	5.864	1.765E-29
6.000	57.0	1.086E-27	3	0.342	0.732	25.03	1.918E-28	1.816E-28	0.2449	1.0280	-0.538	5.841	1.267E-29
6.000	67.0	8.651E-28	3	0.402	0.718	29.21	1.507E-28	1.447E-28	0.2404	0.9695	-0.717	5.825	1.009E-29
6.000	87.0	5.427E-28	3	0.521	0.685	37.28	9.147E-29	9.076E-29	0.2291	0.8482	-1.160	5.787	6.331E-30
6.000	107.0	3.187E-28	3	0.641	0.642	44.93	5.159E-29	5.330E-29	0.2150	0.7280	-1.717	5.739	3.718E-30
6.000	127.0	1.687E-28	3	0.750	0.591	52.11	2.607E-29	2.821E-29	0.1979	0.6133	-2.390	5.680	1.968E-30
6.000	147.0	7.130E-29	4	0.879	0.532	58.83	1.046E-29	1.192E-29	0.1779	0.5055	-3.176	5.610	8.318E-31
8.000	17.0	1.309E-27	3	0.136	1.061	7.31	1.392E-28	1.639E-28	0.3549	1.4749	0.044	5.563	1.144E-29
8.000	27.0	1.059E-27	3	0.216	1.054	11.58	1.120E-28	1.326E-28	0.3528	1.4238	-0.041	5.556	9.252E-30
8.000	37.0	8.696E-28	3	0.296	1.045	15.81	9.129E-29	1.089E-28	0.3496	1.3577	-0.163	5.545	7.527E-30
8.000	47.0	6.669E-28	3	0.376	1.033	20.00	6.933E-29	8.352E-29	0.3455	1.2824	-0.325	5.530	5.827E-30
8.000	57.0	5.042E-28	3	0.456	1.017	24.13	5.179E-29	6.314E-29	0.3405	1.2026	-0.524	5.512	4.405E-30
8.000	67.0	3.757E-28	3	0.536	0.999	28.19	3.805E-29	4.705E-29	0.3344	1.1215	-0.762	5.491	3.282E-30
8.000	87.0	1.740E-28	3	0.695	0.954	36.06	1.703E-29	2.179E-29	0.3194	0.9634	-1.353	5.436	1.520E-30
8.000	107.0	7.609E-29	4	0.854	0.898	43.57	7.142E-30	9.529E-30	0.3005	0.8163	-2.097	5.368	6.648E-31
8.000	127.0	2.991E-29	4	1.013	0.830	50.68	2.676E-30	3.746E-30	0.2777	0.6821	-2.993	5.283	2.613E-31
8.000	147.0	9.713E-30	6	1.172	0.750	57.37	8.239E-31	1.216E-30	0.2510	0.5602	-4.042	5.183	8.486E-32
10.000	17.0	7.637E-28	3	0.170	1.350	7.18	5.283E-29	7.646E-29	0.4517	1.6782	0.103	5.222	5.334E-30
10.000	27.0	5.679E-28	3	0.270	1.342	11.38	3.908E-29	5.686E-29	0.4490	1.6033	-0.003	5.211	3.967E-30
10.000	37.0	3.949E-28	3	0.370	1.332	15.54	2.697E-29	3.954E-29	0.4451	1.5112	-0.156	5.197	2.758E-30
10.000	47.0	2.766E-28	3	0.470	1.315	19.66	1.870E-29	2.769E-29	0.4400	1.4114	-0.358	5.177	1.932E-30

p_3	θ_3	$\dfrac{\mathrm{d}^2\sigma}{\mathrm{d}\Omega_3\,\mathrm{d}p_3}$	Error	p_T	p_L^*	θ_3^*	$\dfrac{1}{p_3^{*\,2}}\cdot\dfrac{\mathrm{d}^2\sigma}{\mathrm{d}\Omega_3^*\,\mathrm{d}p_3^*}$	$\dfrac{E_3}{p_3^2}\cdot\dfrac{\mathrm{d}^2\sigma}{\mathrm{d}\Omega_3\,\mathrm{d}p_3}$	x^*	y_3^*	t	m_4	$\dfrac{\mathrm{d}^2\sigma}{\mathrm{d}t\,\mathrm{d}m_4^2}$
GeV/c	mrad	cm² sr⁻¹ GeV⁻¹ c	%	GeV/c	GeV/c	deg	cm² sr⁻¹ GeV⁻³ c³	cm² sr⁻¹ GeV⁻² c³			GeV²	GeV	cm²/GeV⁴

$p_1 = 24.000$ GeV/c; $p_1^* = 3.290$ GeV/c; $E_1^* + E_2^* = 6.843$ GeV; $s = 46.828$ GeV²; $\gamma_\mathrm{c.m.} = 3.6469$; $\beta_\mathrm{c.m.} = 0.96167$; $y_1^* = 1.9677$; $p_{3,\,\mathrm{max}}^* = 2.988$ GeV/c; [72 A 1] (cont.)

p_3	θ_3	$\dfrac{\mathrm{d}^2\sigma}{\mathrm{d}\Omega_3\,\mathrm{d}p_3}$	Error	p_T	p_L^*	θ_3^*	$\dfrac{1}{p_3^{*\,2}}\cdot\dfrac{\mathrm{d}^2\sigma}{\mathrm{d}\Omega_3^*\,\mathrm{d}p_3^*}$	$\dfrac{E_3}{p_3^2}\cdot\dfrac{\mathrm{d}^2\sigma}{\mathrm{d}\Omega_3\,\mathrm{d}p_3}$	x^*	y_3^*	t	m_4	$\dfrac{\mathrm{d}^2\sigma}{\mathrm{d}t\,\mathrm{d}m_4^2}$
10.000	57.0	1.825E−28	3	0.570	1.296	23.73	1.219E−29	1.827E−29	0.4336	1.3103	−0.607	5.153	1.275E−30
10.000	67.0	1.248E−28	3	0.669	1.273	27.74	8.216E−30	1.250E−29	0.4261	1.2116	−0.905	5.124	8.717E−31
10.000	87.0	4.097E−29	4	0.869	1.217	35.52	2.605E−30	4.102E−30	0.4073	1.0273	−1.643	5.052	2.862E−31
10.000	107.0	1.382E−29	4	1.068	1.146	42.97	8.423E−31	1.384E−30	0.3836	0.8632	−2.573	4.959	9.653E−32
10.000	127.0	4.076E−30	6	1.267	1.061	50.04	2.366E−31	4.081E−31	0.3552	0.7175	−3.694	4.844	2.847E−32
10.000	147.0	1.007E−30	13	1.465	0.962	56.71	5.538E−32	1.008E−31	0.3218	0.5878	−5.005	4.707	7.034E−33
12.000	17.0	3.135E−28	3	0.204	1.635	7.11	1.520E−29	2.615E−29	0.5472	1.8376	0.113	4.850	1.824E−30
12.000	27.0	2.430E−28	3	0.324	1.626	11.27	1.172E−29	2.027E−29	0.5440	1.7373	−0.014	4.837	1.414E−30
12.000	37.0	1.540E−28	3	0.444	1.612	15.40	7.368E−30	1.284E−29	0.5393	1.6199	−0.198	4.818	8.960E−31
12.000	47.0	8.991E−29	3	0.564	1.593	19.49	4.259E−30	7.499E−30	0.5332	1.4986	−0.440	4.793	5.231E−31
12.000	57.0	5.338E−29	3	0.684	1.571	23.52	2.497E−30	4.452E−30	0.5256	1.3804	−0.739	4.762	3.106E−31
12.000	67.0	2.884E−29	3	0.803	1.544	27.50	1.330E−30	2.405E−30	0.5165	1.2683	−1.096	4.724	1.678E−31
12.000	77.0	1.419E−29	4	0.923	1.512	31.40	6.435E−31	1.184E−30	0.5060	1.1635	−1.510	4.680	8.256E−32
12.000	87.0	7.356E−30	5	1.043	1.476	35.23	3.275E−31	6.135E−31	0.4940	1.0659	−1.982	4.629	4.280E−32
12.000	97.0	3.321E−30	6	1.162	1.436	38.98	1.448E−31	2.770E−31	0.4805	0.9752	−2.511	4.572	1.932E−32
12.000	107.0	1.753E−30	8	1.282	1.391	42.65	7.478E−32	1.462E−31	0.4656	0.8906	−3.098	4.507	1.020E−32
12.000	117.0	9.403E−31	10	1.401	1.343	46.22	3.917E−32	7.842E−32	0.4492	0.8118	−3.742	4.435	5.471E−33
12.000	127.0	3.803E−31	16	1.520	1.289	49.69	1.545E−32	3.172E−32	0.4314	0.7379	−4.443	4.355	2.213E−33
12.000	137.0	9.941E−32	44	1.639	1.232	53.07	3.932E−33	8.291E−33	0.4121	0.6686	−5.201	4.268	5.784E−34
14.000	17.0	1.453E−28	3	0.238	1.919	7.07	5.204E−30	1.039E−29	0.6421	1.9659	0.095	4.445	7.245E−31
14.000	27.0	9.304E−29	3	0.378	1.908	11.21	3.314E−30	6.650E−30	0.6384	1.8397	−0.052	4.428	4.639E−31
14.000	37.0	4.234E−29	3	0.518	1.891	15.31	1.496E−30	3.026E−30	0.6329	1.6992	−0.267	4.404	2.111E−31
14.000	47.0	2.078E−29	4	0.658	1.870	19.38	7.270E−31	1.485E−30	0.6257	1.5597	−0.549	4.372	1.036E−31
14.000	57.0	1.026E−29	3	0.798	1.843	23.40	3.545E−31	7.333E−31	0.6169	1.4280	−0.899	4.332	5.116E−32
14.000	67.0	4.458E−30	6	0.937	1.812	27.35	1.518E−31	3.186E−31	0.6063	1.3060	−1.315	4.283	2.223E−32
14.000	77.0	2.213E−30	8	1.077	1.775	31.24	7.411E−32	1.582E−31	0.5940	1.1938	−1.799	4.227	1.103E−32
14.000	87.0	8.284E−31	13	1.216	1.733	35.06	2.723E−32	5.921E−32	0.5800	1.0907	−2.349	4.161	4.130E−33
14.000	97.0	2.545E−31	24	1.356	1.686	38.80	8.196E−33	1.819E−32	0.5643	0.9958	−2.966	4.086	1.269E−33
14.000	107.0	1.774E−31	29	1.495	1.634	42.45	5.587E−33	1.268E−32	0.5469	0.9080	−3.651	4.001	8.845E−34
14.000	117.0	8.010E−32	30	1.634	1.577	46.02	2.463E−33	5.725E−33	0.5278	0.8265	−4.402	3.906	3.994E−34
14.000	127.0	2.876E−32	50	1.773	1.515	49.49	8.622E−34	2.056E−33	0.5070	0.7507	−5.219	3.800	1.434E−34
16.000	17.0	2.724E−29	3	0.272	2.201	7.04	7.496E−31	1.703E−30	0.7366	2.0714	0.061	3.997	1.188E−31
16.000	27.0	1.494E−29	4	0.432	2.188	11.17	4.089E−31	9.342E−31	0.7323	1.9196	−0.108	3.975	6.517E−32
16.000	37.0	5.332E−30	5	0.592	2.170	15.26	1.448E−31	3.334E−31	0.7261	1.7583	−0.354	3.944	2.326E−32
16.000	47.0	2.435E−30	7	0.752	2.145	19.31	6.545E−32	1.523E−31	0.7179	1.6038	−0.677	3.903	1.062E−32
16.000	57.0	9.772E−31	11	0.912	2.115	23.31	2.594E−32	6.110E−32	0.7077	1.4615	−1.076	3.852	4.263E−33
16.000	67.0	2.903E−31	20	1.071	2.079	27.26	7.595E−33	1.815E−32	0.6956	1.3320	−1.552	3.789	1.266E−33
16.000	77.0	1.302E−31	30	1.231	2.037	31.14	3.350E−33	8.141E−33	0.6816	1.2145	−2.104	3.716	5.680E−34
16.000	87.0	5.435E−32	37	1.390	1.989	34.95	1.372E−33	3.398E−33	0.6656	1.1075	−2.733	3.630	2.371E−34
16.000	97.0	1.527E−32	70	1.550	1.935	38.68	3.777E−34	9.548E−34	0.6476	1.0096	−3.439	3.532	6.661E−35

Diddens/Schlüpmann

p_3	θ_3	$\dfrac{\mathrm{d}^2\sigma}{\mathrm{d}\Omega_3\,\mathrm{d}p_3}$	Error	p_T	p_L^*	θ_3^*	$\dfrac{1}{p_3^{*2}}\cdot\dfrac{\mathrm{d}^2\sigma}{\mathrm{d}\Omega_3^*\,\mathrm{d}p_3^*}$	$\dfrac{E_3}{p_3^2}\cdot\dfrac{\mathrm{d}^2\sigma}{\mathrm{d}\Omega_3\,\mathrm{d}p_3}$	x^*	y_3^*	t	m_4	$\dfrac{\mathrm{d}^2\sigma}{\mathrm{d}t\,\mathrm{d}m_4^2}$
GeV/c	mrad	cm² sr⁻¹ GeV⁻¹ c	%	GeV/c	GeV/c	deg	cm² sr⁻¹ GeV⁻³ c³	cm² sr⁻¹ GeV⁻² c³			GeV²	GeV	cm²/GeV⁴

$p_1 = 24.000$ GeV/c; $p_1^* = 3.290$ GeV/c; $E_1^* + E_2^* = 6.843$ GeV; $s = 46.828$ GeV²; $\gamma_{\mathrm{c.m.}} = 3.6469$; $\beta_{\mathrm{c.m.}} = 0.96167$; $y_1^* = 1.9677$; $p_{3,\max}^* = 2.988$ GeV/c; [72 A 1] (cont.)

p_3	θ_3	$\dfrac{\mathrm{d}^2\sigma}{\mathrm{d}\Omega_3\,\mathrm{d}p_3}$	Error	p_T	p_L^*	θ_3^*	$\dfrac{1}{p_3^{*2}}\cdot\dfrac{\mathrm{d}^2\sigma}{\mathrm{d}\Omega_3^*\,\mathrm{d}p_3^*}$	$\dfrac{E_3}{p_3^2}\cdot\dfrac{\mathrm{d}^2\sigma}{\mathrm{d}\Omega_3\,\mathrm{d}p_3}$	x^*	y_3^*	t	m_4	$\dfrac{\mathrm{d}^2\sigma}{\mathrm{d}t\,\mathrm{d}m_4^2}$
18.000	17.0	2.097E−30	9	0.306	2.483	7.03	4.571E−32	1.165E−31	0.8308	2.1591	0.014	3.489	8.130E−33
18.000	27.0	7.457E−31	11	0.485	2.468	11.14	1.619E−32	4.150E−32	0.8260	1.9828	−0.176	3.462	2.895E−33
18.000	37.0	3.204E−31	14	0.666	2.447	15.22	6.891E−33	1.781E−32	0.8189	1.8032	−0.453	3.422	1.242E−33
18.000	47.0	1.164E−31	21	0.846	2.420	19.26	2.478E−33	6.469E−33	0.8097	1.6364	−0.815	3.368	4.513E−34
18.000	57.0	4.026E−32	57	1.025	2.386	23.26	8.465E−34	2.238E−33	0.7983	1.4858	−1.264	3.301	1.561E−34
18.000	67.0	1.982E−32	99	1.205	2.345	27.20	4.107E−34	1.102E−33	0.7847	1.3507	−1.800	3.219	7.684E−35
18.000	77.0	9.517E−33	70	1.385	2.298	31.07	1.939E−34	5.289E−34	0.7689	1.2292	−2.421	3.121	3.690E−35
20.000	17.0	7.627E−32	44	0.340	2.764	7.01	1.349E−33	3.815E−33	0.9248	2.2329	−0.041	2.893	2.661E−34
16.000	22.0	1.801E−29	3	0.352	2.196	9.11	4.944E−31	1.126E−30	0.7347	1.9984	−0.014	3.987	7.856E−32
20.000	22.0	1.800E−32	19	0.440	2.757	9.07	3.176E−34	9.003E−34	0.9224	2.1347	−0.134	2.877	6.280E−35
20.500	22.0	3.400E−33	37	0.451	2.827	9.06	5.712E−35	1.659E−34	0.9458	2.1483	−0.151	2.705	1.157E−35
21.000	22.0	4.520E−34	99	0.462	2.896	9.06	7.238E−36	2.153E−35	0.9692	2.1613	−0.169	2.523	1.502E−36
21.500	22.0	8.402E−35	70	0.473	2.966	9.06	1.284E−36	3.909E−36	0.9926	2.1737	−0.186	2.326	2.727E−37

3.6.4 pp→K⁺X

$p_1 = 3.203$ GeV/c; $p_1^* = 1.061$ GeV/c; $E_1^* + E_2^* = 2.833$ GeV; $s = 8.023$ GeV²; $\gamma_{\mathrm{c.m.}} = 1.5095$; $\beta_{\mathrm{c.m.}} = 0.74910$; $y_1^* = 0.9709$; $p_{3,\max}^* = 0.516$ GeV/c; [68 R 1]

p_3	θ_3	$\dfrac{\mathrm{d}^2\sigma}{\mathrm{d}\Omega_3\,\mathrm{d}p_3}$	Error	p_T	p_L^*	θ_3^*	$\dfrac{1}{p_3^{*2}}\cdot\dfrac{\mathrm{d}^2\sigma}{\mathrm{d}\Omega_3^*\,\mathrm{d}p_3^*}$	$\dfrac{E_3}{p_3^2}\cdot\dfrac{\mathrm{d}^2\sigma}{\mathrm{d}\Omega_3\,\mathrm{d}p_3}$	x^*	y_3^*	t	m_4	$\dfrac{\mathrm{d}^2\sigma}{\mathrm{d}t\,\mathrm{d}m_4^2}$
0.800	0.0	7.100E−29	40	0.000	0.145	0.00	2.027E−28	1.043E−28	0.2798	0.2887	−0.027	2.313	5.452E−29
0.900	0.0	1.020E−28	30	0.000	0.198	0.00	2.430E−28	1.293E−28	0.3829	0.3905	0.037	2.292	6.757E−29
1.050	0.0	1.230E−28	20	0.000	0.273	0.00	2.294E−28	1.295E−28	0.5284	0.5279	0.105	2.252	6.767E−29
1.100	0.0	1.380E−28	15	0.000	0.297	0.00	2.386E−28	1.375E−28	0.5751	0.5701	0.122	2.237	7.188E−29
1.200	0.0	1.440E−28	10	0.000	0.344	0.00	2.156E−28	1.298E−28	0.6662	0.6501	0.149	2.204	6.783E−29
1.260	0.0	1.520E−28	10	0.000	0.372	0.00	2.096E−28	1.296E−28	0.7197	0.6953	0.162	2.183	6.773E−29
1.280	0.0	1.440E−28	10	0.000	0.381	0.00	1.934E−28	1.206E−28	0.7373	0.7100	0.166	2.176	6.303E−29
1.300	0.0	1.530E−28	10	0.000	0.390	0.00	2.001E−28	1.259E−28	0.7549	0.7245	0.169	2.169	6.581E−29
1.350	0.0	1.510E−28	10	0.000	0.412	0.00	1.851E−28	1.191E−28	0.7984	0.7599	0.177	2.150	6.226E−29
1.380	0.0	1.220E−28	10	0.000	0.426	0.00	1.440E−28	9.390E−29	0.8243	0.7805	0.181	2.139	4.908E−29
1.400	0.0	1.150E−28	10	0.000	0.435	0.00	1.324E−28	8.710E−29	0.8415	0.7941	0.183	2.131	4.553E−29
1.450	0.0	8.200E−29	12	0.000	0.457	0.00	8.882E−29	5.974E−29	0.8842	0.8272	0.188	2.111	3.123E−29
1.480	0.0	7.900E−29	12	0.000	0.470	0.00	8.256E−29	5.627E−29	0.9097	0.8466	0.190	2.099	2.941E−29
1.500	0.0	8.000E−29	12	0.000	0.479	0.00	8.165E−29	5.615E−29	0.9266	0.8594	0.192	2.091	2.935E−29
1.520	0.0	8.200E−29	12	0.000	0.487	0.00	8.177E−29	5.672E−29	0.9434	0.8720	0.193	2.083	2.965E−29
1.530	0.0	9.600E−29	12	0.000	0.492	0.00	9.462E−29	6.593E−29	0.9518	0.8782	0.194	2.078	3.446E−29
1.540	0.0	9.000E−29	12	0.000	0.496	0.00	8.769E−29	6.137E−29	0.9602	0.8844	0.194	2.074	3.208E−29
1.560	0.0	1.030E−28	12	0.000	0.505	0.00	9.810E−29	6.925E−29	0.9769	0.8967	0.195	2.066	3.620E−29
1.580	0.0	7.700E−29	12	0.000	0.513	0.00	7.169E−29	5.106E−29	0.9936	0.9089	0.196	2.057	2.669E−29
1.600	0.0	3.800E−29	20	0.000	0.522	0.00	3.460E−29	2.486E−29	1.0102	0.9209	0.196	2.049	1.299E−29
0.640	296.7	5.300E−29	30	0.187	0.010	87.00	1.980E−28	1.046E−28	0.0190	0.0186	−0.351	2.297	5.467E−29
0.750	296.7	8.200E−29	25	0.219	0.067	72.95	2.404E−28	1.309E−28	0.1302	0.1242	−0.275	2.277	6.843E−29

p_3	θ_3	$\dfrac{\mathrm{d}^2\sigma}{\mathrm{d}\Omega_3\,\mathrm{d}p_3}$	Error	p_T	p_L^*	θ_3^*	$\dfrac{1}{p_3^{*2}}\cdot\dfrac{\mathrm{d}^2\sigma}{\mathrm{d}\Omega_3^*\,\mathrm{d}p_3^*}$	$\dfrac{E_3}{p_3^2}\cdot\dfrac{\mathrm{d}^2\sigma}{\mathrm{d}\Omega_3\,\mathrm{d}p_3}$	x^*	y_3^*	t	m_4	$\dfrac{\mathrm{d}^2\sigma}{\mathrm{d}t\,\mathrm{d}m_4^2}$
GeV/c	mrad	cm² sr⁻¹ GeV⁻¹ c	%	GeV/c	GeV/c	deg	cm² sr⁻¹ GeV⁻³ c³	cm² sr⁻¹ GeV⁻² c³			GeV²	GeV	cm²/GeV⁴
$p_1 = 3.203$ GeV/c; $p_1^* = 1.061$ GeV/c; $E_1^* + E_2^* = 2.833$ GeV; $s = 8.023$ GeV²; $\gamma_\mathrm{c.m.} = 1.5095$; $\beta_\mathrm{c.m.} = 0.74910$; $y_1^* = 0.9709$; $p_{3,\max}^* = 0.516$ GeV/c; [68 R 1] (cont.)													
0.840	296.7	8.400E-29	20	0.246	0.111	65.72	2.062E-28	1.160E-28	0.2145	0.1995	-0.234	2.254	6.063E-29
0.930	296.7	8.600E-29	15	0.272	0.152	60.82	1.793E-28	1.047E-28	0.2940	0.2662	-0.207	2.227	5.473E-29
0.980	296.7	9.800E-29	10	0.287	0.174	58.76	1.876E-28	1.120E-28	0.3365	0.2999	-0.198	2.210	5.853E-29
0.990	296.7	1.040E-28	10	0.289	0.178	58.39	1.958E-28	1.174E-28	0.3449	0.3064	-0.196	2.207	6.136E-29
1.100	296.7	9.600E-29	10	0.322	0.224	55.09	1.517E-28	9.566E-29	0.4346	0.3723	-0.186	2.167	5.000E-29
1.140	296.7	9.000E-29	10	0.333	0.241	54.15	1.339E-28	8.604E-29	0.4663	0.3940	-0.185	2.151	4.497E-29
1.200	296.7	7.600E-29	10	0.351	0.265	52.94	1.036E-28	6.849E-29	0.5130	0.4245	-0.186	2.126	3.580E-29
1.220	296.7	7.200E-29	10	0.357	0.273	52.58	9.539E-29	6.367E-29	0.5283	0.4342	-0.188	2.118	3.328E-29
1.250	296.7	4.400E-29	15	0.365	0.285	52.08	5.590E-29	3.785E-29	0.5512	0.4482	-0.190	2.105	1.978E-29
1.280	296.7	3.200E-29	15	0.374	0.296	51.62	3.901E-29	2.680E-29	0.5739	0.4618	-0.192	2.092	1.401E-29
1.320	296.7	3.100E-29	15	0.386	0.312	51.06	3.582E-29	2.507E-29	0.6038	0.4791	-0.197	2.074	1.311E-29
1.350	296.7	2.700E-29	15	0.395	0.323	50.68	2.999E-29	2.130E-29	0.6260	0.4915	-0.201	2.060	1.113E-29
1.360	296.7	2.400E-29	15	0.398	0.327	50.55	2.632E-29	1.877E-29	0.6334	0.4955	-0.203	2.056	9.814E-30
1.370	296.7	2.400E-29	20	0.401	0.331	50.43	2.598E-29	1.862E-29	0.6408	0.4995	-0.204	2.051	9.734E-30
1.400	296.7	1.300E-29	25	0.409	0.342	50.09	1.354E-29	9.846E-30	0.6628	0.5111	-0.209	2.037	5.147E-30
1.420	296.7	4.000E-30	50	0.415	0.350	49.88	4.064E-30	2.982E-30	0.6773	0.5187	-0.212	2.027	1.559E-30
$p_1 = 3.670$ GeV/c; $p_1^* = 1.156$ GeV/c; $E_1^* + E_2^* = 2.978$ GeV; $s = 8.868$ GeV²; $\gamma_\mathrm{c.m.} = 1.5871$; $\beta_\mathrm{c.m.} = 0.77652$; $y_1^* = 1.0365$; $p_{3,\max}^* = 0.657$ GeV/c; [68 R 1]													
0.800	0.0	1.600E-28	20	0.000	0.111	0.00	4.644E-28	2.350E-28	0.1691	0.2230	-0.126	2.469	1.072E-28
0.950	0.0	1.660E-28	20	0.000	0.188	0.00	3.727E-28	1.969E-28	0.2867	0.3725	-0.014	2.442	8.984E-29
1.050	0.0	1.700E-28	15	0.000	0.236	0.00	3.268E-28	1.789E-28	0.3601	0.4622	0.040	2.419	8.162E-29
1.100	0.0	1.910E-28	15	0.000	0.260	0.00	3.411E-28	1.903E-28	0.3957	0.5045	0.063	2.406	8.683E-29
1.200	0.0	1.860E-28	10	0.000	0.305	0.00	2.887E-28	1.676E-28	0.4650	0.5844	0.101	2.378	7.646E-29
1.300	0.0	1.900E-28	10	0.000	0.349	0.00	2.585E-28	1.563E-28	0.5321	0.6589	0.131	2.347	7.132E-29
1.400	0.0	1.890E-28	8	0.000	0.392	0.00	2.270E-28	1.432E-28	0.5976	0.7284	0.153	2.314	6.531E-29
1.450	0.0	1.880E-28	7	0.000	0.414	0.00	2.127E-28	1.370E-28	0.6298	0.7616	0.162	2.297	6.248E-29
1.500	0.0	1.860E-28	7	0.000	0.434	0.00	1.985E-28	1.305E-28	0.6616	0.7937	0.170	2.279	5.956E-29
1.530	0.0	1.960E-28	7	0.000	0.447	0.00	2.021E-28	1.346E-28	0.6806	0.8126	0.174	2.268	6.141E-29
1.550	0.0	1.890E-28	7	0.000	0.455	0.00	1.906E-28	1.280E-28	0.6932	0.8249	0.177	2.261	5.838E-29
1.570	0.0	1.760E-28	7	0.000	0.463	0.00	1.735E-28	1.175E-28	0.7057	0.8372	0.179	2.254	5.361E-29
1.590	0.0	1.770E-28	7	0.000	0.472	0.00	1.707E-28	1.166E-28	0.7183	0.8492	0.181	2.246	5.318E-29
1.610	0.0	1.820E-28	7	0.000	0.480	0.00	1.717E-28	1.182E-28	0.7307	0.8612	0.183	2.239	5.394E-29
1.630	0.0	1.680E-28	7	0.000	0.488	0.00	1.551E-28	1.077E-28	0.7432	0.8730	0.185	2.231	4.913E-29
1.650	0.0	1.700E-28	7	0.000	0.496	0.00	1.536E-28	1.075E-28	0.7556	0.8847	0.187	2.223	4.906E-29
1.670	0.0	1.750E-28	7	0.000	0.504	0.00	1.548E-28	1.093E-28	0.7679	0.8962	0.188	2.216	4.985E-29
1.690	0.0	1.640E-28	7	0.000	0.512	0.00	1.421E-28	1.011E-28	0.7803	0.9077	0.190	2.208	4.612E-29
1.710	0.0	1.470E-28	7	0.000	0.520	0.00	1.247E-28	8.948E-29	0.7925	0.9189	0.191	2.200	4.082E-29
1.730	0.0	1.560E-28	7	0.000	0.528	0.00	1.297E-28	9.377E-29	0.8048	0.9301	0.192	2.192	4.278E-29
1.750	0.0	1.500E-28	7	0.000	0.536	0.00	1.221E-28	8.906E-29	0.8170	0.9412	0.193	2.184	4.063E-29
1.770	0.0	1.500E-28	7	0.000	0.544	0.00	1.197E-28	8.798E-29	0.8292	0.9521	0.194	2.176	4.014E-29
1.790	0.0	1.340E-28	7	0.000	0.552	0.00	1.048E-28	7.766E-29	0.8414	0.9630	0.195	2.168	3.543E-29
1.810	0.0	1.310E-28	8	0.000	0.560	0.00	1.004E-28	7.502E-29	0.8535	0.9737	0.196	2.159	3.422E-29

$p_1 = 3.670$ GeV/c; $p_1^* = 1.156$ GeV/c; $E_1^* + E_2^* = 2.978$ GeV; $s = 8.868$ GeV²; $\gamma_{\text{c.m.}} = 1.5871$; $\beta_{\text{c.m.}} = 0.77652$; $y_1^* = 1.0365$; $p^*_{3,\max} = 0.657$ GeV/c; [68 R 1] (cont.)

p_3 (GeV/c)	θ_3 (mrad)	Error (%)	$\dfrac{d^2\sigma}{d\Omega_3\,dp_3}$ (cm² sr⁻¹ GeV⁻¹ c)	p_T (GeV/c)	p_L^* (GeV/c)	θ_3^* (deg)	$\dfrac{1}{p_3^{*2}}\dfrac{d^2\sigma}{d\Omega_3^*\,dp_3^*}$ (cm² sr⁻¹ GeV⁻³ c³)	$\dfrac{E_3}{p_3^{*2}}\dfrac{d^2\sigma}{d\Omega_3\,dp_3}$ (cm² sr⁻¹ GeV⁻² c³)	x^*	y_3^*	t (GeV²)	m_4 (GeV)	$\dfrac{d^2\sigma}{dt\,dm_4^2}$ (cm²/GeV⁴)
1.830	0.0	8	1.270E-28	0.000	0.568	0.00	9.547E-29	7.188E-29	0.8657	0.9843	0.196	2.151	3.279E-29
1.850	0.0	8	1.210E-28	0.000	0.576	0.00	8.920E-29	6.770E-29	0.8777	0.9948	0.197	2.143	3.088E-29
1.900	0.0	10	8.600E-29	0.000	0.596	0.00	6.042E-29	4.677E-29	0.9078	1.0206	0.197	2.122	2.134E-29
1.920	0.0	10	7.700E-29	0.000	0.604	0.00	5.308E-29	4.141E-29	0.9198	1.0307	0.197	2.113	1.889E-29
1.940	0.0	10	7.100E-29	0.000	0.612	0.00	4.803E-29	3.776E-29	0.9318	1.0407	0.197	2.105	1.723E-29
1.950	0.0	10	7.500E-29	0.000	0.616	0.00	5.027E-29	3.968E-29	0.9378	1.0457	0.197	2.100	1.810E-29
1.960	0.0	10	7.400E-29	0.000	0.620	0.00	4.914E-29	3.893E-29	0.9438	1.0507	0.197	2.096	1.776E-29
1.980	0.0	10	7.300E-29	0.000	0.628	0.00	4.759E-29	3.800E-29	0.9557	1.0605	0.197	2.087	1.733E-29
2.000	0.0	10	6.700E-29	0.000	0.635	0.00	4.288E-29	3.451E-29	0.9676	1.0703	0.197	2.078	1.574E-29
2.020	0.0	10	7.400E-29	0.000	0.643	0.00	4.651E-29	3.771E-29	0.9795	1.0799	0.197	2.069	1.720E-29
2.040	0.0	10	8.200E-29	0.000	0.651	0.00	5.062E-29	4.136E-29	0.9914	1.0895	0.196	2.061	1.887E-29
2.050	0.0	10	9.200E-29	0.000	0.655	0.00	5.628E-29	4.616E-29	0.9973	1.0943	0.196	2.056	2.106E-29
2.060	0.0	10	9.100E-29	0.000	0.659	0.00	5.518E-29	4.543E-29	1.0032	1.0990	0.196	2.052	2.072E-29
2.080	0.0	15	6.700E-29	0.000	0.666	0.00	3.991E-29	3.311E-29	1.0150	1.1084	0.195	2.042	1.510E-29
2.100	0.0	15	4.400E-29	0.000	0.674	0.00	2.575E-29	2.152E-29	1.0269	1.1177	0.194	2.033	9.819E-30
2.120	0.0	15	2.500E-29	0.000	0.682	0.00	1.438E-29	1.211E-29	1.0386	1.1269	0.194	2.024	5.524E-30
2.140	0.0	20	1.900E-29	0.000	0.690	0.00	1.074E-29	9.112E-30	1.0504	1.1361	0.193	2.015	4.157E-30
2.160	0.0	20	1.300E-29	0.000	0.697	0.00	7.225E-30	6.174E-30	1.0622	1.1452	0.192	2.006	2.817E-30
2.180	0.0	20	1.200E-29	0.000	0.705	0.00	6.556E-30	5.644E-30	1.0739	1.1541	0.191	1.996	2.575E-30
2.200	0.0	25	8.000E-30	0.000	0.713	0.00	4.298E-30	3.727E-30	1.0856	1.1631	0.190	1.987	1.700E-30
0.716	264.5	40	1.105E-28	0.187	0.025	82.41	3.546E-28	1.875E-28	0.0380	0.0472	-0.393	2.442	8.553E-29
0.900	296.7	10	1.380E-28	0.263	0.101	69.03	3.076E-28	1.749E-28	0.1536	0.1792	-0.336	2.393	7.979E-29
0.990	296.7	10	1.420E-28	0.289	0.139	64.33	2.721E-28	1.603E-28	0.2119	0.2407	-0.308	2.367	7.312E-29
1.100	296.7	10	1.380E-28	0.322	0.184	60.29	2.228E-28	1.375E-28	0.2795	0.3066	-0.290	2.332	6.274E-29
1.140	296.7	8	1.300E-28	0.333	0.199	59.14	1.978E-28	1.243E-28	0.3033	0.3283	-0.286	2.318	5.669E-29
1.220	296.7	8	1.160E-28	0.357	0.230	57.23	1.576E-28	1.026E-28	0.3497	0.3686	-0.284	2.288	4.680E-29
1.260	296.7	8	1.060E-28	0.368	0.245	56.42	1.363E-28	9.036E-29	0.3724	0.3872	-0.284	2.273	4.122E-29
1.320	296.7	8	9.800E-29	0.386	0.267	55.37	1.164E-28	7.927E-29	0.4059	0.4134	-0.288	2.249	3.616E-29
1.370	296.7	10	9.000E-29	0.401	0.285	54.61	1.002E-28	6.983E-29	0.4334	0.4338	-0.292	2.228	3.186E-29
1.420	296.7	10	8.300E-29	0.415	0.302	53.93	8.686E-29	6.188E-29	0.4605	0.4531	-0.298	2.206	2.823E-29
1.460	296.7	10	7.400E-29	0.427	0.316	53.45	7.376E-29	5.351E-29	0.4820	0.4676	-0.304	2.189	2.441E-29
1.520	296.7	15	6.100E-29	0.444	0.337	52.80	5.663E-29	4.220E-29	0.5138	0.4882	-0.315	2.162	1.925E-29
1.560	296.7	15	4.600E-29	0.456	0.351	52.41	4.078E-29	3.093E-29	0.5347	0.5011	-0.322	2.144	1.411E-29
1.590	296.7	15	3.430E-29	0.465	0.361	52.14	2.914E-29	2.239E-29	0.5503	0.5104	-0.329	2.130	1.021E-29
1.620	296.7	20	2.500E-29	0.474	0.372	51.89	2.072E-29	1.613E-29	0.5659	0.5194	-0.335	2.115	7.360E-30
1.640	296.7	20	2.200E-29	0.479	0.378	51.73	1.784E-29	1.401E-29	0.5762	0.5252	-0.340	2.106	6.391E-30
1.650	296.7	20	1.800E-29	0.482	0.382	51.65	1.444E-29	1.139E-29	0.5813	0.5280	-0.342	2.101	5.195E-30
1.680	296.7	20	1.700E-29	0.491	0.392	51.42	1.320E-29	1.055E-29	0.5967	0.5364	-0.350	2.086	4.812E-30
1.700	296.7	20	1.600E-29	0.497	0.398	51.28	1.216E-29	9.801E-30	0.6069	0.5418	-0.355	2.076	4.471E-30
1.710	296.7	20	1.400E-29	0.500	0.402	51.21	1.053E-29	8.522E-30	0.6120	0.5445	-0.357	2.071	3.888E-30
1.720	296.7	20	1.100E-29	0.503	0.405	51.14	8.185E-30	6.654E-30	0.6170	0.5471	-0.360	2.066	3.035E-30
1.740	296.7	25	7.000E-30	0.509	0.412	51.01	5.101E-30	4.182E-30	0.6272	0.5523	-0.365	2.056	1.908E-30
1.760	296.7	50	4.000E-30	0.515	0.418	50.88	2.855E-30	2.360E-30	0.6373	0.5574	-0.371	2.046	1.077E-30

p_3	θ_3	$\dfrac{d^2\sigma}{d\Omega_3\,dp_3}$	Error	p_T	p_L^*	θ_3^*	$\dfrac{1}{p_3^{*2}}\cdot\dfrac{d^2\sigma}{d\Omega_3^*\,dp_3^*}$	$\dfrac{E_3}{p_3^2}\cdot\dfrac{d^2\sigma}{d\Omega_3\,dp_3}$	x^*	y_3^*	t	m_4	$\dfrac{d^2\sigma}{dt\,dm_4^2}$
GeV/c	mrad	cm² sr⁻¹ GeV⁻¹ c	%	GeV/c	GeV/c	deg	cm² sr⁻¹ GeV⁻³ c³	cm² sr⁻¹ GeV⁻² c³			GeV²	GeV	cm²/GeV⁴

$p_1 = 3.670$ GeV/c; $p_1^* = 1.156$ GeV/c; $E_1^* + E_2^* = 2.978$ GeV; $s = 8.868$ GeV²; $\gamma_{c.m.} = 1.5871$; $\beta_{c.m.} = 0.77652$; $y_1^* = 1.0365$; $p_{3,\,max}^* = 0.657$ GeV/c; [68 R 1] (cont.)

p_3	θ_3	$\dfrac{d^2\sigma}{d\Omega_3\,dp_3}$	Error	p_T	p_L^*	θ_3^*	$\dfrac{1}{p_3^{*2}}\cdot\dfrac{d^2\sigma}{d\Omega_3^*\,dp_3^*}$	$\dfrac{E_3}{p_3^2}\cdot\dfrac{d^2\sigma}{d\Omega_3\,dp_3}$	x^*	y_3^*	t	m_4	$\dfrac{d^2\sigma}{dt\,dm_4^2}$
1.070	250.2	7.375E-29	50	0.265	0.193	53.93	1.281E-28	7.591E-29	0.2939	0.3379	-0.195	2.363	3.463E-29
1.038	306.1	9.733E-29	10	0.313	0.154	63.76	1.718E-28	1.038E-28	0.2349	0.2608	-0.320	2.348	4.737E-29
1.024	392.9	6.445E-29	8	0.392	0.100	75.65	1.094E-28	6.988E-29	0.1527	0.1584	-0.545	2.304	3.188E-29
1.049	472.1	3.848E-29	10	0.477	0.054	83.56	5.887E-29	4.054E-29	0.0821	0.0784	-0.802	2.238	1.850E-29
1.078	514.0	2.077E-29	15	0.530	0.029	86.92	2.923E-29	2.119E-29	0.0434	0.0394	-0.969	2.190	9.668E-30
1.115	550.3	1.320E-29	20	0.583	0.005	89.46	1.694E-29	1.295E-29	0.0084	0.0072	-1.139	2.136	5.907E-30
1.150	558.5	1.000E-29	25	0.609	0.005	89.49	1.206E-29	9.463E-30	0.0083	0.0069	-1.199	2.107	4.317E-30
1.200	558.5	3.000E-30	50	0.636	0.016	88.57	3.357E-30	2.703E-30	0.0242	0.0198	-1.237	2.077	1.233E-30

$p_1 = 3.350$ GeV/c; $p_1^* = 1.092$ GeV/c; $E_1^* + E_2^* = 2.879$ GeV; $s = 8.288$ GeV²; $\gamma_{c.m.} = 1.5343$; $\beta_{c.m.} = 0.75842$; $y_1^* = 0.9925$; $p_{3,\,max}^* = 0.563$ GeV/c; [68 H 1]

p_3	θ_3	$\dfrac{d^2\sigma}{d\Omega_3\,dp_3}$	Error	p_T	p_L^*	θ_3^*	$\dfrac{1}{p_3^{*2}}\cdot\dfrac{d^2\sigma}{d\Omega_3^*\,dp_3^*}$	$\dfrac{E_3}{p_3^2}\cdot\dfrac{d^2\sigma}{d\Omega_3\,dp_3}$	x^*	y_3^*	t	m_4	$\dfrac{d^2\sigma}{dt\,dm_4^2}$
0.714	291.4	4.957E-29	3	0.205	0.039	79.20	1.574E-28	8.441E-29	0.0695	0.0731	-0.334	2.333	4.219E-29
0.687	329.6	5.042E-29	3	0.222	0.013	86.70	1.668E-28	9.038E-29	0.0228	0.0237	-0.407	2.327	4.517E-29
0.700	349.0	5.310E-29	1	0.239	0.012	87.03	1.691E-28	9.283E-29	0.0221	0.0227	-0.429	2.318	4.640E-29
0.750	349.0	5.430E-29	1	0.256	0.036	81.91	1.555E-28	8.668E-29	0.0647	0.0655	-0.402	2.307	4.332E-29
0.800	349.0	5.740E-29	1	0.274	0.059	77.73	1.485E-28	8.432E-29	0.1056	0.1052	-0.380	2.294	4.214E-29
0.850	349.0	6.340E-29	1	0.291	0.082	74.31	1.490E-28	8.626E-29	0.1449	0.1420	-0.364	2.280	4.311E-29
0.900	349.0	6.470E-29	1	0.308	0.103	71.48	1.388E-28	8.200E-29	0.1830	0.1762	-0.352	2.265	4.098E-29
0.950	349.0	6.810E-29	1	0.325	0.124	69.13	1.338E-28	8.079E-29	0.2198	0.2080	-0.344	2.248	4.038E-29
0.975	349.0	6.970E-29	2	0.333	0.134	68.10	1.312E-28	8.013E-29	0.2379	0.2230	-0.342	2.240	4.005E-29
1.000	349.0	7.010E-29	1	0.342	0.144	67.16	1.266E-28	7.818E-29	0.2557	0.2375	-0.340	2.231	3.907E-29
1.025	349.0	6.860E-29	1	0.351	0.154	66.29	1.189E-28	7.429E-29	0.2733	0.2515	-0.339	2.221	3.713E-29
1.050	349.0	6.780E-29	1	0.359	0.164	65.49	1.129E-28	7.136E-29	0.2906	0.2650	-0.338	2.212	3.566E-29
1.100	349.0	6.340E-29	1	0.376	0.183	64.07	9.763E-29	6.318E-29	0.3247	0.2906	-0.340	2.192	3.157E-29
1.150	349.0	5.590E-29	1	0.393	0.202	62.84	7.982E-29	5.290E-29	0.3581	0.3144	-0.343	2.172	2.644E-29
1.200	349.0	4.910E-29	1	0.410	0.220	61.78	6.519E-29	4.425E-29	0.3909	0.3366	-0.349	2.150	2.211E-29
1.250	349.0	3.160E-29	5	0.427	0.238	60.86	3.910E-29	2.718E-29	0.4231	0.3573	-0.357	2.128	1.358E-29
1.300	349.0	1.900E-29	4	0.445	0.256	60.05	2.196E-29	1.563E-29	0.4548	0.3766	-0.367	2.105	7.814E-30
1.005	200.5	8.598E-29	4	0.200	0.208	43.89	1.666E-28	9.532E-29	0.3694	0.3812	-0.068	2.289	4.764E-29
0.967	234.9	7.788E-29	3	0.225	0.179	51.43	1.582E-28	9.043E-29	0.3186	0.3249	-0.130	2.289	4.520E-29
0.938	270.0	7.512E-29	2	0.250	0.154	58.47	1.575E-28	9.050E-29	0.2726	0.2739	-0.195	2.286	4.523E-29
0.924	291.1	7.354E-29	2	0.265	0.139	62.35	1.563E-28	9.024E-29	0.2467	0.2454	-0.235	2.282	4.510E-29
0.916	305.2	7.484E-29	2	0.275	0.130	64.79	1.600E-28	9.282E-29	0.2300	0.2272	-0.263	2.279	4.639E-29
0.906	326.2	6.985E-29	2	0.290	0.116	68.20	1.502E-28	8.780E-29	0.2061	0.2013	-0.305	2.273	4.388E-29
0.900	340.1	6.812E-29	2	0.300	0.107	70.35	1.469E-28	8.633E-29	0.1903	0.1845	-0.334	2.269	4.315E-29
0.894	357.3	6.798E-29	1	0.313	0.097	72.83	1.466E-28	8.687E-29	0.1715	0.1645	-0.370	2.263	4.342E-29
0.891	367.6	6.661E-29	2	0.320	0.090	74.24	1.435E-28	8.547E-29	0.1604	0.1529	-0.393	2.259	4.272E-29
0.889	374.3	6.626E-29	1	0.325	0.086	75.15	1.427E-28	8.526E-29	0.1530	0.1453	-0.408	2.257	4.261E-29
0.888	381.0	6.520E-29	2	0.330	0.082	75.98	1.401E-28	8.401E-29	0.1463	0.1383	-0.423	2.254	4.199E-29
0.887	387.7	6.104E-29	2	0.335	0.079	76.81	1.308E-28	7.876E-29	0.1395	0.1313	-0.438	2.251	3.936E-29
0.886	394.4	6.086E-29	1	0.340	0.075	77.62	1.301E-28	7.864E-29	0.1327	0.1243	-0.453	2.248	3.930E-29
0.885	400.9	6.043E-29	1	0.345	0.071	78.40	1.289E-28	7.819E-29	0.1259	0.1174	-0.468	2.245	3.908E-29

p_3	θ_3	$\dfrac{d^2\sigma}{d\Omega_3\,dp_3}$	Error	p_T	p_L^*	θ_3^*	$\dfrac{1}{p_3^{*2}}\cdot\dfrac{d^2\sigma}{d\Omega_3^*\,dp_3^*}$	$\dfrac{E_3}{p_3^2}\cdot\dfrac{d^2\sigma}{d\Omega_3\,dp_3}$	x^*	y_3^*	t	m_4	$\dfrac{d^2\sigma}{dt\,dm_4^2}$
GeV/c	mrad	cm² sr⁻¹ GeV⁻¹ c	%	GeV/c	GeV/c	deg	cm² sr⁻¹ GeV⁻³ c³	cm² sr⁻¹ GeV⁻² c³			GeV²	GeV	cm²/GeV⁴

$p_1 = 3.350$ GeV/c; $p_1^* = 1.092$ GeV/c; $E_1^* + E_2^* = 2.879$ GeV; $s = 8.288$ GeV²; $\gamma_{c.m.} = 1.5343$; $\beta_{c.m.} = 0.75842$; $y_1^* = 0.9925$; $p_{3,max}^* = 0.563$ GeV/c; [68 H 1] (cont.)

p_3	θ_3	$\dfrac{d^2\sigma}{d\Omega_3\,dp_3}$	Error	p_T	p_L^*	θ_3^*	$\dfrac{1}{p_3^{*2}}\cdot\dfrac{d^2\sigma}{d\Omega_3^*\,dp_3^*}$	$\dfrac{E_3}{p_3^2}\cdot\dfrac{d^2\sigma}{d\Omega_3\,dp_3}$	x^*	y_3^*	t	m_4	$\dfrac{d^2\sigma}{dt\,dm_4^2}$
0.884	407.5	6.191E-29	1	0.350	0.067	79.17	1.317E-28	8.022E-29	0.1189	0.1104	-0.483	2.242	4.009E-29
0.883	420.4	5.916E-29	1	0.360	0.060	80.61	1.250E-28	7.676E-29	0.1058	0.0973	-0.514	2.235	3.837E-29
0.882	426.8	5.227E-29	2	0.365	0.056	81.34	1.312E-28	8.091E-29	0.0988	0.0905	-0.530	2.232	4.044E-29
0.882	433.1	6.078E-29	2	0.370	0.052	81.99	1.275E-28	7.898E-29	0.0924	0.0843	-0.545	2.228	3.947E-29
0.882	439.4	5.899E-29	1	0.375	0.048	82.64	1.232E-28	7.665E-29	0.0860	0.0781	-0.561	2.225	3.831E-29
0.882	445.6	5.763E-29	2	0.380	0.045	83.27	1.199E-28	7.488E-29	0.0797	0.0720	-0.577	2.221	3.743E-29
0.883	451.7	5.669E-29	2	0.385	0.042	83.83	1.172E-28	7.356E-29	0.0740	0.0665	-0.592	2.218	3.676E-29
0.883	457.8	5.274E-29	1	0.390	0.038	84.43	1.085E-28	6.843E-29	0.0675	0.0604	-0.608	2.214	3.420E-29
0.884	469.8	5.228E-29	2	0.400	0.031	85.55	1.065E-28	6.774E-29	0.0552	0.0489	-0.640	2.206	3.386E-29
0.887	487.3	4.889E-29	3	0.415	0.021	87.08	9.772E-29	6.308E-29	0.0376	0.0329	-0.688	2.194	3.153E-29
0.889	498.6	4.623E-29	2	0.425	0.015	88.04	9.127E-29	5.949E-29	0.0259	0.0224	-0.720	2.186	2.973E-29
0.892	509.7	4.220E-29	3	0.435	0.008	88.92	8.215E-29	5.407E-29	0.0146	0.0125	-0.753	2.178	2.703E-29
0.900	524.0	3.690E-29	1	0.450	0.001	89.87	6.998E-29	4.677E-29	0.0018	0.0016	-0.798	2.164	2.337E-29
0.950	524.0	2.600E-29	2	0.475	0.016	88.06	4.499E-29	3.084E-29	0.0286	0.0235	-0.814	2.141	1.542E-29
1.000	524.0	1.350E-29	3	0.500	0.031	86.49	2.140E-29	1.506E-29	0.0544	0.0436	-0.835	2.117	7.525E-30
1.050	524.0	9.640E-30	5	0.525	0.045	85.14	1.404E-29	1.015E-29	0.0793	0.0619	-0.858	2.091	5.071E-30
1.100	524.0	5.050E-30	8	0.550	0.058	83.96	6.785E-30	5.032E-30	0.1034	0.0787	-0.884	2.064	2.515E-30
1.171	210.1	8.288E-29	5	0.244	0.278	41.26	1.245E-28	7.681E-29	0.4941	0.4859	-0.045	2.231	3.839E-29
1.162	223.2	7.727E-29	3	0.257	0.269	43.67	1.168E-28	7.225E-29	0.4784	0.4668	-0.068	2.229	3.611E-29
1.144	255.5	7.036E-29	2	0.289	0.248	49.34	1.074E-28	6.699E-29	0.4409	0.4214	-0.130	2.223	3.348E-29
1.135	287.0	7.023E-29	1	0.321	0.230	54.42	1.067E-28	6.748E-29	0.4081	0.3809	-0.195	2.212	3.372E-29
1.133	302.3	6.875E-29	2	0.337	0.221	56.73	1.038E-28	6.619E-29	0.3930	0.3622	-0.228	2.205	3.308E-29
1.133	317.2	6.651E-29	1	0.353	0.213	58.87	9.949E-29	6.404E-29	0.3789	0.3447	-0.263	2.197	3.200E-29
1.135	331.7	5.915E-29	2	0.370	0.206	60.84	8.739E-29	5.683E-29	0.3660	0.3284	-0.298	2.188	2.840E-29
1.138	345.9	5.766E-29	1	0.386	0.199	62.70	8.400E-29	5.523E-29	0.3535	0.3126	-0.334	2.179	2.760E-29
1.142	359.5	5.384E-29	2	0.402	0.192	64.41	7.724E-29	5.136E-29	0.3415	0.2978	-0.370	2.169	2.567E-29
1.147	372.8	4.803E-29	1	0.418	0.186	66.02	6.774E-29	4.559E-29	0.3299	0.2835	-0.408	2.158	2.279E-29
1.153	385.6	4.321E-29	2	0.434	0.180	67.50	5.984E-29	4.077E-29	0.3189	0.2700	-0.445	2.147	2.038E-29
1.161	397.9	3.530E-29	1	0.450	0.174	68.85	4.786E-29	3.304E-29	0.3090	0.2577	-0.483	2.135	1.651E-29
1.179	421.3	1.604E-29	3	0.482	0.163	71.28	2.080E-29	1.475E-29	0.2900	0.2345	-0.561	2.109	7.372E-30
1.200	442.8	1.095E-29	3	0.514	0.154	73.37	1.353E-29	9.867E-30	0.2727	0.2139	-0.640	2.082	4.932E-30
1.212	452.9	8.607E-30	4	0.530	0.149	74.29	1.036E-29	7.668E-30	0.2649	0.2045	-0.680	2.067	3.832E-30
1.224	462.6	4.726E-30	4	0.546	0.145	75.16	5.548E-30	4.163E-30	0.2570	0.1953	-0.720	2.052	2.081E-30
1.237	471.9	1.417E-30	18	0.562	0.141	75.96	1.620E-30	1.233E-30	0.2496	0.1868	-0.761	2.037	6.164E-31
1.251	480.8	5.897E-31	32	0.579	0.137	76.70	6.557E-31	5.068E-31	0.2428	0.1789	-0.802	2.020	2.533E-31

$p_1 = 3.700$ GeV/c; $p_1^* = 1.162$ GeV/c; $E_1^* + E_2^* = 2.987$ GeV; $s = 8.923$ GeV²; $\gamma_{c.m.} = 1.5919$; $\beta_{c.m.} = 0.77808$; $y_1^* = 1.0405$; $p_{3,max}^* = 0.665$ GeV/c; [68 H 1]

p_3	θ_3	$\dfrac{d^2\sigma}{d\Omega_3\,dp_3}$	Error	p_T	p_L^*	θ_3^*	$\dfrac{1}{p_3^{*2}}\cdot\dfrac{d^2\sigma}{d\Omega_3^*\,dp_3^*}$	$\dfrac{E_3}{p_3^2}\cdot\dfrac{d^2\sigma}{d\Omega_3\,dp_3}$	x^*	y_3^*	t	m_4	$\dfrac{d^2\sigma}{dt\,dm_4^2}$
0.877	196.3	1.153E-28	3	0.171	0.123	54.36	2.811E-28	1.509E-28	0.1845	0.2326	-0.194	2.441	6.827E-29
0.834	227.6	1.099E-28	2	0.188	0.093	63.73	2.854E-28	1.531E-28	0.1397	0.1749	-0.263	2.442	6.930E-29
0.797	260.2	1.017E-28	2	0.205	0.065	72.48	2.787E-28	1.501E-28	0.0974	0.1208	-0.334	2.439	6.793E-29
0.757	294.0	1.018E-28	2	0.222	0.039	80.12	2.908E-28	1.579E-28	0.0582	0.0714	-0.408	2.434	7.143E-29
0.742	328.6	9.381E-29	1	0.239	0.014	86.65	2.766E-28	1.519E-28	0.0211	0.0255	-0.483	2.426	6.872E-29
0.750	349.0	9.400E-29	1	0.256	0.010	87.83	2.696E-28	1.501E-28	0.0146	0.0174	-0.516	2.417	6.790E-29

p_3	θ_3	$\dfrac{\mathrm{d}^2\sigma}{\mathrm{d}\Omega_3\,\mathrm{d}p_3}$	Error	p_T	p_L^*	θ_3^*	$\dfrac{1}{p_3^{*2}}\cdot\dfrac{\mathrm{d}^2\sigma}{\mathrm{d}\Omega_3^*\,\mathrm{d}p_3^*}$	$\dfrac{E_3}{p_3^2}\cdot\dfrac{\mathrm{d}^2\sigma}{\mathrm{d}\Omega_3\,\mathrm{d}p_3}$	x^*	y_3^*	t	m_4	$\dfrac{\mathrm{d}^2\sigma}{\mathrm{d}t\,\mathrm{d}m_4^2}$
GeV/c	mrad	cm^2 sr^{-1} GeV^{-1} c	%	GeV/c	GeV/c	deg	cm^2 sr^{-1} GeV^{-3} c^3	cm^2 sr^{-1} GeV^{-2} c^3			GeV^2	GeV	cm^2/GeV^4

$p_1 = 3.700$ GeV/c; $p_1^* = 1.162$ GeV/c; $E_1^* + E_2^* = 2.987$ GeV; $s = 8.923$ GeV^2; $\gamma_\mathrm{c.m.} = 1.5919$; $\beta_\mathrm{c.m.} = 0.77808$; $y_1^* = 1.0405$; $p_{3,\max}^* = 0.665$ GeV/c; [68 H 1] (cont.)

p_3	θ_3	$\dfrac{\mathrm{d}^2\sigma}{\mathrm{d}\Omega_3\,\mathrm{d}p_3}$	Error	p_T	p_L^*	θ_3^*	$\dfrac{1}{p_3^{*2}}\cdots$	$\dfrac{E_3}{p_3^2}\cdots$	x^*	y_3^*	t	m_4	$\dfrac{\mathrm{d}^2\sigma}{\mathrm{d}t\,\mathrm{d}m_4^2}$
0.800	349.0	1.004E-28	1	0.274	0.032	83.27	2.608E-28	1.475E-28	0.0486	0.0572	-0.490	2.406	6.674E-29
0.850	349.0	1.059E-28	1	0.291	0.054	79.49	2.504E-28	1.441E-28	0.0811	0.0940	-0.470	2.393	6.520E-29
0.900	349.0	1.138E-28	1	0.308	0.075	76.34	2.458E-28	1.442E-28	0.1125	0.1282	-0.454	2.379	6.526E-29
0.950	349.0	1.208E-28	1	0.325	0.095	73.70	2.394E-28	1.433E-28	0.1429	0.1600	-0.443	2.364	6.485E-29
0.975	349.0	1.201E-28	1	0.333	0.105	72.55	2.282E-28	1.381E-28	0.1577	0.1750	-0.439	2.356	6.248E-29
1.000	349.0	1.225E-28	1	0.342	0.115	71.48	2.234E-28	1.366E-28	0.1723	0.1895	-0.436	2.348	6.182E-29
1.025	349.0	1.187E-28	1	0.351	0.124	70.50	2.080E-28	1.285E-28	0.1866	0.2035	-0.434	2.340	5.817E-29
1.050	349.0	1.204E-28	1	0.359	0.134	69.60	2.028E-28	1.267E-28	0.2008	0.2170	-0.432	2.331	5.734E-29
1.100	349.0	1.138E-28	1	0.376	0.152	67.99	1.774E-28	1.134E-28	0.2287	0.2426	-0.432	2.313	5.131E-29
1.150	349.0	1.018E-28	1	0.393	0.170	66.60	1.474E-28	9.634E-29	0.2559	0.2664	-0.433	2.294	4.359E-29
1.200	349.0	9.630E-29	1	0.410	0.188	65.40	1.297E-28	8.678E-29	0.2825	0.2886	-0.438	2.274	3.927E-29
1.250	349.0	8.210E-29	2	0.427	0.205	64.36	1.032E-28	7.062E-29	0.3086	0.3092	-0.444	2.253	3.196E-29
1.300	349.0	7.220E-29	2	0.445	0.222	63.44	8.480E-29	5.941E-29	0.3343	0.3286	-0.452	2.232	2.688E-29
1.115	180.4	1.418E-28	2	0.200	0.236	40.32	2.387E-28	1.391E-28	0.3545	0.4291	-0.068	2.385	6.294E-29
1.073	211.3	1.448E-28	2	0.225	0.207	47.38	2.558E-28	1.486E-28	0.3115	0.3729	-0.130	2.387	6.722E-29
1.041	242.8	1.353E-28	2	0.250	0.181	54.06	2.469E-28	1.439E-28	0.2729	0.3222	-0.194	2.385	6.509E-29
1.025	261.8	1.316E-28	3	0.265	0.167	57.83	2.437E-28	1.425E-28	0.2510	0.2934	-0.235	2.382	6.449E-29
1.016	274.4	1.252E-28	2	0.275	0.158	60.20	2.334E-28	1.370E-28	0.2371	0.2754	-0.263	2.379	6.200E-29
1.004	293.3	1.218E-28	2	0.290	0.144	63.59	2.289E-28	1.352E-28	0.2168	0.2491	-0.305	2.374	6.118E-29
0.997	305.8	1.327E-28	1	0.300	0.135	65.72	2.503E-28	1.485E-28	0.2037	0.2322	-0.334	2.371	6.721E-29
0.990	321.3	1.252E-28	2	0.313	0.125	68.20	2.365E-28	1.413E-28	0.1880	0.2123	-0.371	2.366	6.395E-29
0.985	336.6	1.257E-28	1	0.325	0.115	70.49	2.388E-28	1.439E-28	0.1734	0.1937	-0.408	2.360	6.511E-29
0.980	354.6	1.226E-28	1	0.340	0.104	73.04	2.302E-28	1.401E-28	0.1561	0.1722	-0.453	2.352	6.339E-29
0.977	366.5	1.191E-28	1	0.350	0.096	74.65	2.229E-28	1.366E-28	0.1445	0.1581	-0.483	2.346	6.181E-29
0.976	378.1	1.180E-28	1	0.360	0.089	76.10	2.193E-28	1.355E-28	0.1341	0.1453	-0.514	2.340	6.131E-29
0.974	395.3	1.124E-28	1	0.375	0.078	78.20	2.070E-28	1.294E-28	0.1178	0.1260	-0.561	2.331	5.855E-29
0.975	412.0	1.007E-28	1	0.390	0.069	80.05	1.828E-28	1.158E-28	0.1030	0.1086	-0.608	2.320	5.239E-29
0.975	422.9	1.009E-28	1	0.400	0.062	81.24	1.817E-28	1.160E-28	0.0927	0.0969	-0.640	2.313	5.249E-29
0.978	438.7	9.040E-29	1	0.415	0.052	82.81	1.599E-28	1.035E-28	0.0788	0.0811	-0.688	2.302	4.686E-29
0.980	449.1	9.070E-29	1	0.425	0.046	83.81	1.586E-28	1.036E-28	0.0694	0.0707	-0.721	2.294	4.690E-29
0.987	473.8	8.176E-29	1	0.450	0.031	86.05	1.384E-28	9.263E-29	0.0468	0.0466	-0.802	2.274	4.191E-29
0.997	497.2	6.716E-29	2	0.476	0.017	87.97	1.096E-28	7.517E-29	0.0254	0.0246	-0.885	2.252	3.402E-29
1.009	519.1	5.486E-29	1	0.501	0.003	89.63	8.609E-29	6.053E-29	0.0048	0.0046	-0.969	2.228	2.739E-29

$p_1 = 3.860$ GeV/c; $p_1^* = 1.193$ GeV/c; $E_1^* + E_2^* = 3.035$ GeV; $s = 9.214$ GeV^2; $\gamma_\mathrm{c.m.} = 1.6177$; $\beta_\mathrm{c.m.} = 0.78606$; $y_1^* = 1.0610$; $p_{3,\max}^* = 0.708$ GeV/c; [68 H 1]

p_3	θ_3	$\dfrac{\mathrm{d}^2\sigma}{\mathrm{d}\Omega_3\,\mathrm{d}p_3}$	Error	p_T	p_L^*	θ_3^*	$\dfrac{1}{p_3^{*2}}\cdots$	$\dfrac{E_3}{p_3^2}\cdots$	x^*	y_3^*	t	m_4	$\dfrac{\mathrm{d}^2\sigma}{\mathrm{d}t\,\mathrm{d}m_4^2}$
1.087	232.1	1.795E-28	4	0.250	0.193	52.32	3.094E-28	1.814E-28	0.2729	0.3422	-0.195	2.429	7.867E-29
1.061	262.3	1.763E-28	4	0.275	0.170	58.36	3.106E-28	1.833E-28	0.2395	0.2956	-0.263	2.424	7.950E-29
1.042	292.3	1.749E-28	2	0.300	0.148	63.78	3.114E-28	1.857E-28	0.2089	0.2532	-0.334	2.416	8.057E-29
1.028	321.8	1.676E-28	1	0.325	0.127	68.60	2.991E-28	1.809E-28	0.1801	0.2139	-0.408	2.405	7.845E-29
1.020	350.4	1.583E-28	1	0.350	0.109	72.75	2.804E-28	1.724E-28	0.1537	0.1787	-0.483	2.393	7.479E-29
1.018	361.6	1.508E-28	2	0.360	0.102	74.25	2.657E-28	1.646E-28	0.1435	0.1654	-0.514	2.387	7.141E-29

p_3	θ_3	$\dfrac{d^2\sigma}{d\Omega_3\,dp_3}$	Error	p_T	p_L^*	θ_3^*	$\dfrac{1}{p_3^{*2}}\cdot\dfrac{d^2\sigma}{d\Omega_3^*\,dp_3^*}$	$\dfrac{E_3}{p_3^2}\cdot\dfrac{d^2\sigma}{d\Omega_3\,dp_3}$	x^*	y_3^*	t	m_4	$\dfrac{d^2\sigma}{dt\,dm_4^2}$
GeV/c	mrad	$cm^2\,sr^{-1}\,GeV^{-1}c$	%	GeV/c	GeV/c	deg	$cm^2\,sr^{-1}\,GeV^{-3}c^3$	$cm^2\,sr^{-1}\,GeV^{-2}c^3$			GeV^2	GeV	cm^2/GeV^4

$p_1 = 3.860$ GeV/c; $p_1^* = 1.193$ GeV/c; $E_1^* + E_2^* = 3.035$ GeV; $s = 9.214$ GeV^2; $\gamma_{c.m.} = 1.6177$; $\beta_{c.m.} = 0.78606$; $y_1^* = 1.0610$; $p_{3,max}^* = 0.708$ GeV/c; [68 H 1] (cont.)

p_3	θ_3	$\dfrac{d^2\sigma}{d\Omega_3\,dp_3}$	Error	p_T	p_L^*	θ_3^*			x^*	y_3^*	t	m_4	$\dfrac{d^2\sigma}{dt\,dm_4^2}$
1.017	378.0	1.540E-28	1	0.375	0.091	76.31	2.685E-28	1.683E-28	0.1292	0.1469	-0.561	2.377	7.301E-29
1.017	394.0	1.403E-28	2	0.390	0.082	78.20	2.416E-28	1.534E-28	0.1152	0.1292	-0.608	2.367	6.652E-29
1.017	404.4	1.425E-28	1	0.400	0.075	79.40	2.434E-28	1.558E-28	0.1058	0.1176	-0.640	2.361	6.756E-29
1.019	419.7	1.255E-28	1	0.415	0.065	81.04	2.110E-28	1.369E-28	0.0925	0.1013	-0.688	2.350	5.936E-29
1.021	429.6	1.270E-28	1	0.425	0.059	82.05	2.112E-28	1.382E-28	0.0839	0.0910	-0.721	2.342	5.993E-29
1.028	453.4	1.171E-28	1	0.450	0.045	84.32	1.887E-28	1.264E-28	0.0633	0.0669	-0.802	2.322	5.481E-29
1.038	475.8	9.860E-29	2	0.475	0.031	86.27	1.533E-28	1.052E-28	0.0438	0.0452	-0.885	2.300	4.563E-29
1.050	496.9	7.446E-29	2	0.501	0.018	87.97	1.114E-28	7.836E-29	0.0250	0.0252	-0.969	2.278	3.399E-29

$p_1 = 12.500$ GeV/c; $p_1^* = 2.332$ GeV/c; $E_1^* + E_2^* = 5.028$ GeV; $s = 25.281$ GeV^2; $\gamma_{c.m.} = 2.6796$; $\beta_{c.m.} = 0.92776$; $y_1^* = 1.6420$; $p_{3,max}^* = 2.060$ GeV/c; [71 A 1]

p_3	θ_3	$\dfrac{d^2\sigma}{d\Omega_3\,dp_3}$	Error	p_T	p_L^*	θ_3^*			x^*	y_3^*	t	m_4	$\dfrac{d^2\sigma}{dt\,dm_4^2}$
4.009	137.1	8.826E-28	15	0.548	0.600	42.40	2.333E-28	2.218E-28	0.2912	0.7431	-0.858	3.995	2.971E-29
5.262	226.8	2.065E-29	30	1.183	0.600	63.11	2.784E-30	3.942E-30	0.2912	0.4524	-3.194	3.360	5.279E-31
5.447	234.3	1.274E-29	30	1.265	0.600	64.61	1.582E-30	2.348E-30	0.2913	0.4288	-3.540	3.256	3.146E-31
5.795	246.5	3.948E-30	45	1.414	0.600	67.00	4.237E-31	6.837E-31	0.2913	0.3907	-4.189	3.049	9.158E-32
3.001	212.3	4.090E-28	18	0.632	0.300	64.61	1.612E-28	1.381E-28	0.1456	0.3658	-1.783	4.112	1.850E-29
3.361	189.3	4.738E-28	18	0.632	0.400	57.68	1.589E-28	1.425E-28	0.1942	0.4799	-1.518	4.063	1.908E-29
3.744	169.7	5.237E-28	27	0.632	0.500	51.66	1.492E-28	1.411E-28	0.2427	0.5887	-1.297	4.002	1.890E-29

$p_1 = 14.250$ GeV/c; $p_1^* = 2.502$ GeV/c; $E_1^* + E_2^* = 5.344$ GeV; $s = 28.557$ GeV^2; $\gamma_{c.m.} = 2.8479$; $\beta_{c.m.} = 0.93635$; $y_1^* = 1.7074$; $p_{3,max}^* = 2.246$ GeV/c; [72 A 1]

p_3	θ_3	$\dfrac{d^2\sigma}{d\Omega_3\,dp_3}$	Error	p_T	p_L^*	θ_3^*			x^*	y_3^*	t	m_4	$\dfrac{d^2\sigma}{dt\,dm_4^2}$
4.500	12.0	2.955E-27	5	0.054	0.743	4.16	7.391E-28	6.606E-28	0.3308	1.1925	0.066	4.387	7.762E-29
4.750	12.0	2.889E-27	4	0.057	0.792	4.12	6.539E-28	6.115E-28	0.3526	1.2456	0.090	4.337	7.185E-29
5.000	12.0	2.817E-27	4	0.060	0.841	4.08	5.795E-28	5.661E-28	0.3743	1.2959	0.111	4.285	6.652E-29
5.120	12.0	2.651E-27	4	0.061	0.864	4.07	5.217E-28	5.202E-28	0.3846	1.3191	0.119	4.260	6.112E-29
5.250	12.0	2.876E-27	4	0.063	0.889	4.05	5.400E-28	5.502E-28	0.3958	1.3437	0.128	4.232	6.465E-29
5.500	12.0	2.695E-27	3	0.066	0.937	4.03	4.635E-28	4.920E-28	0.4172	1.3893	0.142	4.178	5.780E-29
5.750	12.0	2.724E-27	4	0.069	0.985	4.01	4.307E-28	4.755E-28	0.4385	1.4327	0.153	4.123	5.587E-29
6.000	12.0	2.696E-27	4	0.072	1.033	3.99	3.931E-28	4.509E-28	0.4597	1.4743	0.162	4.067	5.297E-29
6.250	12.0	2.411E-27	4	0.075	1.080	3.97	3.252E-28	3.870E-28	0.4808	1.5141	0.169	4.010	4.547E-29
6.500	12.0	2.235E-27	3	0.078	1.127	3.96	2.796E-28	3.448E-28	0.5019	1.5523	0.175	3.952	4.052E-29
6.750	12.0	2.165E-27	3	0.081	1.175	3.94	2.519E-28	3.216E-28	0.5229	1.5889	0.179	3.893	3.779E-29
7.000	12.0	2.030E-27	3	0.084	1.222	3.93	2.202E-28	2.907E-28	0.5438	1.6242	0.181	3.833	3.416E-29
7.340	12.0	1.858E-27	3	0.088	1.285	3.92	1.839E-28	2.537E-28	0.5722	1.6702	0.182	3.749	2.981E-29
7.520	12.0	1.754E-27	3	0.090	1.319	3.91	1.656E-28	2.337E-28	0.5871	1.6936	0.182	3.704	2.746E-29
7.710	12.0	1.693E-27	3	0.093	1.354	3.91	1.523E-28	2.200E-28	0.6029	1.7176	0.181	3.655	2.585E-29
7.900	12.0	1.637E-27	3	0.095	1.390	3.90	1.405E-28	2.075E-28	0.6187	1.7411	0.180	3.606	2.439E-29
8.090	12.0	1.558E-27	3	0.097	1.425	3.90	1.277E-28	1.929E-28	0.6345	1.7639	0.178	3.556	2.267E-29
8.290	12.0	1.466E-27	3	0.099	1.462	3.89	1.145E-28	1.772E-28	0.6510	1.7874	0.176	3.503	2.081E-29
8.490	12.0	1.375E-27	3	0.102	1.500	3.89	1.025E-28	1.622E-28	0.6675	1.8102	0.173	3.449	1.906E-29
8.700	12.0	1.331E-27	3	0.104	1.539	3.88	9.464E-29	1.532E-28	0.6849	1.8336	0.169	3.390	1.800E-29
8.920	12.0	1.250E-27	3	0.107	1.579	3.88	8.464E-29	1.403E-28	0.7030	1.8575	0.165	3.329	1.649E-29

p_3	θ_3	$\dfrac{d^2\sigma}{d\Omega_3\,dp_3}$	Error	p_T	p_L^*	θ_3^*	$\dfrac{1}{p_3^{*2}}\cdot\dfrac{d^2\sigma}{d\Omega_3^*\,dp_3^*}$	$\dfrac{E_3}{p_3^2}\cdot\dfrac{d^2\sigma}{d\Omega_3\,dp_3}$	x^*	y_3^*	t	m_4	$\dfrac{d^2\sigma}{dt\,dm_4^2}$
GeV/c	mrad	$cm^2\,sr^{-1}\,GeV^{-1}\,c$	%	GeV/c	GeV/c	deg	$cm^2\,sr^{-1}\,GeV^{-2}\,c^3$	$cm^2\,sr^{-1}\,GeV^{-2}\,c^3$			GeV^2	GeV	cm^2/GeV^4

$p_1 = 14.250$ GeV/c; $p_1^* = 2.502$ GeV/c; $E_1^* + E_2^* = 5.344$ GeV; $s = 28.557$ GeV^2; $\gamma_{c.m.} = 2.8479$; $\beta_{c.m.} = 0.93633$; $y_1^* = 1.7074$; $p_{3,max}^* = 2.246$ GeV/c; [72 A 1] (cont.)

p_3	θ_3	$\dfrac{d^2\sigma}{d\Omega_3\,dp_3}$	Error	p_T	p_L^*	θ_3^*	$\dfrac{1}{p_3^{*2}}\cdot\dfrac{d^2\sigma}{d\Omega_3^*\,dp_3^*}$	$\dfrac{E_3}{p_3^2}\cdot\dfrac{d^2\sigma}{d\Omega_3\,dp_3}$	x^*	y_3^*	t	m_4	$\dfrac{d^2\sigma}{dt\,dm_4^2}$
9.140	12.0	1.133E-27	3	0.110	1.620	3.87	7.315E-29	1.241E-28	0.7212	1.8807	0.161	3.265	1.459E-29
9.360	12.0	1.045E-27	3	0.112	1.661	3.87	6.446E-29	1.119E-28	0.7393	1.9033	0.156	3.201	1.315E-29
9.590	12.0	9.560E-28	3	0.115	1.703	3.87	5.617E-29	9.982E-29	0.7582	1.9263	0.150	3.132	1.173E-29
9.830	12.0	7.856E-28	4	0.118	1.747	3.86	4.397E-29	8.002E-29	0.7779	1.9497	0.143	3.058	9.402E-30
10.320	12.0	6.201E-28	4	0.124	1.838	3.85	3.154E-29	6.016E-29	0.8181	1.9955	0.129	2.902	7.068E-30
10.570	12.0	4.657E-28	5	0.127	1.884	3.85	2.260E-29	4.411E-29	0.8386	2.0180	0.121	2.818	5.182E-30
10.830	12.0	3.700E-28	10	0.130	1.932	3.85	1.712E-29	3.420E-29	0.8599	2.0407	0.112	2.729	4.018E-30
11.100	12.0	2.800E-28	13	0.133	1.981	3.85	1.234E-29	2.525E-29	0.8820	2.0637	0.103	2.633	2.967E-30
11.370	12.0	2.200E-28	15	0.136	2.031	3.84	9.247E-30	1.937E-29	0.9041	2.0860	0.093	2.533	2.276E-30
11.510	12.0	1.800E-28	15	0.138	2.057	3.84	7.385E-30	1.565E-29	0.9155	2.0974	0.088	2.479	1.839E-30
11.650	12.0	1.500E-28	15	0.140	2.082	3.84	6.009E-30	1.289E-29	0.9270	2.1086	0.083	2.425	1.514E-30
11.800	12.0	1.400E-28	20	0.142	2.110	3.84	5.469E-30	1.187E-29	0.9392	2.1204	0.077	2.365	1.395E-30
11.940	12.0	1.050E-28	20	0.143	2.136	3.84	4.007E-30	8.801E-30	0.9506	2.1313	0.071	2.308	1.034E-30
12.090	12.0	6.700E-29	25	0.145	2.163	3.84	2.495E-30	5.546E-30	0.9629	2.1427	0.065	2.244	6.517E-31
12.240	12.0	4.500E-29	30	0.147	2.191	3.84	1.635E-30	3.679E-30	0.9751	2.1541	0.059	2.180	4.323E-31
12.390	12.0	2.500E-29	50	0.149	2.218	3.83	8.868E-31	2.019E-30	0.9874	2.1652	0.053	2.113	2.373E-31

$p_1 = 19.200$ GeV/c; $p_1^* = 2.929$ GeV/c; $E_1^* + E_2^* = 6.151$ GeV; $s = 37.830$ GeV^2; $\gamma_{c.m.} = 3.2779$; $\beta_{c.m.} = 0.95233$; $y_1^* = 1.8562$; $p_{3,max}^* = 2.708$ GeV/c; [70 A 1]

p_3	θ_3	$\dfrac{d^2\sigma}{d\Omega_3\,dp_3}$	Error	p_T	p_L^*	θ_3^*	$\dfrac{1}{p_3^{*2}}\cdot\dfrac{d^2\sigma}{d\Omega_3^*\,dp_3^*}$	$\dfrac{E_3}{p_3^2}\cdot\dfrac{d^2\sigma}{d\Omega_3\,dp_3}$	x^*	y_3^*	t	m_4	$\dfrac{d^2\sigma}{dt\,dm_4^2}$
4.500	12.5	6.106E-27	3	0.056	0.618	5.20	1.722E-27	1.365E-27	0.2281	1.0431	-0.134	5.322	1.190E-28
4.500	20.0	5.741E-27	3	0.090	0.616	8.31	1.615E-27	1.283E-27	0.2275	1.0332	-0.155	5.320	1.119E-28
4.500	30.0	5.515E-27	3	0.135	0.612	12.43	1.545E-27	1.233E-27	0.2261	1.0134	-0.198	5.316	1.075E-28
4.500	40.0	5.109E-27	3	0.180	0.607	16.51	1.422E-27	1.142E-27	0.2242	0.9869	-0.259	5.310	9.960E-29
4.500	50.0	4.801E-27	3	0.225	0.600	20.53	1.326E-27	1.073E-27	0.2218	0.9547	-0.337	5.303	9.359E-29
4.500	60.0	4.515E-27	3	0.270	0.592	24.49	1.235E-27	1.009E-27	0.2188	0.9181	-0.432	5.294	8.802E-29
4.500	70.0	4.099E-27	3	0.315	0.583	28.37	1.109E-27	9.164E-28	0.2152	0.8779	-0.544	5.283	7.991E-29
6.000	12.5	5.039E-27	3	0.075	0.873	4.91	8.380E-28	8.427E-28	0.3223	1.3246	0.051	5.070	7.348E-29
6.000	20.0	5.030E-27	3	0.120	0.870	7.85	8.346E-28	8.412E-28	0.3214	1.3072	0.023	5.067	7.335E-29
6.000	30.0	4.864E-27	3	0.180	0.865	11.75	8.034E-28	8.134E-28	0.3196	1.2734	-0.034	5.062	7.093E-29
6.000	40.0	4.575E-27	3	0.240	0.859	15.61	7.508E-28	7.651E-28	0.3171	1.2296	-0.115	5.054	6.672E-29
6.000	50.0	4.040E-27	3	0.300	0.850	19.44	6.575E-28	6.756E-28	0.3138	1.1784	-0.219	5.043	5.891E-29
6.000	60.0	3.575E-27	3	0.360	0.839	23.21	5.761E-28	5.978E-28	0.3098	1.1222	-0.345	5.031	5.213E-29
6.000	70.0	3.095E-27	3	0.420	0.826	26.93	4.930E-28	5.176E-28	0.3051	1.0630	-0.495	5.016	4.513E-29
8.000	12.5	3.973E-27	3	0.100	1.201	4.76	3.822E-28	4.976E-28	0.4434	1.6028	0.148	4.697	4.339E-29
8.000	20.0	3.880E-27	3	0.160	1.197	7.61	3.724E-28	4.859E-28	0.4422	1.5729	0.111	4.693	4.237E-29
8.000	30.0	3.432E-27	3	0.240	1.191	11.39	3.278E-28	4.298E-28	0.4398	1.5167	0.034	4.684	3.748E-29
8.000	40.0	2.911E-27	3	0.320	1.182	15.15	2.762E-28	3.646E-28	0.4364	1.4473	-0.074	4.673	3.179E-29
8.000	50.0	2.413E-27	5	0.400	1.170	18.87	2.270E-28	3.022E-28	0.4321	1.3702	-0.212	4.658	2.635E-29
8.000	60.0	1.863E-27	3	0.480	1.155	22.55	1.735E-28	2.333E-28	0.4267	1.2897	-0.381	4.640	2.035E-29
8.000	70.0	1.367E-27	3	0.560	1.138	26.18	1.258E-28	1.712E-28	0.4204	1.2088	-0.580	4.618	1.493E-29

Diddens/Schlüpmann

p_3	θ_3	$\dfrac{\mathrm{d}^2\sigma}{\mathrm{d}\Omega_3\,\mathrm{d}p_3}$	Error	p_T	p_L^*	θ_3^*	$\dfrac{1}{p_3^{*2}}\cdot\dfrac{\mathrm{d}^2\sigma}{\mathrm{d}\Omega_3^*\,\mathrm{d}p_3^*}$	$\dfrac{E_3}{p_3^2}\cdot\dfrac{\mathrm{d}^2\sigma}{\mathrm{d}\Omega_3\,\mathrm{d}p_3}$	x^*	y_3^*	t	m_4	$\dfrac{\mathrm{d}^2\sigma}{\mathrm{d}t\,\mathrm{d}m_4^2}$
GeV/c	mrad	cm² sr⁻¹ GeV⁻¹ c	%	GeV/c	GeV/c	deg	cm² sr⁻¹ GeV⁻³ c³	cm² sr⁻¹ GeV⁻² c³			GeV²	GeV	cm²/GeV⁴

$p_1 = 19.200$ GeV/c; $p_1^* = 2.929$ GeV/c; $E_1^* + E_2^* = 6.151$ GeV; $s = 37.830$ GeV²; $\gamma_{\mathrm{c.m.}} = 3.2779$; $\beta_{\mathrm{c.m.}} = 0.95233$; $y_1^* = 1.8562$; $p_{3,\max}^* = 2.708$ GeV/c; [70 A 1] (cont.)

p_3	θ_3	$\dfrac{\mathrm{d}^2\sigma}{\mathrm{d}\Omega_3\,\mathrm{d}p_3}$	Error	p_T	p_L^*	θ_3^*	$\dfrac{1}{p_3^{*2}}\cdot\dfrac{\mathrm{d}^2\sigma}{\mathrm{d}\Omega_3^*\,\mathrm{d}p_3^*}$	$\dfrac{E_3}{p_3^2}\cdot\dfrac{\mathrm{d}^2\sigma}{\mathrm{d}\Omega_3\,\mathrm{d}p_3}$	x^*	y_3^*	t	m_4	$\dfrac{\mathrm{d}^2\sigma}{\mathrm{d}t\,\mathrm{d}m_4^2}$
10.000	12.5	2.900E-27	3	0.125	1.522	4.69	1.809E-28	2.904E-28	0.5621	1.8147	0.167	4.281	2.532E-29
10.000	20.0	2.639E-27	3	0.200	1.518	7.50	1.642E-28	2.642E-28	0.5607	1.7697	0.121	4.276	2.304E-29
10.000	30.0	2.191E-27	3	0.300	1.510	11.24	1.357E-28	2.194E-28	0.5576	1.6885	0.025	4.265	1.913E-29
10.000	40.0	1.660E-27	3	0.400	1.498	14.94	1.021E-28	1.662E-28	0.5534	1.5932	-0.110	4.249	1.449E-29
10.000	50.0	1.181E-27	3	0.500	1.484	18.62	7.203E-29	1.182E-28	0.5480	1.4925	-0.282	4.229	1.031E-29
10.000	60.0	7.685E-28	3	0.600	1.466	22.25	4.639E-29	7.694E-29	0.5413	1.3918	-0.494	4.204	6.710E-30
10.000	70.0	4.950E-28	3	0.699	1.444	25.84	2.952E-29	4.956E-29	0.5334	1.2942	-0.743	4.174	4.322E-30
11.000	12.5	2.301E-27	3	0.137	1.681	4.67	1.191E-28	2.094E-28	0.6210	1.9036	0.161	4.056	1.826E-29
11.000	20.0	1.980E-27	3	0.220	1.677	7.47	1.023E-28	1.802E-28	0.6194	1.8503	0.110	4.050	1.571E-29
11.000	30.0	1.545E-27	3	0.330	1.668	11.19	7.940E-29	1.406E-28	0.6161	1.7562	0.004	4.036	1.226E-29
11.000	40.0	1.114E-27	3	0.440	1.655	14.88	5.687E-29	1.014E-28	0.6114	1.6485	-0.144	4.018	8.840E-30
11.000	50.0	7.099E-28	4	0.550	1.639	18.54	3.593E-29	6.460E-29	0.6054	1.5372	-0.334	3.994	5.633E-30
11.000	60.0	4.338E-28	4	0.660	1.619	22.16	2.173E-29	3.948E-29	0.5981	1.4281	-0.566	3.965	3.442E-30
11.000	70.0	2.656E-28	3	0.769	1.596	25.74	1.314E-29	2.417E-29	0.5895	1.3238	-0.840	3.930	2.108E-30
12.000	12.5	1.886E-27	3	0.150	1.840	4.66	8.230E-29	1.573E-28	0.6797	1.9837	0.148	3.816	1.372E-29
12.000	20.0	1.504E-27	3	0.240	1.836	7.45	6.547E-29	1.254E-28	0.6780	1.9218	0.092	3.809	1.094E-29
12.000	30.0	1.074E-27	3	0.360	1.826	11.15	4.653E-29	8.958E-29	0.6743	1.8146	-0.024	3.794	7.811E-30
12.000	40.0	7.054E-28	3	0.480	1.812	14.83	3.035E-29	5.883E-29	0.6692	1.6950	-0.185	3.772	5.130E-30
12.000	50.0	4.052E-28	3	0.600	1.794	18.48	1.728E-29	3.380E-29	0.6627	1.5741	-0.392	3.745	2.947E-30
12.000	60.0	2.303E-28	3	0.720	1.773	22.09	9.721E-30	1.921E-29	0.6547	1.4575	-0.645	3.711	1.675E-30
12.000	70.0	1.203E-28	3	0.839	1.747	25.66	5.016E-30	1.003E-29	0.6453	1.3476	-0.945	3.670	8.749E-31
13.000	12.5	1.422E-27	3	0.162	1.999	4.65	5.300E-29	1.095E-28	0.7382	2.0564	0.129	3.559	9.545E-30
13.000	20.0	1.030E-27	3	0.260	1.994	7.43	3.830E-29	7.929E-29	0.7363	1.9854	0.068	3.551	6.914E-30
13.000	30.0	6.823E-28	3	0.390	1.983	11.13	2.525E-29	5.252E-29	0.7324	1.8653	-0.057	3.533	4.580E-30
13.000	40.0	4.173E-28	3	0.520	1.968	14.80	1.534E-29	3.212E-29	0.7269	1.7345	-0.231	3.508	2.801E-30
13.000	50.0	1.976E-28	3	0.650	1.949	18.44	7.199E-30	1.521E-29	0.7198	1.6049	-0.456	3.476	1.326E-30
13.000	60.0	1.103E-28	3	0.780	1.925	22.04	3.977E-30	8.491E-30	0.7112	1.4817	-0.730	3.436	7.404E-31
13.000	70.0	5.258E-29	3	0.909	1.898	25.60	1.873E-30	4.048E-30	0.7009	1.3669	-1.055	3.389	3.530E-31
14.000	12.5	8.533E-28	5	0.175	2.157	4.64	2.748E-29	6.099E-29	0.7966	2.1227	0.106	3.282	5.318E-30
14.000	20.0	6.365E-28	3	0.280	2.151	7.42	2.045E-29	4.549E-29	0.7946	2.0424	0.040	3.272	3.967E-30
14.000	30.0	3.909E-28	3	0.420	2.140	11.10	1.250E-29	2.794E-29	0.7903	1.9096	-0.094	3.251	2.436E-30
14.000	40.0	2.194E-28	3	0.560	2.124	14.77	6.966E-30	1.568E-29	0.7844	1.7682	-0.282	3.222	1.367E-30
14.000	50.0	1.053E-28	4	0.700	2.103	18.40	3.314E-30	7.526E-30	0.7768	1.6307	-0.524	3.184	6.563E-31
14.000	60.0	4.721E-29	3	0.839	2.078	22.00	1.470E-30	3.374E-30	0.7675	1.5018	-0.819	3.138	2.942E-31
14.000	70.0	2.011E-29	3	0.979	2.048	25.55	6.187E-31	1.437E-30	0.7565	1.3828	-1.169	3.081	1.253E-31
15.000	12.5	5.380E-28	5	0.187	2.315	4.63	1.511E-29	3.589E-29	0.8549	2.1835	0.079	2.978	3.129E-30
15.000	20.0	3.598E-28	3	0.300	2.309	7.40	1.008E-29	2.400E-29	0.8527	2.0937	0.009	2.966	2.093E-30
15.000	30.0	2.014E-28	3	0.450	2.296	11.09	5.617E-30	1.343E-29	0.8482	1.9484	-0.135	2.942	1.171E-30

Diddens/Schlüpmann

$p_1 = 19.200$ GeV/c; $p_1^* = 2.929$ GeV/c; $E_1^* + E_2^* = 6.151$ GeV; $s = 37.830$ GeV²; $\gamma_{c.m.} = 3.2779$; $\beta_{c.m.} = 0.95233$; $y_1^* = 1.8562$; $p_{3,\,max}^* = 2.708$ GeV/c; [70 A 1] (cont.)

p_3	θ_3	$\dfrac{d^2\sigma}{d\Omega_3\,dp_3}$	Error	p_T	p_L^*	θ_3^*	$\dfrac{1}{p_3^{*2}}\cdot\dfrac{d^2\sigma}{d\Omega_3^*\,dp_3^*}$	$\dfrac{E_3}{p_3^2}\cdot\dfrac{d^2\sigma}{d\Omega_3\,dp_3}$	x^*	y_3^*	t	m_4	$\dfrac{d^2\sigma}{dt\,dm_4^2}$
GeV/c	mrad	cm² sr⁻¹ GeV⁻¹ c	%	GeV/c	GeV/c	deg	cm² sr⁻¹ GeV⁻³ c³	cm² sr⁻¹ GeV⁻² c³			GeV²	GeV	cm²/GeV⁴
15.000	40.0	9.980E−29	3	0.600	2.279	14.74	2.764E−30	6.657E−30	0.8418	1.7972	−0.336	2.907	5.805E−31
15.000	50.0	4.223E−29	3	0.750	2.257	18.37	1.160E−30	2.817E−30	0.8336	1.6525	−0.595	2.852	2.456E−31
15.000	60.0	1.669E−29	3	0.899	2.230	21.97	4.535E−31	1.113E−30	0.8237	1.5186	−0.912	2.807	9.708E−32
15.000	70.0	5.859E−30	3	1.049	2.198	25.51	1.573E−31	3.908E−31	0.8119	1.3960	−1.286	2.739	3.408E−32
16.000	12.5	2.585E−28	10	0.200	2.472	4.62	6.394E−30	1.617E−29	0.9131	2.2394	0.050	2.639	1.410E−30
16.000	20.0	1.457E−28	3	0.320	2.466	7.39	3.594E−30	9.111E−30	0.9108	2.1400	−0.025	2.625	7.945E−31
16.000	30.0	6.840E−29	3	0.480	2.453	11.07	1.679E−30	4.277E−30	0.9059	1.9826	−0.178	2.595	3.730E−31
16.000	40.0	2.936E−29	3	0.640	2.434	14.73	7.157E−31	1.836E−30	0.8991	1.8223	−0.393	2.553	1.601E−31
16.000	50.0	1.003E−29	5	0.800	2.411	18.35	2.424E−31	6.272E−31	0.8904	1.6712	−0.670	2.499	5.469E−32
16.000	60.0	3.161E−30	5	0.959	2.382	21.94	7.558E−32	1.977E−31	0.8798	1.5328	−1.007	2.430	1.724E−32
16.000	70.0	9.175E−31	5	1.119	2.348	25.48	2.167E−32	5.737E−32	0.8672	1.4071	−1.407	2.347	5.003E−33
13.500	12.5	1.022E−27	5	0.169	2.078	4.64	3.536E−29	7.575E−29	0.7675	2.0903	0.118	3.423	6.606E−30
13.800	12.5	9.166E−28	5	0.172	2.125	4.64	3.037E−29	6.646E−29	0.7850	2.1099	0.111	3.339	5.796E−30
14.100	12.5	7.877E−28	5	0.176	2.173	4.64	2.501E−29	5.590E−29	0.8025	2.1290	0.103	3.253	4.875E−30
14.400	12.5	6.895E−28	5	0.180	2.220	4.64	2.100E−29	4.791E−29	0.8200	2.1477	0.096	3.164	4.178E−30
14.700	12.5	5.905E−28	5	0.184	2.267	4.63	1.727E−29	4.020E−29	0.8374	2.1658	0.088	3.072	3.505E−30
15.000	12.5	4.948E−28	5	0.187	2.315	4.63	1.390E−29	3.300E−29	0.8549	2.1835	0.079	2.978	2.878E−30
15.300	12.5	4.021E−28	5	0.191	2.362	4.63	1.086E−29	2.629E−29	0.8724	2.2007	0.071	2.880	2.293E−30
15.600	12.5	3.153E−28	5	0.195	2.409	4.63	8.197E−30	2.022E−29	0.8899	2.2176	0.062	2.780	1.763E−30
15.900	12.5	2.442E−28	5	0.199	2.457	4.63	6.113E−30	1.537E−29	0.9073	2.2340	0.053	2.675	1.340E−30
16.200	12.5	1.744E−28	5	0.202	2.504	4.62	4.207E−30	1.077E−29	0.9248	2.2500	0.044	2.566	9.392E−31
16.500	12.5	1.189E−28	5	0.206	2.551	4.62	2.766E−30	7.209E−30	0.9422	2.2657	0.035	2.452	6.287E−31
16.800	12.5	6.576E−29	5	0.210	2.598	4.62	1.476E−30	3.916E−30	0.9596	2.2810	0.025	2.332	3.415E−31
17.100	12.5	2.682E−29	5	0.214	2.645	4.62	5.812E−31	1.569E−30	0.9771	2.2959	0.015	2.206	1.368E−31
17.400	12.5	4.956E−30	6	0.217	2.693	4.62	1.038E−31	2.849E−31	0.9945	2.3105	0.005	2.072	2.485E−32
17.700	12.5	1.499E−31	21	0.221	2.740	4.62	3.034E−33	8.472E−33	1.0119	2.3248	−0.005	1.929	7.388E−34
13.500	20.0	8.291E−28	5	0.270	2.073	7.42	2.862E−29	6.146E−29	0.7655	2.0147	0.055	3.414	5.359E−30
13.800	20.0	7.079E−28	5	0.276	2.120	7.42	2.340E−29	5.133E−29	0.7829	2.0315	0.046	3.329	4.476E−30
14.100	20.0	6.074E−28	5	0.282	2.167	7.41	1.924E−29	4.310E−29	0.8004	2.0478	0.037	3.243	3.759E−30
14.400	20.0	5.303E−28	5	0.288	2.214	7.41	1.611E−29	3.685E−29	0.8178	2.0636	0.028	3.153	3.213E−30
14.700	20.0	4.457E−28	5	0.294	2.262	7.41	1.303E−29	3.040E−29	0.8353	2.0789	0.019	3.061	2.651E−30
15.000	20.0	3.677E−28	5	0.300	2.309	7.40	1.031E−29	2.453E−29	0.8527	2.0937	0.009	2.966	2.139E−30
15.300	20.0	2.918E−28	5	0.306	2.356	7.40	7.864E−30	1.908E−29	0.8701	2.1081	−0.001	2.868	1.664E−30
15.600	20.0	2.240E−28	5	0.312	2.403	7.40	5.809E−30	1.437E−29	0.8876	2.1220	−0.011	2.766	1.253E−30
15.900	20.0	1.679E−28	5	0.318	2.450	7.39	4.193E−30	1.056E−29	0.9050	2.1356	−0.021	2.661	9.213E−31
16.200	20.0	1.168E−28	5	0.324	2.497	7.39	2.811E−30	7.213E−30	0.9224	2.1487	−0.032	2.551	6.290E−31
16.500	20.0	7.270E−29	5	0.330	2.544	7.39	1.687E−30	4.408E−30	0.9398	2.1614	−0.043	2.436	3.844E−31
16.800	20.0	3.883E−29	5	0.336	2.592	7.39	8.695E−31	2.312E−30	0.9572	2.1738	−0.054	2.315	2.016E−31
17.100	20.0	1.369E−29	5	0.342	2.639	7.38	2.960E−31	8.009E−31	0.9746	2.1858	−0.065	2.188	6.984E−32
17.400	20.0	1.947E−30	7	0.348	2.686	7.38	4.066E−32	1.119E−31	0.9919	2.1975	−0.076	2.052	9.761E−33
17.700	20.0	2.745E−32	45	0.354	2.733	7.38	5.542E−34	1.551E−33	1.0093	2.2089	−0.088	1.907	1.353E−34

p_3	θ_3	$\dfrac{d^2\sigma}{d\Omega_3\,dp_3}$	Error	p_T	p_L^*	θ_3^*	$\dfrac{1}{p_3^{*2}}\cdot\dfrac{d^2\sigma}{d\Omega_3^*\,dp_3^*}$	$\dfrac{E_3}{p_3^2}\cdot\dfrac{d^2\sigma}{d\Omega_3\,dp_3}$	x^*	y_3^*	t	m_4	$\dfrac{d^2\sigma}{dt\,dm_4^2}$
GeV/c	mrad	cm²sr⁻¹GeV⁻¹c	%	GeV/c	GeV/c	deg	cm²sr⁻¹GeV⁻³c³	cm²sr⁻¹GeV⁻²c³			GeV²	GeV	cm²/GeV⁴
$p_1 = 19.200$ GeV/c; $p_1^* = 2.929$ GeV/c; $E_1^* + E_2^* = 6.151$ GeV; $s = 37.830$ GeV²; $\gamma_{\mathrm{c.m.}} = 3.2779$; $\beta_{\mathrm{c.m.}} = 0.95233$; $y_1^* = 1.8562$; $p_{3,\max}^* = 2.708$ GeV/c; [70 A 1] (cont.)													
13.500	30.0	5.354E-28	5	0.405	2.061	11.11	1.839E-29	3.969E-29	0.7614	1.8882	-0.075	3.395	3.461E-30
13.800	30.0	4.539E-28	5	0.414	2.108	11.11	1.493E-29	3.291E-29	0.7788	1.9012	-0.086	3.310	2.870E-30
14.100	30.0	3.678E-28	5	0.423	2.156	11.10	1.159E-29	2.610E-29	0.7961	1.9137	-0.098	3.222	2.276E-30
14.400	30.0	3.105E-28	5	0.432	2.203	11.10	9.388E-30	2.158E-29	0.8135	1.9257	-0.110	3.131	1.881E-30
14.700	30.0	2.450E-28	5	0.441	2.249	11.09	7.141E-30	1.674E-29	0.8308	1.9373	-0.122	3.038	1.460E-30
15.000	30.0	2.007E-28	5	0.450	2.296	11.09	5.598E-30	1.339E-29	0.8482	1.9484	-0.135	2.942	1.167E-30
15.300	30.0	1.552E-28	5	0.459	2.343	11.08	4.162E-30	1.015E-29	0.8655	1.9591	-0.148	2.842	8.850E-31
15.600	30.0	1.132E-28	5	0.468	2.390	11.08	2.921E-30	7.260E-30	0.8828	1.9695	-0.161	2.739	6.331E-31
15.900	30.0	8.295E-29	5	0.477	2.437	11.07	2.061E-30	5.219E-30	0.9002	1.9794	-0.174	2.632	4.551E-31
16.200	30.0	5.421E-29	5	0.486	2.484	11.07	1.298E-30	3.348E-30	0.9175	1.9890	-0.187	2.520	2.919E-31
16.500	30.0	3.211E-29	5	0.495	2.531	11.06	7.415E-31	1.947E-30	0.9348	1.9982	-0.201	2.403	1.698E-31
16.800	30.0	1.434E-29	5	0.504	2.578	11.06	3.195E-31	8.539E-31	0.9521	2.0071	-0.215	2.280	7.446E-32
17.100	30.0	4.551E-30	6	0.513	2.625	11.06	9.791E-32	2.663E-31	0.9694	2.0157	-0.229	2.150	2.322E-32
17.400	30.0	4.600E-31	20	0.522	2.671	11.05	9.560E-33	2.645E-32	0.9867	2.0240	-0.243	2.011	2.306E-33
$p_1 = 24.000$ GeV/c; $p_1^* = 3.290$ GeV/c; $E_1^* + E_2^* = 6.843$ GeV; $s = 46.828$ GeV²; $\gamma_{\mathrm{c.m.}} = 3.6469$; $\beta_{\mathrm{c.m.}} = 0.96167$; $y_1^* = 1.9677$; $p_{3,\max}^* = 3.092$ GeV/c; [72 A 1]													
5.000	17.0	6.324E-27	3	0.085	0.611	7.92	1.609E-27	1.271E-27	0.1976	1.0283	-0.262	6.021	8.866E-29
5.000	27.0	5.651E-27	3	0.135	0.607	12.54	1.430E-27	1.136E-27	0.1963	1.0067	-0.315	6.017	7.923E-29
5.000	37.0	5.430E-27	3	0.185	0.601	17.10	1.365E-27	1.091E-27	0.1944	0.9770	-0.392	6.011	7.613E-29
5.000	47.0	5.130E-27	3	0.235	0.593	21.60	1.278E-27	1.031E-27	0.1919	0.9404	-0.493	6.002	7.192E-29
5.000	57.0	4.698E-27	3	0.285	0.584	26.00	1.157E-27	9.442E-28	0.1889	0.8985	-0.618	5.992	6.587E-29
5.000	67.0	4.220E-27	3	0.335	0.573	30.31	1.026E-27	8.481E-28	0.1852	0.8528	-0.766	5.979	5.916E-29
5.000	77.0	3.646E-27	3	0.385	0.560	34.50	8.728E-28	7.327E-28	0.1810	0.8044	-0.939	5.965	5.129E-29
5.000	87.0	3.088E-27	3	0.434	0.545	38.58	7.268E-28	6.206E-28	0.1761	0.7544	-1.135	5.949	4.329E-29
5.000	97.0	2.551E-27	3	0.484	0.528	42.53	5.916E-28	5.147E-28	0.1707	0.7037	-1.356	5.930	3.591E-29
5.000	107.0	2.156E-27	3	0.534	0.509	46.35	4.880E-28	4.333E-28	0.1647	0.6529	-1.600	5.909	3.023E-29
5.000	117.0	1.743E-27	3	0.584	0.489	50.05	3.860E-28	3.503E-28	0.1581	0.6024	-1.869	5.887	2.444E-29
5.000	127.0	1.505E-27	4	0.633	0.467	53.61	3.256E-28	3.025E-28	0.1510	0.5526	-2.161	5.862	2.110E-29
5.000	137.0	1.180E-27	4	0.683	0.443	57.04	2.491E-28	2.371E-28	0.1432	0.5038	-2.476	5.835	1.654E-29
5.000	147.0	9.420E-28	4	0.732	0.417	60.35	1.938E-28	1.892E-28	0.1348	0.4560	-2.816	5.806	1.321E-29
6.000	17.0	6.252E-27	3	0.102	0.764	7.60	1.142E-27	1.046E-27	0.2472	1.2036	-0.112	5.877	7.294E-29
6.000	27.0	5.744E-27	3	0.162	0.760	12.04	1.044E-27	9.606E-28	0.2457	1.1733	-0.175	5.872	6.701E-29
6.000	37.0	5.338E-27	3	0.222	0.753	16.43	9.629E-28	8.927E-28	0.2434	1.1322	-0.267	5.864	6.227E-29
6.000	47.0	4.817E-27	3	0.282	0.743	20.77	8.607E-28	8.055E-28	0.2404	1.0829	-0.388	5.853	5.620E-29
6.000	57.0	4.152E-27	3	0.342	0.732	25.03	7.333E-28	6.943E-28	0.2367	1.0280	-0.538	5.841	4.844E-29
6.000	67.0	3.713E-27	3	0.402	0.718	29.21	6.469E-28	6.209E-28	0.2324	0.9695	-0.717	5.825	4.332E-29
6.000	77.0	3.081E-27	3	0.462	0.703	33.30	5.284E-28	5.152E-28	0.2273	0.9091	-0.924	5.808	3.594E-29
6.000	87.0	2.409E-27	3	0.521	0.685	37.28	4.060E-28	4.029E-28	0.2215	0.8482	-1.160	5.787	2.810E-29
6.000	97.0	1.951E-27	3	0.581	0.665	41.16	3.225E-28	3.263E-28	0.2150	0.7876	-1.424	5.764	2.276E-29
6.000	107.0	1.512E-27	3	0.641	0.642	44.93	2.448E-28	2.529E-28	0.2078	0.7280	-1.717	5.739	1.764E-29
6.000	117.0	1.164E-27	3	0.700	0.618	48.58	1.842E-28	1.947E-28	0.1999	0.6698	-2.039	5.711	1.358E-29

Diddens/Schlüpmann

p_3	θ_3	$\dfrac{d^2\sigma}{d\Omega_3\,dp_3}$	Error	p_T	p_L^*	θ_3^*	$\dfrac{1}{p_3^{*2}}\cdot\dfrac{d^2\sigma}{d\Omega_3^*\,dp_3^*}$	$\dfrac{E_3}{p_3^2}\cdot\dfrac{d^2\sigma}{d\Omega_3\,dp_3}$	x^*	y_3^*	t	m_4	$\dfrac{d^2\sigma}{dt\,dm_4^2}$
GeV/c	mrad	cm² sr⁻¹ GeV⁻¹ c	%	GeV/c	GeV/c	deg	cm² sr⁻¹ GeV⁻³ c³	cm² sr⁻¹ GeV⁻² c³			GeV²	GeV	cm²/GeV⁴

$p_1 = 24.000$ GeV/c; $p_1^* = 3.290$ GeV/c; $E_1^* + E_2^* = 6.843$ GeV; $s = 46.828$ GeV²; $\gamma_{c.m.} = 3.6469$; $\beta_{c.m.} = 0.96167$; $y_1^* = 1.9677$; $p_{3,\,max}^* = 3.092$ GeV/c; [72 A 1] (cont.)

p_3	θ_3	$\dfrac{d^2\sigma}{d\Omega_3\,dp_3}$	Error	p_T	p_L^*	θ_3^*	$\dfrac{1}{p_3^{*2}}\cdot\dfrac{d^2\sigma}{d\Omega_3^*\,dp_3^*}$	$\dfrac{E_3}{p_3^2}\cdot\dfrac{d^2\sigma}{d\Omega_3\,dp_3}$	x^*	y_3^*	t	m_4	$\dfrac{d^2\sigma}{dt\,dm_4^2}$
6.000	127.0	8.772E-28	4	0.760	0.591	52.11	1.356E-28	1.467E-28	0.1912	0.6133	-2.390	5.680	1.023E-29
6.000	137.0	6.678E-28	4	0.819	0.563	55.53	1.006E-28	1.117E-28	0.1819	0.5585	-2.769	5.646	7.791E-30
6.000	147.0	5.193E-28	4	0.879	0.532	58.83	7.620E-29	8.684E-29	0.1719	0.5055	-3.176	5.610	6.058E-30
7.000	17.0	5.812E-27	3	0.119	0.914	7.42	7.962E-28	8.323E-28	0.2955	1.3499	-0.017	5.724	5.807E-29
7.000	27.0	5.260E-27	3	0.189	0.908	11.75	7.168E-28	7.533E-28	0.2937	1.3097	-0.091	5.718	5.255E-29
7.000	37.0	4.803E-27	3	0.259	0.900	16.05	6.497E-28	6.878E-28	0.2911	1.2564	-0.198	5.708	4.799E-29
7.000	47.0	4.654E-27	3	0.329	0.889	20.30	6.235E-28	6.665E-28	0.2876	1.1941	-0.339	5.696	4.650E-29
7.000	57.0	4.018E-27	3	0.399	0.876	24.48	5.319E-28	5.754E-28	0.2833	1.1264	-0.514	5.680	4.014E-29
7.000	67.0	2.952E-27	3	0.469	0.860	28.58	3.854E-28	4.228E-28	0.2782	1.0560	-0.722	5.662	2.949E-29
7.000	77.0	2.322E-27	3	0.538	0.842	32.61	2.983E-28	3.325E-28	0.2723	0.9850	-0.964	5.641	2.320E-29
7.000	87.0	1.794E-27	3	0.608	0.821	36.54	2.264E-28	2.569E-28	0.2655	0.9148	-1.239	5.616	1.792E-29
7.000	97.0	1.286E-27	3	0.678	0.797	40.37	1.591E-28	1.842E-28	0.2579	0.8462	-1.548	5.589	1.285E-29
7.000	107.0	9.773E-28	3	0.748	0.771	44.10	1.184E-28	1.400E-28	0.2495	0.7796	-1.890	5.558	9.764E-30
7.000	117.0	7.012E-28	3	0.817	0.743	47.72	8.301E-29	1.004E-28	0.2403	0.7155	-2.265	5.524	7.005E-30
7.000	127.0	4.911E-28	4	0.887	0.712	51.24	5.674E-29	7.033E-29	0.2302	0.6538	-2.674	5.487	4.906E-30
7.000	137.0	3.493E-28	4	0.956	0.678	54.65	3.933E-29	5.002E-29	0.2194	0.5947	-3.116	5.446	3.490E-30
7.000	147.0	2.467E-28	5	1.025	0.642	57.94	2.704E-29	3.533E-29	0.2077	0.5379	-3.592	5.403	2.465E-30
8.000	17.0	5.285E-27	3	0.136	1.061	7.31	5.620E-28	6.619E-28	0.3430	1.4749	0.044	5.563	4.617E-29
8.000	27.0	4.795E-27	3	0.215	1.054	11.58	5.072E-28	6.005E-28	0.3410	1.4238	-0.041	5.556	4.189E-29
8.000	37.0	4.322E-27	3	0.296	1.045	15.81	4.537E-28	5.413E-28	0.3379	1.3577	-0.163	5.545	3.776E-29
8.000	47.0	3.923E-27	3	0.376	1.033	20.00	4.078E-28	4.913E-28	0.3340	1.2824	-0.325	5.530	3.427E-29
8.000	57.0	3.100E-27	3	0.456	1.017	24.13	3.184E-28	3.882E-28	0.3291	1.2026	-0.524	5.512	2.708E-29
8.000	67.0	2.276E-27	3	0.536	0.999	28.19	2.305E-28	2.850E-28	0.3232	1.1215	-0.762	5.491	1.988E-29
8.000	77.0	1.658E-27	3	0.615	0.978	32.17	1.652E-28	2.076E-28	0.3164	1.0413	-1.038	5.465	1.449E-29
8.000	87.0	1.173E-27	3	0.695	0.954	36.06	1.148E-28	1.469E-28	0.3087	0.9634	-1.353	5.436	1.025E-29
8.000	97.0	7.934E-28	3	0.775	0.928	39.87	7.610E-29	9.936E-29	0.3000	0.8882	-1.706	5.404	6.932E-30
8.000	107.0	5.648E-28	3	0.854	0.898	43.57	5.302E-29	7.073E-29	0.2904	0.8163	-2.097	5.368	4.935E-30
8.000	117.0	3.751E-28	3	0.934	0.865	47.18	3.440E-29	4.698E-29	0.2799	0.7476	-2.526	5.328	3.277E-30
8.000	127.0	2.543E-28	3	1.013	0.830	50.68	2.275E-29	3.185E-29	0.2684	0.6821	-2.993	5.283	2.222E-30
8.000	137.0	1.667E-28	4	1.093	0.791	54.08	1.453E-29	2.088E-29	0.2560	0.6197	-3.499	5.235	1.456E-30
8.000	147.0	1.124E-28	4	1.172	0.750	57.37	9.535E-30	1.408E-29	0.2426	0.5602	-4.042	5.183	9.820E-31
9.000	17.0	4.828E-27	3	0.153	1.206	7.23	4.095E-28	5.373E-28	0.3900	1.5832	0.081	5.396	3.748E-29
9.000	27.0	4.314E-27	3	0.243	1.199	11.46	3.640E-28	4.801E-28	0.3876	1.5205	-0.014	5.387	3.349E-29
9.000	37.0	3.677E-27	3	0.333	1.188	15.65	3.079E-28	4.092E-28	0.3843	1.4414	-0.152	5.374	2.854E-29
9.000	47.0	2.902E-27	3	0.423	1.174	19.80	2.406E-28	3.229E-28	0.3798	1.3535	-0.333	5.357	2.253E-29
9.000	57.0	2.325E-27	3	0.513	1.157	23.90	1.904E-28	2.587E-28	0.3743	1.2625	-0.558	5.336	1.805E-29
9.000	67.0	1.637E-27	3	0.603	1.137	27.92	1.322E-28	1.822E-28	0.3677	1.1720	-0.825	5.311	1.271E-29
9.000	77.0	1.058E-27	3	0.692	1.113	31.88	8.483E-29	1.188E-28	0.3601	1.0840	-1.136	5.282	8.291E-30
9.000	87.0	7.305E-28	3	0.782	1.086	35.75	5.697E-29	8.129E-29	0.3514	0.9996	-1.490	5.248	5.671E-30
9.000	97.0	4.653E-28	3	0.872	1.056	39.53	3.557E-29	5.178E-29	0.3416	0.9193	-1.887	5.210	3.612E-30

p_3	θ_3	$\dfrac{\mathrm{d}^2\sigma}{\mathrm{d}\Omega_3\,\mathrm{d}p_3}$	Error	p_T	p_L^*	θ_3^*	$\dfrac{1}{p_3^{*2}}\cdot\dfrac{\mathrm{d}^2\sigma}{\mathrm{d}\Omega_3^*\,\mathrm{d}p_3^*}$	$\dfrac{E_3}{p_3^2}\cdot\dfrac{\mathrm{d}^2\sigma}{\mathrm{d}\Omega_3\,\mathrm{d}p_3}$	x^*	y_3^*	t	m_4	$\dfrac{\mathrm{d}^2\sigma}{\mathrm{d}t\,\mathrm{d}m_4^2}$
GeV/c	mrad	cm² sr⁻¹ GeV⁻¹ c	%	GeV/c	GeV/c	deg	cm² sr⁻¹ GeV⁻³ c³	cm² sr⁻¹ GeV⁻² c³			GeV²	GeV	cm²/GeV⁴

$p_1 = 24.000$ GeV/c; $p_1^* = 3.290$ GeV/c; $E_1^* + E_2^* = 6.843$ GeV; $s = 46.828$ GeV²; $\gamma_{\mathrm{c.m.}} = 3.6469$; $\beta_{\mathrm{c.m.}} = 0.96167$; $y_1^* = 1.9677$; $p_{3,\max}^* = 3.092$ GeV/c; [72 A 1] (cont.)

p_3	θ_3	$\dfrac{\mathrm{d}^2\sigma}{\mathrm{d}\Omega_3\,\mathrm{d}p_3}$	Error	p_T	p_L^*	θ_3^*	$\dfrac{1}{p_3^{*2}}\dfrac{\mathrm{d}^2\sigma}{\mathrm{d}\Omega_3^*\mathrm{d}p_3^*}$	$\dfrac{E_3}{p_3^2}\dfrac{\mathrm{d}^2\sigma}{\mathrm{d}\Omega_3\mathrm{d}p_3}$	x^*	y_3^*	t	m_4	$\dfrac{\mathrm{d}^2\sigma}{\mathrm{d}t\,\mathrm{d}m_4^2}$
9.000	107.0	3.065E-28	3	0.961	1.023	43.22	2.292E-29	3.411E-29	0.3308	0.8431	-2.327	5.168	2.379E-30
9.000	117.0	1.821E-28	4	1.051	0.986	46.81	1.330E-29	2.026E-29	0.3189	0.7709	-2.810	5.121	1.414E-30
9.000	127.0	1.093E-28	4	1.140	0.946	50.31	7.789E-30	1.216E-29	0.3060	0.7024	-3.335	5.070	8.485E-31
9.000	137.0	7.814E-29	4	1.229	0.903	53.70	5.424E-30	8.695E-30	0.2921	0.6376	-3.904	5.013	6.066E-31
9.000	147.0	4.438E-29	5	1.318	0.857	56.99	2.997E-30	4.939E-30	0.2770	0.5761	-4.515	4.952	3.445E-31
10.000	17.0	4.340E-27	3	0.170	1.350	7.18	3.002E-28	4.345E-28	0.4366	1.6782	0.103	5.222	3.031E-29
10.000	27.0	3.817E-27	3	0.270	1.342	11.38	2.627E-28	3.822E-28	0.4340	1.6033	-0.003	5.211	2.666E-29
10.000	37.0	3.077E-27	3	0.370	1.330	15.54	2.101E-28	3.081E-28	0.4302	1.5112	-0.156	5.197	2.149E-29
10.000	47.0	2.290E-27	3	0.470	1.315	19.66	1.548E-28	2.293E-28	0.4252	1.4114	-0.358	5.177	1.599E-29
10.000	57.0	1.643E-27	3	0.570	1.296	23.73	1.097E-28	1.645E-28	0.4191	1.3103	-0.607	5.153	1.148E-29
10.000	67.0	1.081E-27	3	0.669	1.273	27.74	7.116E-29	1.082E-28	0.4118	1.2116	-0.905	5.124	7.550E-30
10.000	77.0	6.670E-28	3	0.759	1.247	31.67	4.319E-29	6.678E-29	0.4033	1.1170	-1.250	5.090	4.659E-30
10.000	87.0	3.957E-28	3	0.869	1.217	35.52	2.516E-29	3.962E-29	0.3936	1.0273	-1.643	5.052	2.764E-30
10.000	97.0	2.302E-28	3	0.968	1.184	39.29	1.434E-29	2.305E-29	0.3828	0.9428	-2.084	5.008	1.608E-30
10.000	107.0	1.433E-28	3	1.068	1.146	42.97	8.733E-30	1.435E-29	0.3708	0.8632	-2.573	4.959	1.001E-30
10.000	117.0	8.388E-29	4	1.157	1.106	46.55	4.993E-30	8.398E-30	0.3576	0.7882	-3.109	4.904	5.859E-31
10.000	127.0	5.183E-29	4	1.267	1.061	50.04	3.009E-30	5.189E-30	0.3433	0.7175	-3.694	4.844	3.620E-31
10.000	137.0	2.931E-29	5	1.366	1.013	53.42	1.657E-30	2.935E-30	0.3277	0.6508	-4.325	4.779	2.047E-31
10.000	147.0	1.590E-29	4	1.465	0.962	56.71	8.745E-31	1.592E-30	0.3110	0.5878	-5.005	4.707	1.111E-31
11.000	17.0	3.913E-27	3	0.187	1.493	7.14	2.249E-28	3.561E-28	0.4828	1.7624	0.112	5.040	2.484E-29
11.000	27.0	3.224E-27	3	0.297	1.484	11.32	1.843E-28	2.934E-28	0.4800	1.6749	-0.004	5.028	2.047E-29
11.000	37.0	2.439E-27	3	0.407	1.471	15.46	1.383E-28	2.220E-28	0.4758	1.5700	-0.173	5.011	1.548E-29
11.000	47.0	1.694E-27	3	0.517	1.454	19.56	9.512E-29	1.542E-28	0.4704	1.4591	-0.394	4.989	1.075E-29
11.000	57.0	1.075E-27	3	0.627	1.434	23.61	5.963E-29	9.783E-29	0.4636	1.3489	-0.669	4.962	6.824E-30
11.000	67.0	6.781E-28	3	0.736	1.409	27.60	3.707E-29	6.171E-29	0.4556	1.2430	-0.996	4.929	4.305E-30
11.000	77.0	3.950E-28	3	0.846	1.380	31.52	2.124E-29	3.595E-29	0.4463	1.1429	-1.376	4.890	2.508E-30
11.000	87.0	2.162E-28	3	0.956	1.347	35.36	1.141E-29	1.967E-29	0.4356	1.0489	-1.808	4.845	1.373E-30
11.000	97.0	1.250E-28	3	1.065	1.310	39.12	6.466E-30	1.138E-29	0.4237	0.9609	-2.293	4.795	7.935E-31
11.000	107.0	6.997E-29	4	1.175	1.269	42.78	3.540E-30	6.367E-30	0.4105	0.8786	-2.831	4.739	4.442E-31
11.000	117.0	3.785E-29	4	1.284	1.224	46.36	1.870E-30	3.444E-30	0.3960	0.8014	-3.421	4.676	2.403E-31
11.000	127.0	2.214E-29	4	1.393	1.176	49.84	1.067E-30	2.015E-30	0.3802	0.7290	-4.064	4.607	1.406E-31
11.000	137.0	1.127E-29	6	1.502	1.123	53.22	5.288E-31	1.025E-30	0.3632	0.6609	-4.759	4.531	7.155E-32
11.000	147.0	5.913E-30	7	1.611	1.066	56.51	2.698E-31	5.381E-31	0.3448	0.5966	-5.506	4.448	3.754E-32
12.000	17.0	3.273E-27	3	0.204	1.635	7.11	1.587E-28	2.730E-28	0.5289	1.8376	0.113	4.850	1.904E-29
12.000	27.0	2.594E-27	3	0.324	1.626	11.27	1.251E-28	2.163E-28	0.5258	1.7373	-0.014	4.837	1.509E-29
12.000	37.0	1.811E-27	3	0.444	1.612	15.40	8.665E-29	1.510E-28	0.5213	1.6199	-0.198	4.818	1.054E-29
12.000	47.0	1.188E-27	3	0.564	1.593	19.49	5.627E-29	9.908E-29	0.5153	1.4986	-0.440	4.793	6.912E-30
12.000	57.0	6.970E-28	3	0.684	1.571	23.52	3.261E-29	5.813E-29	0.5080	1.3804	-0.739	4.762	4.055E-30
12.000	67.0	4.387E-28	3	0.803	1.544	27.50	2.023E-29	3.659E-29	0.4992	1.2683	-1.096	4.724	2.553E-30

p_3	θ_3	$\dfrac{\mathrm{d}^2\sigma}{\mathrm{d}\Omega_3\,\mathrm{d}p_3}$	Error	p_T	p_L^*	θ_3^*	$\dfrac{1}{p_3^{*2}}\cdot\dfrac{\mathrm{d}^2\sigma}{\mathrm{d}\Omega_3^*\,\mathrm{d}p_3^*}$	$\dfrac{E_3}{p_3^2}\cdot\dfrac{\mathrm{d}^2\sigma}{\mathrm{d}\Omega_3\,\mathrm{d}p_3}$	x^*	y_3^*	t	m_4	$\dfrac{\mathrm{d}^2\sigma}{\mathrm{d}t\,\mathrm{d}m_4^2}$
GeV/c	mrad	cm^2 sr^{-1} GeV^{-1} c	%	GeV/c	GeV/c	deg	cm^2 sr^{-1} GeV^{-3} c^3	cm^2 sr^{-1} GeV^{-2} c^3			GeV^2	GeV	cm^2/GeV^4

$p_1 = 24.000$ GeV/c; $p_1^* = 3.290$ GeV/c; $E_1^* + E_2^* = 6.843$ GeV; $s = 46.828$ GeV^2; $\gamma_\mathrm{c.m.} = 3.6469$; $\beta_\mathrm{c.m.} = 0.96167$; $y_1^* = 1.9677$; $p_{3,\mathrm{max}}^* = 3.092$ GeV/c; [72 A 1] (cont.)

p_3	θ_3	$\dfrac{\mathrm{d}^2\sigma}{\mathrm{d}\Omega_3\,\mathrm{d}p_3}$	Error	p_T	p_L^*	θ_3^*	$\dfrac{1}{p_3^{*2}}\cdot\dfrac{\mathrm{d}^2\sigma}{\mathrm{d}\Omega_3^*\,\mathrm{d}p_3^*}$	$\dfrac{E_3}{p_3^2}\cdot\dfrac{\mathrm{d}^2\sigma}{\mathrm{d}\Omega_3\,\mathrm{d}p_3}$	x^*	y_3^*	t	m_4	$\dfrac{\mathrm{d}^2\sigma}{\mathrm{d}t\,\mathrm{d}m_4^2}$
12.000	77.0	2.288E-28	3	0.923	1.512	31.40	1.038E-29	1.908E-29	0.4890	1.1635	-1.510	4.680	1.331E-30
12.000	87.0	1.123E-28	4	1.043	1.476	35.23	4.999E-30	9.366E-30	0.4774	1.0659	-1.982	4.629	6.534E-31
12.000	97.0	5.930E-29	4	1.162	1.436	38.98	2.586E-30	4.946E-30	0.4644	0.9752	-2.511	4.572	3.450E-31
12.000	107.0	2.891E-29	4	1.282	1.391	42.65	1.233E-30	2.411E-30	0.4500	0.8906	-3.098	4.507	1.682E-31
12.000	117.0	1.621E-29	5	1.401	1.343	46.22	6.753E-31	1.352E-30	0.4342	0.8118	-3.742	4.435	9.432E-32
12.000	127.0	8.044E-30	6	1.520	1.289	49.69	3.267E-31	6.709E-31	0.4170	0.7379	-4.443	4.355	4.680E-32
12.000	137.0	4.004E-30	7	1.639	1.232	53.07	1.584E-31	3.339E-31	0.3984	0.6686	-5.201	4.268	2.330E-32
12.000	142.0	2.786E-30	12	1.698	1.201	54.73	1.087E-31	2.324E-31	0.3885	0.6356	-5.601	4.220	1.621E-32
13.000	17.0	2.750E-27	3	0.221	1.777	7.09	1.139E-28	2.117E-28	0.5749	1.9050	0.107	4.652	1.477E-29
13.000	27.0	2.035E-27	3	0.351	1.767	11.23	8.386E-29	1.567E-28	0.5715	1.7918	-0.030	4.638	1.093E-29
13.000	37.0	1.340E-27	3	0.481	1.752	15.35	5.479E-29	1.032E-28	0.5666	1.6625	-0.230	4.616	7.196E-30
13.000	47.0	8.023E-28	3	0.611	1.732	19.43	3.248E-29	6.176E-29	0.5601	1.5318	-0.492	4.588	4.308E-30
13.000	57.0	4.488E-28	3	0.741	1.707	23.45	1.794E-29	3.455E-29	0.5522	1.4063	-0.816	4.552	2.410E-30
13.000	67.0	2.273E-28	3	0.870	1.678	27.42	8.956E-30	1.750E-29	0.5427	1.2890	-1.203	4.509	1.221E-30
13.000	77.0	1.073E-28	3	1.000	1.644	31.31	4.158E-30	8.260E-30	0.5316	1.1802	-1.652	4.459	5.762E-31
13.000	87.0	5.157E-29	3	1.130	1.605	35.14	1.962E-30	3.970E-30	0.5191	1.0796	-2.163	4.402	2.769E-31
13.000	97.0	2.357E-29	4	1.259	1.561	38.88	8.784E-31	1.814E-30	0.5050	0.9865	-2.736	4.336	1.266E-31
13.000	107.0	1.176E-29	4	1.388	1.513	42.54	4.286E-31	9.053E-31	0.4894	0.9002	-3.372	4.262	6.315E-32
13.000	117.0	5.511E-30	5	1.518	1.460	46.10	1.961E-31	4.242E-31	0.4722	0.8199	-4.069	4.180	2.959E-32
13.000	127.0	2.372E-30	7	1.647	1.402	49.58	8.230E-32	1.826E-31	0.4536	0.7450	-4.828	4.088	1.274E-32
13.000	137.0	1.214E-30	10	1.775	1.340	52.96	4.101E-32	9.345E-32	0.4334	0.6748	-5.650	3.986	6.519E-33
14.000	17.0	2.320E-27	3	0.238	1.919	7.07	8.308E-29	1.658E-28	0.6206	1.9659	0.095	4.445	1.157E-29
14.000	27.0	1.642E-27	3	0.378	1.908	11.21	5.849E-29	1.174E-28	0.6170	1.8397	-0.052	4.428	8.187E-30
14.000	37.0	9.948E-28	3	0.518	1.891	15.31	3.516E-29	7.110E-29	0.6117	1.6992	-0.267	4.404	4.960E-30
14.000	57.0	2.659E-28	3	0.798	1.843	23.40	9.223E-30	1.908E-29	0.5962	1.4280	-0.899	4.332	1.331E-30
14.000	67.0	1.175E-28	3	0.937	1.812	27.35	4.001F-30	8.398E-30	0.5860	1.3060	-1.315	4.283	5.859E-31
14.000	77.0	5.159E-29	3	1.077	1.775	31.24	1.731E-30	3.694E-30	0.5741	1.1938	-1.799	4.227	2.577E-31
14.000	87.0	2.228E-29	4	1.216	1.733	35.06	7.324E-31	1.592E-30	0.5606	1.0907	-2.349	4.161	1.111E-31
14.000	97.0	1.026E-29	4	1.356	1.686	38.80	3.304E-31	7.333E-31	0.5454	0.9958	-2.966	4.086	5.116E-32
14.000	107.0	4.129E-30	6	1.495	1.634	42.45	1.300E-31	2.951E-31	0.5286	0.9080	-3.651	4.001	2.059E-32
14.000	117.0	2.075E-30	8	1.634	1.577	46.02	6.381E-32	1.483E-31	0.5101	0.8265	-4.402	3.906	1.035E-32
14.000	127.0	8.011E-31	11	1.773	1.515	49.49	2.402E-32	5.726E-32	0.4900	0.7507	-5.219	3.800	3.994E-33
14.000	137.0	3.352E-31	16	1.912	1.448	52.86	9.784E-33	2.396E-32	0.4683	0.6797	-6.104	3.682	1.671E-33
15.000	17.0	1.796E-27	3	0.255	2.060	7.06	5.614E-29	1.198E-28	0.6663	2.0211	0.080	4.227	8.357E-30
15.000	27.0	1.155E-27	3	0.405	2.048	11.18	3.622E-29	7.771E-29	0.6625	1.8821	-0.079	4.208	5.421E-30
15.000	37.0	6.509E-28	3	0.555	2.031	15.28	2.008E-29	4.342E-29	0.6568	1.7308	-0.309	4.181	3.029E-30
15.000	47.0	3.200E-28	3	0.705	2.008	19.34	9.771E-30	2.134E-29	0.6494	1.5834	-0.611	4.144	1.489E-30
15.000	57.0	1.473E-28	3	0.855	1.979	23.35	4.442E-30	9.825E-30	0.6402	1.4461	-0.986	4.099	6.854E-31
15.000	67.0	6.089E-29	3	1.004	1.945	27.30	1.810E-30	4.062E-30	0.6292	1.3201	-1.432	4.044	2.833E-31

p_3	θ_3	$\dfrac{\mathrm{d}^2\sigma}{\mathrm{d}\Omega_3\,\mathrm{d}p_3}$	Error	p_T	p_L^*	θ_3^*	$\dfrac{1}{p_3^{*2}}\cdot\dfrac{\mathrm{d}^2\sigma}{\mathrm{d}\Omega_3^*\,\mathrm{d}p_3^*}$	$\dfrac{E_3}{p_3^2}\cdot\dfrac{c^2\sigma}{\mathrm{d}\Omega_3\,\mathrm{d}p_3}$	x^*	y_3^*	t	m_4	$\dfrac{\mathrm{d}^2\sigma}{\mathrm{d}t\,\mathrm{d}m_4^2}$
GeV/c	mrad	cm^{2}sr^{-1}GeV^{-1}c	%	GeV/c	GeV/c	deg	cm^{2}sr^{-1}GeV^{-3}c^3	cm^{2}sr^{-1}G$_2$V^{-2}c^3			GeV^2	GeV	cm^2/GeV^4

$p_1 = 24.000$ GeV/c; $p_1^* = 3.290$ GeV/c; $E_1^* + E_2^* = 6.843$ GeV; $s = 46.828$ GeV^2; $\gamma_{\mathrm{c.m.}} = 3.6469$; $\beta_{\mathrm{c.m.}} = 0.96167$; $y_1^* = 1.9677$; $p_{3,\max}^* = 3.092$ GeV/c; [72 A 1] (cont.)

p_3	θ_3	$\mathrm{d}^2\sigma/\mathrm{d}\Omega_3\mathrm{d}p_3$	Error	p_T	p_L^*	θ_3^*	(1/p*²)d²σ/dΩ*dp*	(E/p²)c²σ	x^*	y_3^*	t	m_4	d²σ/dt dm₄²
15.000	77.0	2.320E-29	4	1.154	1.906	31.19	6.781E-31	1.548E-30	0.6165	1.2051	-1.950	3.980	1.080E-31
15.000	87.0	8.650E-30	6	1.303	1.861	35.00	2.481E-31	5.770E-31	0.6020	1.0999	-2.539	3.905	4.025E-32
15.000	97.0	3.846E-30	8	1.453	1.811	38.73	1.081E-31	2.565E-31	0.5857	1.0033	-3.201	3.819	1.790E-32
15.000	107.0	1.491E-30	9	1.602	1.755	42.38	4.097E-32	9.945E-32	0.5677	0.9143	-3.934	3.722	6.938E-33
15.000	117.0	6.308E-31	14	1.751	1.694	45.94	1.693E-32	4.208E-32	0.5479	0.8319	-4.739	3.612	2.935E-33
15.000	127.0	1.998E-31	20	1.900	1.628	49.41	5.226E-33	1.333E-32	0.5264	0.7553	-5.615	3.489	9.297E-34
16.000	17.0	1.390E-27	3	0.272	2.201	7.04	3.825E-29	8.692E-29	0.7120	2.0714	0.061	3.997	6.063E-30
16.000	27.0	8.628E-28	3	0.432	2.188	11.17	2.361E-29	5.395E-29	0.7078	1.9196	-0.108	3.975	3.764E-30
16.000	37.0	4.353E-28	3	0.592	2.170	15.26	1.182E-29	2.722E-29	0.7018	1.7583	-0.354	3.944	1.899E-30
16.000	57.0	8.086E-29	3	0.912	2.115	23.31	2.147E-30	5.056E-30	0.6840	1.4615	-1.076	3.852	3.527E-31
16.000	67.0	3.153E-29	4	1.071	2.079	27.26	8.249E-31	1.972E-30	0.6723	1.3320	-1.552	3.789	1.375E-31
16.000	77.0	1.057E-29	4	1.231	2.037	31.14	2.745E-31	6.672E-31	0.6588	1.2145	-2.104	3.716	4.654E-32
16.000	87.0	3.966E-30	5	1.390	1.989	34.95	1.001E-31	2.480E-31	0.6433	1.1075	-2.733	3.630	1.730E-32
16.000	97.0	1.269E-30	7	1.550	1.935	38.68	3.139E-32	7.935E-32	0.6260	1.0096	-3.439	3.532	5.536E-33
16.000	107.0	5.703E-31	12	1.709	1.876	42.33	1.379E-32	3.566E-32	0.6068	0.9196	-4.221	3.419	2.488E-33
16.000	117.0	1.624E-31	26	1.868	1.811	45.89	3.835E-33	1.015E-32	0.5857	0.8364	-5.079	3.291	7.084E-34
16.000	127.0	5.338E-32	44	2.027	1.740	49.35	1.229E-33	3.338E-33	0.5627	0.7591	-6.014	3.146	2.329E-34
17.000	17.0	9.224E-28	3	0.289	2.342	7.03	2.251E-29	5.428E-29	0.7575	2.1172	0.038	3.752	3.787E-30
17.000	27.0	5.171E-28	3	0.459	2.329	11.15	1.255E-29	3.043E-29	0.7531	1.9530	-0.141	3.728	2.123E-30
17.000	37.0	2.482E-28	3	0.629	2.309	15.24	5.978E-30	1.461E-29	0.7467	1.7822	-0.402	3.692	1.019E-30
17.000	47.0	9.623E-29	3	0.799	2.283	19.29	2.294E-30	5.663E-30	0.7383	1.6212	-0.745	3.646	3.951E-31
17.000	57.0	3.442E-29	3	0.968	2.250	23.28	8.105E-31	2.025E-30	0.7278	1.4746	-1.169	3.587	1.413E-31
17.000	67.0	1.182E-29	5	1.138	2.212	27.23	2.743E-31	6.955E-31	0.7154	1.3421	-1.674	3.516	4.852E-32
17.000	77.0	3.582E-30	8	1.308	2.167	31.10	8.173E-32	2.108E-31	0.7010	1.2225	-2.261	3.431	1.471E-32
17.000	87.0	1.330E-30	12	1.477	2.117	34.91	2.978E-32	7.827E-32	0.6846	1.1139	-2.930	3.333	5.460E-33
17.000	97.0	3.439E-31	17	1.646	2.060	38.64	7.544E-33	2.024E-32	0.6661	1.0149	-3.679	3.218	1.412E-33
17.000	107.0	1.731E-31	23	1.816	1.997	42.28	3.713E-33	1.019E-32	0.6457	0.9240	-4.510	3.086	7.106E-34
17.000	117.0	4.457E-32	37	1.984	1.927	45.84	9.334E-34	2.623E-33	0.6233	0.8401	-5.422	2.935	1.830E-34
17.000	127.0	1.247E-32	70	2.153	1.852	49.30	2.546E-34	7.338E-34	0.5989	0.7623	-6.415	2.761	5.119E-35
18.000	17.0	6.302E-28	3	0.306	2.483	7.03	1.374E-29	3.502E-29	0.8030	2.1591	0.014	3.489	2.443E-30
18.000	27.0	3.379E-28	3	0.486	2.468	11.14	7.325E-30	1.878E-29	0.7983	1.9828	-0.176	3.462	1.310E-30
18.000	37.0	1.445E-28	3	0.666	2.447	15.22	3.110E-30	8.056E-30	0.7915	1.8032	-0.453	3.422	5.606E-31
18.000	47.0	5.246E-29	3	0.846	2.420	19.26	1.117E-30	2.916E-30	0.7826	1.6364	-0.815	3.368	2.034E-31
18.000	57.0	1.752E-29	5	1.025	2.386	23.26	3.684E-31	9.737E-31	0.7716	1.4858	-1.264	3.301	6.793E-32
18.000	67.0	4.483E-30	7	1.205	2.345	27.20	9.288E-32	2.491E-31	0.7584	1.3507	-1.800	3.219	1.738E-32
18.000	77.0	1.278E-30	9	1.385	2.298	31.07	2.604E-32	7.103E-32	0.7432	1.2292	-2.421	3.121	4.955E-33
18.000	87.0	4.150E-31	12	1.564	2.244	34.88	8.318E-33	2.312E-32	0.7258	1.1194	-3.129	3.005	1.613E-33
18.000	97.0	1.245E-31	22	1.743	2.184	38.60	2.439E-33	6.919E-33	0.7063	1.0193	-3.923	2.870	4.827E-34

p_3	θ_3	$\dfrac{d^2\sigma}{d\Omega_3\,dp_3}$	Error	p_T	p_L^*	θ_3^*	$\dfrac{1}{p_3^{*2}}\cdot\dfrac{d^2\sigma}{d\Omega_3^*\,dp_3^*}$	$\dfrac{E_3}{p_3^2}\cdot\dfrac{d^2\sigma}{d\Omega_3\,dp_3}$	x^*	y_3^*	t	m_4	$\dfrac{d^2\sigma}{dt\,dm_4^2}$
GeV/c	mrad	cm² sr⁻¹ GeV⁻¹ c	%	GeV/c	GeV/c	deg	cm² sr⁻¹ GeV⁻³ c³	cm² sr⁻¹ GeV⁻² c³			GeV²	GeV	cm²/GeV⁴

$p_1 = 24.000$ GeV/c; $p_1^* = 3.290$ GeV/c; $E_1^* + E_2^* = 6.843$ GeV; $s = 46.828$ GeV²; $\gamma_{c.m.} = 3.6469$; $\beta_{c.m.} = 0.96167$; $y_1^* = 1.9677$; $p_{3,max}^* = 3.092$ GeV/c; [72 A 1] (cont.)

p_3	θ_3	$\dfrac{d^2\sigma}{d\Omega_3\,dp_3}$	Error	p_T	p_L^*	θ_3^*	$\dfrac{1}{p_3^{*2}}\dfrac{d^2\sigma}{d\Omega_3^*\,dp_3^*}$	$\dfrac{E_3}{p_3^2}\dfrac{d^2\sigma}{d\Omega_3\,dp_3}$	x^*	y_3^*	t	m_4	$\dfrac{d^2\sigma}{dt\,dm_4^2}$
19.000	17.0	3.764E-28	3	0.323	2.623	7.02	7.370E-30	1.982E-29	0.8484	2.1975	-0.012	3.205	1.382E-30
19.000	27.0	1.677E-28	3	0.513	2.608	11.13	3.266E-30	8.829E-30	0.8435	2.0096	-0.213	3.174	6.159E-31
19.000	37.0	6.362E-29	3	0.703	2.586	15.21	1.229E-30	3.350E-30	0.8363	1.8218	-0.505	3.127	2.337E-31
19.000	47.0	2.101E-29	3	0.893	2.557	19.25	4.018E-31	1.106E-30	0.8269	1.6496	-0.888	3.065	7.717E-32
19.000	57.0	5.630E-30	5	1.082	2.521	23.24	1.063E-31	2.964E-31	0.8153	1.4955	-1.362	2.987	2.068E-32
19.000	67.0	1.241E-30	8	1.272	2.478	27.17	2.310E-32	6.534E-32	0.8014	1.3581	-1.927	2.891	4.558E-33
19.000	77.0	3.595E-31	10	1.462	2.428	31.05	6.580E-33	1.893E-32	0.7853	1.2350	-2.583	2.775	1.320E-33
19.000	87.0	8.099E-32	18	1.651	2.371	34.85	1.455E-33	4.264E-33	0.7669	1.1240	-3.330	2.637	2.975E-34
19.000	97.0	2.854E-32	35	1.840	2.308	38.57	5.021E-34	1.503E-33	0.7463	1.0231	-4.168	2.473	1.048E-34
19.000	107.0	1.049E-32	70	2.029	2.237	42.21	1.805E-34	5.523E-34	0.7235	0.9308	-5.096	2.278	3.853E-35
4.000	12.0	5.678E-27	4	0.048	0.452	6.07	2.132E-27	1.430E-27	0.1460	0.8164	-0.495	6.156	9.978E-29
4.500	12.0	6.226E-27	4	0.054	0.533	5.78	1.910E-27	1.392E-27	0.1724	0.9322	-0.354	6.091	9.710E-29
5.000	12.0	6.300E-27	3	0.060	0.612	5.60	1.605E-27	1.266E-27	0.1980	1.0356	-0.245	6.023	8.833E-29
5.500	12.0	6.413E-27	3	0.066	0.690	5.47	1.376E-27	1.171E-27	0.2231	1.1290	-0.159	5.952	8.167E-29
6.000	12.0	6.501E-27	3	0.072	0.766	5.37	1.189E-27	1.087E-27	0.2477	1.2140	-0.091	5.879	7.584E-29
6.500	12.0	6.246E-27	3	0.078	0.841	5.30	9.849E-28	9.637E-28	0.2721	1.2920	-0.036	5.803	6.723E-29
7.000	12.0	6.138E-27	3	0.084	0.916	5.24	8.423E-28	8.790E-28	0.2961	1.3639	0.008	5.726	6.132E-29
7.500	12.0	5.796E-27	3	0.090	0.989	5.20	6.981E-28	7.745E-28	0.3200	1.4307	0.043	5.647	5.403E-29
8.000	12.0	5.572E-27	3	0.095	1.063	5.15	5.935E-28	6.978E-28	0.3437	1.4929	0.072	5.566	4.868E-29
8.500	12.0	5.611E-27	3	0.102	1.136	5.13	5.322E-28	6.612E-28	0.3673	1.5511	0.095	5.483	4.613E-29
9.000	12.0	5.299E-27	3	0.108	1.208	5.11	4.502E-28	5.897E-28	0.3908	1.6057	0.113	5.399	4.114E-29
9.500	12.0	5.084E-27	3	0.114	1.280	5.09	3.891E-28	5.359E-28	0.4141	1.6571	0.127	5.313	3.738E-29
10.000	12.0	4.833E-27	3	0.120	1.352	5.07	3.349E-28	4.839E-28	0.4374	1.7056	0.138	5.225	3.376E-29
10.500	12.0	4.576E-27	3	0.125	1.424	5.06	2.884E-28	4.363E-28	0.4606	1.7514	0.145	5.135	3.044E-29
11.000	12.0	4.319E-27	3	0.132	1.496	5.04	2.486E-28	3.930E-28	0.4838	1.7949	0.151	5.044	2.742E-29
11.500	12.0	3.722E-27	3	0.138	1.567	5.03	1.965E-28	3.240E-28	0.5069	1.8363	0.154	4.950	2.260E-29
12.000	12.0	3.459E-27	3	0.144	1.639	5.02	1.680E-28	2.885E-28	0.5300	1.8756	0.155	4.855	2.013E-29
12.500	12.0	3.113E-27	3	0.150	1.710	5.01	1.396E-28	2.492E-28	0.5530	1.9130	0.154	4.757	1.739E-29
13.000	12.0	2.908E-27	3	0.156	1.781	5.01	1.207E-28	2.239E-28	0.5760	1.9488	0.152	4.657	1.562E-29
13.500	12.0	2.673E-27	4	0.162	1.852	5.00	1.030E-28	1.981E-28	0.5989	1.9830	0.149	4.555	1.382E-29
14.000	12.0	2.489E-27	4	0.168	1.923	4.99	8.930E-29	1.779E-28	0.6218	2.0157	0.144	4.450	1.241E-29
14.500	12.0	2.270E-27	4	0.174	1.994	4.99	7.600E-29	1.566E-28	0.6447	2.0470	0.139	4.343	1.093E-29
15.000	12.0	2.093E-27	4	0.180	2.064	4.98	6.554E-29	1.396E-28	0.6676	2.0770	0.132	4.233	9.739E-30
15.500	12.0	1.645E-27	4	0.186	2.135	4.98	4.828E-29	1.062E-28	0.6905	2.1058	0.124	4.120	7.407E-30
16.000	12.0	1.555E-27	4	0.192	2.206	4.98	4.287E-29	9.723E-29	0.7133	2.1335	0.116	4.003	6.783E-30
16.500	12.0	1.334E-27	4	0.198	2.276	4.97	3.460E-29	8.088E-29	0.7361	2.1601	0.107	3.883	5.643E-30
17.000	12.0	1.058E-27	5	0.204	2.347	4.97	2.587E-29	6.226E-29	0.7590	2.1857	0.098	3.759	4.343E-30
17.500	12.0	8.975E-28	5	0.210	2.417	4.97	2.072E-29	5.131E-29	0.7817	2.2103	0.087	3.631	3.579E-30
18.000	12.0	7.648E-28	5	0.216	2.488	4.96	1.670E-29	4.250E-29	0.8045	2.2341	0.077	3.498	2.965E-30

3.6.5 pp→p̄X

p_3	θ_3	$\dfrac{d^2\sigma}{d\Omega_3\,dp_3}$	Error	p_T	p_L^*	θ_3^*	$\dfrac{1}{p_3^{*2}}\cdot\dfrac{d^2\sigma}{d\Omega_3^*\,dp_3^*}$	$\dfrac{E_3}{p_3^2}\cdot\dfrac{d^2\sigma}{d\Omega_3\,dp_3}$	x^*	y_3^*	t	m_4	$\dfrac{d^2\sigma}{dt\,dm_4^2}$
GeV/c	mrad	$cm^2\,sr^{-1}\,GeV^{-1}c$	%	GeV/c	GeV/c	deg	$cm^2\,sr^{-1}\,GeV^{-3}\,c^3$	$cm^2\,sr^{-1}\,GeV^{-2}\,c^3$			GeV^2	GeV	cm^2/GeV^4

$p_1 = 12.500$ GeV/c; $p_1^* = 2.332$ GeV/c; $E_1^* + E_2^* = 5.028$ GeV; $s = 25.281$ GeV^2; $\gamma_{c.m.} = 2.6796$; $\beta_{c.m.} = 0.92776$; $y_1^* = 1.6420$; $p_{3,max}^* = 1.552$ GeV/c; [71 A 1]

p_3	θ_3	$\frac{d^2\sigma}{d\Omega_3 dp_3}$	Error	p_T	p_L^*	θ_3^*	$\frac{1}{p_3^{*2}}\frac{d^2\sigma}{d\Omega_3^* dp_3^*}$	$\frac{E_3}{p_3^2}\frac{d^2\sigma}{d\Omega_3 dp_3}$	x^*	y_3^*	t	m_4	$\frac{d^2\sigma}{dt dm_4^2}$
4.833	131.2	1.519E-30	40	0.632	0.600	46.50	2.500E-31	3.202E-31	0.3866	0.5082	-1.880	3.645	4.288E-32
4.938	143.7	8.930E-31	60	0.707	0.600	49.69	1.395E-31	1.841E-31	0.3865	0.4907	-2.074	3.591	2.466E-32

$p_1 = 19.200$ GeV/c; $p_1^* = 2.929$ GeV/c; $E_1^* + E_2^* = 6.151$ GeV; $s = 37.830$ GeV^2; $\gamma_{c.m.} = 3.2779$; $\beta_{c.m.} = 0.95233$; $y_1^* = 1.8562$; $p_{3,max}^* = 2.320$ GeV/c; [70 A 1]

p_3	θ_3	$\frac{d^2\sigma}{d\Omega_3 dp_3}$	Error	p_T	p_L^*	θ_3^*	$\frac{1}{p_3^{*2}}\frac{d^2\sigma}{d\Omega_3^* dp_3^*}$	$\frac{E_3}{p_3^2}\frac{d^2\sigma}{d\Omega_3 dp_3}$	x^*	y_3^*	t	m_4	$\frac{d^2\sigma}{dt dm_4^2}$
4.500	12.5	2.490E-28	3	0.056	0.400	8.01	5.534E-29	5.652E-29	0.1724	0.4137	-2.179	5.113	4.929E-30
4.500	20.0	2.246E-28	3	0.090	0.398	12.74	4.983E-29	5.098E-29	0.1716	0.4108	-2.200	5.111	4.446E-30
4.500	30.0	2.401E-28	3	0.135	0.394	18.89	5.309E-29	5.450E-29	0.1700	0.4050	-2.244	5.107	4.753E-30
4.500	40.0	2.177E-28	3	0.180	0.389	24.81	4.790E-29	4.942E-29	0.1678	0.3970	-2.304	5.101	4.309E-30
4.500	50.0	2.043E-28	3	0.225	0.383	30.44	4.468E-29	4.638E-29	0.1649	0.3869	-2.382	5.093	4.044E-30
4.500	60.0	1.919E-28	3	0.270	0.375	35.77	4.166E-29	4.356E-29	0.1614	0.3749	-2.477	5.084	3.799E-30
4.500	70.0	1.658E-28	3	0.315	0.365	40.77	3.568E-29	3.764E-29	0.1573	0.3610	-2.589	5.073	3.282E-30
6.000	12.5	1.582E-28	3	0.075	0.708	6.04	2.265E-29	2.669E-29	0.3053	0.6953	-1.335	4.921	2.327E-30
6.000	20.0	1.485E-28	3	0.120	0.706	9.65	2.122E-29	2.505E-29	0.3043	0.6903	-1.364	4.918	2.184E-30
6.000	30.0	1.361E-28	3	0.180	0.701	14.40	1.937E-29	2.295E-29	0.3022	0.6802	-1.421	4.913	2.002E-30
6.000	40.0	1.241E-28	3	0.240	0.694	19.07	1.757E-29	2.093E-29	0.2992	0.6665	-1.502	4.904	1.826E-30
6.000	50.0	1.067E-28	3	0.300	0.685	23.63	1.500E-29	1.800E-29	0.2954	0.6493	-1.605	4.894	1.570E-30
6.000	60.0	9.210E-29	3	0.360	0.675	28.07	1.284E-29	1.554E-29	0.2907	0.6290	-1.732	4.881	1.355E-30
6.000	70.0	7.513E-29	3	0.420	0.662	32.38	1.037E-29	1.267E-29	0.2852	0.6061	-1.882	4.866	1.105E-30
8.000	12.5	5.561E-29	3	0.100	1.077	5.31	4.888E-30	6.999E-30	0.4641	0.9779	-0.738	4.593	6.103E-31
8.000	20.0	5.078E-29	3	0.160	1.074	8.48	4.454E-30	6.391E-30	0.4628	0.9691	-0.775	4.589	5.573E-31
8.000	30.0	4.850E-29	3	0.240	1.067	12.67	4.236E-30	6.104E-30	0.4599	0.9517	-0.852	4.581	5.323E-31
8.000	40.0	3.431E-29	3	0.320	1.058	16.82	2.978E-30	4.318E-30	0.4560	0.9282	-0.960	4.569	3.765E-31
8.000	50.0	2.818E-29	3	0.400	1.046	20.92	2.428E-30	3.547E-30	0.4509	0.8995	-1.098	4.554	3.093E-31
8.000	60.0	2.276E-29	3	0.480	1.032	24.94	1.942E-30	2.864E-30	0.4447	0.8666	-1.267	4.535	2.498E-31
8.000	70.0	1.544E-29	3	0.560	1.015	28.87	1.303E-30	1.943E-30	0.4373	0.8302	-1.466	4.513	1.695E-31
10.000	12.5	1.416E-29	3	0.125	1.423	5.02	8.322E-31	1.422E-30	0.6133	1.1967	-0.416	4.206	1.240E-31
10.000	20.0	1.168E-29	3	0.200	1.419	8.02	6.849E-31	1.173E-30	0.6116	1.1832	-0.463	4.200	1.023E-31
10.000	30.0	8.994E-30	3	0.300	1.411	12.00	5.250E-31	9.033E-31	0.6080	1.1566	-0.559	4.189	7.877E-32
10.000	35.0	6.908E-30	3	0.350	1.405	13.98	4.021E-31	6.938E-31	0.6058	1.1401	-0.621	4.181	6.050E-32
10.000	40.0	6.593E-30	3	0.400	1.399	15.95	3.824E-31	6.622E-31	0.6031	1.1216	-0.693	4.173	5.774E-32
10.000	50.0	4.393E-30	3	0.500	1.385	19.85	2.528E-31	4.412E-31	0.5967	1.0800	-0.866	4.152	3.848E-32
10.000	60.0	2.850E-30	3	0.600	1.367	23.69	1.624E-31	2.863E-31	0.5890	1.0333	-1.077	4.126	2.496E-32
10.000	70.0	1.958E-30	4	0.699	1.345	27.47	1.103E-31	1.967E-31	0.5798	0.9833	-1.326	4.096	1.715E-32
11.000	12.5	5.030E-30	3	0.137	1.591	4.94	2.477E-31	4.589E-31	0.6859	1.2898	-0.312	3.990	4.002E-32
11.000	20.0	4.442E-30	3	0.220	1.587	7.89	2.183E-31	4.053E-31	0.6840	1.2736	-0.363	3.984	3.534E-32
11.000	30.0	3.118E-30	3	0.330	1.578	11.81	1.525E-31	2.845E-31	0.6801	1.2419	-0.469	3.971	2.481E-32

p_3	θ_3	$\dfrac{d^2\sigma}{d\Omega_3\,dp_3}$	Error	p_T	p_L^*	θ_3^*	$\dfrac{1}{p_3^{*2}}\cdot\dfrac{d^2\sigma}{d\Omega_3^*\,dp_3^*}$	$\dfrac{E_3}{p_3^2}\cdot\dfrac{d^2\sigma}{d\Omega_3\,dp_3}$	x^*	y_3^*	t	m_4	$\dfrac{d^2\sigma}{dt\,dm_4^2}$
GeV/c	mrad	$cm^2\,sr^{-1}\,GeV^{-1}\,c$	%	GeV/c	GeV/c	deg	$cm^2\,sr^{-1}\,GeV^{-3}\,c^3$	$cm^2\,sr^{-1}\,GeV^{-2}\,c^3$			GeV^2	GeV	cm^2/GeV^4

$p_1 = 19.200$ GeV/c; $p_1^* = 2.929$ GeV/c; $E_1^* + E_2^* = 6.151$ GeV; $s = 37.830$ GeV^2; $\gamma_{c.m.} = 3.2779$; $\beta_{c.m.} = 0.95233$; $y_1^* = 1.8562$; $p_{3,\,max}^* = 2.320$ GeV/c; [70 A 1] (cont.)

p_3	θ_3	$\dfrac{d^2\sigma}{d\Omega_3\,dp_3}$	Error	p_T	p_L^*	θ_3^*	$\dfrac{1}{p_3^{*2}}\cdot\dfrac{d^2\sigma}{d\Omega_3^*\,dp_3^*}$	$\dfrac{E_3}{p_3^2}\cdot\dfrac{d^2\sigma}{d\Omega_3\,dp_3}$	x^*	y_3^*	t	m_4	$\dfrac{d^2\sigma}{dt\,dm_4^2}$
11.000	35.0	2.350E-30	6	0.385	1.572	13.76	1.146E-31	2.144E-31	0.6776	1.2223	-0.538	3.962	1.870E-32
11.000	40.0	2.085E-30	4	0.440	1.565	15.70	1.013E-31	1.902E-31	0.6747	1.2007	-0.617	3.952	1.659E-32
11.000	50.0	1.255E-30	4	0.550	1.549	19.54	6.050E-32	1.145E-31	0.6677	1.1522	-0.807	3.928	9.985E-33
11.000	60.0	7.660E-31	5	0.660	1.529	23.33	3.656E-32	6.989E-32	0.6591	1.0987	-1.039	3.898	6.094E-33
11.000	70.0	4.186E-31	5	0.759	1.506	27.06	1.975E-32	3.819E-32	0.6490	1.0420	-1.313	3.863	3.330E-33
12.000	12.5	1.425E-30	4	0.150	1.758	4.88	5.961E-32	1.191E-31	0.7576	1.3745	-0.233	3.759	1.039E-32
12.000	20.0	1.242E-30	4	0.240	1.753	7.80	5.184E-32	1.038E-31	0.7555	1.3554	-0.289	3.752	9.053E-33
12.000	30.0	8.368E-31	5	0.360	1.743	11.67	3.476E-32	6.995E-32	0.7513	1.3183	-0.405	3.736	6.099E-33
12.000	35.0	5.130E-31	9	0.420	1.737	13.59	2.125E-32	4.288E-32	0.7485	1.2955	-0.479	3.726	3.739E-33
12.000	40.0	4.233E-31	5	0.480	1.729	15.51	1.747E-32	3.538E-32	0.7454	1.2706	-0.566	3.715	3.085E-33
12.000	50.0	2.340E-31	7	0.600	1.712	19.31	9.579E-33	1.956E-32	0.7377	1.2152	-0.773	3.687	1.706E-33
12.000	60.0	1.453E-31	9	0.720	1.690	23.06	5.888E-33	1.215E-32	0.7284	1.1549	-1.026	3.652	1.059E-33
12.000	70.0	6.380E-32	11	0.839	1.664	26.76	2.555E-33	5.333E-33	0.7174	1.0919	-1.326	3.611	4.650E-34
13.000	12.5	2.630E-31	5	0.162	1.923	4.83	9.454E-33	2.028E-32	0.8286	1.4521	-0.174	3.510	1.769E-33
13.000	20.0	1.971E-31	7	0.260	1.917	7.72	7.069E-33	1.520E-32	0.8264	1.4299	-0.235	3.501	1.326E-33
13.000	30.0	1.030E-31	9	0.390	1.907	11.56	3.677E-33	7.944E-33	0.8218	1.3871	-0.360	3.483	6.927E-34
13.000	35.0	7.330E-32	9	0.455	1.900	13.47	2.609E-33	5.653E-33	0.8188	1.3611	-0.441	3.472	4.930E-34
13.000	40.0	6.413E-32	11	0.520	1.892	15.37	2.274E-33	4.946E-33	0.8154	1.3327	-0.534	3.458	4.313E-34
13.000	50.0	3.080E-32	19	0.650	1.873	19.14	1.083E-33	2.375E-33	0.8071	1.2705	-0.759	3.426	2.071E-34
13.000	60.0	1.030E-32	34	0.780	1.849	22.86	3.586E-34	7.944E-34	0.7970	1.2036	-1.033	3.385	6.927E-35
13.000	70.0	6.948E-33	40	0.909	1.821	26.53	2.390E-34	5.359E-34	0.7851	1.1346	-1.358	3.337	4.673E-35
14.000	12.5	2.060E-32	16	0.175	2.086	4.80	6.429E-34	1.475E-33	0.8991	1.5238	-0.130	3.239	1.286E-34
14.000	20.0	1.281E-32	27	0.280	2.080	7.66	3.988E-34	9.171E-34	0.8967	1.4981	-0.196	3.229	7.997E-35
14.000	30.0	2.958E-33	57	0.420	2.069	11.47	9.166E-35	2.118E-34	0.8917	1.4494	-0.330	3.208	1.847E-35
14.000	35.0	5.090E-33	49	0.490	2.062	13.37	1.572E-34	3.644E-34	0.8885	1.4200	-0.417	3.194	3.178E-35
14.000	40.0	2.438E-33	57	0.560	2.053	15.25	7.505E-35	1.745E-34	0.8848	1.3882	-0.518	3.179	1.522E-35
14.000	50.0	2.481E-33	57	0.700	2.032	19.00	7.573E-35	1.776E-34	0.8759	1.3191	-0.760	3.140	1.549E-35
14.000	60.0	2.359E-33	70	0.839	2.007	22.70	7.128E-35	1.689E-34	0.8650	1.2459	-1.056	3.093	1.473E-35
13.600	12.5	4.997E-32	17	0.170	2.021	4.81	1.648E-33	3.683E-33	0.8710	1.4958	-0.146	3.350	3.212E-34
13.900	12.5	2.911E-32	22	0.174	2.070	4.80	9.210E-34	2.099E-33	0.8921	1.5168	-0.134	3.267	1.830E-34
14.200	12.5	1.023E-32	29	0.177	2.119	4.79	3.107E-34	7.220E-34	0.9131	1.5374	-0.123	3.182	6.296E-35
14.500	12.5	5.700E-33	25	0.181	2.167	4.78	1.663E-34	3.939E-34	0.9342	1.5576	-0.113	3.094	3.435E-35

$p_1 = 24.000$ GeV/c; $p_1^* = 3.290$ GeV/c; $E_1^* + E_2^* = 6.843$ GeV; $s = 46.828$ GeV^2; $\gamma_{c.m.} = 3.6469$; $\beta_{c.m.} = 0.96167$; $y_1^* = 1.9677$; $p_{3,\,max}^* = 2.751$ GeV/c; [72 A 1]

p_3	θ_3	$\dfrac{d^2\sigma}{d\Omega_3\,dp_3}$	Error	p_T	p_L^*	θ_3^*	$\dfrac{1}{p_3^{*2}}\cdot\dfrac{d^2\sigma}{d\Omega_3^*\,dp_3^*}$	$\dfrac{E_3}{p_3^2}\cdot\dfrac{d^2\sigma}{d\Omega_3\,dp_3}$	x^*	y_3^*	t	m_4	$\dfrac{d^2\sigma}{dt\,dm_4^2}$
4.000	17.0	3.906E-28	3	0.068	0.176	21.09	1.048E-28	1.003E-28	0.0641	0.1863	-3.629	5.883	6.997E-30
4.000	37.0	3.550E-28	3	0.148	0.168	41.30	9.450E-29	9.116E-29	0.0612	0.1764	-3.732	5.874	6.359E-30
4.000	57.0	3.011E-28	3	0.228	0.155	55.83	7.907E-29	7.732E-29	0.0562	0.1596	-3.913	5.859	5.394E-30
4.000	67.0	2.710E-28	3	0.268	0.146	61.46	7.054E-29	6.959E-29	0.0529	0.1488	-4.032	5.849	4.855E-30

p_3	θ_3	$\dfrac{\mathrm{d}^2\sigma}{\mathrm{d}\Omega_3\,\mathrm{d}p_3}$	Error	p_T	p_L^*	θ_3^*	$\dfrac{1}{p_3^{*2}}\cdot\dfrac{\mathrm{d}^2\sigma}{\mathrm{d}\Omega_3^*\,\mathrm{d}p_3^*}$	$\dfrac{E_3}{p_3^2}\cdot\dfrac{\mathrm{d}^2\sigma}{\mathrm{d}\Omega_3\,\mathrm{d}p_3}$	x^*	y_3^*	t	m_4	$\dfrac{\mathrm{d}^2\sigma}{\mathrm{d}t\,\mathrm{d}m_4^2}$
GeV/c	mrad	cm² sr⁻¹ GeV⁻¹ c	%	GeV/c	GeV/c	deg	cm² sr⁻¹ GeV⁻³ c³	cm² sr⁻¹ GeV⁻² c³			GeV²	GeV	cm²/GeV⁴

$p_1 = 24.000$ GeV/c; $p_1^* = 3.290$ GeV/c; $E_1^* + E_2^* = 6.843$ GeV; $s = 46.828$ GeV²; $\gamma_{\mathrm{c.m.}} = 3.6469$; $\beta_{\mathrm{c.m.}} = 0.96167$, $y_1^* = 1.9677$; $p_{3,\max}^* = 2.751$ GeV/c; [72 A 1] (cont.)

p_3	θ_3	$\dfrac{\mathrm{d}^2\sigma}{\mathrm{d}\Omega_3\,\mathrm{d}p_3}$	Error	p_T	p_L^*	θ_3^*	$\dfrac{1}{p_3^{*2}}\cdot\dfrac{\mathrm{d}^2\sigma}{\mathrm{d}\Omega_3^*\,\mathrm{d}p_3^*}$	$\dfrac{E_3}{p_3^2}\cdot\dfrac{\mathrm{d}^2\sigma}{\mathrm{d}\Omega_3\,\mathrm{d}p_3}$	x^*	y_3^*	t	m_4	$\dfrac{\mathrm{d}^2\sigma}{\mathrm{d}t\,\mathrm{d}m_4^2}$
4.000	87.0	2.102E−28	3	0.348	0.123	70.48	5.354E−29	5.398E−29	0.0448	0.1229	−4.327	5.823	3.765E−30
4.000	107.0	1.636E−28	3	0.427	0.095	77.47	4.058E−29	4.201E−29	0.0345	0.0920	−4.699	5.791	2.931E−30
4.000	127.0	1.081E−28	3	0.507	0.061	83.14	2.599E−29	2.776E−29	0.0221	0.0571	−5.147	5.753	1.936E−30
4.000	147.0	6.034E−29	3	0.586	0.021	87.94	1.401E−29	1.549E−29	0.0077	0.0190	−5.672	5.707	1.081E−30
6.000	17.0	3.580E−28	3	0.102	0.580	9.98	5.452E−29	6.039E−29	0.2107	0.5811	−2.003	5.705	4.213E−30
6.000	27.0	3.334E−28	3	0.152	0.575	15.73	5.057E−29	5.624E−29	0.2090	0.5722	−2.067	5.700	3.923E−30
6.000	37.0	3.079E−28	3	0.222	0.568	21.34	4.642E−29	5.194E−29	0.2064	0.5595	−2.159	5.692	3.623E−30
6.000	57.0	2.300E−28	3	0.342	0.547	31.98	3.407E−29	3.880E−29	0.1990	0.5239	−2.430	5.668	2.707E−30
6.000	67.0	1.975E−28	3	0.402	0.534	36.96	2.893E−29	3.332E−29	0.1940	0.5018	−2.608	5.652	2.324E−30
6.000	87.0	1.220E−28	3	0.521	0.500	46.18	1.738E−29	2.058E−29	0.1818	0.4506	−3.051	5.613	1.436E−30
6.000	107.0	7.128E−29	3	0.641	0.458	54.45	9.816E−30	1.202E−29	0.1664	0.3928	−3.609	5.563	8.388E−31
6.000	127.0	3.805E−29	3	0.760	0.407	61.84	5.038E−30	6.419E−30	0.1478	0.3308	−4.281	5.502	4.478E−31
6.000	147.0	1.572E−29	4	0.879	0.347	68.45	1.992E−30	2.652E−30	0.1261	0.2667	−5.068	5.430	1.850E−31
8.000	17.0	2.021E−28	3	0.136	0.922	8.39	1.924E−29	2.544E−29	0.3350	0.8616	−1.222	5.442	1.774E−30
8.000	27.0	1.743E−28	3	0.216	0.915	13.28	1.651E−29	2.194E−29	0.3327	0.8461	−1.306	5.434	1.530E−30
8.000	37.0	1.488E−28	3	0.296	0.906	18.09	1.400E−29	1.873E−29	0.3293	0.8243	−1.429	5.423	1.306E−30
8.000	47.0	1.192E−28	3	0.376	0.894	22.81	1.112E−29	1.500E−29	0.3248	0.7971	−1.591	5.408	1.047E−30
8.000	57.0	9.385E−29	3	0.456	0.879	27.42	8.661E−30	1.181E−29	0.3193	0.7654	−1.790	5.389	8.240E−31
8.000	67.0	7.406E−29	3	0.536	0.860	31.90	6.749E−30	9.321E−30	0.3128	0.7299	−2.028	5.367	6.502E−31
8.000	87.0	3.437E−29	4	0.695	0.816	40.44	3.037E−30	4.326E−30	0.2964	0.6514	−2.619	5.312	3.018E−31
8.000	107.0	1.456E−29	5	0.854	0.759	48.38	1.239E−30	1.832E−30	0.2759	0.5673	−3.363	5.241	1.278E−31
8.000	127.0	5.794E−30	5	1.013	0.691	55.71	4.722E−31	7.292E−31	0.2511	0.4816	−4.259	5.155	5.087E−32
8.000	147.0	1.745E−30	7	1.172	0.611	62.45	1.355E−31	2.196E−31	0.2222	0.3967	−5.308	5.052	1.532E−32
10.000	17.0	8.542E−29	3	0.170	1.239	7.82	5.489E−30	8.580E−30	0.4502	1.0778	−0.785	5.130	5.985E−31
10.000	27.0	7.085E−29	3	0.270	1.230	12.37	4.530E−30	7.116E−30	0.4472	1.0541	−0.891	5.120	4.964E−31
10.000	37.0	5.314E−29	3	0.370	1.219	16.88	3.374E−30	5.337E−30	0.4430	1.0214	−1.044	5.105	3.723E−31
10.000	47.0	3.760E−29	3	0.470	1.204	21.32	2.365E−30	3.777E−30	0.4374	0.9816	−1.246	5.085	2.635E−31
10.000	57.0	2.636E−29	4	0.570	1.185	25.68	1.639E−30	2.648E−30	0.4305	0.9363	−1.495	5.060	1.847E−31
10.000	67.0	1.777E−29	4	0.669	1.162	29.95	1.091E−30	1.785E−30	0.4223	0.8871	−1.793	5.031	1.245E−31
10.000	87.0	6.479E−30	5	0.869	1.106	38.16	3.849E−31	6.507E−31	0.4019	0.7825	−2.531	4.957	4.540E−32
10.000	107.0	2.073E−30	6	1.068	1.035	45.89	1.184E−31	2.082E−31	0.3763	0.6756	−3.461	4.862	1.452E−32
10.000	127.0	5.101E−31	10	1.267	0.950	53.13	2.784E−32	5.123E−32	0.3453	0.5712	−4.581	4.746	3.574E−33
10.000	147.0	5.097E−32	33	1.465	0.850	59.86	2.644E−33	5.119E−33	0.3091	0.4713	−5.893	4.605	3.571E−34
12.000	17.0	2.855E−29	4	0.204	1.543	7.53	1.313E−30	2.386E−30	0.5607	1.2525	−0.522	4.779	1.665E−31
12.000	27.0	2.051E−29	4	0.324	1.533	11.93	9.387E−31	1.714E−30	0.5572	1.2192	−0.649	4.766	1.196E−31
12.000	37.0	1.292E−29	4	0.444	1.519	16.29	5.870E−31	1.080E−30	0.5521	1.1743	−0.833	4.747	7.534E−32
12.000	47.0	8.702E−30	3	0.564	1.501	20.59	3.916E−31	7.274E−31	0.5454	1.1210	−1.075	4.721	5.074E−32
12.000	57.0	5.018E−30	4	0.684	1.478	24.82	2.232E−31	4.194E−31	0.5371	1.0619	−1.374	4.689	2.926E−32
12.000	67.0	3.087E−30	5	0.803	1.451	28.98	1.354E−31	2.580E−31	0.5273	0.9996	−1.731	4.651	1.800E−32

p_3	θ_3	$\dfrac{d^2\sigma}{d\Omega_3\,dp_3}$	Error	p_T	p_L^*	θ_3^*	$\dfrac{1}{p_3^{*2}}\cdot\dfrac{d^2\sigma}{d\Omega_3^*\,dp_3^*}$	$\dfrac{E_3}{p_3^2}\cdot\dfrac{d^2\sigma}{d\Omega_3\,dp_3}$	x^*	y_3^*	t	m_4	$\dfrac{d^2\sigma}{dt\,dm_4^2}$
GeV/c	mrad	$cm^2\,sr^{-1}\,GeV^{-1}\,c$	%	GeV/c	GeV/c	deg	$cm^2\,sr^{-1}\,GeV^{-3}\,c^3$	$cm^2\,sr^{-1}\,GeV^{-2}\,c^3$			GeV^2	GeV	cm^2/GeV^4

$p_1 = 24.000$ GeV/c; $p_1^* = 3.290$ GeV/c; $E_1^* + E_2^* = 6.843$ GeV; $s = 46.828$ GeV^2; $\gamma_{c.m.} = 3.6469$; $\beta_{c.m.} = 0.96167$; $y_1^* = 1.9677$; $p_{3,\,max}^* = 2.751$ GeV/c; [72 A 1] (cont.)

p_3	θ_3	$\dfrac{d^2\sigma}{d\Omega_3\,dp_3}$	Error	p_T	p_L^*	θ_3^*	$\dfrac{1}{p_3^{*2}}\cdot\dfrac{d^2\sigma}{d\Omega_3^*\,dp_3^*}$	$\dfrac{E_3}{p_3^2}\cdot\dfrac{d^2\sigma}{d\Omega_3\,dp_3}$	x^*	y_3^*	t	m_4	$\dfrac{d^2\sigma}{dt\,dm_4^2}$
12.000	77.0	1.469E-30	6	0.923	1.419	33.04	6.344E-32	1.228E-31	0.5158	0.9357	-2.145	4.606	8.566E-33
12.000	87.0	7.676E-31	9	1.043	1.383	37.01	3.257E-32	6.416E-32	0.5028	0.8716	-2.617	4.555	4.476E-33
12.000	97.0	3.593E-31	11	1.162	1.343	40.87	1.495E-32	3.003E-32	0.4882	0.8083	-3.146	4.496	2.095E-33
12.000	107.0	1.954E-31	15	1.282	1.299	44.62	7.961E-33	1.633E-32	0.4720	0.7464	-3.733	4.431	1.139E-33
12.000	117.0	9.911E-32	21	1.401	1.250	48.26	3.948E-33	8.284E-33	0.4542	0.6861	-4.377	4.357	5.779E-34
12.000	127.0	8.532E-33	70	1.520	1.196	51.79	3.317E-34	7.132E-34	0.4349	0.6278	-5.078	4.276	4.975E-35
12.000	137.0	1.695E-32	70	1.639	1.139	55.20	6.425E-34	1.417E-33	0.4139	0.5715	-5.836	4.187	9.884E-35
14.000	17.0	6.308E-30	4	0.238	1.839	7.37	2.173E-31	4.516E-31	0.6686	1.3982	-0.358	4.389	3.150E-32
14.000	27.0	4.004E-30	5	0.378	1.828	11.68	1.372E-31	2.866E-31	0.6645	1.3541	-0.506	4.372	2.000E-32
14.000	37.0	2.308E-30	6	0.518	1.812	15.95	7.849E-32	1.652E-31	0.6585	1.2961	-0.721	4.347	1.153E-32
14.000	47.0	1.138E-30	8	0.658	1.790	20.17	3.833E-32	8.147E-32	0.6507	1.2290	-1.003	4.315	5.683E-33
14.000	57.0	6.835E-31	7	0.798	1.764	24.33	2.275E-32	4.893E-32	0.6411	1.1567	-1.353	4.274	3.413E-33
14.000	67.0	3.197E-31	15	0.937	1.732	28.42	1.049E-32	2.289E-32	0.6296	1.0822	-1.769	4.225	1.597E-33
14.000	77.0	1.122E-31	26	1.077	1.696	32.42	3.623E-33	8.032E-33	0.6162	1.0077	-2.252	4.167	5.603E-34
14.000	87.0	2.932E-32	50	1.216	1.654	36.34	9.299E-34	2.099E-33	0.6010	0.9344	-2.803	4.101	1.464E-34
14.000	97.0	3.642E-32	44	1.356	1.607	40.16	1.132E-33	2.607E-33	0.5840	0.8632	-3.420	4.025	1.819E-34
14.000	107.0	2.159E-32	57	1.495	1.555	43.88	6.571E-34	1.546E-33	0.5651	0.7946	-4.104	3.939	1.078E-34
16.000	17.0	6.257E-31	8	0.272	2.132	7.27	1.673E-32	3.924E-32	0.7748	1.5223	-0.257	3.952	2.737E-33
16.000	27.0	3.715E-31	13	0.432	2.119	11.52	9.867E-33	2.326E-32	0.7701	1.4664	-0.426	3.930	1.623E-33
16.000	37.0	1.530E-31	21	0.592	2.100	15.74	4.033E-33	9.579E-33	0.7633	1.3948	-0.672	3.899	6.682E-34
16.000	47.0	8.553E-32	27	0.752	2.076	19.91	2.232E-33	5.355E-33	0.7544	1.3142	-0.994	3.857	3.736E-34
16.000	57.0	4.598E-32	37	0.912	2.045	24.02	1.186E-33	2.879E-33	0.7434	1.2295	-1.394	3.805	2.008E-34
16.000	67.0	1.301E-32	70	1.071	2.009	28.06	3.308E-34	8.145E-34	0.7303	1.1443	-1.869	3.742	5.682E-35
18.000	17.0	2.102E-32	70	0.306	2.421	7.20	4.473E-34	1.169E-33	0.8799	1.6297	-0.198	3.454	8.158E-35

3.6.6 pp→pX

$p_1 = 6.000$ GeV/c; $p_1^* = 1.552$ GeV/c; $E_1^* + E_2^* = 3.627$ GeV; $s = 13.156$ GeV^2; $\gamma_{c.m.} = 1.9330$; $\beta_{c.m.} = 0.85578$; $y_1^* = 1.2774$; $p_{3,\,max}^* = 1.552$ GeV/c; [70 A 3]

p_3	θ_3	$\dfrac{d^2\sigma}{d\Omega_3\,dp_3}$	Error	p_T	p_L^*	θ_3^*	$\dfrac{1}{p_3^{*2}}\cdot\dfrac{d^2\sigma}{d\Omega_3^*\,dp_3^*}$	$\dfrac{E_3}{p_3^2}\cdot\dfrac{d^2\sigma}{d\Omega_3\,dp_3}$	x^*	y_3^*	t	m_4	$\dfrac{d^2\sigma}{dt\,dm_4^2}$
5.401	117.4	1.303E-26	8	0.633	1.300	25.95	1.421E-27	2.449E-27	0.8376	0.9828	-0.456	1.238	6.833E-28
4.779	132.7	1.595E-26	6	0.632	1.100	29.89	2.157E-27	3.404E-27	0.7088	0.8617	-0.549	1.609	9.497E-28
4.179	151.9	1.570E-26	5	0.632	0.900	35.09	2.663E-27	3.850E-27	0.5799	0.7291	-0.690	1.884	1.074E-27
3.610	176.1	1.220E-26	5	0.632	0.700	42.10	2.624E-27	3.492E-27	0.4510	0.5848	-0.892	2.094	9.744E-28
3.079	206.9	9.035E-27	6	0.632	0.500	51.67	2.480E-27	3.068E-27	0.3222	0.4287	-1.174	2.250	8.561E-28
2.594	246.2	7.495E-27	6	0.632	0.300	64.62	2.624E-27	3.072E-27	0.1933	0.2622	-1.554	2.355	8.572E-28
2.167	296.2	5.135E-27	6	0.632	0.100	81.01	2.274E-27	2.583E-27	0.0645	0.0883	-2.049	2.408	7.207E-28
5.513	141.0	4.528E-27	20	0.775	1.300	30.79	4.679E-28	8.331E-28	0.8376	0.9289	-0.663	1.058	2.325E-28
4.901	158.7	8.255E-27	8	0.775	1.100	35.15	1.046E-27	1.715E-27	0.7088	0.8120	-0.774	1.462	4.785E-28
4.313	180.6	6.580E-27	8	0.775	0.900	40.72	1.032E-27	1.561E-27	0.5799	0.6848	-0.935	1.749	4.357E-28

p_3	θ_3	$\dfrac{d^2\sigma}{d\Omega_3\,dp_3}$	Error	p_T	p_L^*	θ_3^*	$\dfrac{1}{p_3^{*2}}\cdot\dfrac{d^2\sigma}{d\Omega_3^*\,dp_3^*}$	$\dfrac{E_3}{p_3^2}\cdot\dfrac{d^2\sigma}{d\Omega_3\,dp_3}$	x^*	y_3^*	t	m_4	$\dfrac{d^2\sigma}{dt\,dm_4^2}$
GeV/c	mrad	cm² sr⁻¹ GeV⁻¹ c	%	GeV/c	GeV/c	deg	cm² sr⁻¹ GeV⁻³ c³	cm² sr⁻¹ GeV⁻² c³			GeV²	GeV	cm²/GeV⁴

$p_1 = 6.000$ GeV/c; $p_1^* = 1.552$ GeV/c; $E_1^* + E_2^* = 3.627$ GeV; $s = 13.156$ GeV²; $\gamma_{c.m.} = 1.9330$; $\beta_{c.m.} = 0.85578$; $y_1^* = 1.2774$; $p_{3,\max}^* = 1.552$ GeV/c; [70 A 3] (cont.)

p_3	θ_3	$\dfrac{d^2\sigma}{d\Omega_3\,dp_3}$	Error	p_T	p_L^*	θ_3^*	$\dfrac{1}{p_3^{*2}}\cdot\dfrac{d^2\sigma}{d\Omega_3^*\,dp_3^*}$	$\dfrac{E_3}{p_3^2}\cdot\dfrac{d^2\sigma}{d\Omega_3\,dp_3}$	x^*	y_3^*	t	m_4	$\dfrac{d^2\sigma}{dt\,dm_4^2}$
3.756	207.7	7.025E-27	7	0.774	0.700	47.89	1.374E-27	1.928E-27	0.4511	0.5476	-1.158	1.963	5.380E-28
3.237	241.7	5.590E-27	7	0.775	0.500	57.16	1.367E-27	1.798E-27	0.3221	0.4002	-1.459	2.120	5.018E-28
2.764	284.1	4.107E-27	7	0.775	0.300	68.83	1.252E-27	1.569E-27	0.1933	0.2441	-1.853	2.224	4.379E-28
2.344	336.7	3.243E-27	7	0.775	0.100	82.64	1.221E-27	1.490E-27	0.0645	0.0822	-2.357	2.276	4.158E-28
5.019	179.2	2.896E-27	30	0.895	1.100	39.12	3.453E-28	5.870E-28	0.7087	0.7701	-0.992	1.305	1.638E-28
4.441	202.8	3.317E-27	12	0.894	0.900	44.83	4.837E-28	7.634E-28	0.5799	0.6480	-1.170	1.609	2.130E-28
3.894	231.8	3.577E-27	10	0.895	0.700	51.96	6.413E-28	9.448E-28	0.4510	0.5166	-1.411	1.830	2.636E-28
3.385	267.4	3.354E-27	9	0.894	0.500	60.79	7.400E-28	1.028E-27	0.3222	0.3768	-1.727	1.989	2.869E-28
2.921	311.2	2.503E-27	10	0.894	0.300	71.46	6.764E-28	9.000E-28	0.1933	0.2294	-2.134	2.094	2.511E-28
2.509	364.5	2.178E-27	10	0.894	0.100	83.62	7.130E-28	9.269E-28	0.0645	0.0771	-2.644	2.146	2.586E-28

$p_1 = 12.400$ GeV/c; $p_1^* = 2.322$ GeV/c; $E_1^* + E_2^* = 5.009$ GeV; $s = 25.094$ GeV²; $\gamma_{c.m.} = 2.6697$; $\beta_{c.m.} = 0.92720$; $y_1^* = 1.6380$; $p_{3,\max}^* = 2.322$ GeV/c; [71 A 1]

p_3	θ_3	$\dfrac{d^2\sigma}{d\Omega_3\,dp_3}$	Error	p_T	p_L^*	θ_3^*	$\dfrac{1}{p_3^{*2}}\cdot\dfrac{d^2\sigma}{d\Omega_3^*\,dp_3^*}$	$\dfrac{E_3}{p_3^2}\cdot\dfrac{d^2\sigma}{d\Omega_3\,dp_3}$	x^*	y_3^*	t	m_4	$\dfrac{d^2\sigma}{dt\,dm_4^2}$
3.665	4.0	3.972E-26	7	0.015	0.420	2.00	1.088E-26	1.119E-26	0.1807	0.4336	-1.439	3.959	1.510E-27
3.638	26.0	2.874E-26	7	0.094	0.409	13.00	7.936E-27	8.158E-27	0.1762	0.4214	-1.488	3.959	1.101E-27
3.522	59.7	2.340E-26	7	0.210	0.364	30.00	6.689E-27	6.876E-27	0.1566	0.3699	-1.699	3.959	9.284E-28
3.364	86.8	1.846E-26	7	0.292	0.302	44.00	5.543E-27	5.697E-27	0.1301	0.3028	-1.986	3.959	7.693E-28
3.159	113.0	1.410E-26	7	0.356	0.223	58.00	4.530E-27	4.656E-27	0.0958	0.2200	-2.355	3.959	6.287E-28
2.918	137.3	1.041E-26	8	0.399	0.130	72.00	3.645E-27	3.747E-27	0.0559	0.1269	-2.786	3.959	5.059E-28
2.637	159.8	8.758E-27	9	0.420	0.022	87.01	3.430E-27	3.526E-27	0.0094	0.0213	-3.287	3.959	4.760E-28
4.519	4.9	6.220E-26	6	0.022	0.640	2.00	1.238E-26	1.406E-26	0.2754	0.6375	-0.958	3.820	1.898E-27
4.484	29.7	4.601E-26	6	0.133	0.626	12.00	9.229E-27	1.048E-26	0.2696	0.6201	-1.021	3.820	1.415E-27
4.412	52.0	3.494E-26	6	0.229	0.597	21.00	7.129E-27	8.096E-27	0.2573	0.5847	-1.154	3.820	1.093E-27
4.303	74.4	2.693E-26	6	0.320	0.554	30.00	5.640E-27	6.406E-27	0.2387	0.5335	-1.354	3.820	8.649E-28
4.159	97.0	2.002E-26	6	0.403	0.497	39.00	4.345E-27	4.935E-27	0.2142	0.4697	-1.619	3.820	6.663E-28
3.983	119.7	1.387E-26	7	0.476	0.428	48.00	3.150E-27	3.573E-27	0.1844	0.3966	-1.940	3.820	4.831E-28
3.756	145.0	1.001E-26	7	0.543	0.339	58.00	2.419E-27	2.747E-27	0.1460	0.3080	-2.353	3.820	3.709E-28
3.528	167.8	7.365E-27	7	0.589	0.250	67.01	1.902E-27	2.160E-27	0.1076	0.2238	-2.768	3.820	2.916E-28
3.256	192.7	5.422E-27	7	0.624	0.144	77.00	1.526E-27	1.733E-27	0.0620	0.1275	-3.260	3.821	2.340E-28
2.999	214.5	4.619E-27	8	0.638	0.045	86.00	1.421E-27	1.614E-27	0.0192	0.0394	-3.721	3.821	2.179E-28
6.240	32.1	8.985E-26	6	0.200	1.031	11.00	1.034E-26	1.456E-26	0.4438	0.9330	-0.506	3.445	1.966E-27
6.187	46.8	6.984E-26	6	0.289	1.009	16.00	8.108E-27	1.142E-26	0.4346	0.9010	-0.605	3.445	1.542E-27
6.114	61.6	5.017E-26	6	0.376	0.980	21.00	5.896E-27	8.302E-27	0.4221	0.8598	-0.740	3.445	1.121E-27
6.042	73.5	3.721E-26	6	0.444	0.952	25.00	4.426E-27	6.232E-27	0.4098	0.8214	-0.873	3.445	8.414E-28
5.936	88.6	2.845E-26	6	0.525	0.909	30.00	3.447E-27	4.854E-27	0.3916	0.7680	-1.070	3.445	6.553E-28
5.813	103.8	2.089E-26	6	0.602	0.860	35.00	2.585E-27	3.640E-27	0.3704	0.7102	-1.298	3.445	4.915E-28
5.673	119.3	1.508E-26	6	0.675	0.804	40.01	1.913E-27	2.694E-27	0.3463	0.6492	-1.558	3.445	3.638E-28
5.550	131.8	1.062E-26	6	0.729	0.755	44.00	1.378E-27	1.941E-27	0.3252	0.5991	-1.785	3.445	2.620E-28
5.383	147.7	7.518E-27	6	0.792	0.689	48.99	1.007E-27	1.418E-27	0.2967	0.5352	-2.093	3.445	1.914E-28
5.203	164.0	5.323E-27	6	0.849	0.617	54.00	7.383E-28	1.040E-27	0.2658	0.4701	-2.427	3.445	1.404E-28

$p_1 = 12.400$ GeV/c; $p_1^* = 2.322$ GeV/c; $E_1^* + E_2^* = 5.009$ GeV; $s = 25.094$ GeV^2; $\gamma_{c.m.} = 2.6697$; $\beta_{c.m.} = 0.92720$; $y_1^* = 1.6380$; $p_{3,\max}^* = 2.322$ GeV/c; [71 A 1] (cont.)

p_3	θ_3	$\dfrac{d^2\sigma}{d\Omega_3\,dp_3}$	Error	p_T	p_L^*	θ_3^*	$\dfrac{1}{p_3^{*2}}\cdot\dfrac{d^2\sigma}{d\Omega_3^*\,dp_3^*}$	$\dfrac{E_3}{p_3^*}\cdot\dfrac{d^2\sigma}{d\Omega_3\,dp_3}$	x^*	y_3^*	t	m_4	$\dfrac{d^2\sigma}{dt\,dm_4^2}$
GeV/c	mrad	cm² sr⁻¹ GeV⁻¹ c	%	GeV/c	GeV/c	deg	cm² sr⁻¹ GeV⁻³ c³	cm² sr⁻¹ GeV⁻² c³			GeV²	GeV	cm²/GeV⁴
5.050	177.3	3.716E−27	6	0.891	0.556	58.01	5.315E−28	7.484E−28	0.2395	0.4178	−2.710	3.445	1.011E−28
4.849	194.1	2.827E−27	6	0.935	0.477	62.99	4.217E−28	5.938E−28	0.2053	0.3526	−3.078	3.445	8.017E−29
4.639	211.4	2.183E−27	6	0.973	0.393	67.99	3.410E−28	4.801E−28	0.1694	0.2871	−3.465	3.445	6.483E−29
4.421	229.1	1.795E−27	6	1.004	0.307	72.99	2.948E−28	4.151E−28	0.1322	0.2217	−3.866	3.445	5.605E−29
4.196	247.3	1.485E−27	7	1.027	0.218	78.00	2.575E−28	3.626E−28	0.0940	0.1563	−4.280	3.445	4.897E−29
4.013	262.1	1.327E−27	7	1.040	0.146	82.00	2.412E−28	3.396E−28	0.0629	0.1042	−4.614	3.445	4.586E−29
3.781	281.0	1.167E−27	8	1.048	0.055	86.99	2.259E−28	3.181E−28	0.0237	0.0391	−5.037	3.445	4.294E−29
7.040	48.6	8.256E−26	6	0.342	1.192	16.00	7.609E−27	1.183E−26	0.5133	1.0120	−0.493	3.224	1.598E−27
6.954	54.0	5.296E−26	6	0.444	1.158	21.00	4.942E−27	7.685E−27	0.4985	0.9605	−0.652	3.224	1.038E−27
6.869	76.4	3.555E−26	6	0.524	1.124	25.00	3.369E−27	5.238E−27	0.4839	0.9134	−0.809	3.224	7.072E−28
6.744	92.1	2.459E−26	6	0.620	1.074	30.00	2.367E−27	3.681E−27	0.4624	0.8491	−1.041	3.224	4.970E−28
6.599	108.0	1.645E−26	6	0.711	1.016	35.00	1.619E−27	2.518E−27	0.4374	0.7809	−1.311	3.224	3.400E−28
6.435	124.2	1.088E−26	6	0.797	0.950	40.00	1.099E−27	1.709E−27	0.4090	0.7103	−1.617	3.224	2.307E−28
6.252	140.7	6.752E−27	6	0.877	0.877	44.99	7.024E−28	1.092E−27	0.3776	0.6386	−1.956	3.224	1.475E−28
6.052	157.6	4.318E−27	6	0.950	0.797	50.00	4.643E−28	7.220E−28	0.3432	0.5663	−2.327	3.224	9.749E−29
5.881	171.4	2.854E−27	6	1.003	0.729	54.00	3.161E−28	4.914E−28	0.3139	0.5085	−2.643	3.224	6.635E−29
5.655	189.1	1.974E−27	6	1.063	0.639	59.00	2.276E−28	3.539E−28	0.2750	0.4364	−3.063	3.224	4.778E−29
5.416	207.3	1.365E−27	6	1.115	0.543	64.01	1.645E−28	2.558E−28	0.2340	0.3648	−3.505	3.224	3.454E−29
5.167	226.0	1.080E−27	6	1.158	0.444	69.01	1.366E−28	2.124E−28	0.1913	0.2939	−3.966	3.224	2.869E−29
4.908	245.3	8.532E−28	6	1.192	0.342	74.00	1.138E−28	1.770E−28	0.1472	0.2234	−4.442	3.224	2.389E−29
6.000	209.4	7.034E−28	5	1.247	0.636	62.99	7.042E−29	1.187E−28	0.2737	0.3969	−3.728	3.015	1.602E−29
5.607	236.9	4.450E−28	5	1.316	0.479	70.01	4.775E−29	8.048E−29	0.2061	0.2920	−4.459	3.015	1.087E−29
5.255	261.5	2.918E−28	5	1.358	0.339	76.00	3.347E−29	5.641E−29	0.1458	0.2037	−5.109	3.015	7.617E−30
5.014	278.6	2.432E−28	5	1.379	0.243	80.01	2.928E−29	4.935E−29	0.1046	0.1452	−5.554	3.015	6.663E−30
8.035	53.2	7.082E−26	6	0.427	1.396	17.00	5.114E−27	8.874E−27	0.6012	1.1113	−0.448	2.930	1.198E−27
7.952	65.8	4.539E−26	6	0.523	1.363	21.00	3.312E−27	5.748E−27	0.5869	1.0593	−0.602	2.930	7.761E−28
7.879	75.4	2.803E−26	6	0.594	1.334	24.00	2.064E−27	3.583E−27	0.5743	1.0168	−0.738	2.930	4.837E−28
7.768	88.4	1.865E−26	6	0.685	1.289	28.00	1.394E−27	2.418E−27	0.5551	0.9567	−0.946	2.930	3.265E−28
7.641	101.4	1.284E−26	6	0.773	1.238	31.99	9.756E−28	1.693E−27	0.5332	0.8943	−1.182	2.931	2.286E−28
7.499	114.7	8.602E−27	6	0.858	1.181	36.00	6.662E−28	1.156E−27	0.5086	0.8302	−1.447	2.930	1.561E−28
7.342	128.2	5.585E−27	6	0.939	1.118	40.01	4.419E−28	7.669E−28	0.4816	0.7657	−1.739	2.930	1.035E−28
7.172	141.9	3.561E−27	6	1.014	1.050	44.00	2.885E−28	5.008E−28	0.4522	0.7012	−2.055	2.930	6.761E−29
7.036	152.4	2.271E−27	6	1.068	0.996	47.01	1.876E−28	3.256E−28	0.4287	0.6529	−2.310	2.930	4.397E−29
6.844	166.6	1.501E−27	6	1.135	0.919	51.01	1.276E−28	2.214E−28	0.3956	0.5892	−2.667	2.930	2.989E−29
6.640	181.1	1.022E−27	6	1.196	0.837	55.00	8.957E−29	1.554E−28	0.3606	0.5263	−3.044	2.930	2.099E−29
6.426	196.0	7.455E−28	6	1.252	0.752	59.00	6.755E−29	1.172E−28	0.3238	0.4639	−3.441	2.930	1.583E−29
6.203	211.3	5.693E−28	6	1.301	0.663	63.01	5.348E−29	9.282E−29	0.2854	0.4022	−3.855	2.930	1.253E−29
9.862	68.0	2.585E−26	6	0.670	1.746	21.00	1.258E−27	2.633E−27	0.7517	1.2026	−0.611	2.239	3.555E−28
9.769	77.9	1.361E−26	6	0.761	1.708	24.00	6.690E−28	1.400E−27	0.7356	1.1463	−0.785	2.239	1.890E−28

p_3	θ_3	$\dfrac{\mathrm{d}^2\sigma}{\mathrm{d}\Omega_3\,\mathrm{d}p_3}$	Error	p_T	p_L^{*}	θ_3^{*}	$\dfrac{1}{p_3^{*2}}\cdot\dfrac{\mathrm{d}^2\sigma}{\mathrm{d}\Omega_3^{*}\,\mathrm{d}p_3^{*}}$	$\dfrac{E_3}{p_3^2}\cdot\dfrac{\mathrm{d}^2\sigma}{\mathrm{d}\Omega_3\,\mathrm{d}p_3}$	x^{*}	y_3^{*}	t	m_4	$\dfrac{\mathrm{d}^2\sigma}{\mathrm{d}t\,\mathrm{d}m_4^2}$
GeV/c	mrad	$\mathrm{cm^2\,sr^{-1}\,GeV^{-1}\,c}$	%	GeV/c	GeV/c	deg	$\mathrm{cm^2\,sr^{-1}\,GeV^{-3}\,c^3}$	$\mathrm{cm^2\,sr^{-1}\,GeV^{-2}\,c^3}$			$\mathrm{GeV^2}$	GeV	$\mathrm{cm^2/GeV^4}$

$p_1 = 12.400\ \mathrm{GeV/c};\ p_1^{*} = 2.322\ \mathrm{GeV/c};\ E_1^{*} + E_2^{*} = 5.009\ \mathrm{GeV};\ s = 25.094\ \mathrm{GeV^2};\ \gamma_{\mathrm{c.m.}} = 2.6697;\ \beta_{\mathrm{c.m.}} = 0.92720;\ y_1^{*} = 1.6380;\ p_{3,\max}^{*} = 2.322\ \mathrm{GeV/c};\ [71\,\mathrm{A}\,1]\ \text{(cont.)}$

p_3	θ_3	$\dfrac{\mathrm{d}^2\sigma}{\mathrm{d}\Omega_3\,\mathrm{d}p_3}$	Error	p_T	p_L^{*}	θ_3^{*}	$\dfrac{1}{p_3^{*2}}\cdot\dfrac{\mathrm{d}^2\sigma}{\mathrm{d}\Omega_3^{*}\,\mathrm{d}p_3^{*}}$	$\dfrac{E_3}{p_3^2}\cdot\dfrac{\mathrm{d}^2\sigma}{\mathrm{d}\Omega_3\,\mathrm{d}p_3}$	x^{*}	y_3^{*}	t	m_4	$\dfrac{\mathrm{d}^2\sigma}{\mathrm{d}t\,\mathrm{d}m_4^2}$
9.627	91.3	7.338E-27	6	0.878	1.651	28.00	3.661E-28	7.659E-28	0.7110	1.0693	-1.051	2.239	1.034E-28
9.465	104.9	4.085E-27	6	0.991	1.586	32.00	2.073E-28	4.337E-28	0.6829	0.9915	-1.354	2.239	5.856E-29
9.283	118.7	2.227E-27	6	1.099	1.513	36.00	1.152E-28	2.411E-28	0.6514	0.9141	-1.694	2.239	3.256E-29
9.083	132.7	1.124E-27	6	1.202	1.433	39.99	5.947E-29	1.244E-28	0.6169	0.8380	-2.066	2.239	1.680E-29
8.866	147.1	5.960E-28	6	1.299	1.345	44.01	3.231E-29	6.760E-29	0.5792	0.7630	-2.474	2.239	9.128E-30
8.692	158.0	3.324E-28	6	1.368	1.275	47.00	1.839E-29	3.847E-29	0.5492	0.7083	-2.796	2.239	5.194E-30
8.447	172.9	1.979E-28	6	1.453	1.177	51.00	1.127E-29	2.357E-29	0.5068	0.6365	-3.253	2.239	3.183E-30
8.187	188.2	1.236E-28	6	1.532	1.073	55.00	7.264E-30	1.520E-29	0.4619	0.5664	-3.738	2.239	2.052E-30
8.052	196.0	1.024E-28	6	1.568	1.019	56.99	6.120E-30	1.280E-29	0.4386	0.5320	-3.989	2.239	1.729E-30
10.525	68.5	1.951E-26	6	0.720	1.877	21.00	8.390E-28	1.861E-27	0.8080	1.2418	-0.636	1.937	2.513E-28
10.425	78.5	9.178E-27	6	0.818	1.836	24.00	3.985E-28	8.839E-28	0.7907	1.1811	-0.823	1.937	1.194E-28
10.272	92.0	4.813E-27	6	0.944	1.775	28.00	2.121E-28	4.705E-28	0.7642	1.0989	-1.108	1.937	6.353E-29
10.098	105.7	2.385E-27	6	1.065	1.705	32.01	1.069E-28	2.372E-28	0.7340	1.0164	-1.435	1.937	3.203E-29
9.903	119.6	1.164E-27	6	1.182	1.626	36.00	5.323E-29	1.181E-28	0.7002	0.9353	-1.799	1.937	1.594E-29
9.688	133.8	5.506E-28	6	1.292	1.540	40.01	2.574E-29	5.710E-29	0.6629	0.8557	-2.201	1.936	7.710E-30
9.454	148.2	2.800E-28	6	1.395	1.446	43.99	1.342E-29	2.976E-29	0.6226	0.7786	-2.635	1.937	4.018E-30
9.203	163.0	1.397E-28	6	1.494	1.345	47.99	6.879E-30	1.526E-29	0.5792	0.7032	-3.104	1.937	2.060E-30
8.936	178.2	7.514E-29	7	1.584	1.237	52.00	3.812E-30	8.455E-30	0.5328	0.6297	-3.604	1.937	1.142E-30
8.653	193.8	4.865E-29	7	1.667	1.124	56.01	2.549E-30	5.655E-30	0.4839	0.5582	-4.132	1.936	7.635E-31
8.529	219.1	2.352E-29	5	1.854	0.986	61.99	1.206E-30	2.774E-30	0.4246	0.4584	-5.181	1.712	3.746E-31
8.284	231.8	1.473E-29	5	1.903	0.888	65.00	7.780E-31	1.789E-30	0.3822	0.4070	-5.639	1.712	2.416E-31
7.863	253.7	7.859E-30	5	1.973	0.718	70.00	4.377E-31	1.007E-30	0.3093	0.3230	-6.426	1.712	1.359E-31
7.690	262.7	5.653E-30	5	1.997	0.649	71.99	3.220E-31	7.406E-31	0.2795	0.2901	-6.745	1.713	1.000E-31
6.980	300.8	2.868E-30	6	2.058	0.365	80.00	1.802E-31	4.146E-31	0.1570	0.1599	-8.068	1.712	5.598E-32
11.300	72.3	1.434E-26	6	0.817	2.021	22.00	5.365E-28	1.273E-27	0.8704	1.2622	-0.740	1.482	1.719E-28
11.227	79.1	9.107E-27	6	0.887	1.992	24.00	3.430E-28	8.140E-28	0.8575	1.2183	-0.878	1.482	1.099E-28
11.147	85.8	5.713E-27	6	0.956	1.959	26.00	2.167E-28	5.143E-28	0.8437	1.1741	-1.028	1.482	6.945E-29
11.061	92.7	3.501E-27	6	1.023	1.925	28.00	1.338E-28	3.177E-28	0.8288	1.1300	-1.188	1.482	4.289E-29
10.969	99.5	2.123E-27	6	1.090	1.888	30.00	8.185E-29	1.942E-28	0.8129	1.0862	-1.359	1.482	2.623E-29
10.872	106.5	1.255E-27	6	1.156	1.849	32.01	4.882E-29	1.159E-28	0.7960	1.0425	-1.543	1.482	1.564E-29
10.769	113.4	7.715E-28	6	1.219	1.807	33.99	3.030E-29	7.191E-29	0.7783	1.0000	-1.733	1.483	9.710E-30
10.661	120.5	4.959E-28	6	1.281	1.764	36.00	1.968E-29	4.670E-29	0.7594	0.9574	-1.937	1.482	6.305E-30
10.547	127.6	3.053E-28	6	1.342	1.718	38.00	1.225E-29	2.906E-29	0.7397	0.9157	-2.149	1.482	3.924E-30
10.428	134.8	1.975E-28	6	1.401	1.570	40.00	8.017E-30	1.903E-29	0.7191	0.8744	-2.372	1.482	2.569E-30
10.304	142.0	1.430E-28	6	1.458	1.520	41.99	5.872E-30	1.394E-29	0.6977	0.8342	-2.602	1.483	1.882E-30
10.175	149.4	9.613E-29	6	1.514	1.558	44.00	3.998E-30	9.488E-30	0.6752	0.7941	-2.845	1.482	1.281E-30
10.041	156.8	6.890E-29	6	1.558	1.514	45.99	2.904E-30	6.892E-30	0.6521	0.7550	-3.094	1.482	9.306E-31
9.902	164.3	4.835E-29	6	1.620	1.459	47.99	2.067E-30	4.904E-30	0.6282	0.7165	-3.352	1.483	6.622E-31
9.832	168.1	3.930E-29	6	1.645	1.430	48.99	1.692E-30	4.015E-30	0.6159	0.6974	-3.484	1.483	5.422E-31
9.687	175.8	2.740E-29	7	1.694	1.372	51.00	1.197E-30	2.842E-30	0.5908	0.6596	-3.756	1.482	3.837E-31
9.538	183.6	2.147E-29	7	1.741	1.312	53.01	9.530E-31	2.262E-30	0.5649	0.6223	-4.036	1.482	3.054E-31

p_3	θ_3	$\dfrac{d^2\sigma}{d\Omega_3\,dp_3}$	Error	p_T	p_L^*	θ_3^*	$\dfrac{1}{p_3^{*2}}\cdot\dfrac{d^2\sigma}{d\Omega_3^*\,dp_3^*}$	$\dfrac{E_3}{p_3^2}\cdot\dfrac{d^2\sigma}{d\Omega_3\,dp_3}$	x^*	y_3^*	t	m_4	$\dfrac{d^2\sigma}{dt\,dm_4^2}$
GeV/c	mrad	cm² sr⁻¹ GeV⁻¹ c	%	GeV/c	GeV/c	deg	cm² sr⁻¹ GeV⁻³ c³	cm² sr⁻¹ GeV⁻² c³			GeV²	GeV	cm²/GeV⁴

$p_1 = 14.250$ GeV/c; $p_1^* = 2.502$ GeV/c; $E_1^* + E_2^* = 5.344$ GeV; $s = 28.557$ GeV²; $\gamma_{c.m.} = 2.8479$; $\beta_{c.m.} = 0.93633$; $y_1^* = 1.7074$; $p_{3,\,max}^* = 2.502$ GeV/c; [72 A 1]

p_3	θ_3	$\dfrac{d^2\sigma}{d\Omega_3\,dp_3}$	Error	p_T	p_L^*	θ_3^*	$\dfrac{1}{p_3^{*2}}\cdot\dfrac{d^2\sigma}{d\Omega_3^*\,dp_3^*}$	$\dfrac{E_3}{p_3^2}\cdot\dfrac{d^2\sigma}{d\Omega_3\,dp_3}$	x^*	y_3^*	t	m_4	$\dfrac{d^2\sigma}{dt\,dm_4^2}$
4.500	12.0	4.323E-26	3	0.054	0.557	5.54	8.983E-27	9.813E-27	0.2227	0.5626	-1.290	4.214	1.153E-27
4.750	12.0	4.911E-26	3	0.057	0.616	5.29	9.379E-27	1.054E-26	0.2461	0.6154	-1.163	4.175	1.238E-27
5.000	12.0	5.460E-26	3	0.060	0.673	5.09	9.610E-27	1.111E-26	0.2690	0.6656	-1.051	4.133	1.305E-27
5.120	12.0	5.758E-26	3	0.061	0.700	5.02	9.754E-27	1.143E-26	0.2798	0.6888	-1.001	4.112	1.343E-27
5.250	12.0	6.139E-26	3	0.063	0.729	4.94	9.983E-27	1.188E-26	0.2915	0.7134	-0.950	4.089	1.396E-27
5.500	12.0	6.880E-26	3	0.066	0.784	4.81	1.036E-26	1.269E-26	0.3135	0.7590	-0.859	4.043	1.491E-27
5.750	12.0	7.841E-26	3	0.069	0.839	4.70	1.096E-26	1.382E-26	0.3353	0.8026	-0.778	3.996	1.623E-27
6.000	12.0	8.669E-26	3	0.072	0.892	4.61	1.128E-26	1.462E-26	0.3567	0.8444	-0.704	3.947	1.718E-27
6.250	12.0	9.125E-26	3	0.075	0.945	4.54	1.107E-26	1.476E-26	0.3779	0.8845	-0.638	3.896	1.735E-27
6.500	12.0	9.970E-26	3	0.078	0.998	4.47	1.130E-26	1.550E-26	0.3988	0.9230	-0.578	3.844	1.821E-27
6.750	12.0	1.124E-25	3	0.081	1.050	4.41	1.192E-26	1.681E-26	0.4195	0.9601	-0.523	3.790	1.975E-27
7.000	12.0	1.223E-25	3	0.084	1.101	4.36	1.217E-26	1.763E-26	0.4401	0.9959	-0.474	3.735	2.071E-27
7.340	12.0	1.419E-25	3	0.088	1.170	4.30	1.297E-26	1.949E-26	0.4678	1.0425	-0.413	3.658	2.290E-27
7.520	12.0	1.503E-25	3	0.090	1.207	4.28	1.315E-26	2.014E-26	0.4823	1.0663	-0.384	3.616	2.366E-27
7.710	12.0	1.605E-25	3	0.093	1.245	4.25	1.343E-26	2.097E-26	0.4976	1.0909	-0.356	3.570	2.464E-27
7.900	12.0	1.808E-25	3	0.095	1.283	4.23	1.447E-26	2.305E-26	0.5128	1.1148	-0.329	3.524	2.708E-27
8.090	12.0	1.955E-25	3	0.097	1.321	4.20	1.499E-26	2.433E-26	0.5279	1.1382	-0.304	3.477	2.858E-27
8.290	12.0	2.067E-25	3	0.099	1.360	4.18	1.516E-26	2.509E-26	0.5438	1.1622	-0.280	3.427	2.948E-27
8.490	12.0	2.161E-25	3	0.102	1.400	4.16	1.517E-26	2.561E-26	0.5596	1.1856	-0.257	3.375	3.009E-27
8.700	12.0	2.246E-25	3	0.104	1.441	4.14	1.507E-26	2.597E-26	0.5761	1.2096	-0.235	3.320	3.051E-27
8.920	12.0	2.405E-25	3	0.107	1.485	4.12	1.541E-26	2.711E-26	0.5934	1.2341	-0.214	3.261	3.185E-27
9.140	12.0	2.581E-25	3	0.110	1.527	4.11	1.581E-26	2.839E-26	0.6106	1.2580	-0.194	3.200	3.335E-27
9.360	12.0	2.590E-25	3	0.112	1.570	4.09	1.517E-26	2.781E-26	0.6277	1.2813	-0.176	3.138	3.267E-27
9.590	12.0	2.714E-25	3	0.115	1.615	4.08	1.520E-26	2.844E-26	0.6455	1.3052	-0.159	3.072	3.341E-27
9.830	12.0	2.816E-25	3	0.118	1.661	4.06	1.505E-26	2.878E-26	0.6641	1.3294	-0.142	3.001	3.381E-27
10.320	12.0	3.250E-25	3	0.124	1.756	4.03	1.585E-26	3.162E-26	0.7018	1.3770	-0.113	2.849	3.715E-27
10.570	12.0	3.422E-25	3	0.127	1.804	4.02	1.595E-26	3.250E-26	0.7210	1.4004	-0.100	2.769	3.819E-27
10.830	12.0	3.787E-25	3	0.130	1.854	4.01	1.686E-26	3.510E-26	0.7409	1.4242	-0.089	2.682	4.124E-27
11.100	12.0	4.023E-25	3	0.133	1.905	4.00	1.709E-26	3.637E-26	0.7615	1.4483	-0.078	2.588	4.273E-27
11.370	12.0	4.041E-25	3	0.136	1.956	3.99	1.640E-26	3.566E-26	0.7820	1.4717	-0.068	2.490	4.190E-27
11.510	12.0	4.229E-25	3	0.138	1.983	3.98	1.677E-26	3.686E-26	0.7926	1.4837	-0.064	2.438	4.331E-27
11.650	12.0	4.415E-25	3	0.140	2.010	3.98	1.711E-26	3.802E-26	0.8033	1.4955	-0.060	2.385	4.467E-27
11.800	12.0	4.479E-25	3	0.142	2.038	3.97	1.694E-26	3.808E-26	0.8146	1.5079	-0.055	2.326	4.474E-27
11.940	12.0	4.745E-25	3	0.143	2.065	3.97	1.754E-26	3.986E-26	0.8252	1.5194	-0.052	2.270	4.684E-27
12.090	12.0	4.622E-25	3	0.145	2.093	3.97	1.668E-26	3.834E-26	0.8366	1.5316	-0.049	2.208	4.505E-27
12.240	12.0	4.751E-25	3	0.147	2.121	3.96	1.675E-26	3.893E-26	0.8479	1.5436	-0.045	2.144	4.574E-27
12.390	12.0	4.789E-25	3	0.149	2.150	3.96	1.649E-26	3.876E-26	0.8592	1.5554	-0.043	2.078	4.554E-27
12.690	12.0	4.885E-25	3	0.152	2.206	3.95	1.607E-26	3.860E-26	0.8819	1.5787	-0.038	1.940	4.535E-27
12.840	12.0	5.059E-25	3	0.154	2.234	3.94	1.627E-26	3.951E-26	0.8931	1.5901	-0.036	1.866	4.642E-27
13.000	12.0	5.312E-25	3	0.156	2.265	3.94	1.668E-26	4.097E-26	0.9052	1.6021	-0.034	1.785	4.813E-27

Diddens/Schlüpmann

p_3	θ_3	$\dfrac{d^2\sigma}{d\Omega_3\,dp_3}$	Error	p_T	p_L^*	θ_3^*	$\dfrac{1}{p_3^{*2}}\cdot\dfrac{d^2\sigma}{d\Omega_3^*\,dp_3^*}$	$\dfrac{E_3}{p_3^2}\cdot\dfrac{d^2\sigma}{d\Omega_3\,dp_3}$	x^*	y_3^*	t	m_4	$\dfrac{d^2\sigma}{dt\,dm_4^2}$
GeV/c	mrad	cm² sr⁻¹ GeV⁻¹ c	%	GeV/c	GeV/c	deg	cm² sr⁻¹ GeV⁻³ c³	cm² sr⁻¹ GeV⁻² c³			GeV²	GeV	cm²/GeV⁴

$p_1 = 19.200$ GeV/c; $p_1^* = 2.929$ GeV/c; $E_1^* + E_2^* = 6.151$ GeV; $s = 37.830$ GeV²; $\gamma_{\text{c.m.}} = 3.2779$; $\beta_{\text{c.m.}} = 0.95233$; $\mathbf{y}_1^* = 1.8562$; $p_{3,\,\text{max}}^* = 2.929$ GeV/c; [70 A 1]

p_3	θ_3	$\dfrac{d^2\sigma}{d\Omega_3\,dp_3}$	Error	p_T	p_L^*	θ_3^*	$\dfrac{1}{p_3^{*2}}\cdot\dfrac{d^2\sigma}{d\Omega_3^*\,dp_3^*}$	$\dfrac{E_3}{p_3^2}\cdot\dfrac{d^2\sigma}{d\Omega_3\,dp_3}$	x^*	y_3^*	t	m_4	$\dfrac{d^2\sigma}{dt\,dm_4^2}$
4.500	12.5	3.363E-26	10	0.056	0.400	8.01	7.474E-27	7.634E-27	0.1366	0.4137	-2.179	5.113	6.657E-28
4.500	20.0	3.542E-26	10	0.090	0.398	12.74	7.858E-27	8.040E-27	0.1360	0.4108	-2.200	5.111	7.011E-28
4.500	30.0	3.174E-26	10	0.135	0.394	18.89	7.018E-27	7.205E-27	0.1347	0.4050	-2.244	5.107	6.283E-28
4.500	40.0	2.953E-26	10	0.180	0.389	24.81	6.498E-27	6.703E-27	0.1329	0.3970	-2.304	5.101	5.845E-28
4.500	50.0	2.941E-26	10	0.225	0.383	30.44	6.432E-27	6.676E-27	0.1307	0.3869	-2.382	5.093	5.822E-28
4.500	60.0	2.555E-26	10	0.270	0.375	35.77	5.547E-27	5.800E-27	0.1279	0.3749	-2.477	5.084	5.058E-28
4.500	70.0	2.607E-26	10	0.315	0.365	40.77	5.611E-27	5.918E-27	0.1246	0.3610	-2.589	5.073	5.160E-28
6.000	12.5	7.129E-26	7	0.075	0.708	6.04	1.021E-26	1.203E-26	0.2419	0.6953	-1.335	4.921	1.049E-27
6.000	20.0	6.570E-26	7	0.120	0.706	9.65	9.390E-27	1.108E-26	0.2411	0.6903	-1.364	4.918	9.665E-28
6.000	30.0	6.058E-26	7	0.180	0.701	14.40	8.624E-27	1.022E-26	0.2394	0.6802	-1.421	4.913	8.911E-28
6.000	40.0	5.451E-26	7	0.240	0.694	19.07	7.717E-27	9.195E-27	0.2370	0.6665	-1.502	4.904	8.019E-28
6.000	50.0	4.803E-26	7	0.300	0.685	23.63	6.752E-27	8.102E-27	0.2340	0.6493	-1.605	4.894	7.065E-28
6.000	60.0	4.210E-26	7	0.360	0.675	28.07	5.868E-27	7.102E-27	0.2303	0.6290	-1.732	4.881	6.193E-28
6.000	70.0	3.631E-26	7	0.420	0.662	32.38	5.011E-27	6.125E-27	0.2260	0.6061	-1.882	4.866	5.341E-28
8.000	12.5	1.180E-25	5	0.100	1.077	5.31	1.037E-26	1.485E-26	0.3677	0.9779	-0.738	4.593	1.295E-27
8.000	20.0	1.075E-25	5	0.160	1.074	8.48	9.430E-27	1.353E-26	0.3666	0.9691	-0.775	4.589	1.180E-27
8.000	30.0	9.014E-26	5	0.240	1.067	12.67	7.873E-27	1.134E-26	0.3644	0.9517	-0.852	4.581	9.893E-28
8.000	40.0	7.569E-26	5	0.320	1.058	16.82	6.571E-27	9.526E-27	0.3612	0.9282	-0.960	4.569	8.307E-28
8.000	50.0	6.128E-26	5	0.400	1.046	20.92	5.279E-27	7.712E-27	0.3572	0.8995	-1.098	4.554	6.725E-28
8.000	60.0	4.755E-26	5	0.480	1.032	24.94	4.058E-27	5.984E-27	0.3523	0.8666	-1.267	4.535	5.219E-28
8.000	70.0	3.590E-26	5	0.560	1.015	28.87	3.030E-27	4.518E-27	0.3465	0.8302	-1.466	4.513	3.940E-28
10.000	12.5	2.065E-25	3	0.125	1.423	5.02	1.214E-26	2.074E-26	0.4859	1.1967	-0.416	4.206	1.809E-27
10.000	20.0	1.722E-25	3	0.200	1.419	8.02	1.010E-26	1.730E-26	0.4845	1.1832	-0.463	4.200	1.508E-27
10.000	30.0	1.254E-25	3	0.300	1.411	12.00	7.320E-27	1.260E-26	0.4817	1.1566	-0.559	4.189	1.098E-27
10.000	40.0	8.923E-26	3	0.400	1.399	15.95	5.176E-27	8.962E-27	0.4778	1.1216	-0.693	4.173	7.815E-28
10.000	50.0	6.233E-26	3	0.500	1.385	19.85	3.586E-27	6.260E-27	0.4728	1.0800	-0.866	4.152	5.459E-28
10.000	60.0	4.200E-26	3	0.600	1.367	23.69	2.393E-27	4.218E-27	0.4666	1.0333	-1.077	4.126	3.679E-28
10.000	70.0	2.808E-26	3	0.699	1.345	27.47	1.582E-27	2.820E-27	0.4593	0.9833	-1.326	4.096	2.459E-28
11.000	12.5	2.632E-25	3	0.137	1.591	4.94	1.296E-26	2.401E-26	0.5434	1.2898	-0.312	3.990	2.094E-27
11.000	20.0	2.006E-25	3	0.220	1.587	7.89	9.858E-27	1.830E-26	0.5419	1.2736	-0.363	3.984	1.596E-27
11.000	30.0	1.328E-25	3	0.330	1.578	11.81	6.496E-27	1.212E-26	0.5388	1.2419	-0.469	3.971	1.057E-27
11.000	40.0	8.624E-26	3	0.440	1.565	15.70	4.191E-27	7.868E-27	0.5345	1.2007	-0.617	3.952	6.861E-28
11.000	50.0	5.304E-26	4	0.550	1.549	19.54	2.557E-27	4.839E-27	0.5290	1.1522	-0.807	3.928	4.220E-28
11.000	60.0	3.409E-26	4	0.660	1.529	23.33	1.627E-27	3.110E-27	0.5222	1.0987	-1.039	3.898	2.712E-28
11.000	70.0	2.224E-26	3	0.769	1.506	27.06	1.049E-27	2.029E-27	0.5142	1.0420	-1.313	3.863	1.769E-28
12.000	12.5	3.116E-25	3	0.150	1.758	4.88	1.304E-26	2.605E-26	0.6002	1.3745	-0.233	3.759	2.271E-27
12.000	20.0	2.274E-25	3	0.240	1.753	7.80	9.491E-27	1.901E-26	0.5985	1.3554	-0.289	3.752	1.658E-27
12.000	30.0	1.360E-25	3	0.360	1.743	11.67	5.650E-27	1.137E-26	0.5952	1.3183	-0.405	3.736	9.913E-28

p_3	θ_3	$\dfrac{d^2\sigma}{d\Omega_3\,dp_3}$	Error	p_T	p_L^*	θ_3^*	$\dfrac{1}{p_3^{*2}}\cdot\dfrac{d^2\sigma}{d\Omega_3^*\,dp_3^*}$	$\dfrac{E_3}{p_3^2}\cdot\dfrac{d^2\sigma}{d\Omega_3\,dp_3}$	x^*	y_3^*	t	m_4	$\dfrac{d^2\sigma}{dt\,dm_4^2}$
GeV/c	mrad	$cm^2\,sr^{-1}\,GeV^{-1}\,c$	%	GeV/c	GeV/c	deg	$cm^2\,sr^{-1}\,GeV^{-3}\,c^3$	$cm^2\,sr^{-1}\,GeV^{-2}\,c^3$			GeV^2	GeV	cm^2/GeV^4

$p_1 = 19.200$ GeV/c; $p_1^* = 2.929$ GeV/c; $E_1^* + E_2^* = 6.151$ GeV; $s = 37.830$ GeV^2; $\gamma_{c.m.} = 3.2779$; $\beta_{c.m.} = 0.95233$; $y_1^* = 1.8562$; $p_{3,\,max}^* = 2.929$ GeV/c; [70 A 1] (cont.)

p_3	θ_3	$\dfrac{d^2\sigma}{d\Omega_3\,dp_3}$	Error	p_T	p_L^*	θ_3^*	$\dfrac{1}{p_3^{*2}}\cdot\dfrac{d^2\sigma}{d\Omega_3^*\,dp_3^*}$	$\dfrac{E_3}{p_3^2}\cdot\dfrac{d^2\sigma}{d\Omega_3\,dp_3}$	x^*	y_3^*	t	m_4	$\dfrac{d^2\sigma}{dt\,dm_4^2}$
12.000	40.0	8.080E-26	3	0.480	1.729	15.51	3.335E-27	6.754E-27	0.5905	1.2706	-0.566	3.715	5.889E-28
12.000	50.0	4.681E-26	3	0.600	1.712	19.31	1.916E-27	3.913E-27	0.5844	1.2152	-0.773	3.687	3.412E-28
12.000	60.0	2.782E-26	3	0.720	1.690	23.06	1.127E-27	2.325E-27	0.5771	1.1549	-1.026	3.652	2.028E-28
12.000	70.0	1.584E-26	3	0.839	1.664	26.76	6.344E-28	1.324E-27	0.5683	1.0919	-1.326	3.611	1.155E-28
13.000	12.5	3.413E-25	3	0.162	1.923	4.83	1.227E-26	2.632E-26	0.6564	1.4521	-0.174	3.510	2.295E-27
13.000	20.0	2.424E-25	3	0.260	1.917	7.72	8.694E-27	1.869E-26	0.6547	1.4299	-0.235	3.501	1.630E-27
13.000	30.0	1.326E-25	3	0.390	1.907	11.56	4.733E-27	1.023E-26	0.6510	1.3871	-0.360	3.483	8.918E-28
13.000	40.0	7.328E-26	3	0.520	1.892	15.37	2.599E-27	5.652E-27	0.6459	1.3327	-0.534	3.458	4.928E-28
13.000	50.0	3.386E-26	3	0.650	1.873	19.14	1.191E-27	2.611E-27	0.6394	1.2705	-0.759	3.426	2.277E-28
13.000	60.0	2.128E-26	3	0.780	1.849	22.86	7.409E-28	1.641E-27	0.6314	1.2036	-1.033	3.385	1.431E-28
13.000	70.0	1.104E-26	3	0.909	1.821	26.53	3.798E-28	8.514E-28	0.6219	1.1346	-1.358	3.337	7.425E-29
14.000	12.5	3.854E-25	5	0.175	2.086	4.80	1.203E-26	2.759E-26	0.7123	1.5238	-0.130	3.239	2.406E-27
14.000	20.0	2.580E-25	3	0.280	2.080	7.66	8.033E-27	1.847E-26	0.7104	1.4981	-0.196	3.229	1.611E-27
14.000	30.0	1.274E-25	3	0.420	2.069	11.47	3.948E-27	9.120E-27	0.7065	1.4494	-0.330	3.208	7.953E-28
14.000	40.0	6.379E-26	3	0.560	2.053	15.25	1.964E-27	4.567E-27	0.7010	1.3882	-0.518	3.179	3.982E-28
14.000	50.0	3.041E-26	4	0.700	2.032	19.00	9.283E-28	2.177E-27	0.6939	1.3191	-0.760	3.140	1.898E-28
14.000	60.0	1.560E-26	3	0.839	2.007	22.70	4.714E-28	1.117E-27	0.6853	1.2459	-1.056	3.093	9.739E-29
14.000	70.0	7.372E-27	3	0.979	1.977	26.35	2.201E-28	5.278E-28	0.6751	1.1713	-1.405	3.036	4.602E-29
15.000	12.5	4.431E-25	5	0.187	2.249	4.77	1.211E-26	2.960E-26	0.7678	1.5901	-0.099	2.941	2.581E-27
15.000	20.0	2.696E-25	3	0.300	2.243	7.62	7.352E-27	1.801E-26	0.7657	1.5610	-0.169	2.929	1.570E-27
15.000	30.0	1.196E-25	3	0.450	2.230	11.41	3.246E-27	7.989E-27	0.7615	1.5060	-0.313	2.905	6.966E-28
15.000	40.0	5.432E-26	3	0.600	2.213	15.17	1.465E-27	3.628E-27	0.7557	1.4380	-0.514	2.870	3.164E-28
15.000	50.0	2.420E-26	3	0.750	2.191	18.89	6.470E-28	1.616E-27	0.7481	1.3622	-0.774	2.824	1.410E-28
15.000	60.0	1.091E-26	3	0.899	2.164	22.57	2.887E-28	7.288E-28	0.7389	1.2829	-1.090	2.767	6.355E-29
15.000	70.0	4.616E-27	3	1.049	2.132	26.20	1.207E-28	3.083E-28	0.7280	1.2029	-1.464	2.699	2.689E-29
16.000	12.5	5.214E-25	5	0.200	2.410	4.74	1.258E-26	3.264E-26	0.8230	1.6519	-0.077	2.607	2.847E-27
16.000	20.0	2.800E-25	3	0.320	2.404	7.58	6.742E-27	1.753E-26	0.8208	1.6190	-0.152	2.593	1.529E-27
16.000	30.0	1.159E-25	3	0.480	2.391	11.35	2.777E-27	7.256E-27	0.8163	1.5577	-0.306	2.563	6.327E-28
16.000	40.0	4.824E-26	3	0.640	2.372	15.09	1.148E-27	3.020E-27	0.8101	1.4828	-0.521	2.521	2.634E-28
16.000	50.0	1.850E-26	5	0.800	2.349	18.80	4.366E-28	1.158E-27	0.8020	1.4005	-0.797	2.466	1.010E-28
16.000	60.0	7.362E-27	5	0.959	2.320	22.47	1.720E-28	4.609E-28	0.7922	1.3154	-1.135	2.396	4.019E-29
16.000	70.0	2.737E-27	3	1.119	2.286	26.08	6.317E-29	1.714E-28	0.7805	1.2305	-1.534	2.311	1.494E-29
13.500	12.5	3.488E-25	3	0.169	2.004	4.81	1.167E-26	2.590E-26	0.6844	1.4887	-0.150	3.377	2.258E-27
13.800	12.5	3.659E-25	3	0.172	2.053	4.80	1.174E-26	2.658E-26	0.7011	1.5099	-0.138	3.295	2.317E-27
14.100	12.5	3.785E-25	3	0.176	2.102	4.79	1.165E-26	2.690E-26	0.7178	1.5306	-0.127	3.211	2.346E-27
14.400	12.5	3.939E-25	3	0.180	2.151	4.78	1.165E-26	2.741E-26	0.7345	1.5509	-0.116	3.124	2.390E-27
14.700	12.5	4.094E-25	3	0.184	2.200	4.77	1.163E-26	2.791E-26	0.7512	1.5707	-0.107	3.034	2.434E-27
15.000	12.5	4.200E-25	3	0.187	2.249	4.77	1.148E-26	2.805E-26	0.7678	1.5901	-0.099	2.941	2.446E-27

p_3	θ_3	$\dfrac{\mathrm{d}^2\sigma}{\mathrm{d}\Omega_3\,\mathrm{d}p_3}$	Error	p_{T}	p_{L}^*	θ_3^*	$\dfrac{1}{p_3^{*2}}\cdot\dfrac{\mathrm{d}^2\sigma}{\mathrm{d}\Omega_3^*\,\mathrm{d}p_3^*}$	$\dfrac{E_3}{p_3^2}\cdot\dfrac{\mathrm{d}^2\sigma}{\mathrm{d}\Omega_3\,\mathrm{d}p_3}$	x^*	y_3^*	t	m_4	$\dfrac{\mathrm{d}^2\sigma}{\mathrm{d}t\,\mathrm{d}m_4^2}$
GeV/c	mrad	$\mathrm{cm^2\,sr^{-1}\,GeV^{-1}\,c}$	%	GeV/c	GeV/c	deg	$\mathrm{cm^2\,sr^{-1}\,GeV^{-3}\,c^3}$	$\mathrm{cm^2\,sr^{-1}\,GeV^{-2}\,c^3}$			GeV^2	GeV	$\mathrm{cm^2/GeV^4}$

$p_1 = 19.200$ GeV/c; $p_1^* = 2.929$ GeV/c; $E_1^* + E_2^* = 6.151$ GeV; $s = 37.830$ GeV^2; $\gamma_{\mathrm{c.m.}} = 3.2779$; $\beta_{\mathrm{c.m.}} = 0.95233$; $y_1^* = 1.8562$; $p_{3,\,\mathrm{max}}^* = 2.929$ GeV/c; [70 A 1] (cont.)

p_3	θ_3	$\dfrac{\mathrm{d}^2\sigma}{\mathrm{d}\Omega_3\,\mathrm{d}p_3}$	Error	p_{T}	p_{L}^*	θ_3^*	$\dfrac{1}{p_3^{*2}}\cdot\dfrac{\mathrm{d}^2\sigma}{\mathrm{d}\Omega_3^*\,\mathrm{d}p_3^*}$	$\dfrac{E_3}{p_3^2}\cdot\dfrac{\mathrm{d}^2\sigma}{\mathrm{d}\Omega_3\,\mathrm{d}p_3}$	x^*	y_3^*	t	m_4	$\dfrac{\mathrm{d}^2\sigma}{\mathrm{d}t\,\mathrm{d}m_4^2}$
15.300	12.5	4.358E-25	3	0.191	2.297	4.76	1.147E-26	2.854E-26	0.7844	1.6091	-0.091	2.845	2.488E-27
15.600	12.5	4.475E-25	3	0.195	2.346	4.75	1.134E-26	2.874E-26	0.8009	1.6277	-0.085	2.746	2.506E-27
15.900	12.5	4.616E-25	3	0.199	2.394	4.75	1.128E-26	2.908E-26	0.8175	1.6459	-0.079	2.643	2.536E-27
16.200	12.5	4.749E-25	3	0.202	2.443	4.74	1.119E-26	2.936E-26	0.8340	1.6638	-0.074	2.535	2.561E-27
16.500	12.5	4.891E-25	3	0.206	2.491	4.73	1.112E-26	2.969E-26	0.8505	1.6812	-0.070	2.423	2.589E-27
16.800	12.5	5.028E-25	3	0.210	2.539	4.73	1.104E-26	2.998E-26	0.8670	1.6984	-0.066	2.305	2.614E-27
17.100	12.5	5.178E-25	3	0.214	2.587	4.72	1.099E-26	3.033E-26	0.8835	1.7152	-0.063	2.180	2.644E-27
17.400	12.5	5.242E-25	3	0.217	2.636	4.72	1.075E-26	3.017E-26	0.8999	1.7317	-0.061	2.048	2.631E-27
17.700	12.5	5.488E-25	3	0.221	2.684	4.71	1.089E-26	3.105E-26	0.9164	1.7479	-0.059	1.906	2.708E-27
18.000	12.5	7.277E-25	3	0.225	2.732	4.71	1.397E-26	4.048E-26	0.9328	1.7638	-0.058	1.753	3.530E-27
18.300	12.5	9.767E-25	3	0.229	2.780	4.70	1.816E-26	5.344E-26	0.9492	1.7794	-0.057	1.584	4.660E-27
18.600	12.5	1.198E-24	5	0.232	2.828	4.70	2.158E-26	6.449E-26	0.9656	1.7947	-0.057	1.396	5.624E-27
18.900	12.5	8.790E-25	5	0.236	2.876	4.70	1.535E-26	4.657E-26	0.9820	1.8097	-0.057	1.177	4.061E-27
13.500	20.0	2.388E-25	3	0.270	1.999	7.69	7.970E-27	1.773E-26	0.6826	1.4647	-0.214	3.368	1.546E-27
13.800	20.0	2.421E-25	3	0.276	2.048	7.68	7.748E-27	1.758E-26	0.6993	1.4849	-0.203	3.285	1.533E-27
14.100	20.0	2.448E-25	3	0.282	2.097	7.66	7.519E-27	1.740E-26	0.7159	1.5047	-0.193	3.200	1.517E-27
14.400	20.0	2.471E-25	3	0.288	2.145	7.65	7.289E-27	1.720E-26	0.7325	1.5239	-0.184	3.113	1.500E-27
14.700	20.0	2.485E-25	3	0.294	2.194	7.63	7.045E-27	1.694E-26	0.7492	1.5427	-0.176	3.022	1.477E-27
15.000	20.0	2.499E-25	3	0.300	2.243	7.62	6.815E-27	1.669E-26	0.7657	1.5610	-0.169	2.929	1.456E-27
15.300	20.0	2.496E-25	3	0.306	2.291	7.61	6.552E-27	1.634E-26	0.7823	1.5789	-0.163	2.833	1.425E-27
15.600	20.0	2.498E-25	3	0.312	2.339	7.60	6.316E-27	1.604E-26	0.7988	1.5964	-0.158	2.733	1.399E-27
15.900	20.0	2.501E-25	3	0.318	2.388	7.59	6.095E-27	1.576E-26	0.8153	1.6134	-0.153	2.629	1.374E-27
16.200	20.0	2.519E-25	3	0.324	2.436	7.58	5.921E-27	1.558E-26	0.8318	1.6301	-0.150	2.520	1.358E-27
16.500	20.0	2.554E-25	3	0.330	2.484	7.57	5.794E-27	1.550E-26	0.8483	1.6465	-0.147	2.407	1.352E-27
16.800	20.0	2.625E-25	3	0.336	2.532	7.56	5.750E-27	1.565E-26	0.8647	1.6624	-0.145	2.288	1.365E-27
17.100	20.0	2.721E-25	3	0.342	2.581	7.55	5.759E-27	1.594E-26	0.8811	1.6781	-0.143	2.162	1.390E-27
17.400	20.0	2.822E-25	3	0.348	2.629	7.54	5.775E-27	1.624E-26	0.8975	1.6934	-0.142	2.028	1.416E-27
17.700	20.0	3.040E-25	3	0.354	2.677	7.53	6.017E-27	1.720E-26	0.9139	1.7083	-0.142	1.884	1.500E-27
18.000	20.0	4.436E-25	3	0.360	2.725	7.53	8.498E-27	2.468E-26	0.9303	1.7230	-0.142	1.728	2.152E-27
18.300	20.0	4.689E-25	3	0.366	2.773	7.52	8.698E-27	2.566E-26	0.9467	1.7374	-0.143	1.557	2.237E-27
13.500	30.0	1.254E-25	3	0.405	1.988	11.51	4.166E-27	9.311E-27	0.6788	1.4190	-0.343	3.349	8.120E-28
13.800	30.0	1.228E-25	3	0.414	2.037	11.49	3.912E-27	8.919E-27	0.6954	1.4374	-0.335	3.265	7.778E-28
14.100	30.0	1.205E-25	3	0.423	2.085	11.47	3.683E-27	8.565E-27	0.7120	1.4553	-0.328	3.179	7.469E-28
14.400	30.0	1.177E-25	3	0.432	2.134	11.44	3.455E-27	8.191E-27	0.7285	1.4727	-0.322	3.090	7.143E-28
14.700	30.0	1.149E-25	3	0.441	2.182	11.42	3.242E-27	7.832E-27	0.7450	1.4896	-0.317	2.999	6.830E-28
15.000	30.0	1.118E-25	3	0.450	2.230	11.41	3.034E-27	7.468E-27	0.7615	1.5060	-0.313	2.905	6.512E-28
15.300	30.0	1.099E-25	3	0.459	2.279	11.39	2.871E-27	7.196E-27	0.7780	1.5220	-0.310	2.807	6.275E-28
15.600	30.0	1.074E-25	3	0.468	2.327	11.37	2.703E-27	6.897E-27	0.7944	1.5376	-0.308	2.705	6.014E-28
15.900	30.0	1.050E-25	3	0.477	2.375	11.36	2.571E-27	6.678E-27	0.8109	1.5528	-0.306	2.599	5.824E-28
16.200	30.0	1.049E-25	3	0.486	2.423	11.34	2.454E-27	6.486E-27	0.8273	1.5675	-0.305	2.489	5.656E-28
16.500	30.0	1.048E-25	3	0.495	2.471	11.33	2.366E-27	6.362E-27	0.8436	1.5819	-0.305	2.374	5.548E-28

Diddens/Schlüpmann

p_3	θ_3	$\dfrac{d^2\sigma}{d\Omega_3\,dp_3}$	Error	p_T	p_L^*	θ_3^*	$\dfrac{1}{p_3^{*2}}\cdot\dfrac{d^2\sigma}{d\Omega_3^*\,dp_3^*}$	$\dfrac{E_3}{p_3^2}\cdot\dfrac{d^2\sigma}{d\Omega_3\,dp_3}$	x^*	y_3^*	t	m_4	$\dfrac{d^2\sigma}{dt\,dm_4^2}$
GeV/c	mrad	cm^2 sr^{-1} GeV^{-1} c	%	GeV/c	GeV/c	deg	cm^2 sr^{-1} GeV^{-3} c^3	cm^2 sr^{-1} GeV^{-2} c^3			GeV^2	GeV	cm^2/GeV^4

$p_1 = 19.200$ GeV/c; $p_1^* = 2.929$ GeV/c; $E_1^* + E_2^* = 6.151$ GeV; $s = 37.830$ GeV^2; $\gamma_{c.m.} = 3.2779$; $\beta_{c.m.} = 0.95233$; $y_1^* = 1.8562$; $p_{3,\max}^* = 2.929$ GeV/c; [70 A 1] (cont.)

p_3	θ_3	$\dfrac{d^2\sigma}{d\Omega_3\,dp_3}$	Error	p_T	p_L^*	θ_3^*	$\dfrac{1}{p_3^{*2}}\cdot\dfrac{d^2\sigma}{d\Omega_3^*\,dp_3^*}$	$\dfrac{E_3}{p_3^2}\cdot\dfrac{d^2\sigma}{d\Omega_3\,dp_3}$	x^*	y_3^*	t	m_4	$\dfrac{d^2\sigma}{dt\,dm_4^2}$
16.800	30.0	1.070E-25	3	0.504	2.519	11.31	2.333E-27	6.379E-27	0.8600	1.5959	-0.306	2.252	5.563E-28
17.100	30.0	1.093E-25	3	0.513	2.567	11.30	2.302E-27	6.401E-27	0.8763	1.6096	-0.307	2.123	5.582E-28
17.400	30.0	1.077E-25	3	0.522	2.614	11.29	2.193E-27	6.199E-27	0.8927	1.6229	-0.309	1.986	5.405E-28
17.700	30.0	1.227E-25	3	0.531	2.662	11.28	2.417E-27	6.942E-27	0.9090	1.6359	-0.312	1.838	6.053E-28
18.000	30.0	1.650E-25	3	0.540	2.710	11.27	3.146E-27	9.179E-27	0.9253	1.6485	-0.315	1.678	8.004E-28
16.517	40.0	4.530E-26	3	0.661	2.455	15.06	1.014E-27	2.747E-27	0.8381	1.5042	-0.527	2.320	2.395E-28
16.817	40.0	4.530E-26	3	0.673	2.502	15.04	9.791E-28	2.698E-27	0.8543	1.5162	-0.532	2.194	2.353E-28
17.082	40.0	4.340E-26	3	0.683	2.544	15.03	9.100E-28	2.545E-27	0.8687	1.5264	-0.537	2.077	2.219E-28
17.382	40.0	4.160E-26	3	0.695	2.592	15.01	8.432E-28	2.397E-27	0.8849	1.5377	-0.543	1.935	2.090E-28
17.681	40.0	5.710E-26	3	0.707	2.639	15.00	1.120E-27	3.234E-27	0.9010	1.5486	-0.549	1.783	2.820E-28
17.820	40.0	7.180E-26	3	0.713	2.661	14.99	1.387E-27	4.035E-27	0.9085	1.5536	-0.552	1.707	3.518E-28
17.981	40.0	5.410E-26	3	0.719	2.686	14.99	1.027E-27	3.013E-27	0.9172	1.5593	-0.556	1.615	2.627E-28
18.117	40.0	5.940E-26	3	0.724	2.708	14.98	1.111E-27	3.283E-27	0.9245	1.5640	-0.559	1.533	2.863E-28
18.281	40.0	4.520E-26	3	0.731	2.734	14.97	8.305E-28	2.476E-27	0.9334	1.5696	-0.564	1.428	2.159E-28
18.499	40.0	3.130E-26	3	0.740	2.768	14.96	5.619E-28	1.694E-27	0.9451	1.5769	-0.569	1.275	1.477E-28
18.657	40.0	2.080E-26	3	0.746	2.793	14.96	3.673E-28	1.116E-27	0.9536	1.5821	-0.574	1.151	9.734E-29
18.729	40.0	3.500E-26	3	0.749	2.804	14.95	6.134E-28	1.871E-27	0.9575	1.5845	-0.576	1.090	1.632E-28
18.763	40.0	6.880E-26	3	0.750	2.810	14.95	1.202E-27	3.671E-27	0.9593	1.5856	-0.577	1.059	3.201E-28
18.780	40.0	1.250E-25	3	0.751	2.812	14.95	2.197E-27	6.718E-27	0.9602	1.5862	-0.577	1.044	5.858E-28
16.417	50.0	1.720E-26	3	0.821	2.414	18.77	3.862E-28	1.049E-27	0.8244	1.4152	-0.809	2.299	9.151E-29
16.717	50.0	1.650E-26	3	0.836	2.462	18.75	3.577E-28	9.886E-28	0.8405	1.4253	-0.819	2.171	8.620E-29
17.017	50.0	1.540E-26	3	0.850	2.509	18.73	3.225E-28	9.064E-28	0.8566	1.4352	-0.829	2.035	7.904E-29
17.420	50.0	1.430E-26	3	0.871	2.572	18.70	2.862E-28	8.221E-28	0.8782	1.4478	-0.844	1.836	7.169E-29
17.654	50.0	2.090E-26	3	0.882	2.609	18.69	4.075E-28	1.186E-27	0.8907	1.4549	-0.853	1.710	1.034E-28
17.803	50.0	1.710E-26	3	0.890	2.632	18.68	3.280E-28	9.618E-28	0.8986	1.4594	-0.859	1.624	8.387E-29
17.955	50.0	1.910E-26	3	0.897	2.656	18.67	3.604E-28	1.065E-27	0.9068	1.4638	-0.866	1.532	9.289E-29
18.181	50.0	1.370E-26	3	0.909	2.691	18.66	2.523E-28	7.545E-28	0.9188	1.4703	-0.875	1.384	6.580E-29
18.326	50.0	8.930E-27	3	0.916	2.714	18.65	1.619E-28	4.879E-28	0.9266	1.4744	-0.881	1.279	4.255E-29
18.483	50.0	5.140E-27	3	0.924	2.738	18.64	9.165E-29	2.785E-28	0.9349	1.4787	-0.888	1.156	2.428E-29
18.527	50.0	3.300E-27	3	0.925	2.745	18.64	5.857E-29	1.783E-28	0.9373	1.4799	-0.890	1.119	1.555E-29
18.610	50.0	1.140E-26	3	0.930	2.758	18.64	2.006E-28	6.134E-28	0.9417	1.4822	-0.894	1.045	5.349E-29
16.483	60.0	6.360E-27	3	0.988	2.395	22.42	1.402E-28	3.865E-28	0.8178	1.3296	-1.159	2.194	3.370E-29
16.882	60.0	5.410E-27	3	1.012	2.457	22.39	1.139E-28	3.210E-28	0.8390	1.3408	-1.181	2.011	2.799E-29
17.182	60.0	4.690E-27	3	1.030	2.504	22.37	9.541E-29	2.734E-28	0.8549	1.3488	-1.198	1.861	2.384E-29
17.481	60.0	5.550E-27	3	1.048	2.550	22.35	1.092E-28	3.179E-28	0.8707	1.3566	-1.216	1.699	2.773E-29
17.663	60.0	4.450E-27	3	1.059	2.578	22.33	8.579E-29	2.523E-28	0.8803	1.3611	-1.227	1.592	2.200E-29
17.781	60.0	4.790E-27	3	1.055	2.596	22.33	9.115E-29	2.698E-28	0.8865	1.3640	-1.234	1.518	2.352E-29
18.002	60.0	3.130E-27	3	1.079	2.631	22.31	5.815E-29	1.741E-28	0.8982	1.3694	-1.248	1.370	1.518E-29
18.155	60.0	1.820E-27	3	1.089	2.654	22.30	3.326E-29	1.004E-28	0.9063	1.3730	-1.257	1.258	8.753E-30
18.285	60.0	1.020E-27	3	1.096	2.674	22.29	1.838E-29	5.586E-29	0.9131	1.3760	-1.266	1.153	4.871E-30

p_3	θ_3	$\dfrac{d^2\sigma}{d\Omega_3\,dp_3}$	Error	p_T	p_L^*	θ_3^*	$\dfrac{1}{p_3^{*2}}\cdot\dfrac{d^2\sigma}{d\Omega_3^*\,dp_3^*}$	$\dfrac{E_3}{p_3^2}\cdot\dfrac{d^2\sigma}{d\Omega_3\,dp_3}$	x^*	y_3^*	t	m_4	$\dfrac{d^2\sigma}{dt\,dm_4^2}$
GeV/c	mrad	cm² sr⁻¹ GeV⁻¹ c	%	GeV/c	GeV/c	deg	cm² sr⁻¹ GeV⁻³ c³	cm² sr⁻¹ GeV⁻² c³			GeV²	GeV	cm²/GeV⁴

$p_1 = 19.200$ GeV/c; $p_1^* = 2.929$ GeV/c; $E_1^* + E_2^* = 6.151$ GeV; $s = 37.830$ GeV²; $\gamma_{c.m.} = 3.2779$; $\beta_{c.m.} = 0.95233$; $y_1^* = 1.8562$; $p_{3,max}^* = 2.929$ GeV/c; [70 A 1] (cont.)

p_3	θ_3	$\dfrac{d^2\sigma}{d\Omega_3\,dp_3}$	Error	p_T	p_L^*	θ_3^*	$\dfrac{1}{p_3^{*2}}\cdot\dfrac{d^2\sigma}{d\Omega_3^*\,dp_3^*}$	$\dfrac{E_3}{p_3^2}\cdot\dfrac{d^2\sigma}{d\Omega_3\,dp_3}$	x^*	y_3^*	t	m_4	$\dfrac{d^2\sigma}{dt\,dm_4^2}$
18.370	60.0	4.750E-28	3	1.102	2.687	22.29	8.483E-30	2.589E-29	0.9176	1.3780	-1.271	1.079	2.258E-30
18.450	60.0	2.700E-27	3	1.106	2.700	22.28	4.781E-29	1.465E-28	0.9218	1.3798	-1.276	1.005	1.278E-29
16.533	70.0	1.430E-27	3	1.156	2.368	26.03	3.097E-29	8.663E-29	0.8085	1.2437	-1.574	2.074	7.555E-30
16.869	70.0	1.160E-27	3	1.180	2.419	26.00	2.416E-29	6.887E-29	0.8260	1.2516	-1.601	1.910	6.006E-30
17.127	70.0	1.030E-27	3	1.198	2.459	25.98	2.083E-29	6.023E-29	0.8395	1.2574	-1.622	1.773	5.252E-30
17.282	70.0	1.060E-27	3	1.209	2.482	25.96	2.106E-29	6.143E-29	0.8476	1.2607	-1.635	1.685	5.356E-30
17.424	70.0	7.840E-28	3	1.219	2.504	25.95	1.533E-29	4.506E-29	0.8550	1.2638	-1.647	1.600	3.929E-30
17.581	70.0	7.520E-28	3	1.230	2.528	25.94	1.445E-29	4.283E-29	0.8632	1.2671	-1.660	1.501	3.735E-30
17.721	70.0	5.090E-28	3	1.239	2.549	25.93	9.633E-30	2.876E-29	0.8705	1.2700	-1.672	1.407	2.508E-30
17.803	70.0	3.700E-28	3	1.245	2.562	25.92	6.940E-30	2.081E-29	0.8748	1.2717	-1.679	1.349	1.815E-30
17.955	70.0	2.520E-28	4	1.256	2.585	25.91	4.649E-30	1.405E-29	0.8827	1.2747	-1.692	1.233	1.226E-30
18.110	70.0	1.600E-28	3	1.267	2.609	25.90	2.903E-30	8.847E-30	0.8907	1.2778	-1.706	1.103	7.714E-31
18.151	70.0	3.130E-28	3	1.270	2.615	25.90	5.653E-30	1.727E-29	0.8929	1.2786	-1.710	1.066	1.506E-30
18.192	70.0	1.100E-27	3	1.272	2.621	25.89	1.978E-29	6.055E-29	0.8950	1.2794	-1.713	1.028	5.280E-30

$p_1 = 24.000$ GeV/c; $p_1^* = 3.290$ GeV/c; $E_1^* + E_2^* = 6.843$ GeV; $s = 46.828$ GeV²; $\gamma_{c.m.} = 3.6469$; $\beta_{c.m.} = 0.96167$; $y_1^* = 1.9677$; $p_{3,max}^* = 3.290$ GeV/c; [72 A 1]

p_3	θ_3	$\dfrac{d^2\sigma}{d\Omega_3\,dp_3}$	Error	p_T	p_L^*	θ_3^*	$\dfrac{1}{p_3^{*2}}\cdot\dfrac{d^2\sigma}{d\Omega_3^*\,dp_3^*}$	$\dfrac{E_3}{p_3^2}\cdot\dfrac{d^2\sigma}{d\Omega_3\,dp_3}$	x^*	y_3^*	t	m_4	$\dfrac{d^2\sigma}{dt\,dm_4^2}$
4.000	17.0	2.142E-26	3	0.068	0.176	21.09	5.747E-27	5.500E-27	0.0536	0.1863	-3.629	5.883	3.837E-28
4.000	27.0	2.018E-26	3	0.108	0.173	31.96	5.397E-27	5.182E-27	0.0526	0.1823	-3.671	5.879	3.615E-28
4.000	37.0	1.878E-26	3	0.148	0.168	41.30	4.999E-27	4.822E-27	0.0512	0.1764	-3.732	5.874	3.364E-28
4.000	47.0	1.786E-26	3	0.188	0.162	49.19	4.726E-27	4.586E-27	0.0493	0.1688	-3.813	5.867	3.199E-28
4.000	57.0	1.676E-26	3	0.228	0.155	55.83	4.401E-27	4.304E-27	0.0470	0.1596	-3.913	5.859	3.002E-28
4.000	67.0	1.572E-26	3	0.268	0.146	61.46	4.092E-27	4.037E-27	0.0443	0.1488	-4.032	5.849	2.816E-28
4.000	77.0	1.411E-26	3	0.308	0.135	66.28	3.636E-27	3.623E-27	0.0411	0.1365	-4.170	5.837	2.528E-28
4.000	87.0	1.275E-26	3	0.348	0.123	70.48	3.248E-27	3.274E-27	0.0375	0.1229	-4.327	5.823	2.284E-28
4.000	97.0	1.156E-26	3	0.387	0.110	74.17	2.907E-27	2.968E-27	0.0334	0.1080	-4.503	5.808	2.071E-28
4.000	107.0	1.041E-26	3	0.427	0.095	77.47	2.582E-27	2.673E-27	0.0289	0.0920	-4.699	5.791	1.865E-28
4.000	117.0	9.134E-27	3	0.467	0.079	80.44	2.232E-27	2.345E-27	0.0239	0.0750	-4.913	5.773	1.636E-28
4.000	127.0	8.190E-27	3	0.507	0.061	83.14	1.969E-27	2.103E-27	0.0185	0.0571	-5.147	5.753	1.467E-28
4.000	137.0	7.255E-27	3	0.546	0.042	85.63	1.715E-27	1.863E-27	0.0127	0.0384	-5.400	5.731	1.300E-28
4.000	147.0	6.182E-27	3	0.586	0.021	87.94	1.435E-27	1.587E-27	0.0064	0.0190	-5.672	5.707	1.107E-28
5.000	17.0	3.396E-26	3	0.085	0.390	12.29	6.777E-27	6.911E-27	0.1186	0.4032	-2.649	5.810	4.821E-28
5.000	27.0	3.102E-26	3	0.135	0.386	19.26	6.167E-27	6.312E-27	0.1174	0.3970	-2.702	5.805	4.404E-28
5.000	37.0	2.880E-26	3	0.185	0.380	25.93	5.695E-27	5.861E-27	0.1156	0.3880	-2.779	5.799	4.088E-28
5.000	47.0	2.626E-26	3	0.235	0.373	32.22	5.155E-27	5.344E-27	0.1133	0.3764	-2.880	5.790	3.728E-28
5.000	57.0	2.356E-26	3	0.285	0.363	38.10	4.585E-27	4.794E-27	0.1104	0.3625	-3.004	5.779	3.345E-28
5.000	67.0	2.059E-26	3	0.335	0.352	43.57	3.966E-27	4.190E-27	0.1070	0.3463	-3.153	5.766	2.923E-28
5.000	77.0	1.781E-26	3	0.385	0.339	48.62	3.390E-27	3.624E-27	0.1030	0.3282	-3.326	5.751	2.528E-28
5.000	87.0	1.530E-26	3	0.434	0.324	53.29	2.874E-27	3.113E-27	0.0984	0.3084	-3.522	5.734	2.172E-28
5.000	97.0	1.304E-26	3	0.484	0.307	57.61	2.413E-27	2.654E-27	0.0933	0.2870	-3.743	5.715	1.851E-28
5.000	107.0	1.095E-26	3	0.534	0.289	61.61	1.994E-27	2.228E-27	0.0877	0.2642	-3.987	5.693	1.554E-28

p_3	θ_3	$\dfrac{d^2\sigma}{d\Omega_3\,dp_3}$	Error	p_T	p_L^*	θ_3^*	$\dfrac{1}{p_3^{*2}}\cdot\dfrac{d^2\sigma}{d\Omega_3^*\,dp_3^*}$	$\dfrac{E_3}{p_3^2}\cdot\dfrac{d^2\sigma}{d\Omega_3\,dp_3}$	x^*	y_3^*	t	m_4	$\dfrac{d^2\sigma}{dt\,dm_4^2}$
GeV/c	mrad	cm² sr⁻¹ GeV⁻¹ c	%	GeV/c	GeV/c	deg	cm² sr⁻¹ GeV⁻³ c³	cm² sr⁻¹ GeV⁻² c³			GeV²	GeV	cm²/GeV⁴
$p_1 = 24.000$ GeV/c; $p_1^* = 3.290$ GeV/c; $E_1^* + E_2^* = 6.843$ GeV; $s = 46.828$ GeV²; $\gamma_{c.m.} = 3.6469$; $\beta_{c.m.} = 0.96167$; $y_1^* = 1.9677$; $p_{3,\,max}^* = 3.290$ GeV/c; [72 A 1] (cont.)													
5.000	117.0	9.130E−27	3	0.584	0.268	65.32	1.634E−27	1.858E−27	0.0815	0.2404	−4.255	5.670	1.296E−28
5.000	127.0	7.607E−27	3	0.633	0.246	68.77	1.336E−27	1.548E−27	0.0748	0.2157	−4.547	5.644	1.080E−28
5.000	137.0	6.177E−27	3	0.683	0.222	71.99	1.064E−27	1.257E−27	0.0675	0.1902	−4.863	5.616	8.769E−29
5.000	147.0	5.036E−27	3	0.732	0.196	75.00	8.496E−28	1.025E−27	0.0596	0.1641	−5.203	5.586	7.149E−29
6.000	17.0	5.027E−26	3	0.102	0.580	9.98	7.656E−27	8.480E−27	0.1762	0.5811	−2.003	5.705	5.916E−28
6.000	27.0	4.586E−26	3	0.162	0.575	15.73	6.956E−27	7.736E−27	0.1747	0.5722	−2.067	5.700	5.397E−28
6.000	37.0	4.156E−26	3	0.222	0.568	21.34	6.265E−27	7.011E−27	0.1726	0.5595	−2.159	5.692	4.891E−28
6.000	47.0	3.668E−26	3	0.282	0.559	26.77	5.486E−27	6.188E−27	0.1698	0.5433	−2.280	5.681	4.317E−28
6.000	57.0	3.180E−26	3	0.342	0.547	31.98	4.711E−27	5.364E−27	0.1664	0.5239	−2.430	5.668	3.742E−28
6.000	67.0	2.624E−26	3	0.402	0.534	36.96	3.843E−27	4.426E−27	0.1623	0.5018	−2.608	5.652	3.088E−28
6.000	77.0	2.147E−26	3	0.462	0.518	41.69	3.104E−27	3.622E−27	0.1575	0.4772	−2.815	5.634	2.527E−28
6.000	87.0	1.740E−26	3	0.521	0.503	46.18	2.479E−27	2.935E−27	0.1520	0.4506	−3.051	5.613	2.048E−28
6.000	97.0	1.381E−26	3	0.581	0.480	50.44	1.936E−27	2.330E−27	0.1459	0.4224	−3.316	5.589	1.625E−28
6.000	107.0	1.079E−26	3	0.641	0.458	54.45	1.486E−27	1.820E−27	0.1391	0.3928	−3.609	5.563	1.270E−28
6.000	117.0	8.395E−27	3	0.700	0.433	58.25	1.134E−27	1.416E−27	0.1317	0.3622	−3.931	5.534	9.879E−29
6.000	127.0	6.441E−27	3	0.750	0.407	61.84	8.528E−28	1.087E−27	0.1236	0.3308	−4.281	5.502	7.580E−29
6.000	137.0	4.863E−27	3	0.819	0.378	65.24	6.302E−28	8.203E−28	0.1149	0.2989	−4.660	5.467	5.723E−29
6.000	147.0	3.720E−27	3	0.879	0.347	68.45	4.713E−28	6.275E−28	0.1055	0.2667	−5.068	5.430	4.378E−29
7.000	17.0	6.555E−26	3	0.119	0.755	8.95	7.818E−27	9.462E−27	0.2295	0.7316	−1.552	5.581	6.601E−28
7.000	27.0	5.922E−26	3	0.189	0.750	14.15	7.021E−27	8.536E−27	0.2278	0.7196	−1.625	5.574	5.955E−28
7.000	37.0	5.226E−26	3	0.259	0.741	19.25	6.156E−27	7.532E−27	0.2253	0.7026	−1.733	5.564	5.255E−28
7.000	47.0	4.848E−26	3	0.329	0.731	24.23	5.663E−27	6.988E−27	0.2221	0.6811	−1.874	5.552	4.875E−28
7.000	57.0	3.982E−26	3	0.399	0.717	29.07	4.604E−27	5.739E−27	0.2180	0.6558	−2.049	5.536	4.004E−28
7.000	67.0	2.952E−26	3	0.469	0.702	33.74	3.372E−27	4.255E−27	0.2132	0.6271	−2.257	5.517	2.968E−28
7.000	77.0	2.298E−26	3	0.538	0.683	38.24	2.589E−27	3.312E−27	0.2077	0.5958	−2.499	5.495	2.311E−28
7.000	87.0	1.741E−26	3	0.608	0.662	42.56	1.931E−27	2.509E−27	0.2013	0.5623	−2.774	5.470	1.751E−28
7.000	97.0	1.282E−26	3	0.678	0.639	46.70	1.398E−27	1.848E−27	0.1942	0.5272	−3.082	5.442	1.289E−28
7.000	107.0	9.327E−27	3	0.748	0.613	50.65	9.979E−28	1.344E−27	0.1863	0.4910	−3.425	5.410	9.378E−29
7.000	117.0	6.712E−27	3	0.817	0.584	54.43	7.038E−28	9.674E−28	0.1776	0.4540	−3.800	5.375	6.749E−29
7.000	127.0	4.818E−27	3	0.887	0.553	58.03	4.945E−28	6.944E−28	0.1682	0.4165	−4.209	5.337	4.844E−29
7.000	137.0	3.368E−27	3	0.956	0.520	61.47	3.379E−28	4.854E−28	0.1580	0.3789	−4.651	5.296	3.387E−29
7.000	147.0	2.362E−27	3	1.025	0.484	64.75	2.314E−28	3.404E−28	0.1470	0.3413	−5.127	5.251	2.375E−29
8.000	17.0	8.158E−26	3	0.136	0.922	8.39	7.765E−27	1.027E−26	0.2801	0.8616	−1.222	5.442	7.163E−28
8.000	27.0	7.207E−26	3	0.216	0.915	13.28	6.828E−27	9.070E−27	0.2782	0.8461	−1.306	5.434	6.328E−28
8.000	37.0	6.243E−26	3	0.296	0.906	18.09	5.875E−27	7.857E−27	0.2753	0.8243	−1.429	5.423	5.481E−28
8.000	47.0	5.458E−26	3	0.376	0.894	22.81	5.091E−27	6.869E−27	0.2716	0.7971	−1.591	5.408	4.792E−28
8.000	57.0	4.255E−26	3	0.455	0.879	27.42	3.927E−27	5.355E−27	0.2670	0.7654	−1.790	5.389	3.736E−28
8.000	67.0	3.091E−26	3	0.536	0.860	31.90	2.817E−27	3.890E−27	0.2615	0.7299	−2.028	5.367	2.714E−28
8.000	77.0	2.227E−26	3	0.615	0.840	36.24	2.000E−27	2.803E−27	0.2551	0.6917	−2.304	5.341	1.955E−28
8.000	87.0	1.548E−26	3	0.695	0.816	40.44	1.368E−27	1.948E−27	0.2479	0.6514	−2.619	5.312	1.359E−28

p_3	θ_3	$\dfrac{d^2\sigma}{d\Omega_3\,dp_3}$	Error	p_T	p_L^*	θ_3^*	$\dfrac{1}{p_3^{*\,2}}\cdot\dfrac{d^2\sigma}{d\Omega_3^*\,dp_3^*}$	$\dfrac{E_3}{p_3^2}\cdot\dfrac{d^2\sigma}{d\Omega_3\,dp_3}$	x^*	y_3^*	t	m_4	$\dfrac{d^2\sigma}{dt\,dm_4^2}$
GeV/c	mrad	cm² sr⁻¹ GeV⁻¹ c	%	GeV/c	GeV/c	deg	cm² sr⁻¹ GeV⁻³ c³	cm² sr⁻¹ GeV⁻² c³			GeV²	GeV	cm²/GeV⁴

$p_1 = 24.000$ GeV/c; $p_1^* = 3.290$ GeV/c; $E_1^* + E_2^* = 6.843$ GeV; $s = 46.828$ GeV²; $\gamma_{c.m.} = 3.6469$; $\beta_{c.m.} = 0.96167$; $y_1^* = 1.9677$; $p_{3,\max}^* = 3.290$ GeV/c; [72 A 1] (cont.)

p_3	θ_3	$\dfrac{d^2\sigma}{d\Omega_3\,dp_3}$	Error	p_T	p_L^*	θ_3^*	$\dfrac{1}{p_3^{*\,2}}\cdot\dfrac{d^2\sigma}{d\Omega_3^*\,dp_3^*}$	$\dfrac{E_3}{p_3^2}\cdot\dfrac{d^2\sigma}{d\Omega_3\,dp_3}$	x^*	y_3^*	t	m_4	$\dfrac{d^2\sigma}{dt\,dm_4^2}$
8.000	97.0	1.053E−26	3	0.775	0.789	44.49	9.226E−28	1.338E−27	0.2397	0.6098	−2.972	5.278	9.333E−29
8.000	107.0	7.100E−27	3	0.854	0.759	48.38	6.043E−28	8.936E−28	0.2307	0.5673	−3.363	5.241	6.234E−29
8.000	117.0	4.684E−27	3	0.934	0.726	52.12	3.904E−28	5.895E−28	0.2208	0.5244	−3.792	5.200	4.113E−29
8.000	127.0	3.089E−27	3	1.013	0.691	55.71	2.518E−28	3.888E−28	0.2100	0.4816	−4.259	5.155	2.712E−29
8.000	137.0	2.045E−27	3	1.093	0.653	59.15	1.629E−28	2.575E−28	0.1983	0.4389	−4.765	5.106	1.796E−29
8.000	147.0	1.335E−27	3	1.172	0.611	62.45	1.037E−28	1.680E−28	0.1858	0.3967	−5.308	5.052	1.172E−29
9.000	17.0	9.755E−26	3	0.153	1.082	8.05	7.566E−27	1.090E−26	0.3289	0.9760	−0.975	5.291	7.602E−28
9.000	27.0	8.472E−26	3	0.243	1.075	12.74	6.539E−27	9.464E−27	0.3267	0.9565	−1.070	5.282	6.602E−28
9.000	37.0	7.037E−26	3	0.333	1.065	17.37	5.394E−27	7.861E−27	0.3235	0.9295	−1.208	5.269	5.484E−28
9.000	47.0	5.497E−26	3	0.423	1.051	21.92	4.175E−27	6.141E−27	0.3193	0.8962	−1.389	5.252	4.284E−28
9.000	57.0	4.277E−26	3	0.513	1.034	26.38	3.213E−27	4.778E−27	0.3141	0.8577	−1.614	5.230	3.333E−28
9.000	67.0	2.968E−26	3	0.603	1.013	30.74	2.201E−27	3.316E−27	0.3080	0.8154	−1.881	5.205	2.313E−28
9.000	77.0	1.985E−26	3	0.692	0.990	34.97	1.450E−27	2.218E−27	0.3008	0.7703	−2.192	5.175	1.547E−28
9.000	87.0	1.269E−26	3	0.782	0.963	39.08	9.115E−28	1.418E−27	0.2926	0.7235	−2.546	5.140	9.890E−29
9.000	97.0	8.002E−27	3	0.872	0.933	43.06	5.643E−28	8.939E−28	0.2835	0.6757	−2.943	5.102	6.236E−29
9.000	107.0	4.961E−27	3	0.961	0.899	46.91	3.429E−28	5.542E−28	0.2733	0.6275	−3.383	5.058	3.866E−29
9.000	117.0	3.035E−27	3	1.051	0.863	50.61	2.053E−28	3.390E−28	0.2621	0.5794	−3.866	5.010	2.365E−29
9.000	127.0	1.854E−27	3	1.140	0.823	54.18	1.225E−28	2.071E−28	0.2500	0.5318	−4.391	4.958	1.445E−29
9.000	137.0	1.132E−27	3	1.229	0.779	57.62	7.303E−29	1.265E−28	0.2369	0.4848	−4.960	4.900	8.822E−30
9.000	147.0	5.904E−28	3	1.318	0.733	60.92	4.342E−29	7.713E−29	0.2228	0.4388	−5.571	4.837	5.380E−30
10.000	17.0	1.172E−25	3	0.170	1.239	7.82	7.531E−27	1.177E−26	0.3764	1.0778	−0.785	5.130	8.212E−28
10.000	27.0	9.613E−26	3	0.270	1.230	12.37	6.147E−27	9.655E−27	0.3740	1.0541	−0.891	5.120	6.736E−28
10.000	37.0	7.600E−26	3	0.370	1.219	16.88	4.825E−27	7.633E−27	0.3704	1.0214	−1.044	5.105	5.325E−28
10.000	47.0	5.628E−26	3	0.470	1.204	21.32	3.540E−27	5.653E−27	0.3658	0.9816	−1.246	5.085	3.943E−28
10.000	57.0	4.090E−26	3	0.570	1.185	25.68	2.544E−27	4.108E−27	0.3600	0.9363	−1.495	5.060	2.866E−28
10.000	67.0	2.685E−26	3	0.669	1.162	29.95	1.648E−27	2.697E−27	0.3531	0.8871	−1.793	5.031	1.881E−28
10.000	77.0	1.646E−26	3	0.769	1.136	34.11	9.948E−28	1.653E−27	0.3452	0.8355	−2.138	4.996	1.153E−28
10.000	87.0	9.721E−27	3	0.869	1.106	38.16	5.775E−28	9.764E−28	0.3361	0.7825	−2.531	4.957	6.811E−29
10.000	97.0	5.574E−27	3	0.968	1.072	42.09	3.250E−28	5.598E−28	0.3259	0.7290	−2.972	4.912	3.906E−29
10.000	107.0	3.160E−27	3	1.068	1.035	45.89	1.805E−28	3.174E−28	0.3146	0.6756	−3.461	4.862	2.214E−29
10.000	117.0	1.794E−27	3	1.167	0.994	49.57	1.002E−28	1.802E−28	0.3022	0.6229	−3.997	4.807	1.257E−29
10.000	127.0	1.020E−27	3	1.267	0.950	53.13	5.567E−29	1.024E−28	0.2887	0.5712	−4.581	4.746	7.147E−30
10.000	137.0	5.926E−28	3	1.366	0.902	56.55	3.155E−29	5.952E−29	0.2741	0.5206	−5.213	4.679	4.152E−30
10.000	147.0	3.355E−28	3	1.465	0.850	59.86	1.740E−29	3.370E−29	0.2585	0.4713	−5.893	4.605	2.351E−30
11.000	17.0	1.386E−25	3	0.187	1.392	7.65	7.488E−27	1.265E−26	0.4230	1.1694	−0.638	4.959	8.822E−28
11.000	27.0	1.072E−25	3	0.297	1.383	12.12	5.763E−27	9.781E−27	0.4203	1.1410	−0.754	4.948	6.823E−28
11.000	37.0	7.950E−26	3	0.407	1.370	16.54	4.248E−27	7.263E−27	0.4164	1.1024	−0.923	4.931	5.066E−28
11.000	47.0	5.569E−26	3	0.517	1.353	20.90	2.944E−27	5.081E−27	0.4113	1.0559	−1.144	4.908	3.545E−28
11.000	57.0	3.680E−26	3	0.627	1.332	25.19	1.923E−27	3.358E−27	0.4049	1.0037	−1.419	4.880	2.342E−28
11.000	67.0	2.290E−26	3	0.735	1.308	29.39	1.181E−27	2.089E−27	0.3974	0.9478	−1.746	4.846	1.458E−28

p_3	θ_3	$\dfrac{\mathrm{d}^2\sigma}{\mathrm{d}\Omega_3\,\mathrm{d}p_3}$	Error	p_T	p_L^*	θ_3^*	$\dfrac{1}{p_3^{*2}}\cdot\dfrac{\mathrm{d}^2\sigma}{\mathrm{d}\Omega_3^*\,\mathrm{d}p_3^*}$	$\dfrac{E_3}{p_3^2}\cdot\dfrac{\mathrm{d}^2\sigma}{\mathrm{d}\Omega_3\,\mathrm{d}p_3}$	x^*	y_3^*	t	m_4	$\dfrac{\mathrm{d}^2\sigma}{\mathrm{d}t\,\mathrm{d}m_4^2}$
GeV/c	mrad	$\mathrm{cm^2\,sr^{-1}\,GeV^{-1}\,c}$	%	GeV/c	GeV/c	deg	$\mathrm{cm^2\,sr^{-1}\,GeV^{-3}\,c^3}$	$\mathrm{cm^2\,sr^{-1}\,GeV^{-2}\,c^3}$			$\mathrm{GeV^2}$	GeV	$\mathrm{cm^2/GeV^4}$

$p_1 = 24.000$ GeV/c; $p_1^* = 3.290$ GeV/c; $E_1^* + E_2^* = 6.843$ GeV; $s = 46.828$ GeV^2; $\gamma_{\mathrm{c.m.}} = 3.6469$; $\beta_{\mathrm{c.m.}} = 0.96167$; $y_1^* = 1.9677$; $p_{3,\max}^* = 3.290$ GeV/c; [72 A 1] (cont.)

p_3	θ_3	$\dfrac{\mathrm{d}^2\sigma}{\mathrm{d}\Omega_3\,\mathrm{d}p_3}$	Error	p_T	p_L^*	θ_3^*	$\dfrac{1}{p_3^{*2}}\cdot\dfrac{\mathrm{d}^2\sigma}{\mathrm{d}\Omega_3^*\,\mathrm{d}p_3^*}$	$\dfrac{E_3}{p_3^2}\cdot\dfrac{\mathrm{d}^2\sigma}{\mathrm{d}\Omega_3\,\mathrm{d}p_3}$	x^*	y_3^*	t	m_4	$\dfrac{\mathrm{d}^2\sigma}{\mathrm{d}t\,\mathrm{d}m_4^2}$
11.000	77.0	1.303E-26	3	0.846	1.279	33.50	6.614E-28	1.189E-27	0.3886	0.8899	-2.126	4.807	8.294E-29
11.000	87.0	7.076E-27	3	0.956	1.246	37.50	3.530E-28	6.456E-28	0.3786	0.8311	-2.558	4.762	4.504E-29
11.000	97.0	3.768E-27	3	1.065	1.209	41.39	1.844E-28	3.438E-28	0.3674	0.7725	-3.043	4.711	2.398E-29
11.000	107.0	1.972E-27	3	1.175	1.168	45.16	9.450E-29	1.799E-28	0.3550	0.7146	-3.581	4.653	1.255E-29
11.000	117.0	1.047E-27	3	1.284	1.123	48.82	4.906E-29	9.553E-29	0.3414	0.6578	-4.171	4.589	6.664E-30
11.000	127.0	5.565E-28	3	1.393	1.074	52.36	2.546E-29	5.077E-29	0.3265	0.6025	-4.814	4.519	3.542E-30
11.000	137.0	2.960E-28	3	1.502	1.022	55.78	1.321E-29	2.701E-29	0.3105	0.5488	-5.509	4.441	1.884E-30
11.000	147.0	1.581E-28	3	1.611	0.965	59.08	6.871E-30	1.442E-29	0.2932	0.4968	-6.256	4.356	1.006E-30
12.000	17.0	1.621E-25	3	0.204	1.543	7.53	7.457E-27	1.355E-26	0.4688	1.2525	-0.522	4.779	9.452E-28
12.000	27.0	1.159E-25	3	0.324	1.533	11.93	5.305E-27	9.688E-27	0.4659	1.2192	-0.649	4.766	6.758E-28
12.000	37.0	7.985E-26	3	0.444	1.519	16.29	3.628E-27	6.674E-27	0.4616	1.1743	-0.833	4.747	4.656E-28
12.000	47.0	5.193E-26	3	0.564	1.501	20.59	2.337E-27	4.341E-27	0.4560	1.1210	-1.075	4.721	3.028E-28
12.000	57.0	3.203E-26	3	0.684	1.478	24.82	1.425E-27	2.677E-27	0.4491	1.0619	-1.374	4.689	1.868E-28
12.000	67.0	1.921E-26	3	0.803	1.451	28.98	8.427E-28	1.606E-27	0.4409	0.9996	-1.731	4.651	1.120E-28
12.000	77.0	9.976E-27	3	0.923	1.419	33.04	4.308E-28	8.339E-28	0.4313	0.9357	-2.145	4.606	5.817E-29
12.000	87.0	4.914E-27	3	1.043	1.383	37.01	2.085E-28	4.107E-28	0.4204	0.8716	-2.617	4.555	2.865E-29
12.000	97.0	2.366E-27	3	1.162	1.343	40.87	9.845E-29	1.973E-28	0.4082	0.8083	-3.146	4.496	1.380E-29
12.000	107.0	1.151E-27	3	1.282	1.299	44.62	4.689E-29	9.621E-29	0.3947	0.7464	-3.733	4.431	6.712E-30
12.000	117.0	5.631E-28	3	1.401	1.250	48.26	2.243E-29	4.707E-29	0.3798	0.6861	-4.377	4.357	3.284E-30
12.000	127.0	2.758E-28	3	1.520	1.196	51.79	1.072E-29	2.305E-29	0.3636	0.6278	-5.078	4.276	1.608E-30
12.000	137.0	1.350E-28	3	1.639	1.139	55.20	5.117E-30	1.123E-29	0.3461	0.5715	-5.836	4.187	7.872E-31
12.000	142.0	9.621E-29	3	1.698	1.108	56.87	3.599E-30	8.042E-30	0.3369	0.5441	-6.236	4.138	5.610E-31
13.000	17.0	2.011E-25	3	0.221	1.692	7.44	7.966E-27	1.551E-26	0.5141	1.3284	-0.430	4.589	1.082E-27
13.000	27.0	1.284E-25	3	0.351	1.681	11.79	5.060E-27	9.903E-27	0.5110	1.2898	-0.568	4.574	6.908E-28
13.000	37.0	7.871E-26	3	0.481	1.666	16.10	3.079E-27	6.070E-27	0.5064	1.2385	-0.767	4.552	4.235E-28
13.000	47.0	4.798E-26	3	0.611	1.646	20.36	1.859E-27	3.700E-27	0.5003	1.1783	-1.029	4.524	2.581E-28
13.000	57.0	2.747E-26	3	0.741	1.622	24.55	1.052E-27	2.119E-27	0.4928	1.1125	-1.354	4.488	1.478E-28
13.000	67.0	1.456E-26	3	0.870	1.592	28.66	5.497E-28	1.123E-27	0.4839	1.0439	-1.740	4.444	7.834E-29
13.000	77.0	6.777E-27	3	1.000	1.558	32.69	2.518E-28	5.227E-28	0.4735	0.9745	-2.189	4.393	3.646E-29
13.000	87.0	3.009E-27	3	1.130	1.519	36.63	1.098E-28	2.321E-28	0.4617	0.9057	-2.700	4.335	1.619E-29
13.000	97.0	1.329E-27	3	1.259	1.476	40.47	4.757E-29	1.025E-28	0.4485	0.8382	-3.274	4.268	7.150E-30
13.000	107.0	6.028E-28	3	1.388	1.427	44.21	2.112E-29	4.649E-29	0.4338	0.7727	-3.909	4.193	3.243E-30
13.000	117.0	2.716E-28	3	1.518	1.374	47.83	9.301E-30	2.095E-29	0.4177	0.7094	-4.606	4.109	1.461E-30
13.000	127.0	1.218E-28	3	1.647	1.317	51.35	4.071E-30	9.394E-30	0.4002	0.6484	-5.366	4.016	6.553E-31
13.000	137.0	5.758E-29	3	1.775	1.254	54.76	1.876E-30	4.441E-30	0.3812	0.5898	-6.187	3.912	3.098E-31
14.000	17.0	2.289E-25	3	0.238	1.839	7.37	7.884E-27	1.639E-26	0.5590	1.3982	-0.358	4.389	1.143E-27
14.000	27.0	1.325E-25	3	0.378	1.828	11.68	4.540E-27	9.486E-27	0.5556	1.3541	-0.506	4.372	6.617E-28
14.000	37.0	7.342E-26	3	0.518	1.812	15.95	2.497E-27	5.256E-27	0.5506	1.2961	-0.721	4.347	3.667E-28
14.000	57.0	2.186E-26	3	0.798	1.764	24.33	7.275E-28	1.565E-27	0.5361	1.1567	-1.353	4.274	1.092E-28
14.000	67.0	1.019E-26	3	0.937	1.732	28.42	3.344E-28	7.295E-28	0.5264	1.0822	-1.769	4.225	5.089E-29

p_3	θ_3	$\dfrac{\mathrm{d}^2\sigma}{\mathrm{d}\Omega_3\,\mathrm{d}p_3}$	Error	p_T	p_L^*	θ_3^*	$\dfrac{1}{p_3^{*2}}\cdot\dfrac{\mathrm{d}^2\sigma}{\mathrm{d}\Omega_3^*\,\mathrm{d}p_3^*}$	$\dfrac{E_3}{p_3^2}\cdot\dfrac{\mathrm{d}^2\sigma}{\mathrm{d}\Omega_3\,\mathrm{d}p_3}$	x^*	y_3^*	t	m_4	$\dfrac{\mathrm{d}^2\sigma}{\mathrm{d}t\,\mathrm{d}m_4^2}$
GeV/c	mrad	cm²sr⁻¹GeV⁻¹c	%	GeV/c	GeV/c	deg	cm²sr⁻¹GeV⁻³c³	cm²sr⁻¹GeV⁻²c³			GeV²	GeV	cm²/GeV⁴

$p_1 = 24.000\ \mathrm{GeV/c}$; $p_1^* = 3.290\ \mathrm{GeV/c}$; $E_1^* + E_2^* = 6.843\ \mathrm{GeV}$; $s = 46.828\ \mathrm{GeV^2}$; $\gamma_\mathrm{c.m.} = 3.6469$; $\beta_\mathrm{c.m.} = 0.96167$; $y_1^* = 1.9677$; $p_{3,\,\mathrm{max}}^* = 3.290\ \mathrm{GeV/c}$; [72 A 1] (cont.)

p_3	θ_3	$\dfrac{\mathrm{d}^2\sigma}{\mathrm{d}\Omega_3\,\mathrm{d}p_3}$	Error	p_T	p_L^*	θ_3^*	$\dfrac{1}{p_3^{*2}}\cdot\dfrac{\mathrm{d}^2\sigma}{\mathrm{d}\Omega_3^*\,\mathrm{d}p_3^*}$	$\dfrac{E_3}{p_3^2}\cdot\dfrac{\mathrm{d}^2\sigma}{\mathrm{d}\Omega_3\,\mathrm{d}p_3}$	x^*	y_3^*	t	m_4	$\dfrac{\mathrm{d}^2\sigma}{\mathrm{d}t\,\mathrm{d}m_4^2}$
14.000	77.0	4.309E-27	3	1.077	1.696	32.42	1.391E-28	3.085E-28	0.5153	1.0077	-2.252	4.167	2.152E-29
14.000	87.0	1.715E-27	3	1.216	1.654	36.34	5.439E-29	1.228E-28	0.5026	0.9344	-2.803	4.101	8.565E-30
14.000	97.0	7.005E-28	3	1.356	1.607	40.16	2.178E-29	5.015E-29	0.4883	0.8632	-3.420	4.025	3.498E-30
14.000	107.0	2.903E-28	3	1.495	1.555	43.88	8.835E-30	2.078E-29	0.4725	0.7946	-4.104	3.939	1.450E-30
14.000	117.0	1.237E-28	3	1.634	1.498	47.50	3.679E-30	8.856E-30	0.4552	0.7286	-4.855	3.842	6.178E-31
14.000	127.0	5.247E-29	3	1.773	1.436	51.01	1.523E-30	3.756E-30	0.4363	0.6654	-5.673	3.734	2.620E-31
14.000	137.0	2.470E-29	3	1.912	1.368	54.41	6.985E-31	1.768E-30	0.4159	0.6049	-6.558	3.614	1.234E-31
15.000	17.0	2.479E-25	3	0.255	1.986	7.32	7.489E-27	1.656E-26	0.6036	1.4626	-0.302	4.177	1.155E-27
15.000	27.0	1.276E-25	3	0.405	1.974	11.59	3.835E-27	8.523E-27	0.5999	1.4127	-0.460	4.158	5.946E-28
15.000	37.0	6.571E-26	3	0.555	1.956	15.83	1.960E-27	4.389E-27	0.5946	1.3480	-0.690	4.130	3.062E-28
15.000	47.0	3.355E-26	3	0.705	1.933	20.03	9.909E-28	2.241E-27	0.5876	1.2740	-0.993	4.093	1.563E-28
15.000	57.0	1.677E-26	3	0.855	1.905	24.16	4.894E-28	1.120E-27	0.5790	1.1954	-1.367	4.047	7.815E-29
15.000	67.0	7.103E-27	3	1.004	1.871	28.22	2.044E-28	4.745E-28	0.5687	1.1154	-1.813	3.992	3.310E-29
15.000	77.0	2.700E-27	3	1.154	1.832	32.21	7.644E-29	1.804E-28	0.5567	1.0361	-2.331	3.926	1.258E-29
15.000	87.0	9.865E-28	3	1.303	1.787	36.11	2.743E-29	6.590E-29	0.5431	0.9589	-2.921	3.851	4.597E-30
15.000	97.0	3.704E-28	3	1.453	1.737	39.91	1.009E-29	2.474E-29	0.5278	0.8844	-3.582	3.764	1.726E-30
15.000	107.0	1.437E-28	3	1.602	1.681	43.62	3.833E-30	9.599E-30	0.5109	0.8130	-4.315	3.665	6.696E-31
15.000	117.0	5.608E-29	3	1.751	1.620	47.23	1.461E-30	3.746E-30	0.4923	0.7447	-5.120	3.553	2.613E-31
15.000	127.0	2.247E-29	3	1.900	1.553	50.73	5.713E-31	1.501E-30	0.4721	0.6796	-5.996	3.428	1.047E-31
16.000	17.0	2.692E-25	3	0.272	2.132	7.27	7.188E-27	1.685E-26	0.6478	1.5223	-0.257	3.952	1.176E-27
16.000	27.0	1.296E-25	3	0.432	2.119	11.52	3.442E-27	8.114E-27	0.6439	1.4664	-0.426	3.930	5.660E-28
16.000	37.0	5.917E-26	3	0.592	2.100	15.74	1.560E-27	3.704E-27	0.6383	1.3948	-0.672	3.899	2.584E-28
16.000	57.0	1.256E-26	3	0.912	2.045	24.02	3.239E-28	7.863E-28	0.6216	1.2295	-1.394	3.805	5.486E-29
16.000	67.0	5.061E-27	3	1.071	2.009	28.06	1.287E-28	3.169E-28	0.6106	1.1443	-1.869	3.742	2.210E-29
16.000	77.0	1.722E-27	3	1.231	1.967	32.03	4.307E-29	1.078E-28	0.5979	1.0607	-2.422	3.668	7.521E-30
16.000	87.0	5.798E-28	3	1.390	1.919	35.92	1.424E-29	3.630E-29	0.5833	0.9799	-3.051	3.581	2.532E-30
16.000	97.0	2.014E-28	3	1.550	1.866	39.71	4.849E-30	1.261E-29	0.5670	0.9024	-3.757	3.481	8.796E-31
16.000	107.0	7.285E-29	3	1.709	1.806	43.41	1.716E-30	4.561E-30	0.5490	0.8285	-4.539	3.367	3.182E-31
16.000	117.0	2.326E-29	3	1.868	1.741	47.01	5.353E-31	1.456E-30	0.5292	0.7583	-5.397	3.237	1.016E-31
16.000	127.0	9.312E-30	3	2.027	1.670	50.51	2.091E-31	5.830E-31	0.5076	0.6914	-6.332	3.089	4.067E-32
17.000	17.0	2.694E-25	3	0.289	2.277	7.23	6.402E-27	1.587E-26	0.6919	1.5778	-0.223	3.712	1.107E-27
17.000	27.0	1.184E-25	3	0.459	2.263	11.46	2.799E-27	6.975E-27	0.6877	1.5158	-0.403	3.688	4.866E-28
17.000	37.0	4.981E-26	3	0.629	2.243	15.66	1.168E-27	2.934E-27	0.6817	1.4374	-0.664	3.652	2.047E-28
17.000	47.0	2.147E-26	3	0.799	2.217	19.81	4.987E-28	1.265E-27	0.6738	1.3501	-1.007	3.605	8.824E-29
17.000	57.0	8.630E-27	3	0.968	2.185	23.91	1.980E-28	5.084E-28	0.6640	1.2597	-1.431	3.545	3.547E-29
17.000	67.0	2.958E-27	3	1.138	2.146	27.93	6.691E-29	1.743E-28	0.6523	1.1697	-1.936	3.473	1.216E-29
17.000	77.0	8.944E-28	3	1.308	2.102	31.89	1.990E-29	5.269E-29	0.6388	1.0821	-2.523	3.388	3.676E-30
17.000	87.0	2.762E-28	3	1.477	2.051	35.76	6.035E-30	1.627E-29	0.6233	0.9980	-3.192	3.288	1.135E-30
17.000	97.0	8.974E-29	3	1.646	1.994	39.54	1.922E-30	5.287E-30	0.6060	0.9179	-3.941	3.172	3.688E-31
17.000	107.0	3.014E-29	3	1.815	1.931	43.24	6.315E-31	1.776E-30	0.5868	0.8419	-4.772	3.038	1.239E-31

p_3	θ_3	$\dfrac{d^2\sigma}{d\Omega_3\,dp_3}$	Error	p_T	p_L^*	θ_3^*	$\dfrac{1}{p_3^{*2}}\cdot\dfrac{d^2\sigma}{d\Omega_3^*\,dp_3^*}$	$\dfrac{E_3}{p_3^2}\cdot\dfrac{d^2\sigma}{d\Omega_3\,dp_3}$	x^*	y_3^*	t	m_4	$\dfrac{d^2\sigma}{dt\,dm_4^2}$
GeV/c	mrad	cm² sr⁻¹ GeV⁻¹ c	%	GeV/c	GeV/c	deg	cm² sr⁻¹ GeV⁻³ c³	cm² sr⁻¹ GeV⁻² c³			GeV²	GeV	cm²/GeV⁴

$p_1 = 24.000$ GeV/c; $p_1^* = 3.290$ GeV/c; $E_1^* + E_2^* = 6.843$ GeV; $s = 46.828$ GeV²; $\gamma_{c.m.} = 3.6469$; $\beta_{c.m.} = 0.96167$; $y_1^* = 1.9677$; $p_{3,\,max}^* = 3.290$ GeV/c; [72 A 1] (cont.)

p_3	θ_3	$\dfrac{d^2\sigma}{d\Omega_3\,dp_3}$	Error	p_T	p_L^*	θ_3^*	$\dfrac{1}{p_3^{*2}}\cdot\dfrac{d^2\sigma}{d\Omega_3^*\,dp_3^*}$	$\dfrac{E_3}{p_3^2}\cdot\dfrac{d^2\sigma}{d\Omega_3\,dp_3}$	x^*	y_3^*	t	m_4	$\dfrac{d^2\sigma}{dt\,dm_4^2}$
17.000	117.0	1.003E-29	3	1.984	1.862	46.83	2.053E-31	5.909E-31	0.5658	0.7698	-5.684	2.884	4.122E-32
17.000	127.0	3.759E-30	4	2.153	1.786	50.32	7.505E-32	2.215E-31	0.5429	0.7015	-6.677	2.706	1.545E-32
18.000	17.0	2.839E-25	3	0.306	2.421	7.20	6.041E-27	1.579E-26	0.7357	1.6297	-0.198	3.454	1.102E-27
18.000	27.0	1.116E-25	3	0.486	2.406	11.42	2.362E-27	6.208E-27	0.7313	1.5613	-0.388	3.426	4.331E-28
18.000	37.0	4.365E-26	3	0.666	2.385	15.60	9.169E-28	2.428E-27	0.7250	1.4760	-0.665	3.386	1.694E-28
18.000	47.0	1.730E-26	3	0.846	2.358	19.73	3.598E-28	9.624E-28	0.7166	1.3824	-1.027	3.332	6.714E-29
18.000	57.0	5.063E-27	3	1.025	2.324	23.81	1.246E-28	3.373E-28	0.7062	1.2865	-1.476	3.264	2.353E-29
18.000	67.0	1.880E-27	3	1.205	2.283	27.83	3.808E-29	1.046E-28	0.6938	1.1919	-2.012	3.180	7.296E-30
18.000	77.0	5.177E-28	3	1.385	2.236	31.77	1.031E-29	2.880E-29	0.6795	1.1007	-2.633	3.081	2.009E-30
18.000	87.0	1.434E-28	3	1.564	2.182	35.63	2.805E-30	7.977E-30	0.6632	1.0137	-3.341	2.964	5.565E-31
18.000	97.0	4.572E-29	3	1.743	2.122	39.41	8.765E-31	2.543E-30	0.6448	0.9312	-4.135	2.827	1.774E-31
18.000	107.0	1.419E-29	3	1.922	2.055	43.09	2.661E-31	7.894E-31	0.6245	0.8533	-5.014	2.667	5.507E-32
18.000	117.0	4.382E-30	4	2.101	1.982	46.68	8.028E-32	2.438E-31	0.6022	0.7797	-5.980	2.479	1.701E-32
18.000	127.0	1.696E-30	5	2.280	1.902	50.17	3.030E-32	9.435E-32	0.5779	0.7101	-7.032	2.257	6.582E-33
19.000	17.0	2.803E-25	3	0.323	2.565	7.18	5.371E-27	1.477E-26	0.7794	1.6782	-0.180	3.174	1.030E-27
19.000	27.0	9.657E-26	3	0.513	2.549	11.38	1.841E-27	5.089E-27	0.7748	1.6034	-0.381	3.142	3.550E-28
19.000	37.0	3.402E-26	3	0.703	2.527	15.54	6.435E-28	1.793E-27	0.7680	1.5112	-0.672	3.095	1.251E-28
19.000	47.0	1.201E-26	3	0.893	2.498	19.66	2.249E-28	6.329E-28	0.7592	1.4114	-1.055	3.033	4.415E-29
19.000	57.0	3.741E-27	3	1.082	2.462	23.73	6.921E-29	1.971E-28	0.7483	1.3103	-1.529	2.954	1.375E-29
19.000	67.0	1.035E-27	3	1.272	2.419	27.74	1.887E-29	5.454E-29	0.7352	1.2116	-2.094	2.856	3.805E-30
19.000	77.0	2.645E-28	3	1.462	2.369	31.67	4.745E-30	1.394E-29	0.7201	1.1170	-2.750	2.739	9.723E-31
19.000	87.0	6.707E-29	3	1.651	2.313	35.52	1.181E-30	3.534E-30	0.7028	1.0274	-3.497	2.599	2.466E-31
19.000	97.0	1.937E-29	3	1.840	2.249	39.29	3.343E-31	1.021E-30	0.6835	0.9428	-4.335	2.433	7.121E-32
19.000	107.0	6.326E-30	3	2.029	2.178	42.97	1.068E-31	3.334E-31	0.6620	0.8632	-5.264	2.234	2.326E-32
19.000	117.0	1.579E-30	5	2.218	2.101	46.55	2.604E-32	8.321E-32	0.6385	0.7882	-6.283	1.992	5.805E-33
20.000	17.0	2.897E-25	3	0.340	2.708	7.16	5.025E-27	1.450E-26	0.8230	1.7238	-0.168	2.865	1.012E-27
20.000	27.0	9.697E-26	3	0.540	2.692	11.34	1.673E-27	4.854E-27	0.8181	1.6423	-0.379	2.828	3.386E-28
20.000	37.0	3.018E-26	3	0.740	2.669	15.50	5.167E-28	1.511E-27	0.8110	1.5434	-0.686	2.774	1.054E-28
20.000	47.0	9.512E-27	3	0.940	2.638	19.61	1.612E-28	4.761E-28	0.8017	1.4376	-1.089	2.700	3.321E-29
20.000	57.0	2.551E-27	3	1.139	2.600	23.66	4.271E-29	1.277E-28	0.7902	1.3316	-1.588	2.606	8.908E-30
20.000	67.0	5.914E-28	3	1.339	2.555	27.66	9.760E-30	2.960E-29	0.7764	1.2290	-2.183	2.489	2.065E-30
20.000	77.0	1.365E-28	3	1.538	2.502	31.58	2.216E-30	6.833E-30	0.7605	1.1314	-2.874	2.346	4.766E-31
20.000	87.0	3.380E-29	3	1.738	2.443	35.43	5.386E-31	1.692E-30	0.7423	1.0393	-3.660	2.172	1.180E-31
20.000	97.0	9.691E-30	3	1.937	2.376	39.19	1.513E-31	4.851E-31	0.7220	0.9529	-4.542	1.959	3.384E-32
20.000	107.0	2.997E-30	4	2.136	2.301	42.87	4.578E-32	1.500E-31	0.6994	0.8718	-5.520	1.691	1.047E-32
20.000	117.0	9.632E-31	6	2.335	2.220	46.44	1.437E-32	4.821E-32	0.6746	0.7956	-6.593	1.337	3.363E-33
4.000	12.0	2.049E-26	3	0.048	0.177	15.14	5.504E-27	5.262E-27	0.0539	0.1877	-3.615	5.884	3.670E-28
4.500	12.0	2.714E-26	3	0.054	0.288	10.60	6.267E-27	6.161E-27	0.0877	0.3023	-3.068	5.853	4.298E-28
5.000	12.0	3.418E-26	3	0.060	0.392	8.71	6.830E-27	6.955E-27	0.1190	0.4053	-2.632	5.811	4.852E-28

p_3	θ_3	$\dfrac{d^2\sigma}{d\Omega_3\,dp_3}$	Error	p_T	p_L^*	θ_3^*	$\dfrac{1}{p_3^{*2}}\cdot\dfrac{d^2\sigma}{d\Omega_3^*\,dp_3^*}$	$\dfrac{E_3}{p_3^2}\cdot\dfrac{d^2\sigma}{d\Omega_3\,dp_3}$	x^*	y_3^*	t	m_4	$\dfrac{d^2\sigma}{dt\,dm_4^2}$
GeV/c	mrad	cm² sr⁻¹ GeV⁻¹ c	%	GeV/c	GeV/c	deg	cm² sr⁻¹ GeV⁻³ c³	cm² sr⁻¹ GeV⁻² c³			GeV²	GeV	cm²/GeV⁴

$p_1 = 24.000$ GeV/c; $p_1^* = 3.290$ GeV/c; $E_1^* + E_2^* = 6.843$ GeV; $s = 46.828$ GeV²; $\gamma_{c.m.} = 3.6469$; $\beta_{c.m.} = 0.96167$; $y_1^* = 1.9677$; $p_{3,\,max}^* = 3.290$ GeV/c; [72 A 1] (cont.)

p_3	θ_3	$\dfrac{d^2\sigma}{d\Omega_3\,dp_3}$	Error	p_T	p_L^*	θ_3^*	$\dfrac{1}{p_3^{*2}}\cdot\dfrac{d^2\sigma}{d\Omega_3^*\,dp_3^*}$	$\dfrac{E_3}{p_3^2}\cdot\dfrac{d^2\sigma}{d\Omega_3\,dp_3}$	x^*	y_3^*	t	m_4	$\dfrac{d^2\sigma}{dt\,dm_4^2}$
5.500	12.0	4.242E-26	3	0.066	0.489	7.69	7.382E-27	7.824E-27	0.1485	0.4987	-2.277	5.762	5.458E-28
6.000	12.0	5.122E-26	3	0.072	0.581	7.06	7.812E-27	8.640E-27	0.1767	0.5841	-1.983	5.707	6.028E-28
6.500	12.0	5.989E-26	3	0.078	0.671	6.63	8.054E-27	9.309E-27	0.2038	0.6627	-1.736	5.647	6.494E-28
7.000	12.0	6.877E-26	3	0.084	0.757	6.33	8.202E-27	9.912E-27	0.2301	0.7356	-1.527	5.583	6.915E-28
7.500	12.0	7.834E-26	3	0.090	0.841	6.11	8.332E-27	1.053E-26	0.2557	0.8034	-1.348	5.515	7.344E-28
8.000	12.0	8.800E-26	3	0.096	0.924	5.93	8.389E-27	1.108E-26	0.2808	0.8669	-1.194	5.444	7.726E-28
8.500	12.0	9.774E-26	3	0.102	1.005	5.80	8.392E-27	1.157E-26	0.3054	0.9264	-1.060	5.370	8.070E-28
9.000	12.0	1.082E-25	3	0.108	1.085	5.69	8.405E-27	1.209E-26	0.3296	0.9826	-0.943	5.294	8.432E-28
9.500	12.0	1.189E-25	3	0.114	1.163	5.60	8.391E-27	1.258E-26	0.3536	1.0356	-0.841	5.215	8.774E-28
10.000	12.0	1.307E-25	3	0.120	1.241	5.52	8.412E-27	1.313E-26	0.3772	1.0859	-0.750	5.133	9.158E-28
10.500	12.0	1.425E-25	3	0.126	1.318	5.46	8.396E-27	1.363E-26	0.4006	1.1336	-0.670	5.049	9.505E-28
11.000	12.0	1.554E-25	3	0.132	1.395	5.41	8.464E-27	1.427E-26	0.4238	1.1791	-0.599	4.963	9.955E-28
11.500	12.0	1.767E-25	3	0.138	1.470	5.36	8.811E-27	1.542E-26	0.4469	1.2225	-0.536	4.875	1.075E-27
12.000	12.0	1.953E-25	3	0.144	1.546	5.32	9.000E-27	1.632E-26	0.4698	1.2640	-0.480	4.784	1.139E-27
12.500	12.0	2.127E-25	3	0.150	1.621	5.29	9.083E-27	1.706E-26	0.4925	1.3037	-0.430	4.690	1.190E-27
13.000	12.0	2.319E-25	3	0.156	1.695	5.26	9.201E-27	1.788E-26	0.5152	1.3418	-0.385	4.594	1.248E-27
13.500	12.0	2.501E-25	3	0.162	1.769	5.23	9.243E-27	1.857E-26	0.5377	1.3784	-0.345	4.496	1.296E-27
14.000	12.0	2.729E-25	3	0.168	1.843	5.21	9.415E-27	1.954E-26	0.5601	1.4136	-0.310	4.394	1.363E-27
14.500	12.0	2.966E-25	3	0.174	1.917	5.19	9.574E-27	2.050E-26	0.5825	1.4475	-0.278	4.290	1.430E-27
15.000	12.0	3.104E-25	3	0.180	1.990	5.17	9.393E-27	2.073E-26	0.6048	1.4802	-0.249	4.183	1.446E-27
15.500	12.0	3.248E-25	3	0.186	2.063	5.15	9.232E-27	2.099E-26	0.6270	1.5117	-0.224	4.073	1.465E-27
16.000	12.0	3.397E-25	3	0.192	2.136	5.14	9.086E-27	2.127E-26	0.6491	1.5422	-0.202	3.959	1.484E-27
16.500	12.0	3.549E-25	3	0.198	2.209	5.12	8.948E-27	2.154E-26	0.6712	1.5716	-0.182	3.841	1.503E-27
17.000	12.0	3.716E-25	3	0.204	2.281	5.11	8.846E-27	2.189E-26	0.6932	1.6001	-0.164	3.720	1.527E-27
17.500	12.0	3.843E-25	3	0.210	2.353	5.10	8.651E-27	2.199E-26	0.7152	1.6277	-0.149	3.594	1.534E-27
18.000	12.0	3.951E-25	3	0.216	2.426	5.09	8.422E-27	2.198E-26	0.7372	1.6544	-0.135	3.463	1.533E-27
18.500	12.0	4.254E-25	3	0.222	2.498	5.08	8.600E-27	2.302E-26	0.7591	1.6804	-0.124	3.327	1.606E-27
19.230	12.0	4.425E-25	3	0.231	2.603	5.07	8.298E-27	2.304E-26	0.7910	1.7169	-0.110	3.116	1.607E-27
19.620	12.0	4.521E-25	3	0.235	2.659	5.06	8.154E-27	2.307E-26	0.8080	1.7358	-0.104	2.998	1.609E-27
20.020	12.0	4.641E-25	3	0.240	2.716	5.05	8.048E-27	2.321E-26	0.8254	1.7547	-0.098	2.871	1.619E-27
20.430	12.0	4.722E-25	3	0.245	2.775	5.05	7.872E-27	2.314E-26	0.8433	1.7737	-0.093	2.735	1.614E-27
20.840	12.0	4.782E-25	3	0.250	2.834	5.04	7.669E-27	2.297E-26	0.8611	1.7923	-0.090	2.591	1.602E-27
21.270	12.0	4.936E-25	3	0.255	2.895	5.04	7.606E-27	2.323E-26	0.8798	1.8113	-0.086	2.431	1.620E-27
21.700	12.0	5.344E-25	3	0.260	2.956	5.03	7.919E-27	2.455E-26	0.8985	1.8299	-0.084	2.260	1.720E-27
22.150	12.0	5.668E-25	3	0.266	3.021	5.03	8.069E-27	2.551E-26	0.9180	1.8489	-0.082	2.066	1.787E-27
22.600	12.0	7.108E-25	3	0.271	3.085	5.02	9.728E-27	3.148E-26	0.9375	1.8675	-0.081	1.850	2.196E-27

$p_1 = 9.900$ GeV/c; $p_1^* = 2.055$ GeV/c; $E_1^* + E_2^* = 4.519$ GeV; $s = 20.420$ GeV²; $\gamma_{c.m.} = 2.4083$; $\beta_{c.m.} = 0.90971$; $y_1^* = 1.5259$; $p_{3,\,max}^* = 2.055$ GeV/c; [67 A 1]

p_3	θ_3	$\dfrac{d^2\sigma}{d\Omega_3\,dp_3}$	Error	p_T	p_L^*	θ_3^*	$\dfrac{1}{p_3^{*2}}\cdot\dfrac{d^2\sigma}{d\Omega_3^*\,dp_3^*}$	$\dfrac{E_3}{p_3^2}\cdot\dfrac{d^2\sigma}{d\Omega_3\,dp_3}$	x^*	y_3^*	t	m_4	$\dfrac{d^2\sigma}{dt\,dm_4^2}$
2.300	100.0	1.056E-26	7	0.230	0.069	73.20	5.121E-27	4.959E-27	0.0337	0.0717	-2.330	3.542	8.386E-28
2.892	100.0	1.349E-26	4	0.289	0.269	47.03	4.818E-27	4.904E-27	0.1309	0.2707	-1.733	3.479	8.293E-28
3.495	100.0	1.795E-26	4	0.349	0.447	37.99	4.851E-27	5.318E-27	0.2174	0.4327	-1.356	3.375	8.993E-28
4.092	100.0	2.040E-26	4	0.409	0.608	33.90	4.297E-27	5.115E-27	0.2958	0.5637	-1.119	3.247	8.650E-28

p_3	θ_3	$\dfrac{d^2\sigma}{d\Omega_3\, dp_3}$	Error	p_T	p_L^*	θ_3^*	$\dfrac{1}{p_3^{*2}} \cdot \dfrac{d^2\sigma}{d\Omega_3^*\, dp_3^*}$	$\dfrac{E_3}{p_3^2} \cdot \dfrac{d^2\sigma}{d\Omega_3\, dp_3}$	x^*	y_3^*	t	m_4	$\dfrac{d^2\sigma}{dt\, dm_4^2}$
GeV/c	mrad	cm² sr⁻¹ GeV⁻¹ c	%	GeV/c	GeV/c	deg	cm² sr⁻¹ GeV⁻³ c³	cm² sr⁻¹ GeV⁻² c³			GeV²	GeV	cm²/GeV⁴

$p_1 = 9.900$ GeV/c; $p_1^* = 2.055$ GeV/c; $E_1^* + E_2^* = 4.519$ GeV; $s = 20.420$ GeV²; $\gamma_{c.m.} = 2.4083$; $\beta_{c.m.} = 0.90971$; $y_1^* = 1.5259$; $p_{3,\,max}^* = 2.055$ GeV/c; [67 A 1] (cont.)

p_3	θ_3	$\dfrac{d^2\sigma}{d\Omega_3\, dp_3}$	Error	p_T	p_L^*	θ_3^*	$\dfrac{1}{p_3^{*2}} \cdot \dfrac{d^2\sigma}{d\Omega_3^*\, dp_3^*}$	$\dfrac{E_3}{p_3^2} \cdot \dfrac{d^2\sigma}{d\Omega_3\, dp_3}$	x^*	y_3^*	t	m_4	$\dfrac{d^2\sigma}{dt\, dm_4^2}$
4.702	100.0	2.186E-26	4	0.469	0.763	31.61	3.655E-27	4.741E-27	0.3711	0.6747	-0.965	3.095	8.017E-28
5.293	100.0	2.038E-26	4	0.528	0.906	30.24	2.778E-27	3.910E-27	0.4410	0.7650	-0.874	2.929	6.613E-28
5.889	100.0	1.743E-26	4	0.588	1.047	29.32	1.967E-27	2.997E-27	0.5093	0.8424	-0.822	2.744	5.069E-28
6.502	100.0	1.425E-26	4	0.649	1.188	28.65	1.344E-27	2.214E-27	0.5780	0.9103	-0.799	2.533	3.745E-28
7.104	100.0	1.136E-26	4	0.709	1.324	28.18	9.108E-28	1.613E-27	0.6442	0.9676	-0.799	2.301	2.728E-28
7.709	100.0	9.077E-27	4	0.770	1.459	27.81	6.251E-28	1.186E-27	0.7098	1.0174	-0.817	2.037	2.006E-28
8.301	100.0	6.893E-27	4	0.829	1.589	27.54	4.131E-28	8.357E-28	0.7732	1.0598	-0.848	1.737	1.413E-28
8.907	100.0	5.530E-27	6	0.889	1.722	27.32	2.900E-28	6.243E-28	0.8376	1.0977	-0.891	1.358	1.056E-28
9.514	100.0	3.541E-27	6	0.950	1.853	27.14	1.637E-28	3.740E-28	0.9016	1.1310	-0.942	0.812	6.325E-29

$p_1 = 20.000$ GeV/c; $p_1^* = 2.992$ GeV/c; $E_1^* + E_2^* = 6.271$ GeV; $s = 39.330$ GeV²; $\gamma_{c.m.} = 3.3422$; $\beta_{c.m.} = 0.95419$; $y_1^* = 1.8766$; $p_{3,\,max}^* = 2.992$ GeV/c; [67 A 1]

p_3	θ_3	$\dfrac{d^2\sigma}{d\Omega_3\, dp_3}$	Error	p_T	p_L^*	θ_3^*	$\dfrac{1}{p_3^{*2}} \cdot \dfrac{d^2\sigma}{d\Omega_3^*\, dp_3^*}$	$\dfrac{E_3}{p_3^2} \cdot \dfrac{d^2\sigma}{d\Omega_3\, dp_3}$	x^*	y_3^*	t	m_4	$\dfrac{d^2\sigma}{dt\, dm_4^2}$
3.888	160.0	6.856E-27	4	0.619	0.073	83.24	1.610E-27	1.814E-27	0.0245	0.0653	-4.866	5.107	1.519E-28
4.479	160.0	5.912E-27	4	0.714	0.185	75.50	1.130E-27	1.349E-27	0.0617	0.1559	-4.617	5.024	1.129E-28
5.084	160.0	4.409E-27	4	0.810	0.288	70.45	6.931E-28	8.819E-28	0.0961	0.2300	-4.498	4.924	7.382E-29
6.285	160.0	2.031E-27	4	1.001	0.472	64.77	2.252E-28	3.267E-28	0.1577	0.3375	-4.516	4.691	2.735E-29
6.893	160.0	1.390E-27	5	1.098	0.558	63.05	1.314E-28	2.035E-28	0.1866	0.3776	-4.609	4.559	1.704E-29
7.502	160.0	1.027E-27	5	1.195	0.642	61.76	8.364E-29	1.380E-28	0.2146	0.4109	-4.742	4.418	1.155E-29
8.103	160.0	6.230E-28	6	1.291	0.722	60.78	4.419E-29	7.740E-29	0.2413	0.4383	-4.904	4.271	6.479E-30
8.699	160.0	4.621E-28	7	1.386	0.800	60.02	2.881E-29	5.343E-29	0.2673	0.4613	-5.087	4.116	4.473E-30
9.304	160.0	2.871E-28	9	1.482	0.877	59.39	1.581E-29	3.101E-29	0.2931	0.4811	-5.292	3.951	2.596E-30
9.899	160.0	1.689E-28	12	1.577	0.952	58.89	8.291E-30	1.714E-29	0.3180	0.4977	-5.509	3.779	1.435E-30
10.490	160.0	7.893E-29	15	1.671	1.025	58.48	3.476E-30	7.554E-30	0.3425	0.5120	-5.737	3.599	6.324E-31
11.100	160.0	5.700E-29	20	1.768	1.099	58.13	2.256E-30	5.153E-30	0.3675	0.5248	-5.984	3.400	4.314E-31

$p_1 = 29.700$ GeV/c; $p_1^* = 3.674$ GeV/c; $E_1^* + E_2^* = 7.584$ GeV; $s = 57.517$ GeV²; $\gamma_{c.m.} = 4.0418$; $\beta_{c.m.} = 0.96891$; $y_1^* = 2.0742$; $p_{3,\,max}^* = 3.674$ GeV/c; [67 A 1]

p_3	θ_3	$\dfrac{d^2\sigma}{d\Omega_3\, dp_3}$	Error	p_T	p_L^*	θ_3^*	$\dfrac{1}{p_3^{*2}} \cdot \dfrac{d^2\sigma}{d\Omega_3^*\, dp_3^*}$	$\dfrac{E_3}{p_3^2} \cdot \dfrac{d^2\sigma}{d\Omega_3\, dp_3}$	x^*	y_3^*	t	m_4	$\dfrac{d^2\sigma}{dt\, dm_4^2}$
3.948	14.0	1.736E-26	10	0.055	0.064	40.82	4.798E-27	4.520E-27	0.0174	0.0680	-4.914	6.641	2.548E-28
4.534	14.0	1.871E-26	10	0.063	0.192	18.31	4.391E-27	4.214E-27	0.0522	0.2026	-4.109	6.621	2.376E-28
4.958	14.0	2.061E-26	14	0.069	0.276	14.09	4.315E-27	4.231E-27	0.0753	0.2898	-3.644	6.597	2.385E-28
6.148	14.0	3.809E-26	8	0.086	0.491	9.93	5.898E-27	6.267E-27	0.1337	0.5004	-2.687	6.502	3.533E-28
6.754	14.0	4.304E-26	6	0.095	0.592	9.07	5.778E-27	6.434E-27	0.1611	0.5926	-2.333	6.443	3.627E-28
7.335	14.0	4.254E-26	6	0.103	0.685	8.53	5.014E-27	5.847E-27	0.1864	0.6735	-2.051	6.380	3.296E-28
7.918	14.0	5.770E-26	6	0.111	0.775	8.14	6.006E-27	7.338E-27	0.2109	0.7484	-1.812	6.314	4.137E-28
8.530	14.0	5.921E-26	8	0.119	0.867	7.84	5.443E-27	6.983E-27	0.2360	0.8213	-1.599	6.240	3.937E-28
8.957	14.0	6.923E-26	7	0.125	0.930	7.68	5.856E-27	7.771E-27	0.2531	0.8690	-1.469	6.186	4.381E-28
9.554	14.0	7.607E-26	7	0.134	1.017	7.49	5.756E-27	8.000E-27	0.2768	0.9320	-1.309	6.109	4.510E-28
10.160	14.0	8.091E-26	7	0.142	1.103	7.35	5.495E-27	7.997E-27	0.3003	0.9919	-1.169	6.027	4.508E-28
10.750	14.0	8.520E-26	7	0.150	1.187	7.23	5.234E-27	7.956E-27	0.3230	1.0468	-1.049	5.945	4.485E-28
11.340	14.0	1.002E-25	7	0.159	1.269	7.13	5.590E-27	8.866E-27	0.3453	1.0987	-0.944	5.860	4.998E-28
11.960	14.0	1.029E-25	7	0.167	1.354	7.05	5.211E-27	8.630E-27	0.3686	1.1501	-0.847	5.769	4.865E-28
12.680	14.0	1.186E-25	7	0.178	1.453	6.97	5.395E-27	9.379E-27	0.3954	1.2065	-0.749	5.660	5.287E-28
13.360	14.0	1.301E-25	7	0.187	1.545	6.90	5.373E-27	9.762E-27	0.4204	1.2567	-0.669	5.553	5.503E-28

$p_1 = 29.700$ GeV/c; $p_1^* = 3.674$ GeV/c; $E_1^* + E_2^* = 7.584$ GeV; $s = 57.517$ GeV^2; $\gamma_{\text{c.m.}} = 4.0418$; $\beta_{\text{c.m.}} = 0.96891$; $y_1^* = 2.0742$; $p_{3,\text{max}}^* = 3.674$ GeV/c; [67 A 1] (cont.)

p_3	θ_3	$\dfrac{d^2\sigma}{d\Omega_3\,dp_3}$	Error	p_T	p_L^*	θ_3^*	$\dfrac{1}{p_3^{*2}}\cdot\dfrac{d^2\sigma}{d\Omega_3^*\,dp_3^*}$	$\dfrac{E_3}{p_3^2}\cdot\dfrac{d^2\sigma}{d\Omega_3\,dp_3}$	x^*	y_3^*	t	m_4	$\dfrac{d^2\sigma}{dt\,dm_4^2}$
GeV/c	mrad	cm² sr⁻¹ GeV⁻¹ c	%	GeV/c	GeV/c	deg	cm² sr⁻¹ GeV⁻³ c³	cm² sr⁻¹ GeV⁻² c³			GeV²	GeV	cm²/GeV⁴
13.970	14.0	1.449E-25	7	0.196	1.627	6.86	5.506E-27	1.040E-26	0.4428	1.2995	-0.605	5.455	5.860E-28
14.560	14.0	1.497E-25	7	0.204	1.706	6.81	5.264E-27	1.030E-26	0.4642	1.3390	-0.550	5.358	5.808E-28
15.170	14.0	1.561E-25	7	0.212	1.787	6.78	5.081E-27	1.031E-26	0.4863	1.3780	-0.500	5.255	5.812E-28
15.770	14.0	1.690E-25	4	0.221	1.866	6.75	5.111E-27	1.074E-26	0.5079	1.4148	-0.456	5.152	6.052E-28
16.380	14.0	1.848E-25	4	0.229	1.947	6.72	5.200E-27	1.130E-26	0.5298	1.4506	-0.416	5.043	6.370E-28
16.970	14.0	2.047E-25	4	0.238	2.024	6.69	5.384E-27	1.208E-26	0.5510	1.4839	-0.381	4.936	6.810E-28
17.580	14.0	2.227E-25	4	0.246	2.104	6.67	5.475E-27	1.269E-26	0.5727	1.5169	-0.350	4.822	7.151E-28
17.970	14.0	2.321E-25	6	0.252	2.155	6.66	5.471E-27	1.293E-26	0.5866	1.5374	-0.331	4.748	7.291E-28
20.170	14.0	2.493E-25	6	0.282	2.441	6.60	4.704E-27	1.237E-26	0.6644	1.6441	-0.251	4.301	6.975E-28
20.760	14.0	2.472E-25	4	0.291	2.518	6.59	4.411E-27	1.192E-26	0.6852	1.6704	-0.235	4.172	6.719E-28
21.370	14.0	2.378E-25	4	0.299	2.596	6.57	4.011E-27	1.114E-26	0.7066	1.6968	-0.220	4.035	6.279E-28
21.970	14.0	2.451E-25	4	0.308	2.674	6.56	3.918E-27	1.117E-26	0.7277	1.7218	-0.208	3.894	6.295E-28
22.560	14.0	2.333E-25	4	0.316	2.750	6.55	3.542E-27	1.035E-26	0.7484	1.7457	-0.198	3.751	5.835E-28
23.180	14.0	2.429E-25	4	0.325	2.829	6.54	3.498E-27	1.049E-26	0.7701	1.7699	-0.189	3.594	5.912E-28
23.760	14.0	2.229E-25	4	0.333	2.904	6.53	3.059E-27	9.389E-27	0.7903	1.7919	-0.182	3.440	5.293E-28
24.370	14.0	2.387E-25	4	0.341	2.982	6.53	3.117E-27	9.802E-27	0.8116	1.8143	-0.176	3.271	5.526E-28
24.970	14.0	2.349E-25	4	0.350	3.059	6.52	2.925E-27	9.414E-27	0.8325	1.8358	-0.172	3.095	5.307E-28
25.570	14.0	2.406E-25	4	0.358	3.136	6.51	2.860E-27	9.416E-27	0.8534	1.8565	-0.169	2.908	5.308E-28
26.170	14.0	2.676E-25	4	0.366	3.212	6.51	3.039E-27	1.023E-26	0.8743	1.8767	-0.166	2.708	5.768E-28
26.720	14.0	3.182E-25	4	0.374	3.283	6.50	3.470E-27	1.192E-26	0.8934	1.8947	-0.165	2.511	6.717E-28
27.160	14.0	3.370E-25	4	0.380	3.339	6.50	3.559E-27	1.242E-26	0.9087	1.9087	-0.165	2.341	6.999E-28
27.620	14.0	3.934E-25	4	0.387	3.397	6.49	4.019E-27	1.425E-26	0.9247	1.9231	-0.165	2.148	8.034E-28
28.070	14.0	4.121E-25	4	0.393	3.455	6.49	4.079E-27	1.469E-26	0.9403	1.9369	-0.166	1.942	8.281E-28
28.510	14.0	6.237E-25	4	0.399	3.511	6.49	5.987E-27	2.189E-26	0.9556	1.9501	-0.167	1.716	1.234E-27
28.970	14.0	5.903E-25	4	0.406	3.569	6.48	5.491E-27	2.039E-26	0.9715	1.9636	-0.169	1.442	1.149E-27
21.950	23.0	1.079E-25	7	0.505	2.656	10.76	1.719E-27	4.920E-27	0.7230	1.6447	-0.426	3.871	2.774E-28
22.360	23.0	1.058E-25	4	0.514	2.709	10.75	1.626E-27	4.736E-27	0.7373	1.6590	-0.423	3.771	2.670E-28
22.950	23.0	1.021E-25	4	0.528	2.784	10.73	1.492E-27	4.453E-27	0.7578	1.6789	-0.419	3.622	2.510E-28
23.550	23.0	9.594E-26	4	0.542	2.861	10.72	1.333E-27	4.077E-27	0.7787	1.6984	-0.418	3.463	2.298E-28
24.080	23.0	8.810E-26	4	0.554	2.929	10.71	1.172E-27	3.651E-27	0.7971	1.7149	-0.417	3.317	2.064E-28
24.660	23.0	8.701E-26	4	0.567	3.003	10.70	1.105E-27	3.531E-27	0.8172	1.7325	-0.418	3.149	1.990E-28
25.130	23.0	8.123E-26	4	0.578	3.062	10.69	9.939E-28	3.235E-27	0.8335	1.7462	-0.419	3.005	1.823E-28
25.590	23.0	7.296E-26	4	0.589	3.121	10.68	8.615E-28	2.853E-27	0.8495	1.7593	-0.422	2.858	1.608E-28
26.070	23.0	8.052E-26	4	0.600	3.182	10.67	9.168E-28	3.091E-27	0.8661	1.7725	-0.425	2.695	1.742E-28
26.490	23.0	7.729E-26	4	0.609	3.235	10.66	8.528E-28	2.920E-27	0.8806	1.7838	-0.428	2.544	1.646E-28
26.880	23.0	8.398E-26	4	0.618	3.285	10.66	9.005E-28	3.126E-27	0.8941	1.7940	-0.431	2.395	1.762E-28
27.190	23.0	8.300E-26	4	0.625	3.324	10.65	8.701E-28	3.054E-27	0.9048	1.8020	-0.434	2.270	1.722E-28
27.660	23.0	8.401E-26	4	0.636	3.384	10.65	8.516E-28	3.039E-27	0.9210	1.8138	-0.439	2.066	1.713E-28
28.150	23.0	1.029E-25	4	0.647	3.446	10.64	1.008E-27	3.657E-27	0.9379	1.8257	-0.445	1.828	2.062E-28
28.620	23.0	1.312E-25	4	0.658	3.506	10.63	1.244E-27	4.587E-27	0.9541	1.8369	-0.451	1.567	2.586E-28
29.100	23.0	1.086E-25	4	0.669	3.566	10.63	9.962E-28	3.734E-27	0.9707	1.8480	-0.458	1.244	2.105E-28

$p_1 = 29.700\ \text{GeV/c}$; $p_1^* = 3.674\ \text{GeV/c}$; $E_1^* + E_2^* = 7.584\ \text{GeV}$; $s = 57.517\ \text{GeV}^2$; $\gamma_{\text{c.m.}} = 4.0418$; $\beta_{\text{c.m.}} = 0.96891$; $y_1^* = 2.0742$; $p_{3,\max}^* = 3.674\ \text{GeV/c}$; [67 A 1] (cont.)

p_3	θ_3	$\dfrac{\mathrm{d}^2\sigma}{\mathrm{d}\Omega_3\,\mathrm{d}p_3}$	Error	p_T	p_L^*	θ_3^*	$\dfrac{1}{p_3^{*2}}\cdot\dfrac{\mathrm{d}^2\sigma}{\mathrm{d}\Omega_3^*\,\mathrm{d}p_3^*}$	$\dfrac{E_3}{p_3^2}\cdot\dfrac{\mathrm{d}^2\sigma}{\mathrm{d}\Omega_3\,\mathrm{d}p_3}$	x^*	y_3^*	t	m_4	$\dfrac{\mathrm{d}^2\sigma}{\mathrm{d}t\,\mathrm{d}m_4^2}$
GeV/c	mrad	cm² sr⁻¹ GeV⁻¹ c	%	GeV/c	GeV/c	deg	cm² sr⁻¹ GeV⁻³ c³	cm² sr⁻¹ GeV⁻² c³			GeV²	GeV	cm²/GeV⁴
21.160	43.0	1.659E-26	6	0.910	2.499	20.00	2.783E-28	7.848E-28	0.6800	1.4036	-1.264	3.953	4.424E-29
21.760	43.0	1.469E-26	4	0.935	2.574	19.97	2.334E-28	6.757E-28	0.7006	1.4178	-1.281	3.806	3.809E-29
22.360	43.0	1.202E-26	4	0.961	2.649	19.94	1.811E-28	5.380E-28	0.7210	1.4312	-1.299	3.653	3.033E-29
22.910	43.0	1.040E-26	4	0.985	2.718	19.92	1.495E-28	4.543E-28	0.7398	1.4429	-1.317	3.506	2.561E-29
23.440	43.0	8.429E-27	4	1.008	2.784	19.89	1.159E-28	3.599E-28	0.7578	1.4536	-1.337	3.359	2.029E-29
23.950	43.0	7.192E-27	4	1.030	2.848	19.87	9.479E-29	3.005E-28	0.7752	1.4635	-1.356	3.210	1.694E-29
24.460	43.0	6.229E-27	4	1.051	2.912	19.85	7.878E-29	2.548E-28	0.7925	1.4729	-1.376	3.055	1.437E-29
24.910	43.0	4.867E-27	4	1.071	2.968	19.84	5.940E-29	1.955E-28	0.8078	1.4809	-1.395	2.910	1.102E-29
25.360	43.0	4.484E-27	4	1.090	3.024	19.82	5.284E-29	1.769E-28	0.8231	1.4886	-1.414	2.758	9.974E-30
25.800	43.0	3.555E-27	4	1.109	3.079	19.81	4.051E-29	1.379E-28	0.8380	1.4958	-1.434	2.600	7.775E-30
26.180	43.0	3.135E-27	4	1.125	3.126	19.80	3.471E-29	1.198E-28	0.8509	1.5019	-1.451	2.456	6.755E-30
26.560	43.0	2.917E-27	4	1.142	3.173	19.79	3.139E-29	1.099E-28	0.8637	1.5077	-1.469	2.302	6.195E-30
26.920	43.0	2.671E-27	6	1.157	3.218	19.78	2.800E-29	9.928E-29	0.8759	1.5131	-1.487	2.146	5.597E-30
27.220	43.0	2.272E-27	6	1.170	3.255	19.77	2.330E-29	8.352E-29	0.8861	1.5174	-1.501	2.007	4.708E-30
27.560	43.0	2.122E-27	6	1.185	3.298	19.76	2.124E-29	7.704E-29	0.8976	1.5223	-1.518	1.837	4.343E-30
27.850	43.0	2.411E-27	6	1.197	3.334	19.75	2.364E-29	8.662E-29	0.9074	1.5263	-1.533	1.678	4.883E-30
28.200	43.0	1.684E-27	6	1.212	3.377	19.75	1.611E-29	5.975E-29	0.9192	1.5310	-1.551	1.464	3.368E-30
28.640	43.0	7.463E-28	8	1.231	3.432	19.74	6.925E-30	2.607E-29	0.9340	1.5367	-1.574	1.138	1.470E-30
29.090	43.0	7.229E-28	8	1.250	3.488	19.73	6.505E-30	2.486E-29	0.9492	1.5424	-1.598	0.653	1.402E-30
5.756	60.0	2.327E-26	6	0.345	0.384	41.95	3.825E-27	4.096E-27	0.1045	0.3752	-3.540	6.493	2.309E-28
6.347	50.0	2.295E-26	6	0.381	0.481	38.33	3.261E-27	3.655E-27	0.1310	0.4591	-3.205	6.434	2.061E-28
7.932	60.0	2.661E-26	7	0.476	0.723	33.36	2.647E-27	3.378E-27	0.1967	0.6419	-2.608	6.248	1.904E-28
8.564	60.0	2.707E-26	4	0.514	0.813	32.27	2.367E-27	3.180E-27	0.2213	0.7014	-2.454	6.166	1.793E-28
9.165	60.0	2.560E-25	4	0.550	0.897	31.48	1.992E-27	2.808E-27	0.2443	0.7524	-2.337	6.084	1.583E-28
9.757	50.0	2.615E-26	4	0.585	0.979	30.87	1.823E-27	2.692E-27	0.2664	0.7979	-2.246	6.000	1.518E-28
10.360	60.0	2.524E-26	4	0.621	1.060	30.36	1.582E-27	2.446E-27	0.2886	0.8401	-2.173	5.911	1.379E-28
10.960	60.0	2.247E-26	4	0.657	1.141	29.95	1.273E-27	2.058E-27	0.3104	0.8783	-2.118	5.821	1.160E-28
11.550	60.0	2.140E-26	5	0.693	1.218	29.62	1.102E-27	1.859E-27	0.3316	0.9127	-2.077	5.729	1.048E-28
12.140	50.0	1.945E-26	5	0.728	1.295	29.33	9.149E-28	1.608E-27	0.3526	0.9442	-2.048	5.634	9.063E-29
12.760	60.0	1.880E-26	5	0.765	1.376	29.08	8.062E-28	1.477E-27	0.3744	0.9746	-2.029	5.532	8.328E-29
13.360	60.0	1.641E-26	5	0.801	1.453	28.87	6.460E-28	1.231E-27	0.3954	1.0016	-2.019	5.430	6.941E-29
14.160	60.0	1.428E-26	5	0.849	1.555	28.64	5.042E-28	1.011E-27	0.4232	1.0342	-2.018	5.291	5.697E-29
14.760	60.0	1.198E-26	6	0.885	1.631	28.49	3.912E-28	8.133E-28	0.4438	1.0565	-2.025	5.183	4.585E-29
15.380	60.0	9.727E-27	5	0.922	1.709	28.36	2.938E-28	6.336E-28	0.4651	1.0778	-2.038	5.068	3.572E-29
15.950	60.0	9.145E-27	5	0.955	1.780	28.25	2.578E-28	5.744E-28	0.4846	1.0959	-2.055	4.960	3.238E-29
16.570	60.0	7.675E-27	6	0.994	1.858	28.14	2.012E-28	4.639E-28	0.5056	1.1142	-2.079	4.839	2.615E-29
17.160	60.0	6.474E-27	6	1.029	1.931	28.05	1.587E-28	3.778E-28	0.5256	1.1303	-2.105	4.721	2.130E-29
17.760	60.0	5.423E-27	6	1.065	2.006	27.97	1.245E-28	3.058E-28	0.5459	1.1456	-2.136	4.597	1.724E-29
18.380	60.0	4.158E-27	6	1.102	2.082	27.89	8.933E-29	2.265E-28	0.5667	1.1602	-2.171	4.465	1.277E-29
18.980	60.0	3.435E-27	8	1.138	2.156	27.83	6.936E-29	1.812E-28	0.5869	1.1734	-2.208	4.333	1.021E-29
19.580	60.0	2.187E-27	8	1.174	2.230	27.77	4.158E-29	1.118E-28	0.6070	1.1858	-2.248	4.196	6.304E-30
20.180	50.0	2.037E-27	8	1.210	2.304	27.71	3.653E-29	1.011E-28	0.6270	1.1973	-2.290	4.055	5.696E-30
20.790	60.0	1.570E-27	10	1.247	2.378	27.66	2.658E-29	7.559E-29	0.6473	1.2082	-2.335	3.905	4.261E-30

p_3	θ_3	$\dfrac{d^2\sigma}{d\Omega_3\,dp_3}$	Error	p_T	p_L^*	θ_3^*	$\dfrac{1}{p_3^{*2}}\cdot\dfrac{d^2\sigma}{d\Omega_3^*\,dp_3^*}$	$\dfrac{E_3}{p_3^2}\cdot\dfrac{d^2\sigma}{d\Omega_3\,dp_3}$	x^*	y_3^*	t	m_4	$\dfrac{d^2\sigma}{dt\,dm_4^2}$
GeV/c	mrad	cm² sr⁻¹ GeV⁻¹ c	%	GeV/c	GeV/c	deg	cm² sr⁻¹ GeV⁻³ c³	cm² sr⁻ GeV⁻² c³			GeV²	GeV	cm²/GeV⁴

$p_1 = 29.700$ GeV/c; $p_1^* = 3.674$ GeV/c; $E_1^* + E_2^* = 7.584$ GeV; $s = 57.517$ GeV²; $\gamma_{c.m.} = 4.0418$; $\beta_{c.m.} = 0.96891$; $y_1^* = 2.0742$; $p_{3,\,max}^* = 3.674$ GeV/c; [67 A 1] (cont.)

p_3	θ_3	$\frac{d^2\sigma}{d\Omega_3 dp_3}$	Error	p_T	p_L^*	θ_3^*	$\frac{1}{p_3^{*2}}\frac{d^2\sigma}{d\Omega_3^* dp_3^*}$	$\frac{E_3}{p_3^2}\frac{d^2\sigma}{d\Omega_3 dp_3}$	x^*	y_3^*	t	m_4	$\frac{d^2\sigma}{dt\,dm_4^2}$
14.720	96.0	6.670E-28	10	1.411	1.459	44.04	2.031E-29	4.540E-29	0.3971	0.7796	-4.477	4.948	2.560E-30
15.330	96.0	5.110E-28	10	1.469	1.529	43.87	1.440E-29	3.340E-29	0.4161	0.7916	-4.591	4.819	1.883E-30
15.930	96.0	3.395E-28	13	1.527	1.597	43.71	8.893E-30	2.135E-29	0.4347	0.8023	-4.709	4.689	1.203E-30
16.530	96.0	3.330E-28	12	1.584	1.665	43.57	8.127E-30	2.018E-29	0.4533	0.8121	-4.831	4.554	1.137E-30
17.110	96.0	1.489E-28	20	1.640	1.731	43.45	3.401E-30	8.716E-30	0.4711	0.8208	-4.954	4.419	4.913E-31
17.720	96.0	1.575E-28	16	1.699	1.800	43.34	3.363E-30	8.901E-30	0.4898	0.8292	-5.086	4.273	5.018E-31
18.330	96.0	9.548E-29	20	1.757	1.868	43.24	1.910E-30	5.216E-30	0.5085	0.8368	-5.222	4.120	2.940E-31
18.920	96.0	6.336E-29	25	1.814	1.934	43.15	1.192E-30	3.353E-30	0.5265	0.8437	-5.356	3.967	1.890E-31
19.520	96.0	4.229E-29	35	1.871	2.001	43.07	7.490E-31	2.169E-30	0.5447	0.8501	-5.496	3.804	1.223E-31
7.947	100.0	6.905E-27	4	0.793	0.622	51.90	6.353E-28	8.749E-28	0.1693	0.4868	-4.113	6.124	4.932E-29
8.555	100.0	5.660E-27	5	0.854	0.701	50.60	4.591E-28	6.656E-28	0.1909	0.5280	-4.080	6.034	3.752E-29
9.153	100.0	4.903E-27	5	0.914	0.778	49.60	3.535E-28	5.385E-28	0.2116	0.5634	-4.077	5.941	3.036E-29
9.736	100.0	4.165E-27	6	0.972	0.850	48.82	2.693E-28	4.299E-28	0.2314	0.5938	-4.097	5.847	2.423E-29
10.340	100.0	3.301E-27	6	1.032	0.924	48.16	1.916E-28	3.206E-28	0.2515	0.6217	-4.139	5.746	1.807E-29
10.950	100.0	3.000E-27	7	1.093	0.998	47.61	1.569E-28	2.750E-28	0.2716	0.6466	-4.198	5.641	1.550E-29
11.540	100.0	2.173E-27	8	1.152	1.068	47.17	1.032E-28	1.839E-28	0.2907	0.6680	-4.269	5.536	1.065E-29
12.140	100.0	1.641E-27	9	1.212	1.139	46.79	7.101E-29	1.356E-28	0.3099	0.6875	-4.353	5.426	7.643E-30
12.740	100.0	1.495E-27	10	1.272	1.209	46.46	5.914E-29	1.177E-28	0.3289	0.7049	-4.448	5.312	6.633E-30
5.563	90.7	1.410E-26	9	0.504	0.299	59.32	2.324E-27	2.570E-27	0.0814	0.2772	-4.431	6.451	1.449E-28
5.277	107.0	1.057E-26	5	0.564	0.217	68.94	1.823E-27	2.034E-27	0.0591	0.1971	-5.107	6.440	1.147E-28
4.971	133.5	6.951E-27	5	0.662	0.102	81.22	1.236E-27	1.425E-27	0.0278	0.0889	-6.230	6.396	8.033E-29
4.812	157.9	4.698E-27	5	0.757	0.008	89.40	8.252E-28	9.947E-28	0.0021	0.0066	-7.323	6.334	5.607E-29
5.739	160.0	2.812E-27	4	0.914	0.127	82.12	3.772E-28	4.965E-28	0.0344	0.0965	-7.291	6.200	2.799E-29
6.342	150.0	1.863E-27	5	1.010	0.199	78.84	2.132E-28	2.970E-28	0.0542	0.1440	-7.341	6.105	1.674E-29
6.939	160.0	1.385E-27	4	1.106	0.266	76.45	1.366E-28	2.014E-28	0.0725	0.1828	-7.462	6.003	1.135E-29
7.540	160.0	9.795E-28	4	1.201	0.331	74.62	8.394E-29	1.309E-28	0.0900	0.2152	-7.639	5.894	7.380E-30
8.114	160.0	6.415E-28	5	1.293	0.389	73.25	4.841E-29	7.959E-29	0.1059	0.2412	-7.849	5.785	4.487E-30
8.729	150.0	4.399E-28	5	1.391	0.449	72.09	2.918E-29	5.069E-29	0.1223	0.2648	-8.109	5.662	2.857E-30
9.329	160.0	3.043E-28	6	1.486	0.506	71.19	1.792E-29	3.278E-29	0.1378	0.2843	-8.391	5.537	1.848E-30
9.926	160.0	1.955E-28	7	1.581	0.562	70.45	1.029E-29	1.978E-29	0.1529	0.3009	-8.694	5.407	1.115E-30
10.520	160.0	1.234E-28	7	1.676	0.615	69.84	5.839E-30	1.178E-29	0.1675	0.3151	-9.014	5.273	6.639E-31
11.130	160.0	7.242E-29	8	1.773	0.669	69.32	3.088E-30	6.530E-30	0.1822	0.3278	-9.360	5.130	3.681E-31
11.720	160.0	4.750E-29	10	1.867	0.721	68.89	1.839E-30	4.066E-30	0.1962	0.3385	-9.707	4.987	2.292E-31
12.320	160.0	2.297E-29	15	1.963	0.772	68.52	8.100E-31	1.870E-30	0.2102	0.3480	-10.072	4.835	1.054E-31
12.920	160.0	1.603E-29	16	2.058	0.823	68.20	5.168E-31	1.244E-30	0.2241	0.3564	-10.447	4.678	7.013E-32

3.7 References for 3

46 H 1	Heisenberg, W.: Cosmic radiation. New York: Dover Publications 1946; Z. Phys. **133** (1952) 65.
51 F 1	Fermi, E.: Phys. Rev. **81** (1951) 683; **92** (1953) 452; **93** (1954) 1434.
51 L 1	Landau, L., Lifshitz, E.: The classical theory of fields (translated by M. Hamermesh). Cambridge, USA: Addison-Wesley Press. 1951, p. 31.
52 M 1	Marshak, R. E.: Meson physics. New York: McGraw Hill Book Company 1952.
54 G 1	Gell-Mann, M., Watson, K. M.: Ann. Rev. Nucl. Sci. **4** (1954) 219.
61 B 1	Baker, W. F., Cool, R. L., Jenkins, E. W., Kycia, T. F., Lindenbaum, S. J., Love, W. A., Luers, D., Niederer, J. A., Ozaki, S., Read, A. L., Russell, J. J., Yuan, L. C. L.: Phys. Rev. Letters **7** (1961) 101.
61 C 1	Cocconi, G., Koester, L. J., Perkins, D. H.: University of California, Radiation Laboratory Report UCRL 10022 (1961), p. 167.
	Cocconi, G.: Nucl. Phys. **B 28** (1971) 341.
62 B 1	Bigi, A., Brandt, S., Carrara, R., Cooper, W. A., de Marco, A., Macleod, G. R., Peyrou, Ch., Sosnowski, R., Wroblewski, A.: Proc. Int. Conf. on High-Energy Physics, Geneva, 1962 (CERN, Geneva, 1962), p. 247.
	See also Imaeda, K.: Nuovo Cimento **48 A** (1967) 482.
62 B 2	Bertocchi, L., Fubini, S., Tonin, M.: Nuovo Cimento **25** (1962) 626.
	Amati, D., Stanghellini, A., Fubini, S.: Nuovo Cimento **26** (1962) 896.
62 F 1	Fidecaro, M., Finocchiaro, G., Gatti, G., Giacomelli, G., Middelkop, W. C., Yamagata, T.: Nuovo Cimento **24** (1962) 73.
62 M 1	Melissinos, A. C., Yamanouchi, T., Fazio, G. G., Lindenbaum, S. J., Yuan, L. C. L.: Phys. Rev. **128** (1962) 2373.
64 D 1	Diddens, A. N., Galbraith, W., Lillethun, E., Manning, G., Parham, A. G., Taylor, A. E., Walker, T. G., Wetherell, A. M.: Nuovo Cimento **31** (1964) 961.
64 F 1	Fowler, P. H., Perkins, D. H.: Proc. Roy. Soc. (London) Ser. A, **278** (1964) 401 and references quoted therein.
65 D 1	Dekkers, D., Geibel, J. A., Mermod, R., Weber, G., Willits, T. R., Winter, K., Jordan, B., Vivargent, M., King, N. M., Wilson, E. J. N.: Phys. Rev. **137 B** (1965) 962.
65 H 1	Hagedorn, R.: Nuovo Cimento Suppl. **3** (1965) 147.
	Hagedorn, R., Ranft, J.: Nuovo Cimento Suppl. **6** (1968) 169.
	Grote, H., Hagedorn, R., Ranft, J.: Atlas of particle production spectra (CERN, Geneva, 1970).
	Ranft, J.: Phys. Letters **31 B** (1970) 529; **32 B** (1970) 207; **33 B** (1970) 481; **36 B** (1971) 225.
	Hagedorn, R.: Nucl. Phys. **B 24** (1970) 93.
65 L 1	Lundy, R. A., Novey, T. B., Yovanovitch, D. D., Telegdi, V. L.: Phys. Rev. Letters **14** (1965) 504, 730.
65 W 1	Wu, T. T., Yang, C. N.: Phys. Rev. **137 B** (1965) 709.
66 K 1	Kastrup, H. A.: Phys. Rev. **147** (1966) 1130.
66 L 1	Lamb, R. C., Lundy, R. A., Novey, T. B., Yovanovitch, D. D., Lander, R.: Phys. Rev. Letters **17** (1966) 100.
66 P 1	Pal, Y.: Cosmic rays and their intersections. In: Handbook of Physics (ed. Condon and Odishaw), 2nd edition. New York: McGraw-Hill: 1967.
66 R 1	Reay, N. W., Melissinos, A. C., Reed, J. T., Yamanouchi, T., Yuan, L. C. L.: Phys. Rev. **142** (1966) 918.
66 T 1	Trilling, G.: University of California, Radiation Laboratory Report UCRL 16830 (1966), 25.
67 A 1	Anderson, E. W., Bleser, E. J., Collins, G. B., Fujii, T., Menes, J., Turkot, F., Carrigan, R. A., Edelstein, R. K., Hien, N. C., McMahon, T. J., Nadelhaft, I.: Phys. Rev. Letters **19** (1967) 198, and private communication from R. K. Edelstein.
67 A 2	Anderson, E. W., Collins, G. B.: Phys. Rev. Letters **19** (1967) 201.
67 A 3	Abraham, F., Gierula, J., Levi-Setti, R., Rybicki, K., Isao, C. H., Wolter, W., Fricken, R. L., Hagget, R. W.: Phys. Rev. **159** (1967) 1110.
67 S 1	Sanford, J. R., Wang, C. L.: Brookhaven National Laboratory Reports BNL 11299, BNL 11479 (1967).
	Wang, C. L.: Phys. Rev. Letters **25** (1970) 1068, 1536.
68 E 1	Elbert, J. W., Erwin, A. R., Mikamo, S., Reeder, D., Chen, Y. Y., Walker, W. D., Weinberg, A.: Phys. Rev. Letters **20** (1968) 124.
68 H 1	Hogan, W. J., Piroué, P. A., Smith, A. J. S.: Phys. Rev. **166** (1968) 1472.
68 M 1	Mack, G.: Phys. Letters **26 B** (1968) 515.

Diddens/Schlüpmann

68 R 1   Reed, J. T., Melissinos, A. C., Reay, N. W., Yamanouchi, T., Sacharidis, E. J., Lindenbaum, S. J., Ozaki, S., Yuan, L. C. L.: Phys. Rev. **168** (1968) 1495.

68 T 1   Turkot, F.: In: Proc. Topical Conf. on High-Energy Collision of Hadrons, Geneva, 1968, CERN 68–7 (1968) 316.

68 W 1   Wayland, J. R.: Phys. Rev. **175** (1968) 2106.
         Lapointe, M., Wayland, J. R.: Phys. Rev. **185** (1969) 2041.

69 A 1   Allaby, J. V., Binon, F., Diddens, A. N., Duteil, P., Klovning, A., Meunier, R., Peigneux, J. P., Sacharidis, E. J., Schlüpmann, K., Spighel, M., Stroot, J. P., Thorndike, A. M., Wetherell, A. M.: Phys. Letters **29 B** (1969) 198.

69 B 1   Benecke, J., Chou, T. T., Yang, C. N., Yen, E.: Phys. Rev. **188** (1969) 2159.
         Chou, T. T., Yang, C. N.: Phys. Letters **25** (1970) 1072.

69 B 2   Bushnin, Yu. B., Denisov, S. P., Donskov, S. V., Dunaitsev, A. F., Gorin, Yu. P., Kachanov, V. A., Khodirev, Yu. S., Kotov, V. I., Kutyin, V. M., Petrukhin, A. I., Prokoshkin, Yu. D., Razuvaev, E. A., Shuvalov, R. S., Stoyanova, D. A., Allaby, J. V., Binon, F., Diddens, A. N., Duteil, P., Giacomelli, G., Meunier, R., Peigneux, J. P., Schlüpman, K., Spighel, M., Stahlbrandt, C. A., Stroot, J. P., Wetherell, A. M.: Phys. Letters **29 B** (1969) 48.

69 B 3   Binon, F., Denisov, S. P., Duteil, P., Kachanov, V. A., Kutyin, V. M., Peigneux, J. P., Prokoshkin, Yu. D., Razuvaev, E. A., Shuvalov, R. S., Spighel, M., Stroot, J. P.: Phys. Letters **30 B** (1969) 506.

69 C 1   Caneschi, L., Pignotti, A.: Phys. Rev. Letters **22** (1969) 1219.

69 F 1   Feynman, R. P.: Phys. Rev. Letters **23** (1969) 1415; In: High energy collisions, p. 237. (Eds. C. N. Yang *et al.*) New York: Gordon and Breach 1969.

69 G 1   Garelick, H.: Phys. Rev. Letters **22** (1969) 674.

69 H 1   Honecker, R., Junkman, B., Schulte, R., Steinberg, R., Tsanos, N., Klugow, J., Nowak, S., Ryseck, E., Walter, M., Böckmann, K., Drevermann, H., Johnssen, W., Steinberg, K., Wagini, B., Böttcher, H., Cocconi, V. T., Hansen, J. D., Kellner, C., Morrison, D. R. O., Paler, K., Mihul, A., Moskalev, V., Coghen, T., Czyżewski, O., Eskreys, K., Loskiewicz, J., Zaorska, J., Brandt, S., Braun, O., Lüth, V., Shah, T. P., Wenninger, H., Bardadin-Otwinowska, M., Hofmokl, T., Michejda, L., Otwinowski, S., Sosnowski, R., Szepticka, M., Wojcik, W., Wroblewski, A.: Nucl. Phys. **B 13** (1969) 571.

69 L 1   Liland, A., Pilkuhn, H.: Phys. Letters **29 B** (1969) 663.

69 P 1   Panvini, R.: Proc. 3rd Int. Conf. on High Energy Collisions, Stony Brook, **1969**, p. 461. New York: Gordon and Breach 1969.

69 S 1   Smith, D. B., Sprafka, R. J., Anderson, J. A.: Phys. Rev. Letters **23** (1969) 1064.
         Smith, D. B.: Thesis, University of California Report UCRL-20632 (1971) and private communication from D. B. Smith.

69 W 1   Wang, C. P.: Phys. Rev. **180** (1969) 1463.

70 A 1   Allaby, J. V., Binon, F., Diddens, A. N., Duteil, P., Klovning, A., Meunier, R., Peigneux, J. P., Sacharidis, E. J., Schlüpmann, K., Spighel, M., Stroot, J. P., Thorndike, A. M., Wetherell, A. M.: CERN 70–12 (1970).

70 A 2   Allaby, J. V., Diddens, A. N., Klovning, A., Schlüpmann, K., Wetherell, A. M.: Phys. Letters **33 B** (1970) 429.

70 A 3   Abolins, M. A., Smith, G. A., Ming Ma, Z., Gellert, E., Wickland, A. B.: Phys. Rev. Letters **25** (1970) 126; private communication from Z. Ming Ma.

70 B 1   Bali, N. F., Brown, L. S., Peccei, R. D., Pignotti, A.: Phys. Rev. Letters **25** (1970) 557.

70 B 2   Berman, S. M., Jacob, M.: Phys. Rev. Letters **25** (1970) 1683.

70 B 3   Benary, O., Price, L. R., Alexander, G.: (Particle Data Group), NN and ND interactions (above 0.5 GeV/c) – a compilation, University of California Radiation Laboratory Report UCRL 20000 NN (1970).

70 C 1   Chen, M., Harte, J.: Phys. Rev. **D 1** (1970) 2603.

70 C 2   Czyżewski, O., Rybicki, K.: Inst. Nucl. Res. Cracow, preprint 215/70 RC/344/59 (1970).

70 D 1   Drell, S. D.: In: Proc. Int. Conf. on Expectations for Particle Reactions at the New Accelerators, Madison, **1970** (University of Wisconsin, Madison, Wis., 1970).

70 E 1   Elbert, J. W., Erwin, A. R., Walker, W. D., Walters, J. W.: Nucl. Phys. **19 B** (1970) 85 and Phys. Rev. **D 3** (1971) 2042.

70 E 2   Erwin, A. R.: In: Proc. Int. Conf. on Expectations for Particle Reactions at the New Accelerators, Madison, **1970** (University of Wisconsin, Madison, Wis., 1970).

70 G 1   Goldberg, H.: Phys. Rev. **D 2** (1970) 839.

70 H 1   Hansen, J. D., Morrison, D. R. O., Tovey, N., Flaminio, E.: CERN-HERA 70–2 (1970).

70 J 1	Jones, L. W., Bassian, A. E., De Meester, G. D., Loo, B. W., Lyon, D. E., Ramana Murthy, P. V., Roth, R. F., Learned, J. G., Mills, F. E., Reeder, D. D., Erickson, K. N., Cork, B.: Phys. Rev. Letters **25** (1970) 1679.
70 M 1	Mueller, A. H.: Phys. Rev. **D 2** (1970) 2963.
70 P 1	Piotrowska, H.: Phys. Letters **32 B** (1970) 71.
70 V 1	Van der Velde, J. C.: Phys. Letters **32 B** (1970) 501.
71 A 1	Akerlof, C. W., Crabb, D. C., Day, J. L., Johnson, N. P., Kalbaci, P., Krisch, A. D., Lin, M. T., Marshak, M. L., Randolph, J. K., Schmueser, P., Read, A. L., Edwards, K. W., Asbury, J. G., Marmer, G. J., Ratner, L. G.: Phys. Rev. **D 3** (1971) 645.
71 A 2	Anthony, R. W., Coffin, C. T., Meanley, E. S., Rice, J. E., Terwilliger, K. M., Stanton, N. R.: Phys. Rev. Letters **26** (1971) 38.
71 B 1	Bøggild, H., Hansen, K. H., Suk, M.: Nucl. Phys. **B 27** (1971) 1; private communication from H. Bøggild.
71 B 2	Bøggild, H., Dahl-Jensen, E., Hansen, K. H., Johnstad, J., Lohse, E., Suk, M., Vije, L., Karimäki, V. J., Laurikainen, K. V., Riipinen, E., Jacobsen, T., Sørensen, S. O., Allan, J., Blomquist, G., Danielsen, O., Ekspong, G., Granstrøm, L., Holmgren, S. O., Nilsson, S., Ronne, B. E., Svedin, U., Yamdagni, N. K.: Nucl. Phys. **B 27** (1971) 285.
71 B 3	Biswas, N. N., Cason, N. M., Kenney, V. P., Powers, J. T., Shephard, W. D., Thomas, D. W.: Phys. Rev. Letters **26** (1971) 1589.
71 B 4	Berger, E. L.: Argonne National Laboratory Report ANL/HEP 7134 (1971), and to be published in Proc. Int. Colloquium on Multiparticle Dynamics, Helsinki, 1971.
71 B 5	Beaupré, J. V., Deutschmann, M., Grässler, H., Schmitz, P., Gensch, U., Nowak, W. D., Schreiber, H. J., Walter, M., de Wolf, E., Grard, F., Herquet, P., Pape, L., Peeters, P., Verbeure, F., Windmolders, R., Angelopoulos, A., Chliapnicov, P. V., Cocconi, V. T., Czyžewski, O., Drijard, D., Duinker, P., Dunwoodie, W., Eskreys, A., Goldschmidt-Clermont, Y., Grant, A., Hansen, J. D., Henri, V. P., Kittel, W., Linglin, D., Matsumoto, S., Meyer, J., Morrison, D. R. O., Muller, F., Nielsen, S., Sekera, Z., Stroynowski, R., Dornan, P. J., Grammatikakis, G. A., Kumar, B. R., Mutalib, A., Buschbeck, B., Froehlich, A., Porth, P., Wahl, H.: Nucl. Phys. **B 30** (1971) 381.
71 B 6	Beaupré, J. V., Deutschmann, M., Kirk, H., Lauscher, P., Matziolis, M., Gensch, U., Nowak, H., Schreiber, H. J., Bosson, J. H., Propach, E., Rost, M., Stöcker, U., Besliu, T., Böckmann, K., Cocconi, V. T., Kellner, G., Kittel, W., Morrison, D. R. O., Sotiriou, D., Stroynowski, R., Wahl, H., Coghen, T., Dziunikowska, K., Blaschke, R., Brandt, S., Michejda, L., Otwinowski, S., Piotrowska, H., Wojcik, W.: Phys. Letters **37 B** (1971) 432.
71 B 7	Berger, E. L., Oh, B. Y., Smith, G. A.: Phys. Rev. Letters **28** (1972) 322.
71 B 8	Berger, E. L., Krzywicki, A.: Phys. Letters **36 B** (1971) 380.
71 C 1	Carazza, B., Marchesini, G.: Phys. Letters **35 B** (1971) 436.
71 C 2	Chliapnikov, P. V., Czyžewski, O., Goldschmidt-Clermont, Y.: Phys. Letters **36 B** (1971) 235.
71 C 3	Chen, M. S., Wang, L. L., Wong, T. F.: Phys. Rev. Letters **26** (1971) 280.
71 C 4	Chan, H. M., Hsue, C. S., Quigg, C., Wang, J. M.: Phys. Rev. Letters **26** (1971) 673.
71 C 5	Chen, M. S., Kinsey, R. R., Morris, T. W., Panvini, R. S., Wang, L. L., Wong, T. F., Stone, S. L., Ferbel, T., Slattery, P., Werner, B., Elbert, J. W., Erwin, A. R.: Phys. Rev. Letters **26** (1971) 1585.
71 C 6	Chou, T. T.: Phys. Rev. Letters **27** (1971) 1247.
71 C 7	Chen, M. S., Wang, L. L., Wong, T. F.: Phys. Rev. **D 5** (1972) 1667.
71 C 8	Cho, Y., Derrick, M., Marmer, G., Wangler, T. P., Day, J. L., Kalbaci, P., Marshak, M. L., Randolph, J. K., Key, A. W.: Phys. Rev. **D 4** (1971) 1967.
71 D 1	De Meester, G. D.: University of Michigan Report UM-HE-71-5 (1971). See also Nucl. Phys. **B 43** (1972) 477.
71 D 2	De Tar, C. E.: Phys. Rev. **D 3** (1971) 128.
71 D 3	Deutschmann, M.: Proc. Int. Conf. on Elementary Particles, Amsterdam, 1971, ed. A. G. Tenner and M. J. G. Veltman (North Holland Publishing Co., Amsterdam, 1972) p. 153.
71 E 1	Ellis, J., Finkelstein, J., Frampton, P. H., Jacob, M.: Phys. Letters **35 B** (1971) 227.
71 E 2	Edelstein, R. M., Rittenberg, V., Rubinstein, H. R.: Phys. Letters **35 B** (1971) 408.
71 E 3	Erwin, J., Ko, W., Lander, R. L., Pellett, D. E., Yager, P. M.: Phys. Rev. Letters **27** (1971) 1534.
71 F 1	Foster, M., Cole, J. A., Kim, E., Lee-Franzini, J., Loveless, R. J., Moore, C., Takahashi, K.: Phys. Rev. Letters **27** (1971) 1312.
71 G 1	Gibbard, B. G., Longo, M. J., Jones, L. W., O'Fallon, J. R., Kreisler, M. N., Perl, M. L.: Nucl. Phys. **B 30** (1971) 77.

71 G 2   Goldberg, H.: Northaestern University Report NUB 2088 (1971).

71 G 3   Gorin, Yu. P., Denisov, S. P., Donskov, S. V., Kotov, V. I., Petrukhin, A. I., Prokoshkin, Yu. D., Seleznev, U. S., Stoyanova, D. A., Khodyrev, Yu. S., Shuvalov, R. S.: Yadernaya Fiz. **14** (1971) 994.

71 H 1   Holder, M., Radermacher, E., Staude, A., Barbiellini, G., Darriulat, P., Hansroul, M., Orito, S., Palazzi, P., Santroni, A., Strolin, P., Tittel, K., Pilcher, J., Rubbia, C., De Zorzi, G., Macri, M., Sette, G., Grosso-Pilcher, C., Fainberg, A., Maderni, G.: Phys. Letters **35 B** (1971) 361.

71 H 2   Van Hove, L.: Phys. Rep. **1 C** (1971) 349.

71 H 3   Horn, D.: Phys. Rep. **4 C** (1972) 1.

71 K 1   Ko, W., Lander, R. L.: Phys. Rev. Letters **26** (1971) 1064.

71 K 2   Ko, W., Lander, R. L., Risk, C.: Phys. Rev. Letters **27** (1971) 1476.

71 L 1   von Lindern, L., Panvini, R. S., Hanlon, J., Salant, E. O.: Phys. Rev. Letters **27** (1971) 1745.

71 M 1   Michejda, L.: Nucl. Phys. **B 35** (1971) 287.

71 N 1   Neuhofer, G., Niebergall, F., Penzias, J., Regler, M., Schubert, K. R., Schumacher, P. E., Schmidt-Parzefall, W., Winter, K.: Phys. Letters **37 B** (1971) 438, and **38 B** (1972) 51.

71 P 1   Peccei, R. D., Pignotti, A.: Phys. Rev. Letters **26** (1971) 1076.

71 Q 1   Quigg, C.: Proc. Workshop on Particle Physics at Intermediate Energies, Pasadena, **1971** (Lawrence Radiation Laboratory, Berkeley, 1971), p. 277.
          Quigg, C.: APS Meeting at University of Rochester, Sept. 1971.

71 R 1   Ratner, L. G., Ellis, R. J., Vannini, G., Babcock, B. A., Krisch, A. D., Roberts, J. B.: Phys. Rev. Letters **27** (1971) 68, and to be published.

71 R 2   Risk, C., Friedman, J. H.: Phys. Rev. Letters **27** (1971) 353.

71 R 3   Risk, C.: Phys. Rev. **D 5** (1972) 1685.

71 R 4   Ranft, J.: Phys. Letters **36 B** (1971) 225.

71 R 5   Ranft, G., Ranft, J.: Fortschr. Phys. **19** (1971) 393.

71 R 6   Ranft, J.: CERN LAB II-Ra 71-1 (1971).

71 S 1   Swanson, W. P., Davier, M., Derado, I., Fries, D. C., Liu, F. F., Mozley, R. F., Odian, A. C., Park, J., Villa, F., Yount, D. E.: Phys. Rev. Letters **27** (1971) 1472.

71 S 2   Sims, W. H., Hanlon, J., Salant, E. O., Panvini, R. S., Kinsey, R. R., Morris, T. W., von Lindern, L.: Nucl. Phys. **B 41** (1972) 317.
          Several other publications exist on these exposures; see for instance, Ellis, W. E., Kinsey, R. R., Morris, T. W., Panvini, R. S., Turkot, F.: Phys. Letters **32 B** (1970) 641, and Ellis, W., Miller, D. J., Morris, T. W., Panvini, R. S., Thorndike, A. M.: Phys. Rev. Letters **21** (1968) 697; [69 P 1].

71 S 3   Stone, S. L., Cohen, D., Farber, M., Ferbel, T., Holmes, R., Slattery, P., Werner, B.: Nucl. Phys. **B 32** (1971) 19.

72 A 1   Allaby, J. V., Diddens, A. N., Dobinson, R. W., Klovning, A., Litt, J., Rochester, L. S., Schlüpmann, K., Wetherell, A. M., Amaldi, U., Biancastelli, R., Bosio, C., Matthiae, G.; contribution to appear in the Proceedings of the Fourth International Conference on High-Energy Collisions, Oxford (April 1972).

72 B 1   Bøggild, H., Dahl-Jensen, E., Gravilas, M., Hansen, K. H., Johnstad, H.: Angular and rapidity distributions of the pions emitted in high-energy proton-proton collisions, submitted to Nucl. Phys. **B**.

72 B 2   Badier, J., Bland, R., Chollet, J. C., Devlin, T., Gaillard, J. M., Lefrançois, J., Merkel, B., Meunier, R., Repellin, J. P., Sauvage, G., Vanderhaghen, R.: Phys. Letters **39 B** (1972) 414.

72 B 3   Bertin, A., Capilappi, P., Cristalline, A., D'Agostino-Bruno, M., Ellis, R. J., Giacomelli, G., Maroni, C., Mercatuli, F., Russi, A. M., Vannini, G.: Phys. Letters **38 B** (1972) 260.

72 B 4   Barbiellini, G., Bozzo, M., Darriulat, P., Diambrini-Palazzi, G., De Zorzi, G., Holder, M., McFarland, A., Maderni, G., Mery, P., Orito, S., Pilcher, J., Rubbia, C., Sette, G., Staude, A., Strolin, P., Tittel, K.: Phys. Letters **39 B** (1972) 294.

72 E 1   Engler, J., Flauger, W., Gibbard, B., Mönnig, E., Runge, K., Schopper, H.: Submitted to Nuclear Physics B.

72 M 1   Mück, H. J., Schachter, M., Selonke, F., Wessels, B., Blobel, V., Brandt, A., Drews, G., Fesefeldt, H., Hellwig, B., Mönkemeyer, D., Söding, P., Brandenburg, G. W., Franz, H., Freund, P., Lüers, D., Richter, W., Schrankel, W.: Contribution to appear in the Proceedings of the Fourth International Conference on High-Energy Collisions, Oxford (April 1972) and Phys. Letters **39 B** (1972) 303.

72 O 1   Proceedings of the Fourth International Conference on High-Energy Collisions, Oxford (April 1972).

72 P 1   Panvini, R. S., Kinsey, R. R., Morris, T. W., Hanlon, J., Salant, E. O., Sims, W. H.: Phys. Letters **38 B** (1972) 55.